Texts and
Monographs
in Physics

A. Galindo P. Pascual

Quantum Mechanics II

Translated by L. Alvarez-Gaumé

With 70 Figures

Springer-Verlag
Berlin Heidelberg New York
London Paris Tokyo
Hong Kong Barcelona

Professor Alberto Galindo
Universidad Complutense
Facultad de Ciencias Físicas
Departamento de Física Teórica
E-28040 Madrid, Spain

Professor Pedro Pascual
Universidad de Barcelona
Facultad de Física
Departamento de Física Teórica
E-08070 Barcelona, Spain

Translator:

Professor Luis Alvarez-Gaumé
Theory Division, CERN
CH-1211 Genève 23, Switzerland

Editors

Wolf Beiglböck
Institut für Angewandte Mathematik
Universität Heidelberg
Im Neuenheimer Feld 294
D-6900 Heidelberg 1, FRG

Joseph L. Birman
Department of Physics, The City College
of the City University of New York
New York, NY 10031, USA

Robert P. Geroch
Enrico Fermi Institute
University of Chicago
5640 Ellis Ave.
Chicago, IL 60637, USA

Elliott H. Lieb
Department of Physics
Joseph Henry Laboratories
Princeton University
Princeton, NJ 08540, USA

Tullio Regge
Istituto di Fisica Teorica
Università di Torino, C. so M. d'Azeglio, 46
I-10125 Torino, Italy

Walter Thirring
Institut für Theoretische Physik
der Universität Wien, Boltzmanngasse 5
A-1090 Wien, Austria

Title of the original Spanish edition: Mecánica Cuántica (II)
© by the authors and Editorial Alhambra, Madrid 1978 (1st ed.)
© by the authors and EUDEMA, Madrid 1989 (2nd ed.)

ISBN 3-540-52309-X Springer-Verlag Berlin Heidelberg New York
ISBN 0-387-52309-X Springer-Verlag New York Berlin Heidelberg

Library of Congress Cataloging-in-Publication Data. Galindo, A. (Alberto), 1934– [Mecánica cuántica II. English] Quantum mechanics II / A. Galindo, P. Pascual ; translated by L. Alvarez-Gaumé. p. cm. – (Texts and monographs in physics) Translation of: Mecánica cuántica II. Includes bibliographical references. (p.) ISBN 0-387-52309-X (U.S.: alk. paper) 1. Quantum theory. I. Pascual, Pedro. II. Title. III. Series. QC174.12.G34213 1990 530.1'2–dc20 90-9566

This work is subject to copyright. All rights are reserved, whether the whole or part of the material is concerned, specifically the rights of translation, reprinting, reuse of illustrations, recitation, broadcasting, reproduction on microfilms or in other ways, and storage in data banks. Duplication of this publication or parts thereof is only permitted under the provisions of the German Copyright Law of September 9, 1965, in its current version, and a copyright fee must always be paid. Violations fall under the prosecution act of the German Copyright Law.

© Springer-Verlag Berlin Heidelberg 1991. Printed in Germany

The use of registered names, trademarks, etc. in this publication does not imply, even in the absence of a specific statement, that such names are exempt from the relevant protective laws and regulations and therefore free for general use.

55/3140-543210 – Printed on acid-free paper

Preface

The first edition of this book was published in 1978 and a new Spanish edition in 1989. When the first edition appeared, Professor A. Martin suggested that an English translation would meet with interest. Together with Professor A.S. Wightman, he tried to convince an American publisher to translate the book. Financial problems made this impossible. Later on, Professors E.H. Lieb and W. Thirring proposed to entrust Springer-Verlag with the translation of our book, and Professor W. Beiglböck accepted the plan. We are deeply grateful to all of them, since without their interest and enthusiasm this book would not have been translated.

In the twelve years that have passed since the first edition was published, beautiful experiments confirming some of the basic principles of quantum mechanics have been carried out, and the theory has been enriched with new, important developments. Due reference to all of this has been paid in this English edition, which implies that modifications have been made to several parts of the book. Instances of these modifications are, on the one hand, the neutron interferometry experiments on wave-particle duality and the 2π rotation for fermions, and the crucial experiments of Aspect et al. with laser technology on Bell's inequalities, and, on the other hand, some recent results on level ordering in central potentials, new techniques in the analysis of anharmonic oscillators, and perturbative expansions for the Stark and Zeeman effects.

Major changes have been introduced in presenting the path-integral formalism, owing to its increasing importance in the modern formulation of quantum field theory. Also, the existence of new and more rigorous results has led us to change our treatment of the W.B.K. method. A new section on the inverse scattering problem has been added, because of its relevance in quantum physics and in the theory of integrable systems. Finally, we have tried to repair some omissions in the first edition, such as lower bounds to the ground-state energy and perturbation expansions of higher order for two-electron atoms.

The material appears in a two-volume edition. Volume I contains Chaps. 1 to 7 and Appendices A to G. The physical foundations and basic principles of quantum mechanics are essentially covered in this part. States, observables, uncertainty relations, time evolution, quantum measurements, pictures, representations, path integrals, inverse scattering, angular momentum, symmetries, are among the topics dealt with. Simple dynamics under one-dimensional and central potentials provide useful illustrations. Pertinent mathematical complements are included in the Appendices as well as a brief introduction to some polemical issues as the collapse of quantum states and hidden variables. Volume II comprises Chaps. 8 to 15. Collision theory, W.B.K. approximation, stationary perturbation theory and variational method, time-dependent perturbations, identical particles, and the quantum theory of radiation are presented. As applications the Stark and Zeeman effects, the atomic fine structure, the van der Waals forces, the BCS theory of superconductivity and the interaction of radiation with matter are discussed.

To our acknowledgements we wish to add a special one to G. García Alcaine for providing us with highly valuable bibliographic material for the updating of this book. We also thank Professor L. Alvarez-Gaumé for the interest he showed in translating the manuscript of the second Spanish edition. We are delighted with his excellent work. Finally, we would like to thank Springer-Verlag for all their help.

Madrid, May 1990 *A. Galindo, P. Pascual*

Preface to the Spanish Edition

Quantum mechanics is one of the basic pillars of physics. It is therefore not surprising that new results have been continuously accumulating, not only in a wide variety of applications, but also in the very foundations, which were laid down almost entirely in the spectacular burst of creative activity of the 1920s. Above all else, these efforts were, and are, intended to establish its axiomatic scheme and to make more rigorous many of the methods commonly employed.

Following the classic texts by Dirac and von Neumann, a large number of books on quantum mechanics and its mathematical and physical foundations have appeared, many of which are excellent. Our reasons for adding another one to the list arise from long teaching experience in the subject. First was the necessity of collecting in one book the clarification of many points which, in our opinion, remain obscure in traditional treatments. Second was the desire to incorporate important material which in general can only be found dispersed among scientific journals or in specialized monographs.

Given the huge quantity of information available, we decided to confine ourselves to what is normally termed "nonrelativistic quantum mechanics". The omission of the relativistic aspects of physical systems is due to our conviction that, for reasons of physics rather than textbook space, developments of relativistic quantum theory should be more than simply the last chapter in a textbook on quantum mechanics. Additionally, Chaps. 1 and 14 were included since it seemed appropriate to include a historical perspective of the genesis of quantum mechanics and its physical basis (albeit a short one since our physics majors have already had a course in quantum physics). Also, given that the atom plays a central role in quantum mechanics, we decided to use it to illustrate (Chap. 14) how one deals with a complex system using the approximation techniques developed in previous chapters. In addition to restricting the scope of the book, we sometimes found it necessary (with the length of the work in mind) to summarize or omit altogether some points of unquestionable interest, e.g., in Chap. 8 the study of dispersion relations, Regge poles, the Glauber approximation, etc.

A major problem which presented itself at the outset was that of choosing the mathematical level of the work. Aiming essentially at students of physics and related sciences, we adopted as a base the knowledge which these students acquire in normal undergraduate training. Thus, we started with the supposition of a certain familiarity with the language and at least the elementary techniques of Hilbert spaces, groups, etc., although at times our desire to make concepts more precise or to include (generally without proof) rigorous results have forced us to use more advanced mathematical terminology. With the aim of helping the reader who wishes to pursue such aspects, in Appendix C we summarize the relevant points and provide a pertinent bibliography.

The postulational basis of quantum mechanics is developed in Chap. 2 in the spirit of the orthodox "Copenhagen" interpretation, and the content of (and necessity for) each postulate is discussed. This approach is completed in Chap. 13 with the symmetrization principle for systems of identical particles. Other more controversial aspects, such as the problem of measurement and hidden variable theory, are discussed briefly in Appendices E and F.

In Chaps. 3 to 8 we develop the quantum principles, introducing some basic concepts with applications to simple systems, the majority of which are exactly soluble. Chapter 3 is devoted to the study of the wave function and its time evolution, and ends with a short introduction to Feynman's alternative formulation of quantum mechanics.

The properties of bound states in essentially one-dimensional systems are treated in Chaps. 4 and 6, the collision problem being included in Chap. 4 (scattering states for one-dimensional problems) and comprising all of Chap. 8. In Chap. 5 angular momentum is studied in detail and, given its practical importance, this study is completed in Appendix B with a summary of the most common formulae and tables of Clebsch-Gordan and Racah coefficients. Finally, Chap. 7 contains a discussion of symmetry transformations, the most important invariances of physical systems, and the associated conservation laws.

The dynamical complexity of the majority of interesting physical problems makes it necessary to resort to approximation methods in order to understand their behaviour. In Chaps. 9 and 10 the techniques appropriate to a discussion of stationary states are presented, reserving for Chap. 11 the usual methods for time-dependent perturbations. In Chap. 12 we study charged particles moving in electromagnetic fields, discuss the gauge-invariance of their dynamics and apply the above-mentioned perturbative methods to calculate the fine structure and the Zeeman effect in hydrogenic atoms.

The symmetrization principle is developed in Chap. 13, introducing second quantization formalism and applying it to many-body systems of identical particles displaying quantum behaviour at the macroscopic level (the phenomena of superfluidity and superconductivity). In Chap. 15, the last one of the book, we present a treatment, relatively complete, of the problem of the interaction of radiation with matter, with the necessary quantization of the radiation field.

The book concludes with a collection of appendices which, in addition to those mentioned above, include a summary of the most important special functions (Appendix A), elements of the theory of distributions and Fourier transforms (Appendix D), and properties of certain antiunitary operators (Appendix G). Contrary to the usual custom, we have not included problems or exercises. Our intent is to collect these in a forthcoming book, which will complement the present work in a practical sense.

The mathematical notation utilized in this work is the traditional one, although in some instances, to simplify the formulae, we have omitted symbols where there can be little room for confusion. Thus, for example, we use indistinguishably $\lambda I - A$ and $\lambda - A$. We also omit the limits or domains of integration on many definite integrals when they coincide with the natural ones. Concerning the numbering of equations, (3.37) or (X.37) indicates equation number 37 of Chap. 3 or Appendix X, respectively. Finally, [XY 57] is a reference to the article or work by a single author whose last name begins with XY, or of various authors whose last names start with X and Y, respectively; the number indicates the last two digits of the year in which the publication appeared. When necessary, we use [XY 57n] to distinguish between analogous cases.

To conclude, we wish to express, first and foremost, our gratitude to our families, whose understanding, sacrifice and support have permitted us to dedicate many hours to the realization of this book. We also wish to show our gratitude to various coworkers for their critical reviews of various chapters and appendices: L. Abellanas (Appendices C and D); R. F. Alvarez-Estrada (Chap. 8); R. Guardiola (Chap. 15); A. Morales (Chap. 7); A. F. Rañada (Chaps. 2 and 3); and C. Sánchez del Rio (Chaps. 1 and 14). Their suggestions were extremely valuable. G. García-Alcaine and M. A. Goñi read considerable portions of the original, and we also benefitted from their comments. Lastly, we appreciate Sra. C. Marcos' typing of a first draft of this work, and especially the patience and care with which Srta. M. A. Iglesias typed the text and formulae of the final version.

We do not wish to end this prologue without thanking the editorial staff of Alhambra, who were most cooperative in accepting our suggestions during the preparation of this book.

Contents

8. Scattering Theory ... 1
8.1 Introduction ... 1
8.2 General Description of Scattering Processes ... 1
8.3 Cross Sections ... 3
8.4 Invariance of the Cross Section. Change LAB ↔ C.M. ... 6
8.5 Simple Scattering: Classical Case ... 9
8.6 Simple Scattering: Quantum Case ... 13
8.7 Scattering Operator or S Matrix. Unitarity of the S Matrix ... 16
8.8 Intertwining Property and Energy Conservation ... 18
8.9 $d\sigma/d\Omega$ as a Function of T_F ... 20
8.10 Scattering into Cones ... 25
8.11 The Optical Theorem ... 26
8.12 Computation of the Scattering Amplitude ... 28
8.13 Space-Time Description of Simple Scattering ... 39
8.14 Symmetries of the Scattering Operator ... 43
8.15 Scattering by a Central Potential: Partial Waves and Phase Shifts ... 44
8.16 Computation and Properties of Phase Shifts ... 48
8.17 Scattering by a Central Square Well ... 53
8.18 Analyticity Properties of the Partial Amplitudes ... 56
8.19 Analyticity Properties of the Scattering Amplitude ... 67
8.20 Coulomb Scattering ... 69
8.21 Two Body Elastic Scattering ... 73
8.22 Multichannel Scattering ... 76
8.23 General Form of the Optical Theorem ... 82
8.24 Symmetries in Multichannel Scattering ... 83
8.25 The Optical Potential ... 85

9. The W.B.K. Method ... 88
9.1 Introduction ... 88
9.2 The W.B.K. Method in One-Dimensional Problems ... 88
9.3 Connection Formulae ... 90
9.4 Bound State Energies ... 95
9.5 The Potential Barrier ... 104
9.6 The Miller-Good Method ... 110
9.7 Transmission by Double Potential Barriers ... 113

9.8 Potential Wells: Several Turning Points 115
9.9 Central Potentials 118

10. Time-Independent Perturbation Theory and Variational Method 122
10.1 Introduction 122
10.2 Time-Independent Perturbations. The Non-Degenerate Case . 122
10.3 The Harmonic Oscillator with λx^4 Perturbation 127
10.4 The Harmonic Oscillator with λx^3 Perturbation 131
10.5 Two-Electron Atoms (I) 134
10.6 Van der Waals Forces (I) 136
10.7 Kato's Theory 138
10.8 The Stark Effect 144
10.9 The Variational Method 146
10.10 Two-Electron Atoms (II) 150
10.11 Van der Waals Forces (II) 154
10.12 One-Electron Atoms 155
10.13 Eigenvalues for Large Coupling Constants 157

11. Time-Dependent Perturbation Theory 161
11.1 Introduction 161
11.2 Nuclear Spin Resonance 162
11.3 The Forced Harmonic Oscillator 164
11.4 The Interaction Picture 172
11.5 Transition Probability 174
11.6 Constant Perturbations 176
11.7 Turning on Perturbations Adiabatically 180
11.8 Periodic Perturbations 182
11.9 Sudden Perturbations 184
11.10 The Adiabatic Theorem 186
11.11 The Adiabatic Approximation 190
11.12 The Decay Law for Unstable Quantum Systems 193

12. Particles in an Electromagnetic Field 200
12.1 Introduction 200
12.2 The Schrödinger Equation 202
12.3 Uncertainty Relations 206
12.4 The Aharonov-Bohm Effect 209
12.5 Spin-1/2 Particles in an E.M. Field 211
12.6 A Particle in a Constant Uniform Magnetic Field 215
12.7 The Fine Structure of Hydrogen-Like Atoms 217
12.8 One-Electron Atoms in a Magnetic Field 220

13. Systems of Identical Particles 227
13.1 Introduction 227

13.2 Symmetrization of Wave Functions 227
13.3 Non-Interacting Identical Particles 235
13.4 Fermi Gas 239
13.5 Bose Gas 243
13.6 Creation and Annihilation Operators 246
13.7 Correlation Functions 252
13.8 Superfluidity 258
13.9 Superconductivity 263

14. Atoms 268
14.1 Introduction 268
14.2 The Thomas-Fermi Method 271
14.3 The Hartree-Fock Method 277
14.4 The Central Field Approximation 283
14.5 Perturbative Calculations 286
14.6 Russell-Saunders or LS Coupling 288
14.7 jj Coupling 296

15. Quantum Theory of Radiation 298
15.1 Introduction 298
15.2 Plane Wave Expansions 299
15.3 Quantization of the E.M. Field 301
15.4 Multipole Waves 304
15.5 Interaction Between Matter and Radiation 308
15.6 Transition Probabilities 310
15.7 Emission and Absorption of Photons 318
15.8 Angular Distribution of Multipole Radiation 320
15.9 Electric Dipole Transitions in Atoms 322
15.10 Radiative Transitions in Nuclei 325
15.11 Low Energy Compton Scattering 333

Bibliography 337

Subject Index 353

Contents of **Quantum Mechanics I** 371

Physical Constants

Avogadro number	$N_A = 6.022\,045\,(31) \times 10^{23}$ particles/mol
Speed of light	$c = 2.997\,924\,58\,(1.2) \times 10^{10}\ \mathrm{cm\,s^{-1}}$
Proton charge	$\lvert e\rvert = 4.803\,242\,(14) \times 10^{-10}$ Fr $= 1.602\,189\,2\,(46) \times 10^{-19}$ C
Reduced Planck constant	$\hbar = 6.582\,173\,(17) \times 10^{-22}$ MeV s $= 1.054\,588\,7\,(57) \times 10^{-27}$ erg s $\hbar c = 1.973\,285\,8\,(51) \times 10^{-11}$ MeV cm $= 197.328\,58\,(51)$ MeV fm $= 0.624\,007\,8\,(16)\ \mathrm{GeV\,mb^{1/2}}$
Fine structure constant	$\alpha = e^2/\hbar c = 1/137.036\,04\,(11)$
Boltzmann constant	$k_B = 1.380\,662\,(44) \times 10^{-16}\ \mathrm{erg\,K^{-1}}$ $= 8.617\,35\,(28) \times 10^{-11}\ \mathrm{MeV\,K^{-1}}$
Electron mass	$m_e = 9.109\,534\,(47) \times 10^{-28}$ g $m_e c^2 = 0.511\,003\,4\,(14)$ MeV
Proton mass	$m_p = 1836.151\,52\,(70)\, m_e$ $m_p c^2 = 938.279\,6\,(27)$ MeV
Electron Compton wavelength	$\lambda_e = h/m_e c = 2.426\,308\,9\,(40) \times 10^{-10}$ cm
Bohr radius	$a_{\infty\mathrm{Bohr}} = 0.529\,177\,06\,(44)$ Å
Rydberg (energy)	$R_\infty = \frac{1}{2} m_e(\alpha c)^2 = 13.605\,804\,(36)$ eV
Rydberg (frequency)	$R_\infty = m_e(\alpha c)^2/4\pi\hbar = 3.289\,842\,(17) \times 10^{15}$ Hz
Rydberg (wave number)	$R_\infty = m_e(\alpha c)^2/4\pi\hbar c = 109\,737.32\,(56)\ \mathrm{cm^{-1}}$
Bohr magneton	$\mu_B = \hbar\lvert e\rvert/2m_e c = 0.578\,837\,85\,(95) \times 10^{-14}\ \mathrm{MeV\,G^{-1}}$
Nuclear magneton	$\mu_N = \hbar\lvert e\rvert/2m_p c = 3.152\,451\,5\,(53) \times 10^{-18}\ \mathrm{MeV\,G^{-1}}$
Gravitational constant	$G = 6.672\,0\,(41) \times 10^{-8}\ \mathrm{cm^3\,g^{-1}\,s^{-2}}$
Gravitational fine structure constant	$\alpha_G = \dfrac{G m_p^2}{\hbar c} = 5.904\,2\,(36) \times 10^{-39}$

Note: The values of these constants have been taken from [AC 84]. Numbers within brackets correspond to one standard deviation and affect the last digits. It should be remembered that in October 1983 the General Conference on Weights and Measures adopted a new definition for the meter, i.e. the length traversed by light in vacuum in a time 1/299 792 458 s, and therefore the speed of light is now, by definition, $c = 299\,792.458$ km/s.

After the printing of the Spanish second edition new adjustments of the fundamental constants have appeared which incorporate additional measurements and the definition for the meter quoted above [PD88].

8. Scattering Theory

8.1 Introduction

In this chapter we discuss some of the fundamental aspects of scattering theory in three dimensional problems. It is beyond the scope of this book to present a detailed account of this interesting subject, whose central importance in quantum mechanics cannot be overemphasized. The number of possible applications and approximation methods is so overwhelming, that we have deemed more appropriate to omit many of them and treat in more detail the fundamentals of the theory. Among the topics we will not discuss are the continuation of scattering amplitudes to the complex angular momentum plane, the impulse approximation, the Glauber approximation, analiticity in the multichannel formalism, etc. It should also be mentioned that there is a vast literature on rigorous results which will only be mentioned fragmentarily. The reader interested in learning more about this topic should consult more complete and/or rigorous sources, such as [GW 64], [AR 65], [SI 71], [TA 72], [JO 75], [AJ 77], [RS 79] and [NE 82].

8.2 General Description of Scattering Processes

Scattering experiments play a central role in quantum physics. They were used to discover some of the crucial aspects of the microphysical world (recall for example the Rutherford and Franck-Hertz experiments) and nowadays they are routinely used to explore the fundamental structure of matter and to produce new elementary particles.

In these experiments, represented diagramatically in Fig. 8.1 there is a beam of *projectiles* (P particles) incident upon a beam of *target* particles (B particles). As a result of the collision there will be some outgoing particles $A_1, A_2, \ldots, A_n$:

$$P + B \to A_1 + \ldots + A_n \ . \tag{8.1}$$

For the same initial configuration, for instance $\alpha + {}^{14}\mathrm{N}$, the final outcome can be quite varied:

$$\begin{aligned} \alpha + {}^{14}\mathrm{N} &\to \alpha + {}^{14}\mathrm{N} \\ &\to p + {}^{17}\mathrm{O} \\ &\to \alpha + \alpha + {}^{10}\mathrm{B} \\ &\to \text{etc.} \end{aligned} \tag{8.2}$$

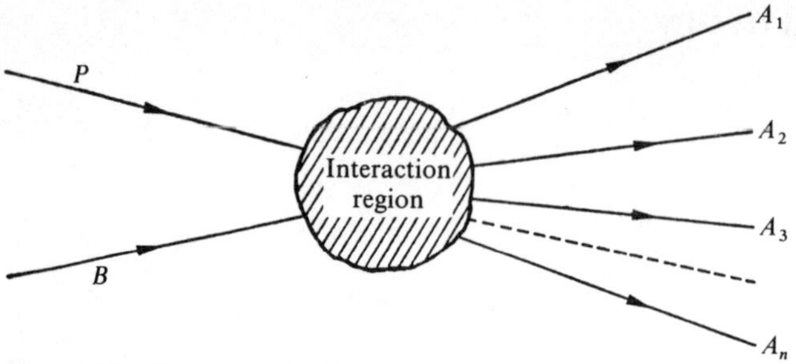

Fig. 8.1. Graphical representation of a scattering process

Each of these possibilities, characterized by the nature of the fragments in the final state and their internal states, is known as a *channel.* Thus the possibilities

$$\begin{aligned} \alpha + {}^{14}\mathrm{N} &\rightarrow \alpha + {}^{14}\mathrm{N} \\ &\rightarrow \alpha + {}^{14}\mathrm{N}^* \end{aligned} \tag{8.3}$$

where in the second case an excited state of ^{14}N is produced, are considered as different channels. However, the final configurations $A_1, A_2, \ldots, A_n$ and $A'_1, A'_2, \ldots, A'_n$ with the same fragments but differing at most in the value of the momentum and spin direction are supposed to be in the same outgoing channel. (Sometimes it is convenient to generalize the concept of channel, by neglecting those internal quantum numbers which do not play an important role in the scattering process. This is the case for example when we speak of the πN channel without specifying the electric charges of these particles.)

In general the scattering process is *multichannel.* It often happens that most of the final channels are *closed* due to energy considerations. The break-up channel $\alpha + {}^{14}\mathrm{N} \rightarrow 2\alpha + {}^{10}\mathrm{B}$ to proceed requires that the CM kinetic energy of the $\alpha + {}^{14}$N system be larger than the *threshold* energy characteristic of the final outgoing channel for a given incoming channel.

If at a given energy only the *elastic* channel is open $P + B \rightarrow P + B$, we say that the process is *simple* or *single-channel* scattering. As the energy increases new channels will open up. This does not mean necessarily that low energy scattering is always simple. Sometimes it is possible to have *exo-energetic* fragmentation:

$$\mathrm{p} + \bar{\mathrm{p}} \rightarrow \pi^+ + \pi^- \tag{8.4}$$

and even if this is not possible, as for the case of e + p, we have for any energy the open channel

$$\mathrm{e} + \mathrm{p} \rightarrow \mathrm{e} + \mathrm{p} + \gamma \tag{8.5}$$

since the vanishing of the photon mass implies that it can come out with an energy as small as we want.

The single-channel scattering is therefore an idealization, but under appropriate conditions, when all other channels have very small probabilities, it is a valid idealization of the physical situation, as in e + p until relatively high energies are reached.

Another idealization often used is to assume that the final fragments are stable. This is certainly not true for neutrons which constitute a very common particle beam in scattering experiments, and similarly if in the final state one finds excited atoms or nuclei. What happens in general is that the lifetimes of these particles are so large compared to the interaction time that for all practical purposes we can take them to be stable. Only when the interaction time and the lifetimes become comparable the idealization has to be called into question.

8.3 Cross Sections

In practice, the results of scattering experiments are expressed in terms of cross sections.

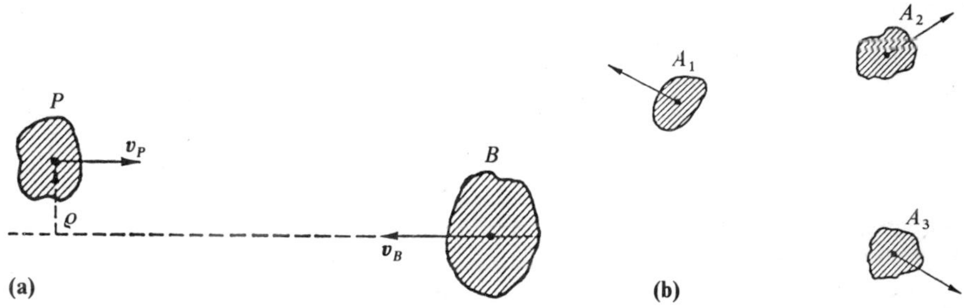

Fig. 8.2. (a) Before the collision. (b) After the collision

Consider a classical collision (Fig. 8.2) where the projectile and the target move with antiparallel velocities. (Incidentally, one of them can be at rest. In practice it is usually the target). After the collision, we observe some fragments (A_1, A_2, A_3 in Fig. 8.2b) coming out with some determined momenta, if the *impact parameter* ϱ (see Fig. 8.2a) allows the collision. Thus, little information is extracted of a single event. But if the experiment is repeated varying only ϱ, from the variety of final configurations for the same type of fragments (same scattering channel) we can extract very important information about the interaction P–B. In general not all final configurations will occur with the same frequency. It is precisely their probability distribution the indirect, although very valuable information that we obtain from scattering experiments.

In practice the repetition of experiments is achieved by sending a beam of incident particles with velocity $\boldsymbol{v}_P$ with randomly distributed impact parameters, onto a beam of targets with velocity $\boldsymbol{v}_B$.

A system of detectors $D_1, \ldots, D_n$ (Fig. 8.3) registers the fragments $A_1, \ldots, A_n$ produced in some individual P–B collision, with some momenta, third component of spin, and final internal configurations f within some interval Δf.

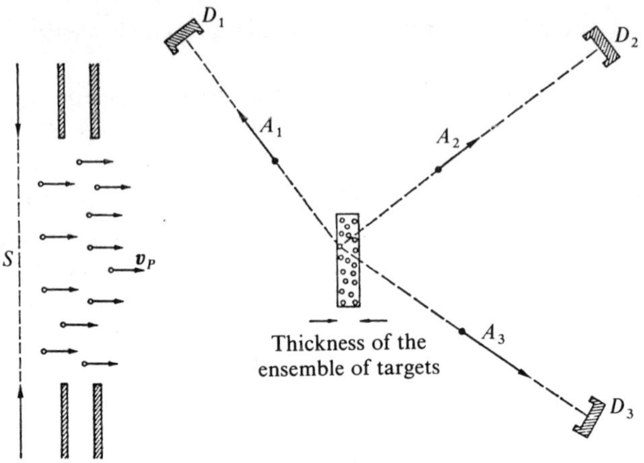

Fig. 8.3. Sketch of a scattering experiment

If $N(i \to \Delta f)$ is the total number of P–B collisions theoretically registered (after correcting the actual number of events registered by the efficiency of the detection device) where i specifies the momenta, spin states and other internal characteristics of the projectiles and targets before the collision, it is observed experimentally under certain conditions to be spelled out presently, that $N(i \to \Delta f) \propto N_B N_P / S$, where N_P, N_B are respectively the number of projectiles and targets, and S is the transverse cross section of the beam of projectiles.

This proportionality requires that each pair P–B acts independently of the presence of the others. For this to hold it is necessary that:

1) The target beam should be sufficiently "thin", and its particles randomly distributed, so that multiple collisions can be neglected, as well as shadowing of one target by another.
2) The incident beam should also be sufficiently "tenuous", so that the interactions between different particles can be neglected.

Finally it is clear that the proportionality alluded to will only take place if

3) The section S is sufficiently large so that all impact parameters capable of producing the reaction (8.1) with $f \in \Delta f$ are (randomly) represented.

The proportionality coefficient $\sigma(i \to \Delta f)$:

$$\boxed{\sigma(i \to \Delta f) = S \frac{N(i \to \Delta f)}{N_P N_B}} \tag{8.6}$$

has dimensions of (length)^2 and it is known as the *cross section* for the process (8.1), where the initial and final particles have momenta, spin states and internal configurations specified by i and $f \in \Delta f$, respectively.

A very convenient, typical unit for cross sections is the *barn* ($\equiv 10^{-24}\,\mathrm{cm}^2$), roughly the geometrical cross section of an object with a diameter of 10^{-12} cm. Calculating in natural units ($\hbar = c = 1$), $1(\mathrm{MeV})^{-2} = 3.893856\,(20) \times 10^2$ barn.

Usually (8.6) is rewritten in the form

$$\boxed{\sigma(i \to \Delta f) = \frac{N(i \to \Delta f)/\text{unit time and target particle}}{\text{flux of projectiles}}} . \tag{8.7}$$

When the scattering is governed by the laws of quantum mechanics, it is not possible to prepare the beams with exactly known momenta, and if we try to make them more precise, the notion of impact parameter looses its meaning. Using various filters and collimation equipment, it is at most possible to obtain a momentum distribution "peaked" around some central momentum. This leaves an enormous arbitrariness in the quantum state of each projectile and target. Assume first the *ideal* case where the projectiles are described by pure states and the state of one of them can be obtained from any other by random displacements transverse to the beam axis. If $|\phi\rangle$ denotes the joint state of a projectile and a target, and $|\phi_{\varrho}\rangle$ the state resulting after translating the projectile by ϱ perpendicularly to the incident direction, then in this ideal case, and with a single target, we have

$$N(\phi \to \Delta f) = \frac{N_P}{S} \int d^2\varrho \, W(\phi_{\varrho} \to \Delta f) \tag{8.8}$$

where $W(\phi_{\varrho} \to \Delta f)$ is the probability for $P + B$ in the state $|\phi_{\varrho}\rangle$ to produce (8.1) with final state $|f\rangle \in \Delta f$. Therefore, the ideal cross section is [TA 72]

$$\boxed{\sigma(\phi \to \Delta f) = \int d^2\varrho \, W(\phi_{\varrho} \to \Delta f)} . \tag{8.9}$$

We will see later (Sect. 8.9) that $\sigma(\phi \to \Delta f)$ depends only (under conditions to be specified below) on the central momenta and internal quantum numbers of the projectile and target, and not on the fine details of their center of mass wave functions.

In a real experiment, we have in principle control over the central momenta and internal quantum states of the projectile and the target, but there is a great deal of arbitrariness in the center of mass wave functions. Therefore, one would have to average over these uncontrollable aspects of the wave functions. The result mentioned previously indicates that this average is unnecessary, and hence the right hand side of (8.9) provides $\sigma(i \to \Delta f)$.

Finally, in the quantum case we have to add that for the proportionality (8.6) to hold, we must require the relative *incoherence* of the contributions from different targets, so that we can ignore the relative phases between individual amplitudes, and therefore sum over probabilities. An extraordinarily important example where this incoherence does not take place, is the scattering of particles with targets in a regular structure (crystal lattice). When the wavelengths of the incoming particles become comparable to the size of the unit cell of the lattice, the coherence between the individual amplitudes $P - B_i$, $i = 1, 2, \ldots$ can become very important, originating diffraction phenomena. Similarly, we have to take into account the possibility that some of the outgoing fragments for example A_1, A_2 are *identical* particles, and the detectors can not distinguish between the alternatives $(A_1 \to D_1,\ A_2 \to D_2)$, $(A_1 \to D_2,\ A_2 \to D_1)$. The detectors are sensitive to the fragment type, and to the quantum numbers imposed by their acceptance window, but not to the fictitious indices 1, 2. Hence, in this case, the transition probability $W(\phi \to f)$ is obtained by first summing over all transition amplitudes corresponding to all possible ways of permuting all the indices in each family of identical fragments, and then computing the absolute square of the quantity obtained. Later on we will study some examples illustrating the previous remarks (Sect. 8.21), as well as the influence of the bosonic or fermionic character of the identical particles on the relative phases due to the permutation of particle indices.

8.4 Invariance of the Cross Section. Change LAB ↔ C.M.

All quantities appearing in the right hand side of (8.6) are invariant under inertial transformations parallel to the incident direction. Thus, we obtain the invariance of the cross section:

$$\sigma(i \to \Delta f) = \sigma(i' \to (\Delta f)') \tag{8.10}$$

where i', $(\Delta f)'$ stand for the central momenta, spin states, etc., initial and final after the inertial transformation considered.

Particularly interesting is the change LAB ↔ C.M. for two body scattering:

$$P + B \to A_1 + A_2\ . \tag{8.11}$$

Given the simplicity of the relativistic kinematics involved in the analysis of this process, we present the discussion in this framework; to obtain the relevant expressions in the non-relativistic situations of interest, it suffices to take the galilean limit of the formulae we will derive.

When studying the effect of physical symmetries on transition amplitudes, we will see that the space-time translational invariance implies the conservation of energy and momentum in the process (8.1) and in particular in (8.11).

To simplify the notation in the rest of this section we will adopt units where $c = 1$.

Let $p \equiv (E, \boldsymbol{p})$, $p^* \equiv (E^*, \boldsymbol{p}^*)$ be the four momenta in the C.M. and LAB frames respectively (Fig. 8.4a,b); introducing the relativistic scalar Mandelstam variables,

$$\begin{aligned} s &\equiv (p_P + p_B)^2 = (p_P^* + p_B^*)^2 \\ t &\equiv (p_{A_1} - p_P)^2 = (p_{A_1}^* - p_P^*)^2 \\ u &\equiv (p_{A_2} - p_P)^2 = (p_{A_2}^* - p_P^*)^2 \end{aligned} \tag{8.12}$$

a simple computation shows that s, t, u are not independent, but they are related by

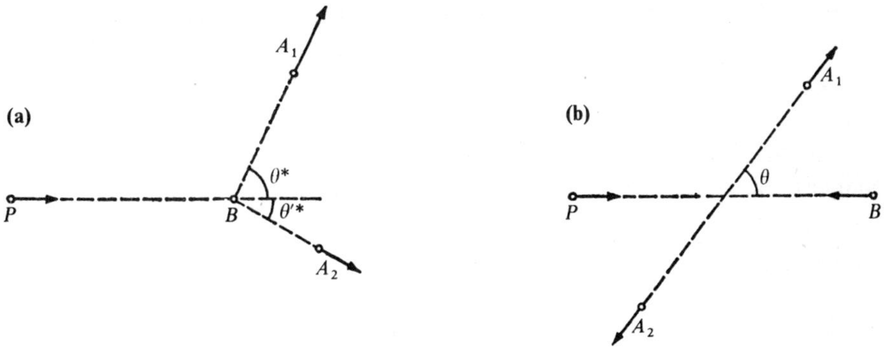

Fig. 8.4. (a) Collision in LAB frame. (b) Collision in the C.M. frame

$$s + t + u = M_P^2 + M_B^2 + M_{A_1}^2 + M_{A_2}^2 \tag{8.13}$$

and furthermore, in the C.M.:

$$\begin{aligned} E_P &= \frac{s + M_P^2 - M_B^2}{2\sqrt{s}}, \quad E_B = \frac{s + M_B^2 - M_P^2}{2\sqrt{s}} \\ |\boldsymbol{p}_P| = |\boldsymbol{p}_B| &= \frac{\lambda^{1/2}(s, M_P^2, M_B^2)}{2\sqrt{s}} \\ \cos\theta &= \frac{s(t-u) + (M_P^2 - M_B^2)(M_{A_1}^2 - M_{A_2}^2)}{\lambda^{1/2}(s, M_P^2, M_B^2)\lambda^{1/2}(s, M_{A_1}^2, M_{A_2}^2)} \end{aligned} \tag{8.14}$$

where

$$\lambda(a, b, c) \equiv a^2 + b^2 + c^2 - 2ab - 2bc - 2ca \,. \tag{8.15}$$

Similar expressions are obtained *mutatis mutandis* for E_{A_1}, E_{A_2} and $|\boldsymbol{p}_{A_1}|$ $(= |\boldsymbol{p}_{A_2}|)$.

Hence, knowing two of the variables s, t, u fixes uniquely the kinematical configuration of momenta in the C.M. up to rotations. The same should certainly occur in the LAB ($\boldsymbol{p}_B^* = 0$) frame, and indeed

$$\begin{aligned} E_P^* &= \frac{s - M_P^2 - M_B^2}{2M_B} \\ |\boldsymbol{p}_P^*| &= \frac{\lambda^{1/2}(s, M_B^2, M_P^2)}{2M_B} \\ \tan\theta &= \frac{(M_B/E_B)\sin\theta}{\cos\theta + (E_{A_1}/|\boldsymbol{p}_{A_1}|)(|\boldsymbol{p}_B|/E_B)} = \frac{\sin\theta}{\gamma\cos\theta + \tau} \,. \end{aligned} \tag{8.16}$$

The first two are derived with elementary manipulations, and the third is obtained using the Lorentz transformation changing the C.M. frame to the LAB frame. The quantities γ, τ are $\gamma \equiv E_B/M_B$, $\tau \equiv \gamma v_B/v_{A_1}$, and v_B, v_{A_1} are the velocities of B and A_1.

In particular, if the scattering is elastic ($P + B \to P + B$) and $M_P = M_B$

$$\tan\theta^* = \frac{M_B}{E_B}\tan(\theta/2)\ ; \tag{8.17}$$

thus, in this case $\theta^* \leq \pi/2$, and $\theta^* \simeq \theta/2$ in the non-relativistic limit $E_B \simeq M_B$. Similarly, in elastic collisions, and even if $M_P \neq M_B$,

$$\tan\theta^{*\prime} = \frac{M_B}{E_B}\cot(\theta/2) \tag{8.18}$$

and therefore in the non-relativistic limit of the elastic scattering of two particles of equal mass $\theta^* + \theta^{*\prime} \simeq \pi/2$.

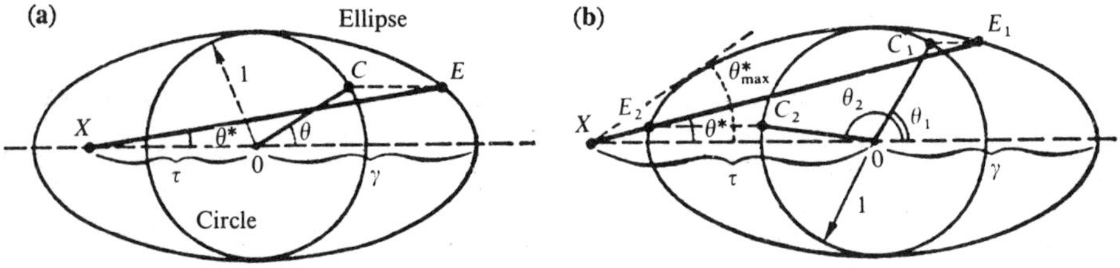

Fig. 8.5. **(a)** $\tau < \gamma$ case. **(b)** $\tau > \gamma$ case

The relation (8.16) is visualized geometrically in Fig. 8.5, where the outer ellipse has semiaxes $1, \gamma$. The point X to the left of the origin is at a distance τ of it. Suppose first $\tau < \gamma$ (Fig. 8.5a) in other words, $v_B < v_{A_1}$. If from X we draw the half-line with slope θ^* cutting the ellipse at the point E, and from there we draw a line parallel to the major axis of the ellipse, it will cut the unit circle at C. The slope of the line OC determines θ. In this case the correspondence $\theta \leftrightarrow \theta^*$ is one-to-one, and the range of θ^* is $[0, \pi]$. If $\tau > \gamma$ (Fig. 8.5b), i.e. $v_B > v_{A_1}$, there are two possible values of θ for each value of θ^*. The range of θ^* is $[0, \theta^*_{\text{max}}]$, with $\theta^*_{\text{max}} < \pi/2$. Notice however that each θ has associated a single θ^*. In the limiting case $\tau = \gamma$ (for example in the elastic collision of two particles of equal mass) X lies on the ellipse and the range of θ^* is $[0, \pi/2]$. Finally for low energy elastic collisions ($\gamma \to 1$) the ellipse and the circle coincide, $\tau \simeq M_P/M_B$ (taking $A_1 = P$), and we have

$$\begin{aligned} &\tau < 1 \Rightarrow \theta/2 \leq \theta^* \leq \theta \\ &\tau = 1 \Rightarrow \theta^* = \theta/2 \\ &\tau > 1 \Rightarrow \theta^* \leq \theta_1/2 \end{aligned} \tag{8.19}$$

where θ_1 is the smaller of the two θ's corresponding to θ^*.

After these kinematical considerations, we are ready to relate the differential cross sections in the C.M. and LAB systems for a collision (8.11). Denote by $\lambda_i = (\lambda_P, \lambda_B)$, $\lambda_f = (\lambda_{A_1}, \lambda_{A_2})$ the internal quantum states (spin etc.) of P, B, A_1, A_2 in the collision (8.11) in the center of mass, and let

$\sigma_{\lambda_f,\lambda_i}(\boldsymbol{p}_P, \boldsymbol{p}_B \to \Delta\boldsymbol{p}_{A_1}, \Delta\boldsymbol{p}_{A_2})$ be the cross section for final fragments with momenta in $\Delta\boldsymbol{p}_{A_1}, \Delta\boldsymbol{p}_{A_2}$. Using the conservation laws of energy and momentum, the kinematical configuration of the final momenta is fixed by the direction of $\boldsymbol{p}_{A_1}$. If the range of final momenta is such that this direction falls within a small solid angle $\Delta\Omega$, we can write

$$\sigma_{\lambda_f,\lambda_i}(\boldsymbol{p}_P, \boldsymbol{p}_B \to \Delta\Omega) \simeq \frac{d\sigma_{\lambda_f,\lambda_i}(\boldsymbol{p}_P, \boldsymbol{p}_B \to \boldsymbol{p}_{A_1}, \boldsymbol{p}_{A_2})}{d\Omega}\Delta\Omega \tag{8.20}$$

and according to (8.10), the differential cross sections $d\sigma/d\Omega$, $d\sigma/d\Omega^*$ in the C.M. and LAB frames respectively are therefore related by

$$\frac{d\sigma_{\lambda_f,\lambda_i}(\boldsymbol{p}_P, \boldsymbol{p}_B \to \boldsymbol{p}_{A_1}, \boldsymbol{p}_{A_2})}{d\Omega} = \frac{d\sigma_{\lambda_f^*,\lambda_i^*}(\boldsymbol{p}_P^*, \boldsymbol{p}_B^* \to \boldsymbol{p}_{A_1}^*, \boldsymbol{p}_{A_2}^*)}{d\Omega^*} \times \left|\frac{d\Omega^*}{d\Omega}\right| . \tag{8.21}$$

With * we indicate the internal states λ and momenta $\boldsymbol{p}$ as seen from the LAB frame.

It is clear that $|d\Omega^*/d\Omega| = |d(\cos\theta^*)/d(\cos\theta)|$, and using (8.16) we can write (8.21) explicitly

$$\frac{d\sigma_{\lambda_f^*,\lambda_i^*}(\boldsymbol{p}_P^*, \boldsymbol{p}_B^* \to \boldsymbol{p}_{A_1}^*, \boldsymbol{p}_{A_2}^*)}{d\Omega^*} = \frac{[(\tau+\gamma\cos\theta)^2+\sin^2\theta]^{3/2}}{|\gamma+\tau\cos\theta|}\frac{d\sigma_{\lambda_f,\lambda_i}(\boldsymbol{p}_P, \boldsymbol{p}_B \to \boldsymbol{p}_{A_1}, \boldsymbol{p}_{A_2})}{d\Omega} . \tag{8.22}$$

Since θ^* in the LAB does not fix the configuration of final momenta when $\tau > \gamma$, the application of (8.22) implicitly assumes the configuration as given. The kinematic proportionality factor relating $d\sigma/d\Omega$ and $d\sigma/d\Omega^*$ can distort considerably the shape of the differential cross section, and its effect is to reinforce $d\sigma/d\Omega^*$ in the forward direction ($\theta^* \simeq 0$).

Finally it is sometimes necessary to analyze collisions where the incident particles do not collide head on (for instance molecules in a gas). In these cases, it is natural to replace S by the area of the normal section of the projectile beam perpendicular to the relative motion of both beams. In this way (8.10) still holds for any inertial coordinate change.

8.5 Simple Scattering: Classical Case

In order to simplify the notation, but without loosing any of the essential features, we now discuss the scattering of a spinless non-relativistic particle by a potential (or very massive target particle whose recoil we neglect). Unless otherwise stated we will consider this situation in the following. It is also of great help in visualizing the process, and in the motivation of later concepts to analyze the classical situation.

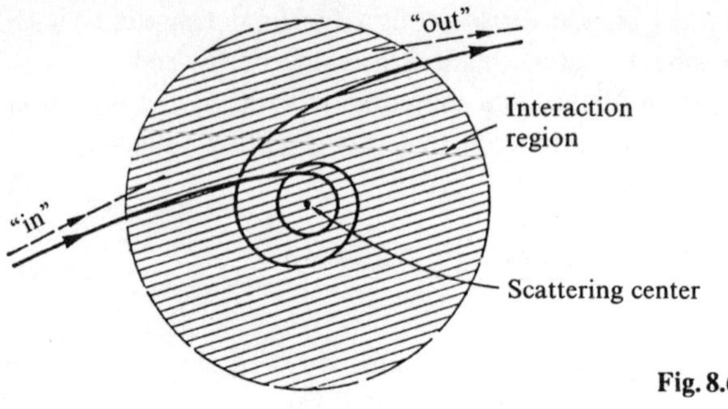

Fig. 8.6. Classical scattering

Suppose a classical particle moving towards an interaction region produced by an external scattering source (Fig. 8.6).

The particle described by a position vector $\boldsymbol{r}(t)$ will move freely in the asymptotic past ($t \to -\infty$) along the "in" asymptote ($\boldsymbol{a}_{\text{in}}, \boldsymbol{v}_{\text{in}}$):

$$\boldsymbol{r}(t) \underset{t\to-\infty}{\sim} \boldsymbol{a}_{\text{in}} + \boldsymbol{v}_{\text{in}} t \equiv \boldsymbol{r}_{\text{in}}(t) \tag{8.23}$$

if the interaction considered is not only sufficiently regular, but it also decays fast at large distances so that its effects do not persist in the asymptotic region. If the scattering center is not "too" attractive, after some finite time, the particle will emerge from the interaction region, escaping to infinity with an asymptotically free motion:

$$\boldsymbol{r}(t) \underset{t\to+\infty}{\sim} \boldsymbol{a}_{\text{out}} + \boldsymbol{v}_{\text{out}} t \equiv \boldsymbol{r}_{\text{out}}(t) \, . \tag{8.24}$$

From the point of view of collision theory we are not so much interested in the analytic details of $\boldsymbol{r}(t)$ as in the relation between its "in" and "out" asymptotes:

$$(\boldsymbol{a}_{\text{in}}, \boldsymbol{v}_{\text{in}}) \xrightarrow{S} (\boldsymbol{a}_{\text{out}}, \boldsymbol{v}_{\text{out}}) \, . \tag{8.25}$$

Under fairly general conditions for the interaction potential [SI 71a] (a particularly important one is the requirement that the force goes to zero as $|\boldsymbol{r}| \to \infty$ faster than $1/r^2$, to guarantee the free asymptotic behavior) it is possible to prove that for each "in" or "out" asymptote $(\boldsymbol{a}, \boldsymbol{v})$ with $\boldsymbol{v} \neq 0$ there are unique initial conditions $\boldsymbol{r}(0)$, $(d\boldsymbol{r}/dt)(0) \equiv \dot{\boldsymbol{r}}(0)$ determining a trajectory $\boldsymbol{r}(t)$ with the asymptotic behavior given by $(\boldsymbol{a}, \boldsymbol{v})$. Writing

$$\begin{aligned} \Omega_+(\boldsymbol{a}_{\text{in}}, \boldsymbol{v}_{\text{in}}) &= (\boldsymbol{r}(0), \dot{\boldsymbol{r}}(0)) \\ \Omega_-(\boldsymbol{a}_{\text{out}}, \boldsymbol{v}_{\text{out}}) &= (\boldsymbol{r}(0), \dot{\boldsymbol{r}}(0)) \end{aligned} \tag{8.26}$$

it is shown [SI 71a] that $\Omega_\pm$ provide maps in phase space preserving the Liouville measure, and whose ranges coincide up to zero measure sets. This leads to the fact that $\Omega_\pm$ are isometries in the space of square integrable functions in phase space with identical ranges, and therefore we obtain the existence of a unitary operator $(\Omega_-)^{-1}\Omega_+ \equiv S$ satisfying (8.25) except perhaps for a zero measure set.

To convince the reader that it is necessary to impose conditions guaranteeing the existence of asymptotic states, assume that the interaction potential is central $V(r)$ vanishing as $r \to \infty$. The equations determining the trajectories are:

$$M^2 \left(\frac{dr}{dt}\right)^2 + \frac{J^2}{r^2} + U(r) = \varepsilon$$
$$M^2 r^4 \left(\frac{d\theta}{dt}\right)^2 = J^2 \tag{8.27}$$

with $U(r) \equiv 2MV(r)$, $\varepsilon \equiv 2ME$, J is the angular momentum (a constant of motion because the interaction is spherically symmetric), and r, θ are the polar coordinates in the plane of the trajectory.

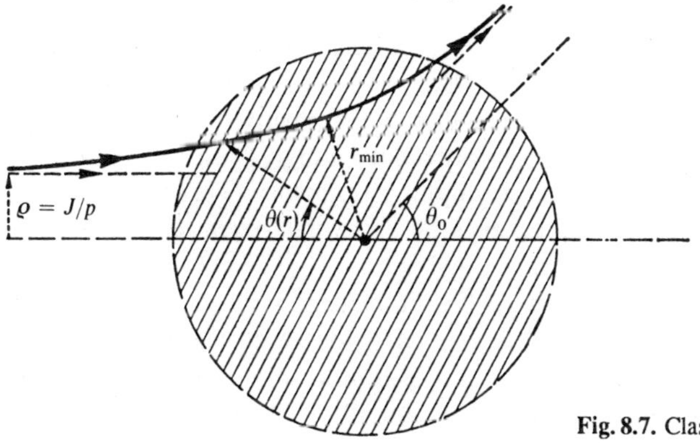

Fig. 8.7. Classical trajectories

Hence, for trajectories approaching the interaction region along the polar axis,

$$\theta(r) = J \int_r^\infty \frac{dr}{r^2} [p^2 - U_{\text{eff}}(r)]^{-1/2} \tag{8.28}$$

with $p^2 \equiv \varepsilon$, $U_{\text{eff}}(r) \equiv U(r) + J^2/r^2$, and r decreasing from ∞ until r_{min}, where r_{min} is the minimum distance of the incoming particle to the scattering center (in other words, the largest value of r solution to $p^2 = U_{\text{eff}}(r)$), the deflection angle θ_0 is (Fig. 8.7):

$$\theta_0 = \pi - 2\theta(r_{\text{min}}) \ . \tag{8.29}$$

Hence $-\infty \le \theta_0 \le \pi$.

If V decreases monotonically to zero (Fig. 8.8) the particle traverses any compact subset of $[r_{\text{min}}, \infty)$ in a finite time, and we have "normal bouncing". In this case we have naturally $0 \le \theta_0 \le \pi$, and θ_0 decreases as the impact parameter ϱ increases.

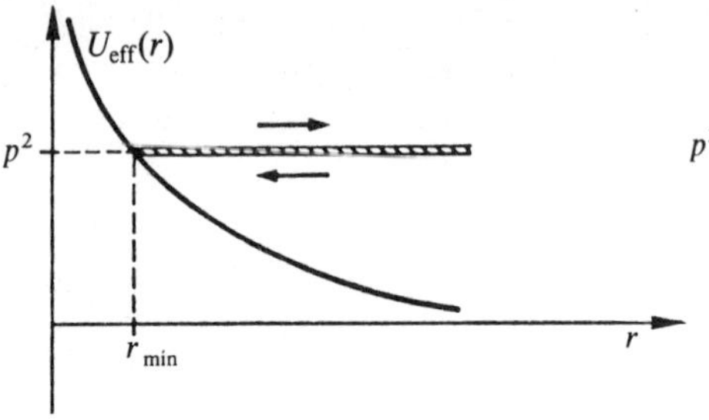

Fig. 8.8. Effective potential without maxima

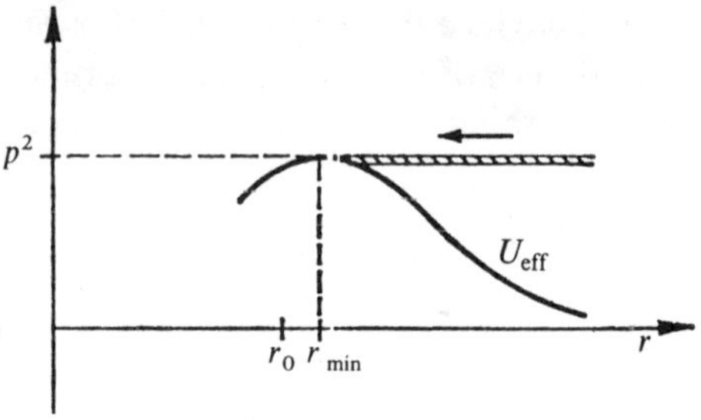

Fig. 8.9. Effective potential with a maximum

If on the other hand the potential V presents a relative maximum $V_0 > 0$ at some $r_0 > 0$ which we will assume as far as possible from the origin, then, for J sufficiently small $U_{\rm eff}$ will have a local maximum $U_{\rm eff,max}$ close to r_0, and for an energy such that $p^2 = U_{\rm eff,max}$ (see Fig. 8.9) there will be no bouncing, the projectile will "spiral" down to a distance $r_{\rm min}$ taking an amount of time which diverges at least logarithmically if $J \neq 0$ ($\theta(r_{\rm min}) = \infty$). [This "orbiting" effect is always possible so long as the potential $U_{\rm eff}$ has a positive local maximum.] Under the conditions stated, the particle does not emerge from the interaction region and there will not be an "out" asymptotic trajectory. This however will only occur for a zero measure set of asymptotic initial states, and it will be irrelevant for the general properties of the collision.

If the potential $V(r)$ is singularly attractive at the origin ($\lim_{r\to 0} r^2 V(r) < 0$), then for a set of initial asymptotic conditions with non-zero measure, there will not be any $r_{\rm min} > 0$, and such particles will reach in a finite time the scattering center if they fall from a finite distance. If the singularity is of the form $-a/r^n$, $a > 0$, then $\theta(r_{\rm min} = 0)$ is finite if $n > 2$, it is logarithmically divergent for $n = 2$, and $(d\theta/dr)(r_{\rm min} = 0)$ is finite for $n \geq 4$, so that the trajectory can be continued through the center unambiguously. On the other hand, if $2 < n < 4$, $(d\theta/dr)(r_{\rm min} = 0)$ is infinite, and it is impossible to continue the trajectory through the center in a unique way.

Not only the singularities at $r = 0$ can generate complications in the assignment of "in" and "out" asymptotic states. It is also essential to consider the behavior of $V(r)$ at large r. A typical example is found in the attractive or repulsive Coulomb potential. In this case it is impossible to assign asymptotes to collision trajectories. The reason is the following: since $U_{\rm eff}(r) = \alpha/r + J^2/r^2$, a simple computation using (8.27) shows that

$$|t| \underset{r\to\infty}{\sim} \frac{r(t)}{v} + \frac{\alpha}{2p^2 v} \ln r(t) , \quad v \equiv p/M \tag{8.30}$$

while if (8.23, 24) were satisfied, we should have $v|t| \sim r + \text{const}$. The same argument can be used to show that when $V(r)$ behaves as λ/r^α for $r \to \infty$, only for $\alpha > 1$ do the asymptotes exist.

Finally we recall that if the scattering angle θ decreases as ϱ increases (as in the case of a repulsive potential), the classical differential cross section is given by

$$\boxed{\frac{d\sigma}{d\Omega} = \varrho\left|\frac{d\varrho}{d(\cos\theta)}\right|} \,. \tag{8.31}$$

Hence the cross section for the projectile to emerge with an angle θ such that $\theta_1 \leq \theta \leq \theta_2$ will be

$$\sigma(\theta_1 \leq \theta \leq \theta_2) = \pi[\varrho_1^2 - \varrho_2^2] \tag{8.32}$$

where ϱ_1, ϱ_2 are the impact parameters corresponding to θ_1 and θ_2. From (8.32) it follows immediately that if the potential is not strictly of finite range (in other words, if it does not vanish from some distance on), the total cross section is infinite. This is because as $\theta_1 \to 0$ we have in this case $\varrho_1 \to \infty$: no matter how large the impact parameter, there is always scattering, and therefore there will be an infinite number of projectiles scattered in a neighborhood of the forward direction. This peculiar feature does not generally occur in the quantum case as we will show later. (For possible generalizations of the classical formula (8.31), see [NE 82].)

Next we discuss simple scattering in the context of quantum mechanics.

8.6 Simple Scattering: Quantum Case

Now the time evolution of the state, or quantum trajectory of the particle satisfies (2.123, 135)

$$|\Psi(t)\rangle = U(t)|\Psi(0)\rangle \tag{8.33}$$

with $U(t) = \exp(-itH/\hbar)$, $H = H_0 + V$ conservative, and $H_0 = p^2/2M$. We generally assume the particle to be spinless and the interaction to be of potential type $V(\boldsymbol{r})$. Outside the interaction region the influence of the potential will be negligible in the situation of interest, and the wave packet will evolve quasi-freely (Fig. 8.10). Hence it is reasonable to expect that there are states $|\psi_{\text{in}}\rangle$, $|\psi_{\text{out}}\rangle$ such that

$$|\Psi(t)\rangle \underset{t\to\mp\infty}{\sim} \mathrm{e}^{-itH_0/\hbar}|\psi_{\substack{\text{in}\\ \text{out}}}\rangle \equiv U_0(t)|\psi_{\substack{\text{in}\\ \text{out}}}\rangle \,. \tag{8.34}$$

More precisely, such that

$$\|\Psi(t) - \mathrm{e}^{-itH_0/\hbar}\psi_{\substack{\text{in}\\ \text{out}}}\| \to 0 \,, \quad t \to \mp\infty \,. \tag{8.35}$$

It is obvious that $|\psi_{\substack{\text{in}\\ \text{out}}}\rangle$ play now the role of $(\boldsymbol{a}_{\substack{\text{in}\\ \text{out}}}, \boldsymbol{v}_{\substack{\text{in}\\ \text{out}}})$.

From (8.33), (8.35), and the unitarity of $U(t)$ we immediately obtain

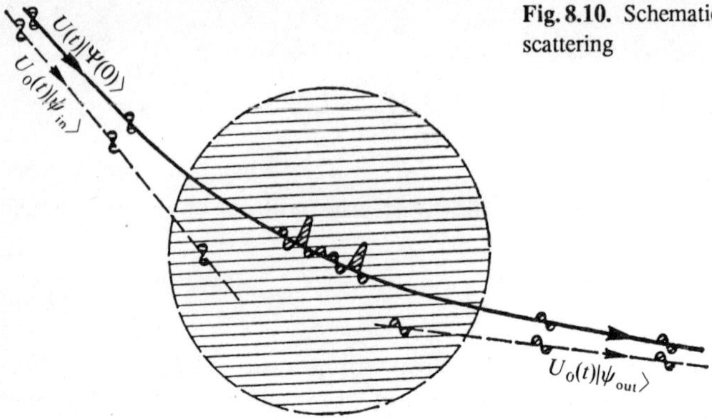

Fig. 8.10. Schematic description of quantum scattering

$$|\Psi(0)\rangle = \lim_{t\to\mp\infty} e^{itH/\hbar}e^{-itH_0/\hbar}|\psi_{\substack{in\\out}}\rangle \equiv \Omega_\pm|\psi_{\substack{in\\out}}\rangle \tag{8.36}$$

where the Möller operators

$$\boxed{\Omega_\pm \equiv \lim_{t\to\mp\infty} U^{-1}(t)U_0(t)} \tag{8.37}$$

as strong limits of unitary ones are isometric in their domain of definition and they extend to partial isometries (i.e. they are defined as zero in the subspace orthogonal to their domain of isometry). [It is physically clear why one cannot expect the limits (8.37) to exist in the uniform or norm topology of operators: the trajectories of two collisions, one taking place today and the other one hundred years from now would not reach their asymptotes at the same time except for trivial cases.

If an analytic argument is preferred, assume that

$$\|U^{-1}(t)U_0(t) - \Omega_+\| \to 0\,, \quad t \to -\infty\,. \tag{8.38}$$

Given $\varepsilon > 0$, there will exist a $T(\varepsilon)$ such that

$$\|U^{-1}(t)U_0(t) - \Omega_+\| < \varepsilon\,, \quad t < T(\varepsilon)\,. \tag{8.39}$$

From (8.38) we derive immediately that $\Omega_+ = U^{-1}(\tau)\Omega_+U_0(\tau)$ for all τ. Using (8.39) we will have

$$\begin{aligned}\|U^{-1}(t+\tau)U_0(t+\tau) - \Omega_+\| &= \|U^{-1}(t+\tau)U_0(t+\tau) - U^{-1}(\tau)\Omega_+U_0(\tau)\|\\ &= \|U^{-1}(t)U_0(t) - \Omega_+\| < \varepsilon\,, \quad t < T(\varepsilon)\end{aligned} \tag{8.40}$$

and since τ can have any value,

$$\|U^{-1}(t)U_0(t) - \Omega_+\| < \varepsilon\,, \quad \forall t\,. \tag{8.41}$$

As ε is arbitrary, $\Omega_+ = U^{-1}(t)U_0(t)$ for all t, namely $\Omega_+ = I$ and $H = H_0$.]

Let $DI(\Omega_\pm)$ be the isometry domains of $\Omega_\pm$. Several important questions arise naturally:

1. Which states are asymptotes of scattering orbits? In other words which are the domains $DI(\Omega_\pm)$?
2. Which quantum orbits admit asymptotes? Or equivalently how do we determine the ranges $R(\Omega_\pm) \equiv \mathcal{H}_{\substack{\text{in}\\\text{out}}}$ of $\Omega_\pm$?

From a physical point of view we expect $DI(\Omega_\pm)$ to fill all of $\mathcal{H}$, the space of all (pure) physical states of the particle. In the case of Ω_+ this is because the projectiles can be sent asymptotically ($t \to -\infty$) with any state which results from the free evolution towards the past of any state. [Ultimately the reason is the asymptotic spreading of free wave packets making the probability of finding a particle in any finite size region decrease as $|t|^{-3}$ for large $|t|$, as shown in Sect. 3.8. Hence for potentials decreasing sufficiently fast at large distances, this shows that the packets are asymptotically not affected by the potential.] For Ω_- the previous argument does not apply. However, if we assume for example that the interaction is time reversal invariant (as in the scattering by $V(\boldsymbol{r})$), then it is easy to see (Sect. 8.14) that $U_T\Omega_\pm = \Omega_\mp U_T$, and therefore $DI(\Omega_+) = \mathcal{H}$ implies $DI(\Omega_-) = \mathcal{H}$.

Furthermore, we should not expect that any quantum orbit under H should have asymptotes. It suffices to consider the bound states of H if any. Nevertheless, it is natural to expect that any quantum trajectory orthogonal to the bound states is a scattering state and possesses "in" and "out" asymptotes.

The questions (1) and (2) pose a well defined mathematical problem, and as in the classical case it is reasonable to expect that certain mathematical conditions must be required on V if we want to see reflected in the formalism the previous intuitive arguments. For local potentials, these conditions will limit the behavior of the potential at large distances and the nature of its singularities.

We refer the reader to [SI 71], [RS 79] for a discussion of the conditions which ensure the existence of $\Omega_\pm$ and other properties necessary to make the formal arguments we will present rigorous. As a guide, suffice it to say that if $V(\boldsymbol{r}) \in L^2(\mathbb{R}^3)$, $|V(\boldsymbol{r}| < C|\boldsymbol{r}|^{-(3+\varepsilon)}$, $\varepsilon > 0$, $|\boldsymbol{r}| \geq R_0$ and $V(\boldsymbol{r})$ is locally Hölder (namely $|V(\boldsymbol{r})-V(\boldsymbol{r}')| = O(|\boldsymbol{r}-\boldsymbol{r}'|^\alpha)$, $\boldsymbol{r}' \to \boldsymbol{r}$, $\alpha > 0$), except perhaps for a finite number of singularities (*Ikebe* conditions [IK 60, 65]) our formal arguments can be justified mathematically, and in particular the following properties are quite important:

1) $DI(\Omega_\pm) = \mathcal{H}$.
2) $\mathcal{H}_{\text{in}} = \mathcal{H}_{\text{out}} = \mathcal{H} \ominus \mathcal{H}_{\text{bound}}$ (*asymptotic completeness*).
3) H does not have any positive energy bound states, and only a finite number of bound states with energy ≤ 0.

Similarly, for central potentials these properties are satisfied under conditions such as: (a) $V(r)$ is continuous in $(0, \infty)$ except perhaps for a finite number of finite discontinuities; (b) $V(r) = O(1/r^{3+\varepsilon})$, $r \to \infty$, $\varepsilon > 0$; (c) $V(r) = O(1/r^{3/2-\varepsilon})$, $r \to 0$, $\varepsilon > 0$. The previous conditions or the Ikebe type conditions will be called *standard* conditions. With regard to property (2) it is interesting to point out that it automatically eliminates the possibility of having a singularly continuous part in the spectrum $\sigma(H)$. [The states in this part of the spectrum, with (2) not satisfied would neither be bound nor scattering states, and therefore would have an obscure physical interpretation.] Finally the orthogonality $\mathcal{H}_{\substack{\text{in}\\\text{out}}} \perp \mathcal{H}_{\text{bound}}$ expresses mathematically the idea that the scattering trajectories are orthogonal to bound trajectories because at large $|t|$ the overlap between a free evolution packet and a bound state wave function vanishes.

The conditions (1) and (2) are satisfied for potentials more general than those considered here. It suffices to have for example $(1 + |\boldsymbol{r}|)^{1+\varepsilon} V(\boldsymbol{r})$ in $L^2(\mathbb{R}^3) + L^\infty(\mathbb{R}^3)$ for some $\varepsilon > 0$; this family nearly contains the Coulomb potential. For more information on these issues see [RS 79]. Other interesting families of potentials for scattering theories have been introduced by *Enss* [EN 78], [EN 79], [EN 85] who also gave a new formulation of scattering theory with great physical and geometrical content based on the notion of localizability in phase space.

As in classical scattering, the asymptotic completeness can be affected by singular behavior at finite distance or slow decrease at large distances. Examples of the first case can be found in [PE 75].

8.7 Scattering Operator or S Matrix. Unitarity of the S Matrix

In a scattering process ordinarily only the asymptotic states are observables. This does not mean that one has to wait for a very long time to observe them. A slow projectile like a thermal neutron passes through the interaction region of a macromolecule in a few tenths of a nanosecond. Outside of this time interval its motion will be essentially free. Therefore what we want in scattering processes is a relation analogous to (8.25) between the "in" and "out" states.

Taking a potential V satisfying the conditions (1) and (2) of the previous section (something we will assume in the following), $\Omega_\pm$ are partial isometries with $DI(\Omega_\pm) = \mathcal{H}$, $R(\Omega_\pm) = \mathcal{H} \ominus \mathcal{H}_{\text{bound}} = \mathcal{H}_{\text{scattering}}$. This leads to

$$\boxed{\Omega_\pm^\dagger \Omega_\pm = I\,, \quad \Omega_\pm \Omega_\pm^\dagger = 1 - P_{\text{bound}} = P_{\text{scattering}}} \tag{8.42}$$

where P_{bound} and $P_{\text{scattering}}$ are the orthogonal projection operators onto $\mathcal{H}_{\text{bound}}$ and $\mathcal{H}_{\text{scattering}}$. Thus, using (8.36)

$$\boxed{|\psi_{\text{out}}\rangle = \Omega_-^\dagger |\Psi(0)\rangle = \Omega_-^\dagger \Omega_+ |\psi_{\text{in}}\rangle \equiv S|\psi_{\text{in}}\rangle} \tag{8.43}$$

where the operator S is given by

$$\boxed{S \equiv \Omega_-^\dagger \Omega_+}\,, \tag{8.44}$$

is *unitary* and it is known as the *scattering operator* or S *matrix*.

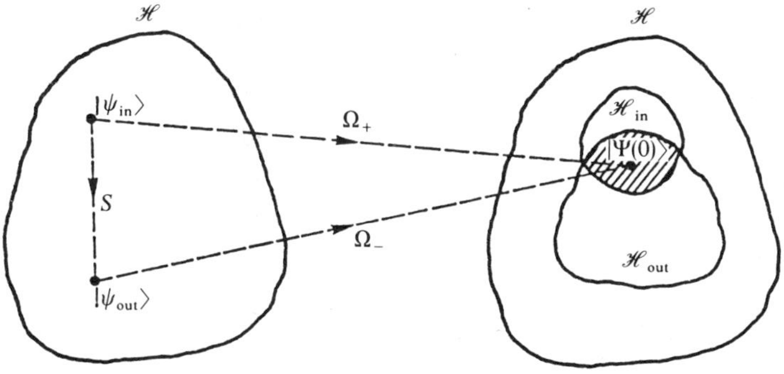

Fig. 8.11. Schematic representation of the domains and ranges of $\Omega_\pm$

The unitarity of S is of fundamental importance, and later we will discuss some of its physical consequences. It is the asymptotic completeness $\mathcal{H}_{in} = \mathcal{H}_{out}$ which guarantees the unitarity of S (Fig. 8.11).

Only if $\Omega_+|\psi_{in}\rangle \in \mathcal{H}_{out} = DI(\Omega_-^\dagger)$ will $\Omega_-^\dagger(\Omega_+|\psi_{in}\rangle)$ have the same norm as $\Omega_+|\psi_{in}\rangle$ and therefore as $|\psi_{in}\rangle$. On the other hand, only if $\mathcal{H}_{out} \subset \mathcal{H}_{in}$ will the range of $\Omega_-^\dagger\Omega_+$ fill $\mathcal{H}$. Therefore S is unitary if and only if $\mathcal{H}_{in} = \mathcal{H}_{out}$, provided that $DI(\Omega_\pm) = \mathcal{H}$. The unitarity of S is not simply a consequence of probability conservation [HU 68]; this is guaranteed by the self-adjoint property of H which in turn is not enough to guarantee asymptotic completeness. With this completeness taken for granted, the unitarity of S follows from elementary arguments; however, proving in a particular case that the asymptotic completeness condition is satisfied is in general a very difficult mathematical problem. If the S operator were not unitary, we would have rather strange situations. For example, we could have a scattering trajectory $|\Psi(t)\rangle$, therefore orthogonal to $\mathcal{H}_{bound}$ of which we would only see one asymptote. This is indeed very counter-intuitive. Actually, when the underlying scattering dynamics is unknown, the unitarity of S is taken as a basic postulate.

The relation (8.43) between $|\Psi(0)\rangle$ and $|\psi_{\substack{in\\out}}\rangle$ is represented schematically in Fig. 8.12.

The operator S relating the asymptotes of a quantum scattering trajectory is clearly very important:

Imagine that a projectile comes out of its source (for example an accelerator) in a state which if evolved freely would equal $|\psi_{in}\rangle$ at $t = 0$. After undergoing the interaction, it moves toward a detector far from the interaction region which is triggered when the "out" asymptote is in $P_{detector}\mathcal{H} \equiv \mathcal{H}_{detector}$. In other cases, the probability to trigger the detector is proportional to the norm of the projection of the asymptote onto $\mathcal{H}_{detector}$ (we are assuming 100% efficiency of the detector). Consequently the detection probability $W(\psi_{in} \to \mathcal{H}_{detector})$ will be

$$W(\psi_{in} \to \mathcal{H}_{detector}) = \|P_{detector}S\psi_{in}\|^2 . \tag{8.45}$$

In particular, for the ideal case $P_{detector} = |\phi\rangle\langle\phi|$, we have

$$\boxed{W(\psi_{in} \to \phi) = |\langle\phi|S|\psi_{in}\rangle|^2} . \tag{8.46}$$

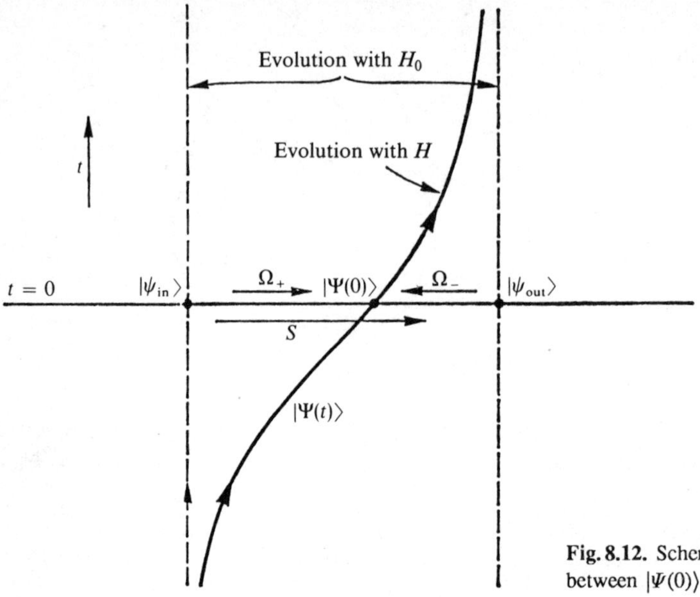

Fig. 8.12. Schematic relation between $|\Psi(0)\rangle$ and $|\psi_{\substack{in \\ out}}\rangle$

Thus the matrix element $\langle\phi|S|\psi_{in}\rangle$ represents the probability amplitude for the transition $\psi_{in} \to \phi$, namely, the probability amplitude for the projectile with incoming asymptote $|\psi_{in}\rangle$ to be found as $t \to \infty$ in the state $\exp(-itH_0/\hbar)|\phi\rangle$. The equations (8.9) and (8.46) explain the importance of the operator S in the computation of scattering cross sections.

The Ikebe conditions, or the more general conditions explained in [RS 79] which guarantee the existence and unitarity of S leave out unfortunately the familiar Coulomb interaction $V(r) = \alpha/r$. As in the classical case its long tail keeps on acting on the state at very long distances so that one never reaches effectively the asymptotic region. We will see when we treat this case explicitly that the Möller operators do not exist in the case of Coulomb interaction, although they can be modified without changing the physics of the asymptotic limit. This redefinition leads to a unitary S matrix [DO 64].

8.8 Intertwining Property and Energy Conservation

Since H is time independent, it is clear that the energy is conserved. This does not mean however that S and H commute. We have to take into account that S relates asymptotes for which the energy is purely kinetic and asymptotically equal to the total energy of the quantum trajectory since at large times the interaction ceases to act. What happens is that

$$\boxed{[S, H_0] = 0} \,. \tag{8.47}$$

The conservation of energy in the collision expressed in (8.47) is easy to prove. From the definition (8.37) of $\Omega_\pm$ and the relation

$$U(t)[U^{-1}(t_0)U_0(t_0)] = [U^{-1}(t_0 - t)U_0(t_0 - t)]U_0(t) \tag{8.48}$$

and taking the limit $t_0 \to \mp\infty$ we obtain

$$\mathrm{e}^{-\mathrm{i}tH/\hbar}\Omega_\pm = \Omega_\pm \mathrm{e}^{-\mathrm{i}tH_0/\hbar} \, . \tag{8.49}$$

Hence $\Omega_\pm$ *intertwines* the total and free time evolution operators. In particular

$$\boxed{H\Omega_\pm = \Omega_\pm H_0} \, . \tag{8.50}$$

Similarly taking adjoints in (8.49, 50):

$$\begin{aligned} \Omega_\pm^\dagger \mathrm{e}^{\mathrm{i}tH/\hbar} &= \mathrm{e}^{\mathrm{i}tH_0/\hbar}\Omega_\pm^\dagger \\ \Omega_\pm^\dagger H &= H_0 \Omega_\pm^\dagger \end{aligned} \tag{8.51}$$

and therefore

$$\begin{aligned} \mathrm{e}^{-\mathrm{i}tH_0/\hbar} S &= \mathrm{e}^{-\mathrm{i}tH_0/\hbar}\Omega_-^\dagger \Omega_+ = \Omega_-^\dagger \mathrm{e}^{-\mathrm{i}tH/\hbar}\Omega_+ \\ &= \Omega_-^\dagger \Omega_+ \mathrm{e}^{-\mathrm{i}tH_0/\hbar} = S\mathrm{e}^{-\mathrm{i}tH_0/\hbar} \end{aligned} \tag{8.52}$$

implying (8.47). [Actually (8.47, 50) and the second equation in (8.51) are formal ways of writing (8.52, 49) and the first equation in (8.51) respectively. They should be interpreted in the sense that the spectral families of each of the operators appearing in the vanishing commutator do commute among each other. We will not insist on this point any further.]

As it is well known, the fact that S and H_0 commute implies that S cannot connect eigenstates (in a generalized sense) of H_0 with different kinetic energies. The operator S is hence defined by its restriction S_E, $E \in \sigma(H_0)$, to the spaces $\mathcal{H}_E$ of "eigenvectors" of H_0 with eigenvalue E. Symbolically we can write

$$\langle E'\hat{\boldsymbol{p}}'|S|E\hat{\boldsymbol{p}}\rangle = \delta(E' - E)\langle E\hat{\boldsymbol{p}}'|S_E|E\hat{\boldsymbol{p}}\rangle \tag{8.53}$$

$|E\hat{\boldsymbol{p}}\rangle$ is the eigenstate of H_0 with eigenvalue E and direction $\hat{\boldsymbol{p}}$ for the momentum. S_E is only defined on the *energy shell* E and it is unitary in $\mathcal{H}_E$. Separating from S the non-interacting part ($S = I$), it is customary to write

$$S_E = I_E - 2\pi\mathrm{i}\varrho_E T_E \, . \tag{8.54}$$

$\varrho_E \equiv p^2/(dE/dp)$ is a kinematical factor chosen for convenience, and related to the change of basis $\langle \boldsymbol{p}'|E\hat{\boldsymbol{p}}\rangle = \varrho_E^{-1/2}\delta(E - E_{p'})\delta(\hat{\boldsymbol{p}} - \hat{\boldsymbol{p}}')$ which guarantees $\langle E'\hat{\boldsymbol{p}}|E\hat{\boldsymbol{p}}\rangle = \delta(E' - E)\delta(\hat{\boldsymbol{p}}' - \hat{\boldsymbol{p}})$. T_E is an operator in $\mathcal{H}_E$. Under quite general conditions on the potential $V(\boldsymbol{r})$ (for instance the standard conditions) it can be shown [SI 71], [JS 72] that T_E is a trace class operator. Hence it defines an integral operator $T_E(\hat{\boldsymbol{p}}', \hat{\boldsymbol{p}}) \equiv \langle E\hat{\boldsymbol{p}}'|T_E|E\hat{\boldsymbol{p}}\rangle$ in the space of square integrable functions on the surface of the unit sphere. We can write (8.53) in the form

$$\boxed{\langle \mathbf{p}'|S|\mathbf{p}\rangle = \delta(\mathbf{p}'-\mathbf{p}) - 2\pi\mathrm{i}\delta(E_{p'}-E_p)T_{E_p}(\hat{\mathbf{p}}',\hat{\mathbf{p}})} \tag{8.55}$$

as a symbolic expression for

$$\langle\phi'|S|\phi\rangle = \langle\phi'|\phi\rangle - 2\pi\mathrm{i}\int d^3p d^3p' \hat{\phi}'^*(\mathbf{p}')\delta(E_{p'}-E_p)T_{E_p}(\hat{\mathbf{p}}',\hat{\mathbf{p}})\hat{\phi}(\mathbf{p})\ . \tag{8.56}$$

Under the standard conditions (8.56) makes sense for all $\phi, \phi' \in \mathcal{H}$ because in this case $T_{E_p}(\hat{\mathbf{p}}',\hat{\mathbf{p}})$ is a continuous function of $\hat{\mathbf{p}}',\hat{\mathbf{p}}, \forall E_p > 0$ [IK 65]. Under more general conditions (8.56) may make sense only on states ϕ, ϕ' sufficiently regular [SI 71]. Since the kernel $T_{E_p}(\hat{\mathbf{p}}',\hat{\mathbf{p}})$ is the part of S_E coming from the interaction it is natural to expect that the differential cross section can be expressed in terms of it. We will prove this in the next section.

8.9 $d\sigma/d\Omega$ as a Function of T_E

If the incident beam consists of particles with "in" states $|\phi_\varrho\rangle$, obtained one from the others by translations orthogonal to the central direction of the beam, the scattering cross section $\sigma(\phi \to \Delta f)$ to a set of final states with momenta in a neighborhood $\Delta \boldsymbol{p}_f$ of the final momentum $\boldsymbol{p}_f$ is given according to (8.9) and (8.45) by

$$\sigma(\phi \to \Delta\boldsymbol{p}_f) = \int d^2\varrho \int_{p\in\Delta p_f} d^3p\ |(S\phi_\varrho)\hat{}(\mathbf{p})|^2\ . \tag{8.57}$$

If the momentum distribution $|\hat{\phi}(\mathbf{p})|^2$ occupies a region $\Delta\boldsymbol{p}_i$ peaked around the incoming momentum $\boldsymbol{p}_i$, and if $\boldsymbol{p}_f \neq \boldsymbol{p}_i$, the only contribution to (8.57) when the neighborhoods $\Delta\boldsymbol{p}_i, \Delta\boldsymbol{p}_f$ are sufficiently small so that there is no overlap comes from $S-1$ (Fig. 8.13). On the other hand, from (8.56) we have

$$\begin{aligned} |((S-1)\phi_\varrho)\hat{}(\mathbf{p})|^2 &= (2\pi)^2 \int d^3p' d^3p''\ \delta(E_p - E_{p'})\delta(E_p - E_{p''}) \\ &\times T^*_{E_p}(\hat{\mathbf{p}},\hat{\mathbf{p}}')T_{E_p}(\hat{\mathbf{p}},\hat{\mathbf{p}}'')\mathrm{e}^{\mathrm{i}(p'-p'')\cdot\varrho/\hbar}\hat{\phi}^*(\mathbf{p}')\hat{\phi}(\mathbf{p}'')\ . \end{aligned} \tag{8.58}$$

Since

$$\begin{aligned} &\int d^2\varrho\ \mathrm{e}^{\mathrm{i}(p'-p'')\cdot\varrho/\hbar} = (2\pi\hbar)^2\delta(\mathbf{p}'_\perp - \mathbf{p}''_\perp) \\ &\delta(E_{p'} - E_{p''})\delta(\mathbf{p}'_\perp - \mathbf{p}''_\perp) \\ &\quad = \frac{1}{v_{p'}}\left[\delta(\mathbf{p}'-\mathbf{p}'') + \delta(\mathbf{p}'_\perp - \mathbf{p}''_\perp)\delta(\mathbf{p}'_\parallel + \mathbf{p}''_\parallel)\right]\frac{|\mathbf{p}'|}{|\mathbf{p}'_\parallel|} \end{aligned} \tag{8.59}$$

where $v_p \equiv |\nabla_p E_p|$ and $\parallel, \perp$ indicate the projections over $\boldsymbol{p}_i$ and over the plane perpendicular to it respectively. Substituting (8.58) in (8.57) and using (8.59) we easily obtain

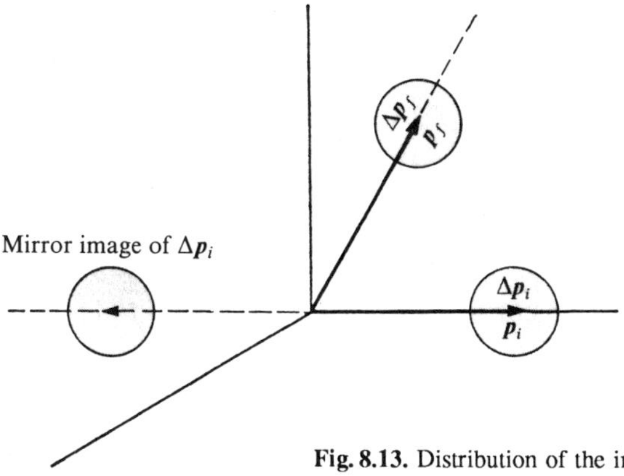

Fig. 8.13. Distribution of the initial and detected momenta

$$\sigma(\phi \to \Delta\boldsymbol{p}_f) = (2\pi)^4\hbar^2 \int d^3p' \int_{\Delta p_f} d^3p \times \delta(E_{p'} - E_p)|T_{E_p}(\hat{\boldsymbol{p}}, \hat{\boldsymbol{p}}')|^2|\hat{\phi}(\boldsymbol{p}')|^2 \frac{|\boldsymbol{p}'|}{v_{p'}|\boldsymbol{p}'_{\|}|} . \qquad (8.60)$$

(The term $\delta(\boldsymbol{p}'_\perp - \boldsymbol{p}''_\perp)\delta(\boldsymbol{p}'_{\|} + \boldsymbol{p}''_{\|})$ in (8.59) does not contribute to (8.57) if $\boldsymbol{p}_i \neq 0$ and if the neighborhood $\Delta\boldsymbol{p}_i$ is sufficiently small so that it does not overlap with its mirror image with respect to the plane $\boldsymbol{p}\cdot\boldsymbol{p}_i = 0$. This will be assumed to be the case).

Admitting now that $|T_{E_p}(\hat{\boldsymbol{p}}, \hat{\boldsymbol{p}}')|$ depends continuously on $|\boldsymbol{p}|, \hat{\boldsymbol{p}}, \hat{\boldsymbol{p}}'$ and that $\Delta\boldsymbol{p}_i, \Delta\boldsymbol{p}_f$ are small enough so that we can neglect the variation of $|T_{E_p}(\hat{\boldsymbol{p}}, \hat{\boldsymbol{p}}')|$ in $\Delta\boldsymbol{p}_f \times \Delta\boldsymbol{p}_i$ and $|\hat{\boldsymbol{p}}'| \simeq |\hat{\boldsymbol{p}}'_{\|}|$ the equation (8.60) can be rewritten as

$$\begin{aligned}\sigma(\phi \to \Delta\boldsymbol{p}_f) &\simeq (2\pi)^4\hbar^2|T_{E_i}(\hat{\boldsymbol{p}}_f, \hat{\boldsymbol{p}}_i)|^2 \int d^3p' \int_{\Delta p_f} d^3p \frac{1}{v_{p'}}\delta(E_{p'} - E_p)|\hat{\phi}(\boldsymbol{p}')|^2 \\ &= (2\pi)^4\hbar^2|T_{E_i}(\hat{\boldsymbol{p}}_f, \hat{\boldsymbol{p}}_i)|^2 \int d^3p' \int_{\Delta\hat{p}_f} d\Omega_{\hat{p}} \left(\frac{p'}{v_{p'}}\right)^2 |\hat{\phi}(\boldsymbol{p}')|^2 \times \begin{cases}1\\0\end{cases} \\ &\simeq (2\pi)^4\hbar^2 \left(\frac{p_i}{v_i}\right)^2 |T_{E_i}(\hat{\boldsymbol{p}}_f, \hat{\boldsymbol{p}}_i)|^2 \Delta\Omega_f \times \begin{cases}1\\0\end{cases} \qquad (8.61)\end{aligned}$$

where $E_i \equiv E_{p_i}$, $\Delta\Omega_f = \int_{\Delta\hat{p}_f} d\Omega_{\hat{p}}$ and we will have 1(0) according to whether $p_i\hat{\boldsymbol{p}}_f$ falls in (out) of $\Delta\boldsymbol{p}_f$. (This was expected as a consequence of energy conservation).

As anticipated in Sect. 8.3 the ideal cross section $\sigma(\phi \to \Delta\boldsymbol{p}_f)$ does not depend on the details of $|\phi\rangle$ but only on its central momentum $\boldsymbol{p}_i$. For this reason we can write

$$\sigma(\boldsymbol{p}_i \to \Delta\boldsymbol{p}_f) \simeq (2\pi\hbar)^4 \left(\frac{k_i}{v_i}\right)^2 |T_{E_i}(\hat{\boldsymbol{p}}_f, \hat{\boldsymbol{p}}_i)|^2 \Delta\Omega_f \times \begin{cases}1\\0\end{cases} \qquad (8.62)$$

with $k_i \equiv p_i/\hbar$, and summing over all final energies:

$$\boxed{\sigma(\boldsymbol{p}_i \to \Delta\hat{\boldsymbol{p}}_f) \simeq (2\pi\hbar)^4 \left(\frac{k_i}{v_i}\right)^2 |T_{E_i}(\hat{\boldsymbol{p}}_f, \hat{\boldsymbol{p}}_i)|^2 \Delta\Omega_f} \,. \tag{8.63}$$

$\sigma(\boldsymbol{p}_i \to \Delta\hat{\boldsymbol{p}}_f)$ means the scattering cross section into a solid angle $\Delta\Omega_f$ around the final direction $\hat{\boldsymbol{p}}_f$. The approximation symbols in (8.61–63) becomes equalities when the dimensions $\Delta\boldsymbol{p}_f$, $\Delta\hat{\boldsymbol{p}}_f$ tend to zero and therefore

$$\boxed{\frac{d\sigma(\boldsymbol{p}_i \to \boldsymbol{p}_j)}{d\Omega_f} = (2\pi\hbar)^4 \left(\frac{k_i}{v_i}\right)^2 |T_{E_i}(\hat{\boldsymbol{p}}_f, \hat{\boldsymbol{p}}_i)|^2} \tag{8.64}$$

with $p_i = p_f$ necessarily. This expression is valid both in relativistic and non-relativistic kinematics.

It is important to keep in mind that (8.62–64) are valid only as long as $|T_{E_p}(\hat{\boldsymbol{p}}, \hat{\boldsymbol{p}}')|$ is practically constant over $\Delta\boldsymbol{p}_f \times \Delta\boldsymbol{p}_i$. Hence (8.63,64) will hold if $|T_{E_f}(\hat{\boldsymbol{p}}_f, \hat{\boldsymbol{p}}')|$ does not present any important oscillations in the region $\Delta\boldsymbol{p}_i$ where ϕ is relevant. Very often this condition is satisfied in practice. However there are many important physical situations where T_E has a strong dependence on the energy and $\Delta\boldsymbol{p}_i$ cannot be made as small as one wants due for instance to the characteristics of the source of projectiles. These cases are virtually always related to the appearance of resonances in the scattering process and for them (8.62–64) are not valid.

In the previous derivation of scattering cross sections we used only $S - 1$ because the observation region always excluded $\Delta\boldsymbol{p}_i$. Thus the differential cross sections are given by (8.64) as long as $\hat{\boldsymbol{p}}_i \neq \hat{\boldsymbol{p}}_f$. The forward differential cross section ($\hat{\boldsymbol{p}}_i = \hat{\boldsymbol{p}}_f$) is not observable directly and it has to be defined by extrapolation from $d\sigma/d\Omega$ in the limit $\hat{\boldsymbol{p}}_f \to \hat{\boldsymbol{p}}_i, \hat{\boldsymbol{p}}_f \neq \hat{\boldsymbol{p}}_i$.

Finally the total elastic cross section is obtained by integrating (8.64) over all directions:

$$\sigma_{\text{el}}(\boldsymbol{p}_i) = (2\pi\hbar)^4 \left(\frac{k_i}{v_i}\right)^2 \int d\Omega_{\hat{\boldsymbol{p}}} |T_{E_i}(\hat{\boldsymbol{p}}, \hat{\boldsymbol{p}}_i)|^2 \,. \tag{8.65}$$

Under the standard conditions, T_{E_i} is a continuous function of its arguments, and therefore $\sigma_{\text{el}}(\boldsymbol{p}_i) < \infty$, $\forall \boldsymbol{p}_i \neq 0$. It would suffice to have T_E Hilbert-Schmidt for $\sigma_{\text{el}}(\boldsymbol{p}_i)$ to be finite for almost all $\boldsymbol{p}_i$. Furthermore if the interaction is spherically symmetric, σ_{el} only depends on p_i i.e. on E_i and in such case [JS 72]:

$$\begin{aligned} \sigma_{\text{el}}(p_i) &= \frac{1}{4\pi}(2\pi\hbar)^4 \left(\frac{k_i}{v_i}\right)^2 \|T_{E_i}\|_2^2 = \frac{\pi}{k_i^2}\|S_{E_i} - I_{E_i}\|_2^2 \\ &\leq \frac{2\pi}{k_i^2}\|S_{E_i} - I_{E_i}\|_1 \,. \end{aligned} \tag{8.66}$$

[The inequality is a consequence of the unitarity of S_{E_i}: the eigenvalues of S_{E_i} have unit modulus, and if $S_{E_i} - I_{E_i}$ is of trace class, the spectrum of S_{E_i}

will be discrete except for an accumulation point at 1. We also use the fact that for a complex number z of unit modulus, the following inequality holds, $|z-1|^2 \leq 2|z-1|$.]

In contrast with the classical case, $\sigma_{\rm el}$ may be finite even for potentials whose tails extend to infinity. This difference is physically explained by the uncertainty principle: if the impact parameter ϱ of the incoming particles is known with a precision $\Delta\varrho$ ($\ll \varrho$ so that it makes sense to talk of an approximate classical trajectory) we will necessarily have an uncertainty in its transverse momentum $\Delta p_\perp$ of the order of $\hbar/\Delta\varrho$. This uncertainty will in turn generate an uncertainty $\Delta\theta \sim (\Delta p_\perp)/p$ in the scattering angle, assuming ϱ large enough so that the change in the trajectory is small. Thus, to be able to say that the particle has deviated from its initial trajectory it is necessary that $\Delta\theta \ll \theta_0$, the scattering angle. Using (8.28, 29) and assuming $V(r) \sim ar^{-\lambda}$, $r \to \infty$, a simple computation shows that $\theta_0 \sim$ const. $\times\, \varrho^{-\lambda}$. Hence for large impact parameters it will make sense to say that the particle has deviated only if $\hbar/p \ll$ const. $\times\, \Delta\varrho/\varrho^\lambda$, namely if $\hbar/p \ll$ const. $\times\, \varrho^{1-\lambda}$. This is not the case for $\lambda > 1$, where the indeterminacy $\Delta\theta$ due to the uncertainty principle is $\gg \theta_0$ for $\varrho \to \infty$; then the classical description is not valid, and there is no reason to expect $\sigma_{\rm el}$ to diverge.

In fact, if $1 < \lambda \leq 2$, $\sigma_{\rm el}$ can be infinite (this is indeed the case when $V = C(1+r)^{-\lambda}$, $C > 0$). For locally integrable potentials decreasing at large distances faster than $|\boldsymbol{x}|^{-2}$, it can be shown that the cross section averaged over small energy intervals is finite [ES 80]. For a standard potential, which decreases therefore faster than $|\boldsymbol{x}|^{-3}$, we have already argued that the differential cross section is continuous for non-vanishing energies, and the total cross section is finite. In [ES 80] as well as in [MA 79], [MA 80] one can find bounds for $\sigma_{\rm el}$.

The classical cross section is finite for sufficiently regular potentials with compact support. It is then sensible to ask whether there exists any connection between the classical and the quantum cross sections computed for the same potential as $\hbar \to 0$. This is a very subtle problem [ES 80], and it can be argued that for positive spherically symmetric potentials with compact support, the quantum differential cross section coincides in this limit with the classical result for scattering angles different from zero. Keeping $\hbar$ fixed, this is equivalent to saying that for a potential λV and energy λE the differential cross section for $\lambda \to \infty$ tends to the corresponding classical cross section for a potential V and energy E. In spite of this, the total quantum cross section $\sigma_{\rm el}$ does not become the classical result, but twice that: $\sigma_{\rm el}(p_i) \to 2\sigma_{\rm cl}(p_i)$ as $\hbar \to 0$. This factor of two is due to the fact that when $\hbar \ll pR$, where R is the range of the potential, the differential cross section presents a sharp peak for angles $\theta \lesssim 1/kR$. This peak, in the $\hbar \to 0$ limit, becomes a delta function $\delta(\Omega_f - \Omega_i)$ which is missing in $\sigma_{\rm cl}$, for this cross section only counts deviated particles. It can be shown generally [ES 80] that this peak always contributes an amount πR^2 to $\sigma_{\rm el}$ in this limit. An illustrative example is provided by scattering by a hard impenetrable sphere of radius R. Classically, $d\sigma/d\Omega = R^2/4$, which is also valid in Q.M. for $kR \gg 1$ and $\theta \neq 0$ [NE 82]. However, the computation of $\sigma_{\rm el}$ has to be carried out by evaluating the radial flux of the scattered wave $\langle \boldsymbol{x}|(\Omega_+ - 1)|\boldsymbol{p}_i\rangle$ through a spherical surface containing the target (see Sect. 8.13). If $\hbar \to 0$, this scattered wave is made out of the rays deviated by the target (contributing an amount πR^2 to σ_{el}) and the rays of the incident beam which due to the "shadow" of the target do not appear in $\langle \boldsymbol{x}|\Omega_+|\boldsymbol{p}_i\rangle$. These last rays clearly acccount for the extra πR^2 contribution. In this discussion we are using the picture of rays, because it is allowed in the $\hbar \to 0$ limit.

The derivation of (8.64) has been quite laborious for the sake of rigor. For purely mnemonic reasons there are many formal derivations of (8.64). The simplest one is the following:

From (8.46) and (8.55) the transition probability $W(\boldsymbol{p}_i \to \boldsymbol{p}_f)$ taking $\boldsymbol{p}_i \neq \boldsymbol{p}_f$ will be

$$W(\boldsymbol{p}_i \to \boldsymbol{p}_f) = 4\pi^2\delta^2(E_f - E_i)|T_{E_i}(\hat{\boldsymbol{p}}_f, \hat{\boldsymbol{p}}_i)|^2 \,. \tag{8.67}$$

Formally

$$\delta^2(E_f - E_i) = \delta(E_f - E_i)\delta(0) \tag{8.68}$$

(although the notation δ^2 makes as little sense as (8.68) itself). Defining the δ-function as a limit

$$\delta(E) = \frac{1}{2\pi\hbar} \lim_{T\to\infty} \int_{-T/2}^{T/2} dt\, e^{iEt/\hbar} \tag{8.69}$$

we can say that

$$\delta(0) = \lim_{T\to\infty} \frac{T}{2\pi\hbar} \,. \tag{8.70}$$

Since we are improperly using stationary states, the interaction lasts an infinite amount of time ($\lim T$, $T \to \infty$) and the transition probability $w(\boldsymbol{p}_i \to \boldsymbol{p}_f)$ per unit time which follows from (8.67, 68 and 70) is

$$\boxed{w(\boldsymbol{p}_i \to \boldsymbol{p}_f) = \frac{2\pi}{\hbar}\delta(E_f - E_i)|T_{E_i}(\hat{\boldsymbol{p}}_f, \hat{\boldsymbol{p}}_i)|^2} \,. \tag{8.71}$$

Integrating (8.71) over a set of final states $\Delta\hat{\boldsymbol{p}}_f$ sufficiently narrow in the angular direction and with arbitrary energies yields

$$\boxed{w(\boldsymbol{p}_i \to \Delta\hat{\boldsymbol{p}}_f) \simeq \frac{2\pi}{\hbar}\varrho_{E_i}|T_{E_i}(\hat{\boldsymbol{p}}_f, \hat{\boldsymbol{p}}_i)|^2\Delta\Omega_f} \,. \tag{8.72}$$

This formula is sometimes known as the "platinum rule". Using (8.72) we immediately obtain (8.64) after we take (8.7) into account and use that $(2\pi\hbar)^{-3}v_i$ is the flux for a plane wave $\langle \boldsymbol{x}|\boldsymbol{p}_i\rangle$.

8.10 Scattering into Cones

We have computed the differential cross section using the probability for the outgoing particle momentum to lie in some element of solid angle $\Delta\Omega$ around a given direction in momentum space. If we look at the experimental set-ups used in scattering experiments, it seems more natural to compute the probability that as $t \to \infty$ the outgoing particle is found within a solid angle $\Delta\Omega$ of the direction joining the target and the detector. Although from a physical point of view it is to be expected that the results will not depend on the method used, we would like to sketch here a derivation for the scattering into cones in coordinate space [DO 69], [JL 72].

Let X be any measurable set in $\mathbb{R}^3$, and

$$P(X) \equiv \lim_{t\to\infty} \int_X d^3x \, |\Psi(\boldsymbol{x};t)|^2 \tag{8.73}$$

with $\Psi(\boldsymbol{x};t)$ the (normalized) wave function of a scattering trajectory. Since (8.35) and (8.43) imply that

$$\|\Psi(t) - \mathrm{e}^{-\mathrm{i}tH_0/\hbar} S\psi_{\mathrm{in}}\| \to 0\,, \quad t \to \infty \tag{8.74}$$

it is clear that

$$P(X) = \lim_{t\to\infty} \int_X d^3x \, |\Phi(\boldsymbol{x};t)|^2 \tag{8.75}$$

where $\Phi(\boldsymbol{x};t)$ is the wave function of $\exp(-\mathrm{i}tH_0/\hbar)S|\psi_{\mathrm{in}}\rangle$. In the non-relativistic case ($H_0 = p^2/2M$) (3.113) gives for $t \neq 0$ a representation of the free evolution operator of the form

$$\mathrm{e}^{-\mathrm{i}tH_0/\hbar} = C_t Q_t \tag{8.76}$$

with the unitary operators C_t, Q_t given for $t \neq 0$ by

$$\begin{aligned} (Q_t\phi)(\boldsymbol{x}) &\equiv \exp(\mathrm{i}Mx^2/2\hbar t)\phi(\boldsymbol{x}) \\ (C_t\phi)(\boldsymbol{x}) &\equiv (M/\mathrm{i}t)^{3/2} \exp(\mathrm{i}Mx^2/2\hbar t)\hat{\phi}(M\boldsymbol{x}/t)\,. \end{aligned} \tag{8.77}$$

The usefulness of the decomposition (8.76) is due to a remarkable property: The probability distribution $|\Phi(\boldsymbol{x};t)|^2$, in position space coincides asymptotically ($|t| \to \infty$) with $|(C_t\phi)(\boldsymbol{x})|^2$, where $\phi(\boldsymbol{x}) \equiv \Phi(\boldsymbol{x};0)$. [It suffices to take into account that $\|\exp(-\mathrm{i}tH_0/\hbar)\phi - C_t\phi\| = \|Q_t\phi - \phi\| \to 0$ if $|t| \to \infty$.] Since $|(C_t\phi)(\boldsymbol{x})|^2 = (M/|t|)^3|\hat{\phi}(M\boldsymbol{x}/t)|^2$ we immediately obtain:

$$P(X) = \lim_{t\to\infty} \left(\frac{M}{t}\right)^3 \int_X d^3x \, |\hat{\phi}(M\boldsymbol{x}/t)|^2\,. \tag{8.78}$$

In particular, if X is a cone C (a set of points of the form $\lambda\hat{\boldsymbol{x}}$, $\lambda \geq 0$, $\hat{\boldsymbol{x}} \in$ measurable set in the surface of the unit sphere), the change of variables $\boldsymbol{x} = (t/M)\boldsymbol{p}$ shows that

$$P(C) = \int_C d^3p\, |\hat{\phi}(\boldsymbol{p})|^2 = \int_C d^3p\, |(S\psi_{\text{in}})^\wedge(\boldsymbol{p})|^2 \,. \tag{8.79}$$

This equality states that the probability for the outgoing momentum to fall asymptotically inside C coincides with the probability for the particle to appear inside the same cone in position space. Incidentally, it is important to notice that $P(C)$ given by (8.73) does not change if we remove from C a compact set. This implies that as long as the detector covers the same solid angle as seen from the target, we can bring it closer or farther away without changing its counts. This applies clearly in the asymptotic region.

A very important consequence of the previous analysis [DO 69] is that the equation (8.73) defining $P(X)$ can make sense when $\Omega_\pm$ do not exist or even when S is not unitary. A trivial example is provided by the interaction V = const. $\neq 0$. A less trivial case is furnished by Coulomb scattering as we will see later in this chapter. Thus it is possible to treat and give meaning to scattering problems which scape the general framework. The basic idea comes from the observation that when S is unitary,

$$\begin{aligned} P(X) &= \lim_{t\to\infty} \int_X d^3x\, |(\mathrm{e}^{-\mathrm{i}tH/\hbar}\Omega_+\psi_{\text{in}})(\boldsymbol{x})|^2 \\ &= \lim_{t\to\infty} \int_X d^3x\, |(\mathrm{e}^{-2\mathrm{i}tH/\hbar}\mathrm{e}^{\mathrm{i}tH_0/\hbar}\psi_{\text{in}})(\boldsymbol{x})|^2 \,. \end{aligned} \tag{8.80}$$

Hence even if $\Omega_\pm$ do not exist, it could happen that for every scattering trajectory $|\Psi(t)\rangle$ there exists a unique $|\psi_{\text{in}}\rangle$ such that $P(X)$ coincides with the limit of the second integral in (8.80). If this were the case it would make sense to think of $|\psi_{\text{in}}\rangle$ as the "in" asymptote of $|\Psi(t)\rangle$ in some generalized sense, and to use $P(C)$ as the basic ingredient $W(\psi_{\text{in}} \to \hat{\boldsymbol{p}}_f \in C)$ to define using (8.9) the differential cross section.

Equation (8.79) still holds in the relativistic case [JL 72].

8.11 The Optical Theorem

From equation (8.54), and using the unitarity of S_E we obtain

$$\operatorname{Im} T_E \equiv \frac{1}{2\mathrm{i}}(T_E - T_E^\dagger) = -\pi\varrho_E T_E^\dagger T_E \tag{8.81}$$

or in integral form

$$\begin{aligned} &\frac{1}{2\mathrm{i}}\left[T_E(\hat{\boldsymbol{p}}', \hat{\boldsymbol{p}}) - T_E^*(\hat{\boldsymbol{p}}, \hat{\boldsymbol{p}}')\right] \\ &\quad = -\pi\varrho_E \int d\Omega_{\hat{\boldsymbol{p}}''}\, T_E^*(\hat{\boldsymbol{p}}'', \hat{\boldsymbol{p}}')T_E(\hat{\boldsymbol{p}}'', \hat{\boldsymbol{p}}) \,. \end{aligned} \tag{8.82}$$

In the forward scattering direction ($\hat{\boldsymbol{p}} = \hat{\boldsymbol{p}}'$):

$$\operatorname{Im} T_E(\hat{\boldsymbol{p}}, \hat{\boldsymbol{p}}) = -\pi \varrho_E \int d\Omega_{\hat{\boldsymbol{p}}''} |T_E(\hat{\boldsymbol{p}}'', \hat{\boldsymbol{p}})|^2 \tag{8.83}$$

and using (8.65) with $v \equiv v_p$

$$\operatorname{Im} T_E(\hat{\boldsymbol{p}}, \hat{\boldsymbol{p}}) = -\frac{\hbar v}{2(2\pi\hbar)^3} \sigma_{\mathrm{el}}(\boldsymbol{p}) \; . \tag{8.84}$$

It is customary to rewrite (8.84) using the *scattering amplitude* $f(\boldsymbol{p}_i \to \boldsymbol{p}_f)$ for $E_i = E_f$ (energy shell E_i):

$$\boxed{f(\boldsymbol{p}_i \to \boldsymbol{p}_f) \equiv -(2\pi\hbar)^2 \frac{k_i}{v_i} T_{E_i}(\hat{\boldsymbol{p}}_f, \hat{\boldsymbol{p}}_i)} \tag{8.85}$$

which will appear naturally in the study of the space dependence of $|\Psi(0)\rangle$ at large distances from the target. Then (8.64) and (8.84) become

$$\boxed{\frac{d\sigma(\boldsymbol{p}_i \to \boldsymbol{p}_f)}{d\Omega_f} = |f(\boldsymbol{p}_i \to \boldsymbol{p}_f)|^2} \tag{8.86}$$

$$\boxed{\operatorname{Im} f(\boldsymbol{p}_i \to \boldsymbol{p}_i) = \frac{k_i}{4\pi} \sigma_{\mathrm{el}}(\boldsymbol{p}_i)} \; . \tag{8.87}$$

This equation relating the imaginary part of the forward scattering amplitude and the total elastic cross section for simple scattering is known as the *optical theorem* or Bohr-Peierls-Placzek relation. Although it was discovered in quantum mechanics by *Feenberg* [FE 32] its origin goes back to Lord Rayleigh. Physically the optical theorem says that the scattering process diminishes the forward intensity of the beam after it crosses the target and that this attenuation, proportional to the cross section, can be explained in terms of the intensity of the wave $\Psi(\boldsymbol{x}; t)$. The latter is a superposition of the incident and the scattered waves, and it is precisely their interference in the forward direction proportional to $\operatorname{Im} f(\boldsymbol{p}_i \to \boldsymbol{p}_i)$ what explains qualitatively and quantitatively this attenuation. This will be explained in more detail in Sect. 8.13. This physical reasoning leads us to think that even though (8.84, 87) have been written only for the case of single channel scattering, a similar relation should hold for multichannel scattering if we replace in (8.84, 87) σ_{el} by σ_{tot}. This is indeed the case as will be shown in Sect. 8.13. Hence

$$\boxed{\operatorname{Im} f(\boldsymbol{p}_i \to \boldsymbol{p}_i) = \frac{k_i}{4\pi} \sigma_{\mathrm{tot}}(\boldsymbol{p}_i)} \tag{8.88}$$

and the experimental measurement of $\sigma_{\mathrm{tot}}(\boldsymbol{p}_i)$ gives the imaginary part of $f(\boldsymbol{p}_i \to \boldsymbol{p}_i)$. The extrapolation of $d\sigma_{\mathrm{el}}(\boldsymbol{p}_i \to \boldsymbol{p}_f)/d\Omega_f$ to $\Omega_i = \Omega_f$ allows us then to obtain $|\operatorname{Re} f(\boldsymbol{p}_i \to \boldsymbol{p}_i)|$. The sign of the real part has to be determined by interference of the elastic scattering produced by V with some other well known elastic scattering (like for instance Coulomb scattering) or using dispersion relations.

It is not easy to verify experimentally the optical theorem due to the difficulty of measuring the absolute phase of the scattering amplitude [NE 76]. Nevertheless there are scattering data obtained via interference with Coulomb scattering verifying the optical theorem with a 3% precision [ET 74].

8.12 Computation of the Scattering Amplitude

The previous arguments indicate that the quantity $T_E(\hat{\boldsymbol{p}}_f, \hat{\boldsymbol{p}}_i)$, or $f(\boldsymbol{p}_i \to \boldsymbol{p}_f)$ contains all the necessary information to compute elastic scattering cross sections. This quantity depends on the two Hamiltonians H, H_0, namely, the total and free energies, of the incoming particles. The aim of this section is to develop computational tools which extract the scattering amplitude from the dynamics represented by H and H_0.

We will proceed formally referring the reader to the literature (for example [SI 71], [RS 79]) for the conditions making the arguments rigorous.

a) The Lippmann-Schwinger Equations

The Möller operators were defined as strong limits in (8.37)

$$\Omega_\pm = \lim_{t \to \mp\infty} \mathrm{e}^{\mathrm{i}tH/\hbar} \mathrm{e}^{-\mathrm{i}tH_0/\hbar} . \tag{8.89}$$

If $|\phi\rangle \in DI(\Omega_\pm)$, then $\exp(\mathrm{i}tH/\hbar)\exp(-\mathrm{i}tH_0/\hbar)|\phi\rangle$ is a bounded continuous function of t, and therefore, its limit coincides with the abelian limit

$$\Omega_\pm|\phi\rangle = \lim_{\varepsilon \downarrow 0} \mp\frac{\varepsilon}{\hbar} \int_0^{\mp\infty} dt\, \mathrm{e}^{-\varepsilon|t|/\hbar} \mathrm{e}^{\mathrm{i}tH/\hbar} \mathrm{e}^{-\mathrm{i}tH_0/\hbar}|\phi\rangle . \tag{8.90}$$

Assume *formally* that $|\phi\rangle$ is an eigenstate of H_0 with eigenvalue E. Then (8.90) leads immediately to

$$|\phi_\pm\rangle \equiv \Omega_\pm|\phi\rangle = \lim_{\varepsilon \downarrow 0} \frac{\pm\varepsilon}{\mathrm{i}(H - E \mp \mathrm{i}\varepsilon)} |\phi\rangle . \tag{8.91}$$

The formal identities

$$(A+B)^{-1} = A^{-1} - A^{-1}B(A+B)^{-1} = A^{-1} - (A+B)^{-1}BA^{-1} \tag{8.92}$$

between the operators A, B applied to the resolvents

$$G(z) \equiv (z - H)^{-1} , \quad G_0(z) \equiv (z - H_0)^{-1} \tag{8.93}$$

lead to equations between the resolvents

$$\begin{aligned} G(z) &= G_0(z) + G_0(z)VG(z) \\ &= G_0(z) + G(z)VG_0(z) . \end{aligned} \tag{8.94}$$

Hence, using $H_0|\phi\rangle = E|\phi\rangle$, (8.91) becomes

$$\boxed{\begin{aligned} |\phi_\pm\rangle &= |\phi\rangle + G_0(E \pm \mathrm{i}0)V|\phi_\pm\rangle \\ &= |\phi\rangle + G(E \pm \mathrm{i}0)V|\phi\rangle \end{aligned}} \tag{8.95}$$

where $f(E \pm \mathrm{i}0)$ is defined as $\lim f(E \pm \mathrm{i}\varepsilon)$, $\varepsilon \downarrow 0$. The first equality in (8.95) provides the *Lippmann-Schwinger integral equations*, and the second gives a formal solution. This solution will be unique if the homogeneous integral equation does not possess any non-trivial solution. The homogeneous equation $|\psi\rangle = G_0(E \pm \mathrm{i}0)V|\psi\rangle$ makes sense also when $E < 0$. Then $|\psi\rangle = G_0(E)V|\psi\rangle$ and therefore formally $H|\psi\rangle = E|\psi\rangle$; thus, for $E < 0$ its solutions in $\mathcal{H}$ are the bound states of H with negative energy (this is the case rigorously if $V(\boldsymbol{x}) \in L^2(\mathbb{R}^3)$). The situation for $E \geq 0$ is more complicated and we refer the reader to the literature for a detailed account. We only mention that for local potentials, and under very general conditions (for instance if $V \in L^1 \cap L^2$), (8.95) has a unique solution for $E \geq 0$ except perhaps for a zero measure closed and bounded set of values of E containing the point spectrum of H in the region $E \geq 0$.

Due to the intertwinning property (8.50), $|\phi_\pm\rangle$ are both (improper) eigenfunctions of H with eigenvalue E. They differ by their boundary conditions: $|\phi_+\rangle(|\phi_-\rangle)$ is the scattering stationary state which in the far past (future) has $|\phi\rangle$ as its "free asymptote". These statements without qualifications are absurd, because the stationary character of $|\phi_\pm\rangle$ implies that its characteristics do not change with time, and therefore the free asymptotic character is strictly unreachable. However, in the same way that an incoming wave packet can be constructed by continuous superposition of the stationary states $|\phi\rangle$ of H_0, a scattering packet will formally be *the same* superposition of the states $|\phi_\pm\rangle$ of H due to the linearity of $\Omega_\pm$. If we construct this superposition with $|\phi_+\rangle$, the scattering state will have as its "in" asymptote the same superposition of the $|\phi\rangle$'s; and if instead we make it for the $|\phi_-\rangle$ states, the same superposition of the $|\phi\rangle$'s will be its "out" asymptote. Hence the boundary conditions of $|\phi_\pm\rangle$ make sense after the mollification due to the integration needed to construct wave packets.

Specifically, if $|\phi\rangle = |\boldsymbol{p}\rangle$ each $|\psi_{\substack{\mathrm{in}\\\mathrm{out}}}\rangle \in \mathcal{H}$ can be written as

$$|\psi_{\substack{\mathrm{in}\\\mathrm{out}}}\rangle = \int d^3p\, \hat{\psi}_{\substack{\mathrm{in}\\\mathrm{out}}}(\boldsymbol{p})|\boldsymbol{p}\rangle \tag{8.96}$$

and therefore

$$\Omega_\pm|\psi_{\substack{\mathrm{in}\\\mathrm{out}}}\rangle = \int d^3p\, \hat{\psi}_{\substack{\mathrm{in}\\\mathrm{out}}}(\boldsymbol{p})|\boldsymbol{p}_\pm\rangle\ . \tag{8.97}$$

b) Transition Operator T

Using (8.95) we immediately obtain

$$|\phi_+\rangle - |\phi_-\rangle = [G(E + \mathrm{i}0) - G(E - \mathrm{i}0)]V|\phi\rangle \tag{8.98}$$

and the formal manipulation of the relations between distributions

$$(x \pm \mathrm{i}0)^{-1} = \mathrm{PV}x^{-1} \mp \mathrm{i}\pi\delta(x) \tag{8.99}$$

leads to

$$G(E + \mathrm{i}0) - G(E - \mathrm{i}0) = -2\pi\mathrm{i}\delta(E - H)\ . \tag{8.100}$$

Hence, if $|\phi\rangle$, $|\phi'\rangle$ are eigenstates of H_0 with eigenvalues E, E',

$$\begin{aligned}\langle\phi'|S|\phi\rangle &= \langle\phi'_-|\phi_+\rangle = \langle\phi'_-|\phi_-\rangle - 2\pi\mathrm{i}\langle\phi'_-|\delta(E - H)V|\phi\rangle \\ &= \langle\phi'_+|\phi_+\rangle - 2\pi\mathrm{i}\langle\phi'|V\delta(E' - H)|\phi_+\rangle\ .\end{aligned} \tag{8.101}$$

The isometry property of $\Omega_\pm$ implies that $\langle\phi'_\pm|\phi_\pm\rangle = \langle\phi'|\phi\rangle$, and since $|\phi_+\rangle$, $|\phi'_-\rangle$ are eigenstates of H with eigenvalues E, E' (8.50), we finally obtain

$$\begin{aligned}\langle\phi'|S|\phi\rangle &= \langle\phi'|\phi\rangle - 2\pi\mathrm{i}\delta(E' - E)\langle\phi'_-|V|\phi\rangle \\ &= \langle\phi'|\phi\rangle - 2\pi\mathrm{i}\delta(E' - E)\langle\phi'|V|\phi_+\rangle\ .\end{aligned} \tag{8.102}$$

Comparing with (8.55), we derive on the energy shell $E_{\boldsymbol{p}} = E_{\boldsymbol{p}'}$

$$T_{E_p}(\hat{\boldsymbol{p}}', \hat{\boldsymbol{p}}) = \langle\boldsymbol{p}'_-|V|\boldsymbol{p}\rangle = \langle\boldsymbol{p}'|V|\boldsymbol{p}_+\rangle \tag{8.103}$$

and using (8.95),

$$T_E(\hat{\boldsymbol{p}}', \hat{\boldsymbol{p}}) = \langle\boldsymbol{p}'|V|\boldsymbol{p}\rangle + \int \langle\boldsymbol{p}'|V|\boldsymbol{p}''\rangle \frac{d^3p''}{E_{\boldsymbol{p}} + \mathrm{i}0 - E_{\boldsymbol{p}''}} \langle\boldsymbol{p}''|V|\boldsymbol{p}_+\rangle\ . \tag{8.104}$$

The previous expression is "almost" an integral equation for T_{E_p}. The problem is that the integration ranges over all $\boldsymbol{p}''$, and therefore $E_{\boldsymbol{p}''} \neq E_{\boldsymbol{p}}$ in general. Nevertheless it suggests the introduction of an operator $T(z)$ as a function of a complex variable z satisfying the equation

$$\boxed{T(z) = V + VG_0(z)T(z)} \tag{8.105}$$

and when $E_{\boldsymbol{p}} = E_{\boldsymbol{p}'}$ we will have

$$\boxed{T_{E_p}(\hat{\boldsymbol{p}}', \hat{\boldsymbol{p}}) = \langle\boldsymbol{p}'|T(E_{\boldsymbol{p}} + \mathrm{i}0)|\boldsymbol{p}\rangle} \tag{8.106}$$

as follows easily from (8.103) and (8.104).

The operator $T(z)$ is known as the *transition operator*. Its matrix elements $\langle\boldsymbol{p}'|T(z)|\boldsymbol{p}\rangle$ for $E_{\boldsymbol{p}} \neq E_{\boldsymbol{p}'}$ provide an off-shell extrapolation of $T_{E_p}(\hat{\boldsymbol{p}}', \hat{\boldsymbol{p}})$. Such extrapolation is very useful in practice in the computation of scattering amplitudes and in the determination of their analytic properties.

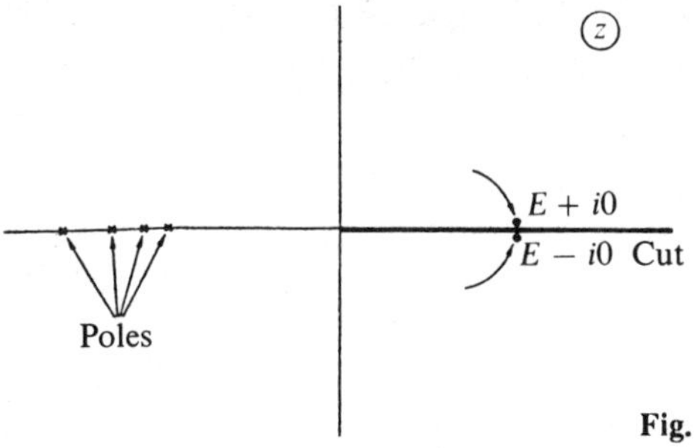

Fig. 8.14. Singularities of $T(z)$

Equation (8.105) is known as the *operator* form of the *Lippmann-Schwinger equation.* Its formal solution is

$$T(z) = (1 - VG_0(z))^{-1}V \tag{8.107}$$

and after power series expansion of $(1 - VG_0)^{-1}$ it leads to the famous *Born series*

$$\boxed{T(z) = V + VG_0(z)V + VG_0(z)VG_0(z)V + \ldots} \tag{8.108}$$

which is the starting point of numerous applications, and will be discussed later in this section. Since $G_0(z) + G_0(z)VG_0(z) + \ldots$ sums formally to $G(z)$ (8.94),

$$T(z) = V + VG(z)V \tag{8.109}$$

and $T(z)$ has essentially the same z-dependence as $G(z)$. Hence we expect $T(z)$ to be analytic in the z complex plane except for the non-negative real axis (continuous spectrum of H) where it will have a cut, and the values of E representing the negative point spectrum of H which will appear as isolated singularities (poles in general, see Fig. 8.14). Since V and H are self-adjoint from (8.109) we obtain

$$\boxed{T(z) = T^\dagger(z^*)} \tag{8.110}$$

hence $T(z)$ is a *Hermitian* analytic function. Its discontinuity across the cut satisfies

$$\begin{aligned}\operatorname{Im} T(E) &\equiv \frac{1}{2\mathrm{i}}[T(E + \mathrm{i}0) - T(E - \mathrm{i}0)] = V(\operatorname{Im} G(E))V \\ &= -\pi V\delta(E - H)V \leq 0\end{aligned} \tag{8.111}$$

due to (8.100). From (8.111) we obtain once again the optical theorem:

$$\begin{aligned}\operatorname{Im} T_{E_p}(\hat{\boldsymbol{p}}, \hat{\boldsymbol{p}}) &= \langle \boldsymbol{p}|\operatorname{Im} T(E_p)|\boldsymbol{p}\rangle = -\pi \int d^3p' \langle \boldsymbol{p}|V|\boldsymbol{p}'_-\rangle \delta(E_p - E_{p'})\langle \boldsymbol{p}'_-|V|\boldsymbol{p}\rangle \\ &= -\pi \varrho_{E_p} \int d\Omega_{\hat{p}} |T_{E_p}(\hat{\boldsymbol{p}}', \hat{\boldsymbol{p}})|^2 \,.\end{aligned} \tag{8.112}$$

When the integral kernel $VG_0(z)$ is compact, equation (8.105) can be analyzed rigorously [HU 68]. This happens for example when $V(\boldsymbol{x}) \in L^2(\mathbb{R}^3)$, since then $\|VG_0(z)\|_2 = M\hbar^{-2}\|V\|_2(2\pi\,\mathrm{Im}\,k)^{-1/2}$ with $\hbar k \equiv (2Mz)^{1/2}$, $\mathrm{Im}\,k > 0$; to derive the previous equality it suffices to use the formula

$$\langle \boldsymbol{x}|G_0(z)|\boldsymbol{y}\rangle = -\frac{M}{2\pi\hbar^2}\frac{\exp(\mathrm{i}k|\boldsymbol{x}-\boldsymbol{y}|)}{|\boldsymbol{x}-\boldsymbol{y}|} \tag{8.113}$$

for the integral kernel obtained from

$$\langle \boldsymbol{x}|G_0(z)|\boldsymbol{y}\rangle = \frac{1}{(2\pi\hbar)^3}\int d^3p\,\frac{\exp[(\mathrm{i}/\hbar)\boldsymbol{p}\cdot(\boldsymbol{x}-\boldsymbol{y})]}{z-(p^2/2M)} \tag{8.114}$$

performing first the angular integral, and then evaluating the radial integral by the method of residues.

When we approach the real axis ($\mathrm{Im}\,k \to 0$), the region of interest in scattering problems, the Hilbert-Schmidt norm diverges $\|VG_0(z)\|_2 \to \infty$. A very ingenious and efficient way of overcoming this difficulty and at the same time of finding additional advantages, is to introduce the "symmetrized" kernel [SW 64a]

$$\tilde{K}(z) \equiv |V|^{1/2}G_0(z)V^{1/2} \tag{8.115}$$

where $|V|^{1/2} \equiv (V^2)^{1/4}$, $V^{1/2} \equiv V/|V|^{1/2}$. In terms of $\tilde{K}(z)$, (8.107) becomes

$$T(z) = V^{1/2}[1-\tilde{K}(z)]^{-1}|V|^{1/2}\,. \tag{8.116}$$

The norm $\|\,.\,\|_2$ of $\tilde{K}(z)$, with $U(\boldsymbol{x}) \equiv 2MV(\boldsymbol{x})/\hbar^2$, is:

$$\begin{aligned}\|\tilde{K}(z)\|_2 &= \frac{1}{4\pi}\left[\int d^3x\,d^3x'\,\frac{|U(\boldsymbol{x})|\exp(-2\mathrm{Im}\,k|\boldsymbol{x}-\boldsymbol{x}'|)U(\boldsymbol{x}')|}{|\boldsymbol{x}-\boldsymbol{x}'|^2}\right]^{1/2}\\ &\le \frac{1}{4\pi}\|U\|_{\mathrm{R}}\end{aligned} \tag{8.117}$$

$\|\,.\,\|_{\mathrm{R}}$ is the Rollnik norm (6.10). Hence if $V \in R$, $\tilde{K}(z)$ will be Hilbert-Schmidt for all z in the *canonically cut* plane (the plane without the non-negative real axis), and the same property holds for its limiting values $\tilde{K}(E\pm\mathrm{i}0)$. Furthermore it can be shown that then $[1-\tilde{K}(z)]^{-1}$ is meromorphic in the cut plane and its poles coincide with the negative eigenvalues of H_0+V (this operator defined in the sense of quadratic forms [SI 71]) which are finite in number and multiplicity. The equation (8.116) provides a solution to (8.105), and (8.106) becomes

$$T_{E_p}(\hat{\boldsymbol{p}}',\hat{\boldsymbol{p}}) = \langle \boldsymbol{p}'|V^{1/2}[1-\tilde{K}(E_p+\mathrm{i}0)]^{-1}|V|^{1/2}|\boldsymbol{p}\rangle\,. \tag{8.118}$$

Assuming that $V \in R\cap L^1$, then $V^{1/2}|\boldsymbol{p}'\rangle$, $|V|^{1/2}|\boldsymbol{p}\rangle$ are normalizable and using (8.118) we can calculate the scattering amplitude through the matrix elements of $[1-\tilde{K}(E_p+\mathrm{i}0)]^{-1}$ between two states of $\mathcal{H}$. Finally, these inverse operators exist as bounded operators except perhaps for a closed, zero measure set of values of E_p [SI 71]. [These are the values of E mentioned after equation (8.95).] To summarize, if we have at least $V \in R\cap L^1$ (for instance, if $V \in L^1\cap L^2$), once the potential is known explicitly, (8.115) and (8.118) can be used to compute the scattering amplitude for almost all incident energies.

c) Born Series. Born Approximation. Applications

We proved in the previous subsection that the Born series

$$T(z) = V + VG_0(z)V + VG_0(z)VG_0(z)V + \dots \tag{8.119}$$

provides a formal solution for the transition operator $T(z)$. The convergence of the series on the right hand side depends on the convergence of

$$[1 - \tilde{K}(z)]^{-1} = 1 + \tilde{K}(z) + \tilde{K}(z)\tilde{K}(z) + \dots \tag{8.120}$$

between normalizable states as mentioned previously. When $\tilde{K}(z)$ is compact ($V \in R \cap L^1$), the series (8.120) converges in norm if the modulus $|\lambda_1(z)|$ of the eigenvalue of $\tilde{K}(z)$ farthest away from the origin satisfies $|\lambda_1(z)| < 1$. Since $|\lambda_1(z)| \leq \|\tilde{K}(z)\| \leq \|\tilde{K}(z)\|_2 \leq \|\tilde{K}(|z| \pm \mathrm{i}0)\|_2 = \|U\|_{\mathrm{R}}/4\pi$, we conclude:

1) The Born series converges for all z (in the canonically cut plane, as well as on the lips of the non-negative real axis) if the potential is sufficiently weak to satisfy $\|U\|_{\mathrm{R}} < 4\pi$; this inequality guarantees the absence of bound states in the spectrum of H (6.13).

On the other hand, defining $\tilde{K}_-(z) \equiv -|V|^{1/2}G_0(z)|V|^{1/2}$ it is clear that for all $\varphi \in L^2(\mathbb{R}^3)$ we have $|\langle\varphi, \tilde{K}_-(z)\varphi\rangle| \leq \langle|\varphi|, \tilde{K}(0)|\varphi|\rangle$, and therefore $|\lambda_1(z)| \leq \lambda_{1-}(0) = \|\tilde{K}_-(0)\|$, as $\tilde{K}_-(0) \geq 0$. Since for $E \leq 0$, $\tilde{K}_-(E)$ is a non-decreasing family of non-negative self-adjoint operators, continuous in the norm $\|\,.\,\|_2$ as a function of E and satisfying $\lim \tilde{K}_-(E) = 0$ when $E \to -\infty$, then necessarily $\lambda_{1-}(0) \leq 1$ if $-|V|$ has no bound states with negative energy. Furthermore, if $\lambda_{1-}(0) > 1$ then it is also true that $|\lambda_{1-}(E + \mathrm{i}0)| > 1$ for $E > 0$ sufficiently small, because the family of compact operators $\tilde{K}_-(E + \mathrm{i}0)$ is continuous in the $\|\,.\,\|_2$ norm on the axis $E \geq 0$. Consequently:

2) The Born series (8.120) converges for all z if the attractive potential $-|V|$ does not have any bound state with energy ≤ 0, and no resonances at zero energy (solutions of $\tilde{K}_-(0)\varphi = \varphi$ in $L^2(\mathbb{R}^3)$ and such that there is no $\psi \in L^2(\mathbb{R}^3)$ with $\varphi = |V|^{1/2}\psi$). If $V = \pm|V|$ this condition is also necessary.
3) If V is attractive and it possesses at least a bound state with negative energy, the Born series necessarily diverges at low energies.

Finally, it can be shown that $\|\tilde{K}(z)\| \to 0$ for z real $\to \pm\infty$; thus:
4) The Born series always converges at sufficiently high energies.

In practice, we rarely use more than the first term in the series (8.108) which yields $T(z)$ when (8.120) converges. With this truncation

$$\boxed{T(z) \simeq T_{\mathrm{B}}(z) \equiv V} \tag{8.121}$$

is the *(first) Born approximation*. We learn from the previous analysis that this approximation will be the better, the higher the energy. It is therefore a high energy approximation. And this holds regardless of the intensity of the potential. If furthermore the potential is weak, the validity of the approximation can be extended to all energies. Heuristic criteria are often given to determine the validity of the Born approximation. They are based on the comparison of the transition amplitude $f^{(2)}$ computed in second order (with the term $VG_0(z)V$) with the first order or Born approximation. When $|f^{(2)}(\boldsymbol{p}_i \to \boldsymbol{p}_f)| \ll |f_{\mathrm{B}}(\boldsymbol{p}_i \to \boldsymbol{p}_f)|$ one should expect the approximation $|f(\boldsymbol{p}_i \to \boldsymbol{p}_f)| \simeq |f_{\mathrm{B}}(\boldsymbol{p}_i \to \boldsymbol{p}_f)|$ to be good, although this is not always the case necessarily. A way of estimating the quotient $|f^{(2)}/f_{\mathrm{B}}| = |\langle \boldsymbol{p}_f|VG_0(E_i+\mathrm{i}0)V|\boldsymbol{p}_i\rangle/\langle \boldsymbol{p}_f|V|\boldsymbol{p}_i\rangle|$ and to make plausible that it is $\ll 1$ is to require that

$$|\langle \boldsymbol{x}|G_0(E_i+\mathrm{i}0)V|\boldsymbol{p}_i\rangle| \ll |\langle \boldsymbol{x}|\boldsymbol{p}_i\rangle| \tag{8.122}$$

over the $\boldsymbol{x}$ region where V is not negligible. In other words, we want the wave function $\langle \boldsymbol{x}|\boldsymbol{p}_{i+}\rangle$, stationary scattering wave with incoming asymptote $\langle \boldsymbol{x}|\boldsymbol{p}_i\rangle$ and satisfying the integral equation (8.95):

$$\langle \boldsymbol{x}|\boldsymbol{p}_{i+}\rangle = \langle \boldsymbol{x}|\boldsymbol{p}_i\rangle + \langle \boldsymbol{x}|G_0(E_i+\mathrm{i}0)V|\boldsymbol{p}_{i+}\rangle \tag{8.123}$$

with formal solution

$$\langle \boldsymbol{x}|\boldsymbol{p}_{i+}\rangle = \langle \boldsymbol{x}|\boldsymbol{p}_i\rangle + \langle \boldsymbol{x}|G_0(E_i+\mathrm{i}0)V|\boldsymbol{p}_i\rangle + \dots \tag{8.124}$$

to be such that its difference with $\langle \boldsymbol{x}|\boldsymbol{p}_i\rangle$ in first order of perturbation theory be negligible with respect to the unperturbed wave in the region of interest. Using (8.113) and assuming that $|V|$ takes its maximum in a neighborhood of $\boldsymbol{x} = 0$, the condition (8.122) becomes essentially

$$\frac{1}{4\pi}\left|\int d^3y\, \frac{\exp[\mathrm{i}(|\boldsymbol{k}_i||\boldsymbol{y}| + \boldsymbol{k}_i\cdot\boldsymbol{y})]}{|\boldsymbol{y}|}\, U(\boldsymbol{y})\right| \ll 1 \tag{8.125}$$

and if V is central

$$\left|\int_0^\infty dr\, U(r)[1-\mathrm{e}^{2\mathrm{i}k_i r}]\right| \ll 2k_i\,. \tag{8.126}$$

Thus, if $|V|$ has a range R and average value $|\overline{V}|$, the previous condition will be satisfied for all incident energies if

$$|\overline{V}| \ll \hbar^2/MR^2 \tag{8.127}$$

namely, if $|\overline{V}|$ is very small compared to the average kinetic energy of the particle confined in the region within the range of the potential; hence $-|V|$ is unable to bound the incoming particle. At high energies ($k_iR \gg 1$) the oscillating term in (8.126) can be neglected, and the condition becomes

$$|\overline{V}| \ll \hbar p_i/MR = \frac{\hbar^2}{MR^2}\left(\frac{R}{\lambda\!\!\!^{-}_i}\right) \tag{8.128}$$

or equivalently, the time R/v_i spent by the incoming particle in the region of influence of the potential is $\ll$ than the time $\hbar/|\overline{V}|$ required for the potential to affect the particle trajectory. Notice that the estimate (8.127) is essentially $\|\tilde{K}(z)\|_2 \ll 1$.

In the Born approximation, and using (8.85) and (8.106) the scattering amplitude becomes

$$f_B(\boldsymbol{p}_i \to \boldsymbol{p}_f) = -\frac{1}{4\pi}\int d^3x\, U(\boldsymbol{x}) e^{-i(\boldsymbol{k}_j - \boldsymbol{k}_i)\cdot \boldsymbol{x}} \tag{8.129}$$

i.e. it is proportional to the Fourier transform of the potential. Defining $\boldsymbol{q} \equiv \boldsymbol{k}_f - \boldsymbol{k}_i$, then $\hbar \boldsymbol{q}$ is the *momentum transfer* and (8.129) becomes for central potentials:

$$f_B(\boldsymbol{p}_i \to \boldsymbol{p}_f) \equiv f_B(p_i, \cos\theta) = -\int_0^\infty r^2\, dr\, U(r) \frac{\sin qr}{qr}\,, \qquad q = 2k_i \sin(\theta/2)\,. \tag{8.130}$$

Several important properties of the Born approximation appear after cursory examination of (8.129):

1) Im $f_B(\boldsymbol{p}_i \to \boldsymbol{p}_i) = 0$. From the optical theorem only when the total cross section $\sigma_{tot}(\boldsymbol{p}_i)$ vanishes, can the forward scattering amplitude have a vanishing imaginary part. Hence the formula used for the Born approximation *violates unitarity*. This has a simple explanation: the optical theorem (8.87) relates two expressions, one linear, and the other quadratic in the amplitude f. If the amplitude appearing in the cross section is computed to order n in the potential, to verify (8.87) we would have to calculate Im $f(\boldsymbol{p}_i \to \boldsymbol{p}_i)$, to order $2n$ in V. Since $f = 0$ to order 0, we must have necessarily that Im $f^{(1)}(\boldsymbol{p}_i \to \boldsymbol{p}_i) = 0$, and $\sigma_B^{el}(\boldsymbol{p}_i)$ will be related to Im $f^{(2)}(\boldsymbol{p}_i \to \boldsymbol{p}_i)$.

   This violation of unitarity in the use of the Born approximation will be negligible whenever the approximation is justified. It is interesting to notice that the criteria (8.127) and (8.128) can be summarized in a single condition:

$$\sigma_B^{el}(p_i) \ll \sigma_{geom} \equiv \pi R^2\,. \tag{8.131}$$

   If this inequality holds for all energies, the Born approximation will be plausibly valid everywhere. Otherwise, (8.131) will hold at high energies where $f \simeq f_B$ will be a good approximation in general.
2) At low energies

$$\lim_{p_i \to 0} f_B(\boldsymbol{p}_i \to \boldsymbol{p}_f) = -\frac{1}{4\pi}\int d^3x\, U(\boldsymbol{x}) \tag{8.132}$$

   hence f_B is isotropic in this limit (see Fig. 8.15).
3) In the forward direction ($\boldsymbol{q} = 0$)

$$f_B(\boldsymbol{p}_i \to \boldsymbol{p}_i) = -\frac{1}{4\pi}\int d^3x\, U(\boldsymbol{x}) \tag{8.133}$$

   and it does not depend on the energy (see Fig. 8.15).

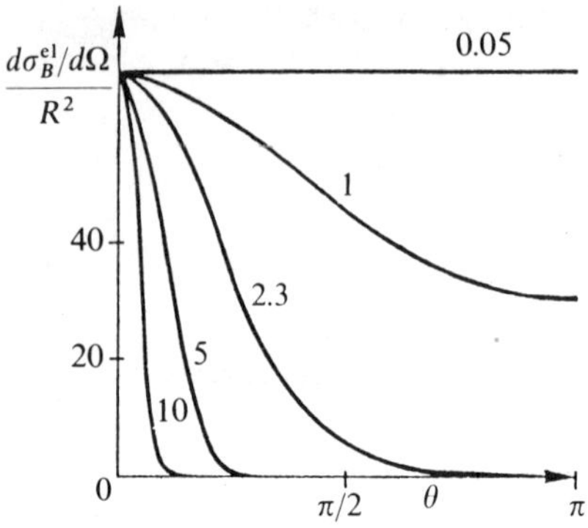

Fig. 8.15. Differential cross section in units of R^2 in the Born approximation for a spherical well of radius R, parameter $x_0 = 5$ and for different values of the momentum x ($\equiv kR$) [(8.134, 135)]

4) At high energies $d\sigma_B^{el}/d\Omega$ is concentrated at small angles, because for $q \to \infty$, $f_B(\boldsymbol{p}_i \to \boldsymbol{p}_f) \to 0$ as a consequence of (8.129). From this expression we also see that the scattering angles θ for which $d\sigma_B^{el}/d\Omega$ is important satisfy $\theta \lesssim 1/k_i R$. Since the Born approximation is good at high energies this concentration into a *forward peak* is also true for $d\sigma_{el}/d\Omega$. The width of this peak becomes narrower as we increase the energy. This remarkable fact is shown in Fig. 8.15.

We now analyze some simple examples where we will be able to discuss more explicitly the previous considerations.

a) Spherical Square Well

Potential $V(r) = \begin{cases} V_0 < 0 & \text{if } r < R \\ 0 & \text{if } r > R \end{cases}$

Defining for convenience the dimensionless parameters

$$x_0 \equiv (-U_0)^{1/2} R \,, \quad x \equiv k_i R \tag{8.134}$$

we obtain from (8.130)

$$f_B(p_i, \cos\theta) = R \left[\frac{x_0}{qR}\right]^2 \left(\frac{\sin(qR)}{qR} - \cos(qR)\right) \tag{8.135}$$

which leads to

$$\sigma_B^{el}(x) = \pi R^2 \left\{\frac{1}{2}\frac{x_0^4}{x^2}\left[1 - \frac{1}{(2x)^2} + \frac{\sin 4x}{(2x)^3} - \frac{\sin^2 2x}{(2x)^4}\right]\right\} . \tag{8.136}$$

The comparison of σ_B^{el} with the *exact* cross section σ_{el} obtained numerically shows (see Figs. 8.16 and 8.17) the validity of the Born approximation at high energies ($x_0^2 \ll 2x$) as well as for weak potentials ($x_0^2 \ll 2$). In this example the estimates can be made more precise: for any central potential we have by integration in (6.10)

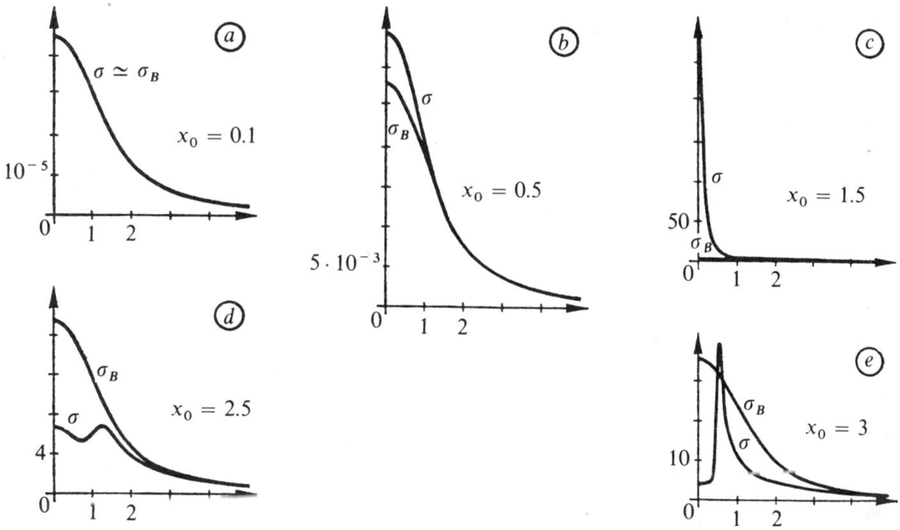

Fig. 8.16a–e. Graphical comparison of σ_{el} and $\sigma_{\mathrm{B}}^{\mathrm{el}}$ for elastic scattering by a spherical well of parameter x_0. The abscissa is x and the ordinates are σ_{el}, $\sigma_{\mathrm{B}}^{\mathrm{el}}$ in units of πR^2

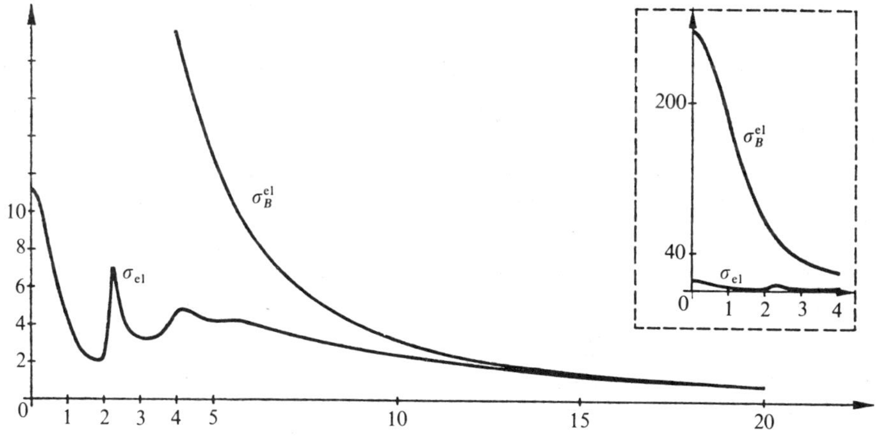

Fig. 8.17. Elastic scattering cross section for a spherical well with $x_0 = 5$. σ_{el} is the exact answer and $\sigma_{\mathrm{B}}^{\mathrm{el}}$ is the Born approximation. Both in units of πR^2. The abscissa represents the incident momentum k in units of $1/R$ ($kR \equiv x$). On the right we have a more detailed view of the low energy region: $\sigma_{\mathrm{el}}(0) = 11.24$, $\sigma_{\mathrm{B}}^{\mathrm{el}}(0) = 277.8$

$$\|V\|_{\mathrm{R}}^2 = 8\pi^2 \int_0^\infty \int_0^\infty dx\, dy\, xy |V(x)|\, |V(y)| \ln \left| \frac{x+y}{x-y} \right| \tag{8.137}$$

and in our case

$$\|U\|_{\mathrm{R}} = 2\pi |U_0| R^2 \tag{8.138}$$

therefore

$$\|\tilde{K}(E_i + \mathrm{i}0)\|_2 = \tfrac{1}{2}x_0^2 \tag{8.139}$$

and the Born approximation will converge whenever $x_0 < \sqrt{2}$ for all energies. The sufficient condition $x_0 < \sqrt{2}$ does not differ very much from the necessary and sufficient condition $x_0 < \pi/2$ obtained from (6.61) by requiring that those x_0 not satisfying this inequality should correspond to potentials with some bound state with energy ≤ 0 or some resonance at $E = 0$. In other words, that $\tilde{K}(0)$ has some eigenvalue ≥ 1. The case (c) in Fig. 8.16 corresponding to $x_0 = 1.5$ is very close to the limit of convergence of the series, and the convergence will be therefore very slow [at zero energy for example $f^{(2)}/f_{\mathrm{B}} = (2/5)x_0^2 = 0.9$ in case (c)], and one should not be surprised by the large discrepancy between σ_{el} and $\sigma_{\mathrm{B}}^{\mathrm{el}}$ at low energies ($\sigma_{\mathrm{el}}/\sigma_{\mathrm{B}}^{\mathrm{el}} = 125.47$, at $x = 0$, $x_0 = 1.5$). The fact that $\sigma_{\mathrm{el}}(0) \geq \sigma_{\mathrm{B}}^{\mathrm{el}}(0)$ for (a), (b) and (c) is easily explained: the contributions of the different terms $f^{(n)}$ in the Born series to the scattering amplitude at zero energy are all non-negative for negative potentials, as can easily be proved. Hence for all those values x_0 where the series converges, $f \geq f_{\mathrm{B}} \geq 0$ at zero energy.

There are approximation techniques which improve the convergence rate of the Born series or even make it convergent when it is not. The basic idea of the *quasi-particle method* [WE 63a, 63b, 64], [SW 64, 64a] consists of subtracting from $\tilde{K}$ the part responsible of its eigenvalues which are on or outside the unit circle, and the new $\tilde{K}_{\mathrm{Q}}$ so obtained will make the Born series convergent. One can also subtract from $\tilde{K}_{\mathrm{Q}}$ the contribution due to the eigenvalues with largest modulus, so that $\tilde{K}_{\mathrm{Q}}$ has a norm smaller than $\tilde{K}$ and this will accelerate the convergence rate of the series. The results obtained using this method are quite satisfactory, even though one often does not subtract the exact quantity described, but only a sensible approximation.

b) Yukawa Potential. With the potential $V(r) = V_0 R \exp(-r/R)/r$, $V_0 < 0$ and using the definitions (8.134), we obtain from (8.130)

$$f_{\mathrm{B}}(p_i, \cos\theta) = R\frac{x_0^2}{1 + (qR)^2} \quad \text{hence} \tag{8.140}$$

$$\sigma_{\mathrm{B}}^{\mathrm{el}}(x) = \pi R^2 \frac{x_0^4}{x^2 + 1/4}\,. \tag{8.141}$$

To compute $\|V\|_{\mathrm{R}}$ instead of using (8.137) it is more convenient in this case, where $V \in R \cap L^1$, to use the equivalent expression [SI 71]

$$\|V\|_{\mathrm{R}}^2 = 2\pi^2 \int \frac{d^3k}{|\boldsymbol{k}|} \||V|\hat{\ }(\boldsymbol{k})|^2 \tag{8.142}$$

where $|V|\hat{\ }(\boldsymbol{k})$ is the Fourier transform of $|V|$. Therefore

$$\|V\|_{\mathrm{R}} = \frac{4\pi}{\sqrt{2}}|V_0|R^2 \quad \text{and} \tag{8.143}$$

$$\|\tilde{K}(E_i + \mathrm{i}0)\|_2 = \frac{1}{\sqrt{2}}x_0^2\,. \tag{8.144}$$

The Born series will converge at all energies so long as $x_0 < 2^{1/4} = 1.189$. This value should be compared with the "exact" estimate $x_0 = 1.2961$ for the minimum intensity of a Yukawa potential in order to have a bound state. Finally, the second order contribution $f^{(2)}$ to the scattering amplitude can be computed with some effort [NE 82] with the result

$$f^{(2)}(p_i, \cos\theta) = R\frac{x_0^4}{2gqR}\left[2\arctan\frac{qR}{2g} + \mathrm{i}\ln\frac{g + xqR}{g - xqR}\right] \tag{8.145}$$

where

$$g \equiv [1 + 4x^2 + x^2(qR)^2]^{1/2} . \tag{8.146}$$

Thus,

$$f^{(2)}(p_i, 1) = R\frac{x_0^4/2}{1 - 2\mathrm{i}x} \quad \text{and} \tag{8.147}$$

$$\mathrm{Im} f^{(2)}(p_i, 1) = \frac{k_i}{4\pi}\sigma_{\mathrm{B}}^{\mathrm{el}} \tag{8.148}$$

as predicted when we discussed the unitarity violation in the Born approximation.

8.13 Space-Time Description of Simple Scattering

In the previous section we discussed the computation of the scattering amplitude using stationary states. This idealization is useful for computational purposes, but as we have remarked several times, in a real scattering experiment the states of the incoming particles are not plane waves but rather wave packets of finite size. Using equation (8.97) we will analyze the behavior of these packets after scattering, at least in regions far from the range of the potential. We begin by discussing the asymptotic form of $\langle \boldsymbol{x}|\boldsymbol{p}_+\rangle$ and next compute the time evolution of a wave packet built out of these states. We will follow the lucid exposition of *Taylor* [TA 72].

a) Asymptotic Form of $\langle \boldsymbol{x}|\boldsymbol{p}_+\rangle$

Using (8.123) and (8.113) we have

$$\langle \boldsymbol{x}|\boldsymbol{p}_+\rangle = \langle \boldsymbol{x}|\boldsymbol{p}\rangle - \frac{1}{4\pi}\int \frac{\mathrm{e}^{\mathrm{i}k|\boldsymbol{x}-\boldsymbol{y}|}}{|\boldsymbol{x}-\boldsymbol{y}|}U(\boldsymbol{y})\, d^3y\, \langle \boldsymbol{y}|\boldsymbol{p}_+\rangle . \tag{8.149}$$

The presence of $U(\boldsymbol{y})$ in the integrand limits the effective region of integration, and it suggests to use the approximation

$$|\boldsymbol{x} - \boldsymbol{y}| \simeq |\boldsymbol{x}|\left[1 - \frac{\boldsymbol{x}\cdot\boldsymbol{y}}{|\boldsymbol{x}|^2}\right] \tag{8.150}$$

so long as $|\boldsymbol{x}| \gg$ the range of R of the potential; hence using (8.85) and (8.103)

$$\begin{aligned}\langle \boldsymbol{x}|\boldsymbol{p}_+\rangle \underset{|\boldsymbol{x}|\to\infty}{\sim} \langle \boldsymbol{x}|\boldsymbol{p}\rangle - \frac{1}{4\pi}\frac{\mathrm{e}^{\mathrm{i}k|\boldsymbol{x}|}}{|\boldsymbol{x}|}\int \mathrm{e}^{-\mathrm{i}k\hat{\boldsymbol{x}}\cdot\boldsymbol{y}}U(\boldsymbol{y})\, d^3y\, \langle \boldsymbol{y}|\boldsymbol{p}_+\rangle \\ = \langle \boldsymbol{x}|\boldsymbol{p}\rangle - \frac{(2\pi\hbar)^{3/2}}{4\pi}\langle p\hat{\boldsymbol{x}}|U|\boldsymbol{p}_+\rangle\frac{\mathrm{e}^{\mathrm{i}k|\boldsymbol{x}|}}{|\boldsymbol{x}|}\end{aligned}$$

$$\Rightarrow \boxed{\langle \boldsymbol{x}|\boldsymbol{p}_+\rangle \underset{|\boldsymbol{x}|\to\infty}{\sim} (2\pi\hbar)^{-3/2}\left[\mathrm{e}^{\mathrm{i}\boldsymbol{k}\cdot\boldsymbol{x}} + f(\boldsymbol{p}\to p\hat{\boldsymbol{x}})\frac{\mathrm{e}^{\mathrm{i}k|\boldsymbol{x}|}}{|\boldsymbol{x}|}\right]} \tag{8.151}$$

with an error of the order of $R/|\boldsymbol{x}|$ and $kR^2/|\boldsymbol{x}|$. The first error comes from approximating $|\boldsymbol{x}-\boldsymbol{y}|^{-1}$ by $|\boldsymbol{x}|^{-1}$, and the second from the approximation of the exponential $\exp(\mathrm{i}k|\boldsymbol{x}-\boldsymbol{y}|)$. The interpretation of (8.151) is simple: When we send a plane wave $\exp(\mathrm{i}\boldsymbol{k}\cdot\boldsymbol{x})$ towards the potential, the resulting scattering state far from the scattering region will consist of two pieces: the same incident plane wave and a scattered wave with radial emergence $\exp(\mathrm{i}k|\boldsymbol{x}|)/|\boldsymbol{x}|$ modulated in amplitude by the factor $f(\boldsymbol{p}\to p\hat{\boldsymbol{x}})$ depending on the direction of observation. The "out" nature of the scattered wave is due to the structure of $|\boldsymbol{p}_+\rangle$. In other words, we have approached E from the upper half plane ($E+\mathrm{i}0$). The asymptotic approximation (8.151) is justifiable rigorously under very general conditions, for example the standard conditions on V. The definition (8.85) of the scattering amplitude is justified a posteriori from (8.151). (Had we used the relativistic definition of the energy, we would have also obtained the last term in (8.151) as part of the asymptotic form of $\langle \boldsymbol{x}|\boldsymbol{p}_+\rangle$, with the same definition (8.85). See for example [GW 64].)

The behavior (8.151) of the outgoing wave in the elastic channel allows us to derive the optical theorem once again but with some new physical insight. The radial probability current J_r associated to $\langle \boldsymbol{x}|\boldsymbol{p}_+\rangle$ at large r is

$$J_r \simeq (2\pi\hbar)^{-3}v\left[\cos\theta + \frac{1}{r}\mathrm{Re}\left(f\mathrm{e}^{\mathrm{i}k(r-z)} + f^*\cos\theta\mathrm{e}^{-\mathrm{i}k(r-z)}\right) + \frac{|f|^2}{r^2}\right] \tag{8.152}$$

where $r=|\boldsymbol{x}|$, $\cos\theta = \hat{\boldsymbol{k}}\cdot\hat{\boldsymbol{x}}$, $z=r\cos\theta$ and we have neglected terms contributing a vanishing radial flux as $r\to\infty$, either because they decrease explicitly as $1/r^3$ or faster, or by taking into account (8.153). The interference terms between the incident and scattered waves oscillate very rapidly when $kr \gg 1$ except in the direction of incidence. Indeed, a simple computation shows that as a distribution in $\boldsymbol{k}$ we have:

$$\mathrm{e}^{\mathrm{i}k(r-z)} \underset{r\to\infty}{\sim} \frac{2\pi\mathrm{i}}{kr}\left[\delta(\hat{\boldsymbol{p}}-\hat{\boldsymbol{x}}) - \mathrm{e}^{2\mathrm{i}kr}\delta(\hat{\boldsymbol{p}}+\hat{\boldsymbol{x}})\right] . \tag{8.153}$$

Therefore

$$J_r \simeq (2\pi\hbar)^{-3}v\left[\cos\theta - \frac{1}{r^2}\frac{4\pi}{k}\{\mathrm{Im} f(\boldsymbol{p}\to\boldsymbol{p})\}\delta(\hat{\boldsymbol{p}}-\hat{\boldsymbol{x}}) + \frac{|f|^2}{r^2}\right] . \tag{8.154}$$

This equation reveals firstly that the radial flux in the direction of incidence after crossing the scattering region and far from the target has changed with respect to the incident flux. Actually it has decreased, because we know that $\mathrm{Im} f(\boldsymbol{p} \to \boldsymbol{p}) > 0$ if there is scattering at all. This attenuation of the flux in the direction of incidence represents the "shadow" produced by the target. Second, the flux of outgoing particles through a spherical surface S of large radius r centered upon the target is

$$\int_S \boldsymbol{J} \cdot d\boldsymbol{s} \underset{r\to\infty}{\sim} \int J_r r^2 \, d\Omega = (2\pi\hbar)^{-3} v \left[-\frac{4\pi}{k} \mathrm{Im} f(\boldsymbol{p} \to \boldsymbol{p}) + \sigma_{\mathrm{el}}(\boldsymbol{p}) \right] . \quad (8.155)$$

In a simple scattering system, $\langle \boldsymbol{x} | \boldsymbol{p}_+ \rangle$ is the total wave function of the scattering state $|\boldsymbol{p}_+\rangle$, and since it is stationary, $\boldsymbol{J}$ satisfies the equation $\nabla \cdot \boldsymbol{J} = 0$ and the total flux through S must vanish. Thus we obtain again (8.87). However, if there are possible inelastic channels, $\langle \boldsymbol{x} | \boldsymbol{p}_+ \rangle$ will only be the component in the elastic channel of the wave function of $|\boldsymbol{p}_+\rangle$, the equation $\nabla \cdot \boldsymbol{J} = 0$ does not hold any longer and part of the incoming flux particles will be used in the production of inelastic reactions. There will be less particles coming out in the elastic channel than came in. Consequently the right hand side of (8.155) will be negative in general, and equal to the number of incoming particles disappearing per unit time in the interior of S, namely it equals $-(2\pi\hbar)^{-3} v \sigma_{\mathrm{inel}}(\boldsymbol{p})$. Writing $\sigma_{\mathrm{tot}} = \sigma_{\mathrm{el}} + \sigma_{\mathrm{inel}}$ we obtain (8.88).

b) Time Evolution of a Scattering Wave Packet

In a realistic situation the in-state $|\psi\rangle$ in a simple scattering system will not be mono-energetic but rather a normalized superposition of plane waves

$$|\psi\rangle = \int d^3p \, \hat{\psi}(\boldsymbol{p}) |\boldsymbol{p}\rangle . \quad (8.156)$$

The associated scattering state (8.97) will be

$$|\psi_+\rangle = \int d^3p \, \hat{\psi}(\boldsymbol{p}) |\boldsymbol{p}_+\rangle . \quad (8.157)$$

The corresponding wave functions, assuming they evolve with H and H_0 respectively are

$$\begin{aligned} \Psi(\boldsymbol{x};t) &= (2\pi\hbar)^{-3/2} \int d^3p \, \hat{\psi}(\boldsymbol{p}) \exp\left[\mathrm{i}(\boldsymbol{p}\cdot\boldsymbol{x} - E_p t)/\hbar\right] \\ \Psi_+(\boldsymbol{x};t) &= \int d^3p \, \hat{\psi}(\boldsymbol{p}) \exp(-\mathrm{i}E_p t/\hbar) \langle \boldsymbol{x} | \boldsymbol{p}_+ \rangle . \end{aligned} \quad (8.158)$$

If $\boldsymbol{p}_0 = \langle \psi | \boldsymbol{p} | \psi \rangle$ is the average momentum of $|\psi\rangle$, we know (Sect. 1.12) that

$$\Psi(\boldsymbol{x};t) \simeq \exp\left[\mathrm{i}(\boldsymbol{p}_0 \cdot \boldsymbol{x} - E_{p_0} t)/\hbar\right] \phi(\boldsymbol{x} - \boldsymbol{v}_0 t) \left[1 + \mathrm{O}\left(\frac{t(\Delta p)^2}{\hbar M} \right) \right] \quad (8.159)$$

where

$$\phi(\boldsymbol{x}) \equiv (2\pi\hbar)^{-3/2} \int d^3p\, \hat{\psi}(\boldsymbol{p}) \exp\left[\mathrm{i}(\boldsymbol{p} - \boldsymbol{p}_0) \cdot \boldsymbol{x}/\hbar\right] . \tag{8.160}$$

If $(t/\hbar M)(\Delta p)^2 \ll 1$ the free wave packet will not show any sign of distortion and it will move, apart from the multiplicative plane wave, as a whole with the group velocity $\boldsymbol{v}_0$.

For $\Psi_+(\boldsymbol{x}; t)$, using (8.151) and therefore within relative errors of $\mathrm{O}(R/|\boldsymbol{x}|)$, $\mathrm{O}(k_0 R^2/|\boldsymbol{x}|)$, we obtain

$$\begin{aligned} \Psi_+(\boldsymbol{x}; t) &\simeq \Psi(\boldsymbol{x}; t) + (2\pi\hbar)^{-3/2} \int d^3p\, f(\boldsymbol{p} \to p\hat{\boldsymbol{x}}) \\ &\quad \times \frac{\exp\left[\mathrm{i}(p|\boldsymbol{x}| - E_p t)/\hbar\right]}{|\boldsymbol{x}|} \hat{\psi}(\boldsymbol{p}) \\ &\equiv \Psi(\boldsymbol{x}; t) + \Psi_{\mathrm{sc}}(\boldsymbol{x}; t) . \end{aligned} \tag{8.161}$$

Assuming now that $\hat{\psi}(\boldsymbol{p})$ is peaked around $\boldsymbol{p}_0$ in a sufficiently small region for $f(\boldsymbol{p} \to p\hat{\boldsymbol{x}})$ to be constant to a good approximation in it (according to the Born approximation (8.129) this is to be expected for $(\Delta k)R \ll 1$), and taking into account that in this range of momenta $p|\boldsymbol{x}| = \boldsymbol{p} \cdot (|\boldsymbol{x}|\hat{\boldsymbol{p}}_0)[1 + \mathrm{O}((\Delta p)^2/p_0^2)]$, we can write

$$\Psi_{\mathrm{sc}}(\boldsymbol{x}; t) = \frac{f(\boldsymbol{p}_0 \to p_0\hat{\boldsymbol{x}})}{|\boldsymbol{x}|} \Psi(|\boldsymbol{x}|\hat{\boldsymbol{p}}_0; t) \left[1 + \mathrm{O}\left(\frac{|\boldsymbol{x}|(\Delta p)^2}{\hbar p_0}\right)\right] . \tag{8.162}$$

Hence up to errors of order $t(\Delta p)^2/\hbar M$, $R/|\boldsymbol{x}|$, $k_0 R^2/|\boldsymbol{x}|$, $(\Delta k)R$, $|\boldsymbol{x}|(\Delta p)^2/\hbar p_0$, we have

$$\Psi_+(\boldsymbol{x}; t) \simeq \Psi(\boldsymbol{x}; t) + \frac{f(\boldsymbol{p}_0 \to p_0\hat{\boldsymbol{x}})}{|\boldsymbol{x}|} \Psi(|\boldsymbol{x}|\hat{\boldsymbol{p}}_0; t) \tag{8.163}$$

and Ψ moves without any important distortion.

In conclusion, the incident wave packet, which moves freely until its wave function "feels" the potential will behave at large distances, after the collision, as a superposition of the initial packet and a scattered wave which is important only in a spherical shell of width similar to that of Ψ in the line of incidence going through the target and moving in phase with Ψ at a speed v_0 (see Fig. 8.18).

If $a \sim 1/\Delta k$ is the width of the incident packet, a detector placed at a distance d from the target and forming an angle θ with the incoming direction will register only the scattered packet without interferences of the incident one if $a \ll d|\sin\theta|$, and the approximation (8.163) applies on the detector if $R \ll a \ll d \ll k_0 a^2$, $k_0 R^2 \ll d$ as follows from an estimate of the errors with $t \sim d/v_0$. That these conditions are generally satisfied in actual experiments can be seen by considering for instance 0.1 keV electrons as projectiles with a collimation of $a \sim 1\,\mathrm{mm}$ scattered by atoms ($R \sim 10^{-8}$ cm) and a detector at a distance $d \sim 1$ m.

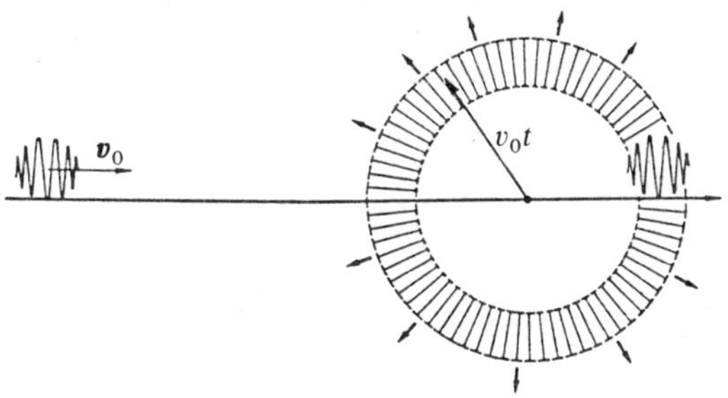

Fig. 8.18. Time evolution of the incident and scattered wave functions

8.14 Symmetries of the Scattering Operator

As in any dynamical problem, symmetry considerations in the scattering process provide useful information about it. Although later we will present a more complete discussion we want to study in this section the consequences of symmetries in single channel scattering.

The free Hamiltonian H_0 is invariant under parity, time reversal and rotations. Let U be the unitary (or antiunitary) operator implementing one of these symmetries. We have

$$UH_0 = H_0U \; . \tag{8.164}$$

If the total Hamiltonian is also invariant under U,

$$UH = HU \tag{8.165}$$

from (8.37) and (8.44) we obtain

$$\boxed{\left.\begin{array}{l} U\Omega_\pm = \Omega_\pm U \\ US = SU \end{array}\right\} \text{ if } U \text{ is unitary}} \quad \text{or} \tag{8.166}$$

$$\boxed{\left.\begin{array}{l} U\Omega_\pm = \Omega_\mp U \\ US = S^\dagger U \end{array}\right\} \text{ if } U \text{ is antiunitary}} \; . \tag{8.167}$$

Therefore, on the energy shell E [see (8.54)], stable under U by (8.164),

$$\boxed{\begin{array}{ll} UT_EU^{-1} = T_E & \text{unitary case} \\ UT_EU^{-1} = T_E^\dagger & \text{antiunitary case .} \end{array}} \tag{8.168}$$

Thus, if H conserves parity

$$\boxed{f(\boldsymbol{p} \to \boldsymbol{p}') = f(-\boldsymbol{p} \to -\boldsymbol{p}')} \; . \tag{8.169}$$

If H is invariant under time reversal,

$$\boxed{f(\boldsymbol{p} \to \boldsymbol{p}') = f(-\boldsymbol{p}' \to -\boldsymbol{p})} \tag{8.170}$$

known as the *principle of microreversibility* or *reciprocity theorem.* If H is spherically symmetric

$$\boxed{f(\boldsymbol{p} \to \boldsymbol{p}') = f(\mathcal{R}\boldsymbol{p} \to \mathcal{R}\boldsymbol{p}')} \ . \tag{8.171}$$

From these expressions we see that invariance under rotations implies automatically (8.169) and (8.170) and therefore the invariance of S under parity and time reversal (in the sense of (8.167)). This implication which holds for simple scattering of spinless particles is not valid in more general cases. From (8.171) we conclude that such scattering amplitude depends only on the magnitude of the incoming momentum and on the scattering angle: $f(\boldsymbol{p} \to \boldsymbol{p}') = f(p, \cos\theta)$. Given the practical importance of central potentials, we will study in detail their properties in simple scattering processes.

8.15 Scattering by a Central Potential: Partial Waves and Phase Shifts

The invariance of H_0 and H under rotations make it useful to expand the scattering states in eigenstates of $\boldsymbol{L}^2$, L_z, and in particular to use the basis $\{|pLM\rangle\}$ (Sect. 5.10) for the asymptotic states. This expansion is known as the *partial wave expansion.* It is not only a very efficient way of fully exploiting spherical symmetry, but also for short range potentials and for a fixed energy, it will allow us to consider only those partial waves more sensitive to the scattering process.

Consider the expansion of the incident plane wave in spherical harmonics (5.129, 136)

$$\langle \boldsymbol{x}|\boldsymbol{p}\rangle = \sum_{LM} \langle \boldsymbol{x}|pLM\rangle Y_L^{M*}(\hat{\boldsymbol{p}}) \tag{8.172}$$

where (5.137)

$$\langle \boldsymbol{x}|pLM\rangle = \sqrt{\frac{2}{\pi\hbar^3}}\, \mathrm{i}^L j_L(kr) Y_L^M(\hat{\boldsymbol{x}}) \ . \tag{8.173}$$

In the discussion of reduced radial functions it is convenient to introduce the notation $\tilde{f}(z) = zf(z)$ where $f(z)$ can be any of the spherical Bessel, Neumann or Hankel functions (Sect. A.9). The relative probability density $|\tilde{j}_L(kr)|^2$ to find the particle at a distance r from the origin when the state is $|pLM\rangle$ vanishes as $(kr)^{2(L+1)}$ when $kr \ll 1$, and it oscillates between 0 and approximately 1 at large distances ($kr \gg 1$) (Sect. A.9). Furthermore, it is possible to show that $|\tilde{j}_L(kr)|^2$ remains small for $kr \lesssim L$, and therefore a particle in the state $|pLM\rangle$ will hardly feel the presence of a potential centered at the origin and with a range

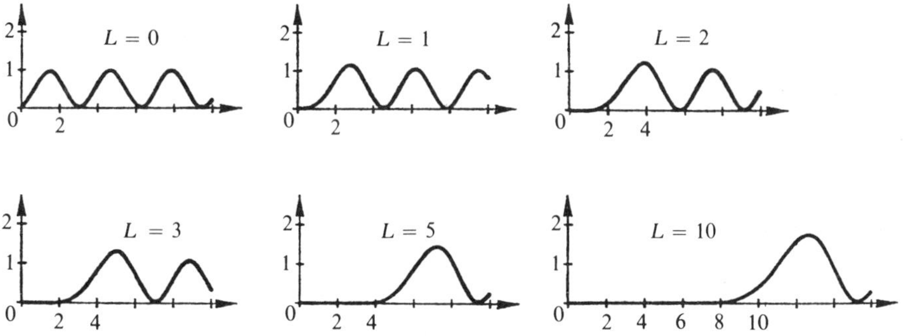

Fig. 8.19. The functions $|\tilde{j}_L(x)|^2$ for different values of L

$\lesssim L/k$. The sensitivity to the potential decreases as we increase L. The physical reason for this behavior is the presence of the centrifugal barrier which makes it difficult for the particle to penetrate in regions of small r. Figure 8.19 illustrates the behavior of these probabilities for different values of L.

Finally, (A.14) and the asymptotic expansion of $j_L(kr)$ lead to the following expression for the asymptotic behavior of a plane wave:

$$e^{i\boldsymbol{k}\cdot\boldsymbol{x}} \underset{|\boldsymbol{x}|\to\infty}{\sim} \sum_{L=0}^{\infty}\left(L+\frac{1}{2}\right)P_L(\hat{\boldsymbol{p}}\cdot\hat{\boldsymbol{x}})\left[(-1)^{L+1}\frac{e^{-ikr}}{ikr}+\frac{e^{ikr}}{ikr}\right]. \tag{8.174}$$

The plane wave appears as a superposition of *partial waves* ($\langle\boldsymbol{x}|pLM\rangle$), of definite angular momentum, which asymptotically are made out of an incoming spherical wave ($\exp(-ikr)/r$ type) and an outgoing one ($\exp(ikr)/r$ type) with relative phases $(-1)^{L+1}$ and modulated by $P_L(\hat{\boldsymbol{p}}\cdot\hat{\boldsymbol{x}})$. As it will be seen the action on it of a central potential in a scattering process will be to preserve the structure (8.174) but producing additional phases for the outgoing waves $\exp(ikr)/r$ of each partial wave L without changing the phase of the incoming ones (this is expected because these waves come towards the target from infinity).

Since H has spherical symmetry, (8.166) implies that

$$\boxed{RS = SR\,, \quad \forall \text{rotation } \mathcal{R}} \tag{8.175}$$

and therefore

$$\boxed{\boldsymbol{L}S = S\boldsymbol{L}} \tag{8.176}$$

i.e. the angular momentum is conserved in the scattering. Since S also conserves the kinetic energy and in particular p, and it is unitary we have

$$S|pLM\rangle = \exp[2i\delta_{LM}(p)]|pLM\rangle\,, \quad \delta_{LM} \text{ real}\,. \tag{8.177}$$

Moreover, (8.176) implies that $L_\pm S = SL_\pm$, hence $\delta_{LM} = \delta_{LM\pm1} \pmod{\pi}$ and δ_{LM} can be chosen independent of M. Thus,

$$\boxed{S|pLM\rangle = \exp[2i\delta_L(p)]|pLM\rangle\,, \quad \delta_L \text{ real}}\,. \tag{8.178}$$

The numbers $\delta_L(p)$ defined modulo π are known as the *phase shifts* produced by the interaction in the partial wave L with momentum p. When there is no interaction, $S = 1$ and $\delta_L(p) = 0$. It is also clear that the result of the scattering will be known as soon as we know the phase shifts. In particular, it must be possible to express the scattering amplitude $f(p, \cos\theta)$ in terms of them. This is indeed the case: computing in the basis $|qLM\rangle$,

$$\begin{aligned}
\langle \mathbf{p}'|S|\mathbf{p}\rangle &= \sum_{L=0}^{\infty}\sum_{M=-L}^{+L}\int_0^\infty q^2\,dq\,\langle \mathbf{p}'|qLM\rangle \exp[2i\delta_L(q)]\langle qLM|\mathbf{p}\rangle \\
&= \frac{\delta(p'-p)}{p^2}\sum_{L=0}^{\infty}\exp[2i\delta_L(p)]\left[\sum_{M=-L}^{+L} Y_L^M(\hat{\mathbf{p}}')Y_L^{M*}(\hat{\mathbf{p}})\right] \quad \text{[use (A.41)]} \\
&= \frac{\delta(p'-p)}{p^2}\frac{1}{4\pi}\sum_{L=0}^{\infty}(2L+1)\exp[2i\delta_L(p)]P_l(\hat{\mathbf{p}}'\cdot\hat{\mathbf{p}})
\end{aligned} \tag{8.179}$$

and using (8.55, 85),

$$\boxed{f(p, \cos\theta) = \sum_{L=0}^{\infty}(2L+1)f_L(p)P_L(\cos\theta)} \tag{8.180}$$

with $\cos\theta = \hat{\mathbf{p}}\cdot\hat{\mathbf{p}}'$ and

$$\boxed{\begin{aligned} f_L(p) &\equiv \frac{\exp[i\delta_L(p)]\sin\delta_L(p)}{k} = \frac{s_L(p)-1}{2ik}\,, \\ s_L(p) &\equiv \exp[2i\delta_L(p)]\,. \end{aligned}} \tag{8.181}$$

The complex functions $f_L(p)$ are known as the *partial (wave) scattering amplitudes*. (The expansion (8.180) of the scattering amplitude in partial amplitudes was introduced by Faxen and Holtsmark imitating its use in acoustics and electromagnetism.)

Once $f(p, \cos\theta)$ is known, $f_L(p)$ can be obtained formally from

$$f_L(p) = \frac{1}{2}\int_{-1}^{+1} dx\, f(p, x)P_L(x) \tag{8.182}$$

using (A.10) and (8.180).

From (8.180),

$$\boxed{\sigma_{\text{el}}(p) = \frac{4\pi}{k^2}\sum_{L=0}^{\infty}(2L+1)\sin^2\delta_L(p) \equiv \sum_{L=0}^{\infty}\sigma_{\text{el}}^{(L)}(p)\,,} \tag{8.183}$$

where

$$\boxed{\sigma_{\mathrm{el}}^{(L)}(p) \equiv \frac{4\pi}{k^2}(2L+1)\sin^2\delta_L(p)} \tag{8.184}$$

is the contribution to $\sigma_{\mathrm{el}}(p)$ of the L-partial wave. Since the unitarity of the S matrix implies that the phase shifts are real, the partial cross sections $\sigma_{\mathrm{el}}^{(L)}$ are bounded:

$$\boxed{\sigma_{\mathrm{el}}^{(L)} \leq \frac{4\pi}{k^2}(2L+1) \quad \text{(unitarity limit)}}\ . \tag{8.185}$$

This bound is saturated when $\delta_L(p) = \pi/2 \pmod{\pi}$.

The relation $|s_L(p)| = 1$ valid for simple scattering in a central potential implies using (8.81):

$$\operatorname{Im} f_L(p) = k|f_L(p)|^2\ . \tag{8.186}$$

From this and $P_L(1) = 1$ we can derive again the optical theorem.

With respect to the convergence of the expansion (8.180), it is clear that if $\sigma_{\mathrm{el}}(p) < \infty$, the series of partial amplitudes converges to $f(p, .)$ in $L^2(-1,+1)$. As in the case of Fourier series, the pointwise convergence is more delicate, and there is a variety of local conditions guaranteeing it. For example, if $f(p, .)$ is piecewise C^1, (8.180) will converge in each point $\cos\theta$ of continuity in $f(p, .)$ to $f(p, \cos\theta)$. Similarly if $|f_L| = \mathrm{O}(L^{-3/2})$. Thus for piecewise continuous central potentials less singular than $r^{-3/2}$ at the origin and such that $V(r) = \mathrm{O}(r^{-\alpha})$, $r \to \infty$, it can be shown [AR 65], [TA 72] that $|f^L(p)| = \mathrm{O}(L^{-(\alpha-1)})$, $L \to \infty$ and since $|P_L(x)| \leq 1$ in $[-1,+1]$ the series of partial amplitudes will converge pointwise and uniformly towards $f(p, .)$, a continuous function, if $\alpha > 3$. Equation (8.186) guarantees the pointwise convergence of the imaginary parts of (8.180) for $\alpha > 2$. If furthermore the potential $V(r)$ is *short range*, a discrete or continuous superposition of Yukawa potentials ($V(r) = \mathrm{O}(\exp(-\mu r))$, $r \to \infty$), the partial amplitudes f_L will decrease exponentially as L increases, and the series (8.180) converges (uniformly over compact sets) towards an analytic function $f(p, z)$ in the interior of an ellipse of focal points $z = \pm 1$ and semimajor axis $1 + \mu^2/2k^2$ (Lehmann's ellipse) [AR 65].

Finally replacing (8.180) in (8.151) we obtain

$$\langle \boldsymbol{x}|\boldsymbol{p}_+\rangle \underset{r\to\infty}{\sim} (2\pi\hbar)^{-3/2} \sum_L \left(L+\frac{1}{2}\right) P_l(\hat{\boldsymbol{p}}\cdot\hat{\boldsymbol{x}}) \times \left[(-1)^{L+1}\frac{\mathrm{e}^{-ikr}}{ikr} + s_L(p)\frac{\mathrm{e}^{ikr}}{ikr}\right] \tag{8.187}$$

and as announced, the effect of the interaction does not change the intensity of the outgoing partial waves (in simple scattering the scattering center does not act as a source or sink of particles). It changes their phases by factors $\exp[2i\delta_L(p)]$.

8.16 Computation and Properties of Phase Shifts

Since H_0 and V (central potential) commute with $\boldsymbol{L}$, the same property is shared by $\Omega_\pm$, and therefore $\langle \boldsymbol{x}|\Omega_\pm|pLM\rangle$ has an angular dependence of the form $Y_L^M(\hat{\boldsymbol{x}})$. In analogy with (8.173) we can write

$$\langle \boldsymbol{x}|\Omega_\pm|pLM\rangle = \left(\frac{2}{\pi\hbar^3}\right)^{1/3} \mathrm{i}^L \frac{u^\pm_{L,p}(r)}{kr} Y_L^M(\hat{\boldsymbol{x}}) \tag{8.188}$$

and the radial wave function $u^\pm_{L,p}(r)$ will be the regular solution to the reduced radial equation

$$-u'' + \left[U(r) + \frac{L(L+1)}{r^2} - k^2\right] u = 0 \tag{8.189}$$

because $\Omega_\pm|pLM\rangle$ is a (generalized) eigenstate of H with eigenvalue $p^2/2M$.

The normalization (5.129) and the isometric property of $\Omega_\pm$ imply:

$$\int_0^\infty dr\, u^{\pm *}_{L,p'}(r) u^\pm_{L,p}(r) = \tfrac{\pi}{2}\delta(k' - k)\,. \tag{8.190}$$

To completely specify $u^\pm_{L,p}$ it suffices to use (8.151, 180, 181) and (5.136) together with the relation

$$\langle \boldsymbol{x}|\Omega_\pm|pLM\rangle = \int \langle \boldsymbol{x}|\Omega_\pm|\boldsymbol{p}'\rangle\, d^3p'\, \langle \boldsymbol{p}'|pLM\rangle \tag{8.191}$$

and $\Omega_- = U_T^{-1}\Omega_+ U_T$. It is now easy to see that

$$u^\pm_{L,p}(r) \underset{r\to\infty}{\sim} \exp\left[\pm \mathrm{i}\delta_L(p)\right] \sin(kr - L\tfrac{\pi}{2} + \delta_L(p)) \tag{8.192}$$

and $u^\pm_{L,p}$ are the regular solutions ($u^\pm_{L,p}(0) = 0$) to (8.189) with the required asymptotic behavior. Comparing with (8.173) and using the asymptotic behaviors we see that up to a phase, $u^\pm_{L,p}(r)$ oscillates asymptotically as in the free case with the difference that the oscillations are *displaced* by an amount $\delta_L(p)$. The numerical or graphical comparison in the asymptotic region between the free solution $\tilde{j}_L(kr)$ and any non-trivial regular solution of (8.189) allows us to determine $\delta_L(p)$ (mod π) in principle.

Figure 8.20 illustrates the comparison between $u_{1,p}(r) \equiv \exp\left[\mp \mathrm{i}\delta_1(p)\right] u^\pm_{1,p}(r)$ and $\tilde{j}_1(kr)$ for attractive and repulsive central square wells. We see that the attractive potential tends to "bend" the wave function towards the axis, with the opposite behavior in the repulsive case. These effects are expected from (8.189) and they are quite general, independent of the concrete form of V.

To determine the phase shifts analytically it suffices to do a partial wave decomposition of the relation

$$f(\boldsymbol{p} \to \boldsymbol{p}') = -\frac{(2\pi\hbar)^3}{4\pi} \langle \boldsymbol{p}'|U|\boldsymbol{p}_+\rangle \tag{8.193}$$

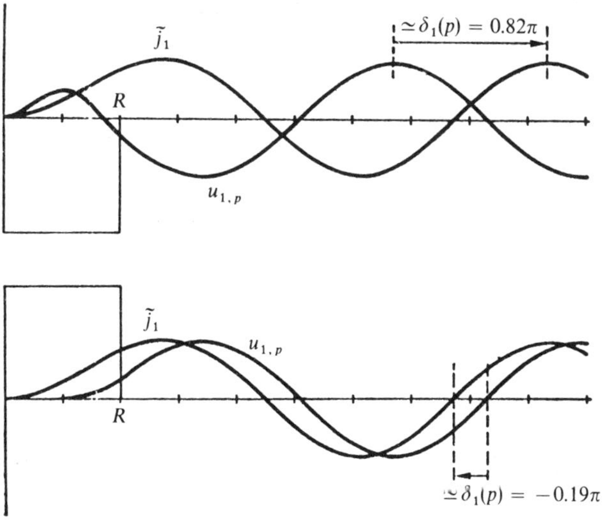

Fig. 8.20. Comparison between $\tilde{j}_1(kr)$ and $u_{1,p}(r)$ for the potentials drawn with parameters $U_0 R^2 = \mp 25$, $kR = 2$

using (8.182, 188), (A.41, 42) and the equality

$$\langle \boldsymbol{p}'|U|\boldsymbol{p}_+\rangle = \sum_{LM} \int \langle \boldsymbol{p}'|\boldsymbol{x}\rangle d^3x\ U(r)\langle \boldsymbol{x}|\Omega_+|p''LM\rangle p''^2 dp'' \langle p''LM|\boldsymbol{p}\rangle \ . \quad (8.194)$$

We easily obtain

$$\boxed{f_L(p) = -\frac{1}{k^2}\int_0^\infty dr\ \tilde{j}_L(kr)U(r)u^+_{L,p}(r)} \quad (8.195)$$

and

$$\boxed{\sin\delta_L(p) = -\frac{1}{k}\int_0^\infty dr\ \tilde{j}_L(kr)U(r)u_{L,p}(r)}\ . \quad (8.196)$$

The last equation determines $\delta_L(p) (\bmod \pi)$ once $u_{L,p}(r) \equiv \exp[-i\delta_L(p)]u^+_{L,p}(r)$ is known. A simpler procedure to arrive at (8.196) is to use the wronskian theorem applied to the functions $u^{(1)}_{L,p}(r)$, $u^{(2)}_{L,p}(r)$ associated to the respective potentials V_1 and V_2; they are both regular and with asymptotic behavior $\sin(kr - L\pi/2 + \delta_L^{(1,2)}(p))$; thus

$$\sin\left(\delta_L^{(2)}(p) - \delta_L^{(1)}(p)\right) = -\frac{1}{k}\int_0^\infty dr\ u^{(2)}_{L,p}(r)[U_2(r) - U_1(r)]u^{(1)}_{L,p}(r) \quad (8.197)$$

and (8.196) is a particular case taking $U_1 \equiv 0$ and therefore $\delta_L^{(1)}(p) = 0$. When $U_2 - U_1 \equiv \Delta U$ is very small, (8.197) shows that the difference $\delta_L^{(2)}(p) - \delta_L^{(1)}(p)$ satisfies approximately

$$\Delta\delta_L(p) \simeq -\frac{1}{k}\int_0^\infty dr\,\left(u^{(1)}_{L,p}(r)\right)^2 \Delta U(r) \tag{8.198}$$

so that the change in the phase shifts is opposite to the change in the potentials: As *the potential increases, the phase shifts decrease and vice versa*. Therefore if we build $U(r)$ by small increments starting with $U \equiv 0$ we can say that

$$\boxed{\begin{array}{lll} U(r) \geq 0 & \text{(repulsive potential)} & \Rightarrow \delta_L(p) \leq 0 \\ U(r) \leq 0 & \text{(attractive potential)} & \Rightarrow \delta_L(p) \geq 0\,. \end{array}} \tag{8.199}$$

This statement presupposes the choice of phase shifts by continuity from the vanishing phase shifts in the case $U \equiv 0$ as functions of the intensity of the potential. Similarly, and with analogous criteria,

$$\boxed{U_1(r) \geq U_2(r) \Rightarrow \delta^{(1)}_L(p) \leq \delta^{(2)}_L(p)}\,. \tag{8.200}$$

The intrinsic ambiguity in the definition of phase shifts is usually solved as follows: at high energies the kinetic energy dominates over the potential energy and the effect of the interaction is negligible: $\delta_L(p) \to 0 \pmod{\pi}$, $p \to \infty$. We choose $\delta_L(\infty) = 0$. After this, we proceed by continuity of δ_L as a function of p. This fixes uniquely a concrete choice for $\delta_L(p)$: the continuous choice such that $\delta_L(p) = 0, p \to \infty$. This is always possible as we will see below under the standard conditions on V. Once the choice is made, it can be shown that $\delta_L(p)$ is a continuous function of the intensity of the potential and vanishing with it. Finally it can also be shown that $\delta_L(p) \to 0$ if $L \to \infty$ (this is quite natural, modulo π, because at large L the centrifugal barrier dominates over the potential and it screens the particle from its effects).

In contrast with (8.196) there exists a non-linear differential equation for a new set of phase shifts incorporating the choice $\delta_L(\infty) = 0$ and which allows a more transparent analysis of the properties of the phase shifts. This equation derived by *Calogero* [CA 67] appears when we consider the variation with r of the phase shifts produced by the potential U truncated after r: $U_r(r') = U(r')\theta(r - r')$. Defining these phase shifts as $\delta_L(p,r)$ and if they are chosen by continuity from $\delta_L(p,0) = 0$ it can be shown that

$$\frac{\partial \delta_L(p,r)}{\partial r} = -\frac{1}{k}U(r)[\tilde{\jmath}_L(kr)\cos\delta_L(p,r) - \tilde{y}_L(kr)\sin\delta_L(p,r)]^2 \tag{8.201}$$

and the $\delta_L(p)$ with the previous choice are nothing but $\delta_L(p) = \delta_L(p,\infty)$. The equation (8.201) is known as the *phase equation* and it can be treated easily using numerical analysis. Furthermore it exhibits quite clearly the statements made before concerning the relation between the sign of δ_L and the attractive or repulsive nature of the potential. Using it (8.200) can be proved rigorously without difficulty. On the other hand, integrating the phase equation for $L = 0$ we obtain

$$\delta_0(p) = -\frac{1}{k}\int_0^\infty dr\, U(r)\sin^2[kr + \delta_0(p,r)] \tag{8.202}$$

and we immediately derive the bounds

$$\int_0^\infty dr\, U_-(r) \le k\delta_0(p) \le \int_0^\infty dr\, U_+(r) \tag{8.203}$$

which can be improved if additional conditions are imposed on the potential [CA 67].

The use of (8.195) in the computation of $f_L(p)$ requires the knowledge of $u^+_{L,p}(r)$. This function, regular solution of (8.189) with the asymptotic behavior (8.192) is determined by the Lippmann-Schwinger integral equation (8.95).

Using (8.113) and (A.135) we can write

$$u^+_{L,p}(r) = \tilde{\jmath}_L(kr) + \int_0^\infty dr' R_0^{(L,p)}(r,r')U(r')u^+_{L,p}(r') \tag{8.204}$$

with

$$R_0^{(L,p)}(r,r') \equiv -\frac{\mathrm{i}}{k}\tilde{\jmath}_L(kr_<)\tilde{h}_L^{(1)}(kr_>) \tag{8.205}$$

and $r_< \equiv \min(r,r')$, $r_> \equiv \max(r,r')$. Writing (8.204) in operator form:

$$u^+_{L,p} = \tilde{\jmath}_L + R_0^{(L,p)}Uu^+_{L,p} \tag{8.206}$$

an iterative computation leads to the formal series

$$u^+_{L,p} = \tilde{\jmath}_L + R_0^{(L,p)}U\tilde{\jmath}_L + R_0^{(L,p)}UR_0^{(L,p)}U\tilde{\jmath}_L + \dots \tag{8.207}$$

which provides upon substitution in (8.195) the *Born series for the partial amplitudes*

$$f_L(p) = -\frac{1}{k^2}\left[\int dr\,\tilde{\jmath}_L U\tilde{\jmath}_L + \int dr\,\tilde{\jmath}_L U R_0^{(L,p)}U\tilde{\jmath}_L + \dots\right] . \tag{8.208}$$

The analysis of the convergence of this series can be carried out in parallel with the arguments of Sect. 8.12. Introducing the symmetrized "partial" kernel

$$\tilde{K}_L(E) \equiv |U|^{1/2}R_0^{(L,p)}U^{1/2} \tag{8.209}$$

it is easy to see that under the conditions assumed for V, $\tilde{K}_L$ is Hilbert-Schmidt, and we have

$$\begin{gathered} \|\tilde{K}(E+\mathrm{i}0)\| = \sup_L \|\tilde{K}_L(E)\| \\ \frac{1}{4\pi}\|U\|_R = \|\tilde{K}(E+\mathrm{i}0)\|_2 = \left[\sum_L (2L+1)\|\tilde{K}_L(E)\|_2^2\right]^{1/2} . \end{gathered} \tag{8.210}$$

Consequently:

1) From some energy on, (8.208) converges for all L.
2) From some L on, (8.208) converges for all incident energies.
3) If $-|V|$ does not have any bound states or zero energy resonances, (8.208) converges for all E, L.
4) The *Born approximation*

$$s_L(p) \simeq s_L^{\mathrm{B}}(p) \equiv 1 - \frac{2\mathrm{i}}{k} \int_0^\infty dr\, U(r)[\tilde{\jmath}_L(kr)]^2 \tag{8.211}$$

can be trusted at either high energies, or for large angular momenta or for weak potentials. For example, using the bound

$$|R_0^{(L,p)}(r,r')|^2 \leq \frac{\pi r r'}{2L+1} \tag{8.212}$$

it can be shown [NE 82] that

$$|s_L(p) - s_L^{\mathrm{B}}(p)| \leq \frac{\pi^{1/2}\alpha^2}{(2L+1)^{1/2} - \alpha} \tag{8.213}$$

so long as

$$\alpha \equiv \pi^{1/2} \int_0^\infty dr\, r|U(r)| < \sqrt{2L+1}\,. \tag{8.214}$$

Using the inequality (8.213) we can estimate in many cases the error in the Born approximation $f_L(p) \simeq f_L^{\mathrm{B}}(p)$.
Since $f_L^{\mathrm{B}}(p)$ is real, this approximation will be good only if $\delta_L(p)$ is small (mod π); then

$$\tan \delta_L^{\mathrm{B}}(p) = -\frac{1}{k} \int_0^\infty dr\, U(r)[\tilde{\jmath}_L(kr)]^2\,. \tag{8.215}$$

Finally at low energies

$$R_0^{(L,p)}(r,r') \underset{p\to 0}{\sim} -\frac{1}{2L+1} r_<^{L+1} r_>^{-L} \tag{8.216}$$

and this together with (8.204) suggests that $u_{L,p}^+(r) \propto p^{L+1}$, $p \to 0$, as with $\tilde{\jmath}_L(kr)$. Using (8.195) one would naively obtain

$$f_L(p) \underset{p\to 0}{\sim} -a_L k^{2L}\,. \tag{8.217}$$

This behavior is not true in general as we will see later on, however (8.217) is valid for short range potentials unless there are exceptional circumstances such as a zero energy resonance with $L = 0$ or a zero energy bound state with $L > 0$.

Assuming the previous equation to hold, we see that at low energies only $f_0(p)$ survives, and (8.180) yields

$$f(p, \cos\theta) \underset{p\to 0}{\sim} -a_0 \tag{8.218}$$

and the differential cross section will be isotropic

$$\frac{d\sigma}{d\Omega} \underset{p\to 0}{\simeq} |a_0|^2 . \tag{8.219}$$

The quantities a_L, called *scattering lengths* (in L-wave) (although only a_0 has dimensions of length), are real: the behavior (8.217) implies that $\delta_L(p) \sim n_L\pi - a_L k^{2L+1}$, $p \to 0$.

Summarizing, the partial wave formalism presents many advantages in the computation of scattering amplitudes at low energies. The behavior (8.217) and the heuristic argument indicating that only those waves of angular momentum $L \lesssim kR$, $R \simeq$ range of the potential, will be affected by the potential imply that a small number of partial amplitudes suffices for the description of the scattering process. In addition, at energies where the Born approximation for the total amplitude $f_B(\boldsymbol{p} \to \boldsymbol{p}')$ is not applicable, the Born approximation $f_L^B(p)$ to $f_L(p)$ is still valid for large enough L, say $L > L_0(p)$, and we can approximate f as:

$$f(p, \cos\theta) \simeq f_B(p, \cos\theta) + \sum_0^{L_0(p)} (2L+1)[f_L(p) - f_L^B(p)]P_L(\cos\theta) \tag{8.220}$$

which requires only the computation of a few partial wave amplitudes. In particular,

$$\sigma_{el}(p) \simeq \sigma_B^{el}(p) + 4\pi \sum_0^{L_0(p)} (2L+1)[|f_L(p)|^2 - f_L^B(p)|^2] . \tag{8.221}$$

8.17 Scattering by a Central Square Well

There are few cases where the radial equation (8.189) can be solved in terms of elementary functions for all p and L. With the conditions imposed on $V(r)$ there is only one: the attractive or repulsive spherical square well, or the trivial generalization to piecewise constant central potentials. For simplicity we present here only the central square well:

$$V(r) = \begin{cases} V_0, & r < R \\ 0, & r > R . \end{cases} \tag{8.222}$$

Define $x \equiv kR$, $\bar{x} \equiv (k^2 - U_0)^{1/2}R$, $\operatorname{Im}\bar{x} \geq 0$; in the interior of the well $u_{L,p}(r) \propto \bar{j}_L(\bar{x}r/R)$, while on the outside it will be a linear combination of

$\tilde{j}_L(xr/R)$ and $\tilde{y}_L(xr/R)$. From (8.192) we can write $u_{L,p}(r) \propto \tilde{j}_L(xr/R) + ikf_L(p)\tilde{h}_L^{(1)}(xr/R)$, and solving the matching conditions at $r = R$ we obtain after a few simplifications:

$$\tan \delta_L(p) = \frac{\bar{x}\tilde{j}_{L-1}(\bar{x})\tilde{j}_L(x) - \tilde{j}_L(\bar{x})x\tilde{j}_{L-1}(x)}{\bar{x}\tilde{j}_{L-1}(\bar{x})\tilde{y}_L(x) - \tilde{j}_L(\bar{x})x\tilde{y}_{L-1}(x)} \tag{8.223}$$

taking $\tilde{j}_{-1} \equiv -\tilde{y}_0$, $\tilde{y}_{-1} \equiv \tilde{j}_0$. For example, for $L = 0$ (8.223) becomes

$$\cot(x + \delta_0) = \frac{\bar{x}}{x} \cot \bar{x} \, . \tag{8.224}$$

A continuous fraction representation of (8.223) obtained using the recursion relations for the Bessel functions, valid for $L \geq 1$, and very useful in numerical computations is

$$\cot(x + \delta_L) = \frac{1}{x} - \frac{1}{\frac{3}{x} - } \cdots \frac{1}{\frac{2L-1}{x} - }$$
$$\times \frac{1}{\frac{2L+1}{x} - \frac{\bar{x}}{x}\left\{\frac{2L+1}{x} - \frac{1}{\frac{2L-1}{x} - } \cdots \frac{1}{\frac{3}{\bar{x}} - } \frac{1}{\frac{1}{\bar{x}} - \cot \bar{x}}\right\}} \, . \tag{8.225}$$

In particular

$$\cot(x + \delta_1) = \frac{\bar{x}^2 - x^2 + x^2\bar{x}\cot\bar{x}}{x\bar{x}^2} \, . \tag{8.226}$$

From (8.223) it is easy to see that the scattering lengths satisfy

$$a_L/R^{2L+1} = -\frac{2L+1}{[(2L+1)!!]^2} \frac{\tilde{j}_{L+1}(x_0)}{\tilde{j}_{L-1}(x_0)} \tag{8.227}$$

with $x_0 \equiv [-U_0]^{1/2}R$, Im $x_0 \geq 0$. Thus, for the $L = 0$ wave

$$\frac{a_0}{R} = 1 - \frac{\tan x_0}{x_0} \, . \tag{8.228}$$

It is important to notice that a_L becomes infinite when the parameters of the potential allow the formation of a bound state at zero energy ($L > 0$) or a zero energy resonance at $L = 0$ (see Sect. 8.12, Sect. 6.3). This feature is not particular to the square well, it is a property of all short range potentials.

It is instructive to analyze the behavior of the phase shifts at low energies, specially for the $L = 0$ wave. Using (8.224) it is not difficult to obtain the expansion

$$k \cot \delta_0(p) = -\frac{1}{a_0} + \frac{1}{2} r_0 k^2 + \ldots \tag{8.229}$$

where a_0 is the scattering length (8.228) and the new parameter r_0, known as the *effective range* in $L = 0$ wave, is given by

$$\frac{r_0}{R} = 1 - \frac{R}{x_0^2 a_0} - \frac{R^2}{3a_0^2} \,. \tag{8.230}$$

Expansions of the type (8.229) apply in general for all short range potentials and are used often in the analysis of low energy scattering data.

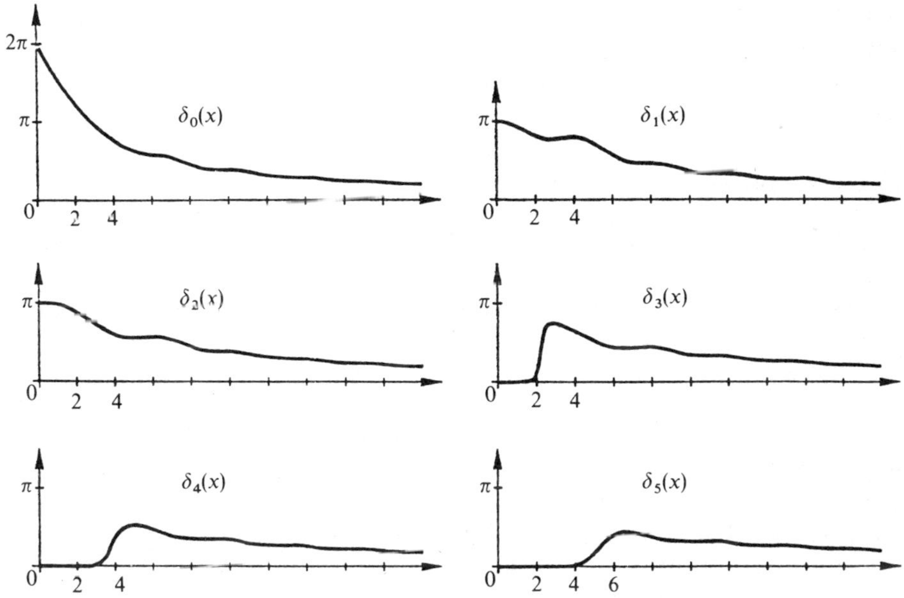

Fig. 8.21. Some phase shifts for the attractive central square well with parameter $x_0 = 5$

To conclude with this example, we include in Fig. 8.21 the lower partial wave phase shifts for a central square well of parameter $x_0 = 5$, the same one used in Fig. 8.17. These phase shifts have been obtained continuously from $\delta_L(\infty) = 0$ and one can see that for all of them $\delta_L(0) = n_L\pi$, where n_L is the number of bound states of the potential in the L wave (Sect. 6.3). This result is a simple illustration of Levinson's theorem which we discuss below.

Another remarkable fact in Figs. 8.17 and 8.21 is that $\delta_3(p)$ in the vicinity of $x \simeq 2.3$ undergoes a rapid increase of the order of π correlated with the resonance peak in $\sigma_{\rm el}(p)$ in the same energy region. This relation between sudden increases in $\delta_L(p)$ and the appearance of resonances is quite general. Finally, according to the previous discussion about the relative sizes of the different partial waves, notice the smallness, mod π, of the phase shifts of large L at low energies.

8.18 Analyticity Properties of the Partial Amplitudes

The study of the analyticity properties of the S matrix has played an important role in the physics of elementary particles, and it has originated very useful technical tools such as dispersion relations, Regge poles, etc. In the scattering by a central potential, these properties can be proved, while in the general case they are taken as postulates of the dynamics underlying the scattering of elementary particles. Since the main purpose of this book is not scattering theory, we will only present the most important results on the analyticity for simple scattering, referring the interested reader to the specialized literature for a more thorough and rigorous treatment [GW 64], [AR 65], [TA 72], [RS 79], [NE 82].

Unless otherwise stated in what follows we will assume that $V(r)$ satisfies the standard conditions.

a) Jost Functions

The differential equation (8.189) has $u^+_{L,p}(r)$ as the regular solution satisfying (8.204). Any other solution regular at the origin is proportional to it. Among them we can single out the real solution $v_{L,p}(r)$ characterized by the condition

$$v_{L,p}(r) \underset{r\to 0}{\sim} \tilde{\jmath}_L(kr) \tag{8.231}$$

namely, it is not only regular at the origin, but it is normalized in such a way that in a neighborhood of the origin it behaves like the free solution $\tilde{\jmath}_L$. The solution $v_{L,p}$ does not satisfy an integral equation of Fredholm type (8.204) but rather a Volterra integral equation to account for the new boundary condition (8.231):

$$\begin{aligned} v_{L,p}(r) &= \tilde{\jmath}_L(kr) + \int_0^r dr' \mathcal{R}_0^{(L,p)}(r,r')U(r')v_{L,p}(r') \\ \mathcal{R}_0^{(L,p)}(r,r') &\equiv \frac{\mathrm{i}}{2k}\left[\tilde{h}_L^{(2)}(kr)\tilde{h}_L^{(1)}(kr') - \tilde{h}_L^{(1)}(kr)\tilde{h}_L^{(2)}(kr')\right] \end{aligned} \tag{8.232}$$

as one can easily prove. The equation (8.232) under the conditions imposed on the potential can be solved by an iterative procedure (Neumann series) even for complex p. This series converges uniformly over compact sets in p and r, and the function obtained $v_{L,p}(r)$ is for each r an entire function of p satisfying (8.189) and (8.231). Using the properties of spherical functions under $k \to -k$ we can easily verify using (8.232) that

$$v_{L,p}(r) = (-1)^{L+1} v_{L,-p}(r) = (v_{L,p^*}(r))^* \ . \tag{8.233}$$

For real p and large r, the influence of the potential is negligible, and $v_{L,p}$ will behave as a linear combination of $\tilde{h}_L^{(1)}(kr)$ and $\tilde{h}_L^{(2)}(kr)$:

$$v_{L,p}(r) \underset{r\to\infty}{\sim} \tfrac{1}{2}\left[F_L(p)\tilde{h}_L^{(2)}(kr) + F_L(-p)\tilde{h}_L^{(1)}(kr)\right] \tag{8.234}$$

where (8.233) has been taken into account. The function $F_L(p)$ appearing in (8.234) is known as the *Jost function* for L waves. Comparing with (8.192) it is clear that

$$\boxed{v_{L,p}(r) = F_L(p)u^+_{L,p}(r)\ , \quad s_L(p) = \frac{F_L(-p)}{F_L(p)}} \tag{8.235}$$

and since we are now assuming p to be real, the reality of $v_{L,p}$ yields

$$F_L(-p) = (F_L(p))^*\ . \tag{8.236}$$

The asymptotic behavior (8.232) and (8.234) implies that

$$F_L(p) = 1 + \frac{\mathrm{i}}{k}\int_0^\infty dr'\tilde{h}^{(1)}_L(kr')U(r')v_{L,p}(r')\ . \tag{8.237}$$

This equation valid for p real is the starting point for the analytic extension of the Jost function to the complex plane. The fact that for $\operatorname{Im} p > 0$ the function $\tilde{h}^{(1)}_L(kr')$ decreases exponentially counters the possible exponential growth of $v_{L,p}$, and it can be shown that the right hand side of (8.237) defines an analytic function $F_L(p)$ in the upper half plane $\operatorname{Im} p > 0$ and continuous for $\operatorname{Im} p \geq 0$. On the other hand, from (8.233) and (8.237) we immediately obtain

$$\boxed{(F_L(-p^*))^* = F_L(p)} \tag{8.238}$$

generalizing (8.236) and implying in particular that the Jost function is real over the positive imaginary axis $\operatorname{Re} p = 0$, $\operatorname{Im} p \geq 0$.

The second equation in (8.235) implies that it is not possible to extend $s_L(p)$ in general to complex momenta as an analytic function. It would require extending $F_L(p)$ to the lower half plane, and (8.237) does not allow such an extension unless $U(r)$ decreases sufficiently fast (together with the standard conditions). If for example $U(r)$ is of compact support, the exponential growth of $v_{L,p}$ and $\tilde{h}^{(1)}_L$ is irrelevant in the integral in (8.237), and this equation defines F_L as an entire function. For these potentials, s_L is a meromorphic function of p. On the other hand, if $U(r) = \mathrm{O}(\exp(-\mu r))$, $r \to \infty$, then $F_L(p)$ is analytic for $\operatorname{Im} k > -\mu/2$ and s_L is meromorphic in the (Bargmann) strip $|\operatorname{Im} k| < \mu/2$. Finally, if the potential is in addition a discrete or continuous superposition of Yukawa or exponential potentials, it can be shown that F_L is analytic except perhaps on the Yukawa cut $\operatorname{Re} k = 0$, $\operatorname{Im} k \leq -\mu/2$, and s_L is meromorphic in the whole complex plane with the possible exclusion of cuts $\operatorname{Re} k = 0$, $|\operatorname{Im} k| \geq \mu/2$.

In general it is not possible to determine explicitly the Jost functions; its knowledge is equivalent to solving the Schrödinger equation. In the case of the central square well (Sect. 8.17), it is not difficult to obtain:

$$F_L(p) = \mathrm{i}\left(\frac{x}{\bar{x}}\right)^L \left[\frac{x}{\bar{x}}\tilde{j}_L(\bar{x})\tilde{h}^{(1)}_{L+1}(x) - \tilde{j}_{L+1}(\bar{x})\tilde{h}^{(1)}_L(x)\right] \tag{8.239}$$

and for s-waves

$$F_0(p) = e^{ix}\left[\cos\bar{x} - i\frac{x}{\bar{x}}\sin\bar{x}\right] . \tag{8.240}$$

As expected from the previous arguments, $F_L(p)$ is an entire function.

For the exponential potential $V(r) \propto \exp(-\mu r)$, $F_0(p)$ is explicitly calculable, and it is meromorphic in the plane with simple poles at $k = -i(\mu/2)n$, $n = 1, 2, \ldots$. The Yukawa cut is reduced in this case to a discrete set of simple poles. On the other hand, for the Yukawa potential which can be considered as a superposition of exponentials,

$$\frac{e^{-\mu r}}{r} = \int_\mu^\infty d\lambda\, e^{-\lambda r} \tag{8.241}$$

the union of pole sets arising from each exponential component fill the Yukawa cut, and give rise to a true cut in $F_0(p)$.

b) Zeroes of the Jost Function and Bound States

Equation (8.234) is generally meaningful only if p is real. For these values of p (8.189) has two solutions, necessarily non-regular, $F_L(\pm p, r)$ characterized by their asymptotic behavior

$$F_L(p,r) \underset{r\to\infty}{\sim} \tilde{h}_L^{(1)}(kr) , \quad F_L(-p,r) \underset{r\to\infty}{\sim} \tilde{h}_L^{(2)}(kr) . \tag{8.242}$$

The behavior (8.234) shows that $v_{L,p}(r)$ is the following linear combination of these other solutions:

$$v_{L,p}(r) = \tfrac{1}{2}[F_L(p)F_L(-p,r) + F_L(-p)F_L(p,r)] \tag{8.243}$$

and in particular, the coefficient $F_L(p)$ is simply the wronskian

$$F_L(p) = \frac{i}{k} W[F_L(p, .), v_{L,p}(.)] . \tag{8.244}$$

As for $v_{L,p}$ the new solutions $F_L(\pm p, r)$ satisfy integral equations of Volterra type but with the integration interval $[r, \infty)$. This makes it difficult to extend them analytically. Nevertheless it is not difficult to prove that $F_L(p,r)$ is analytic for all r if $\mathrm{Im}\, p > 0$, and continuous if $\mathrm{Im}\, p \geq 0$. Consequently, the right hand side of (8.244) makes sense if $\mathrm{Im}\, p > 0$ and coincides in this half plane with the analytic extension of F_L.

Suppose now that $F_L(p) = 0$ for some p with $\mathrm{Im}\, p > 0$. In this case $F_L(p,r)$ and $v_{L,p}(r)$ are linearly dependent and the regular solution $v_{L,p}(r)$ will behave at large distances as const.$\tilde{h}_L^{(1)}(kr)$; it will be exponentially decreasing. Therefore (8.189) will have for this momentum a square integrable regular solution. It is not difficult to verify that this function belongs to the domain of the self-adjoint operator $-d^2/dr^2 + (L(L+1)/r^2 + U(r))$ (with the boundary condition $u(0) = 0$ if $L = 0$); this necessarily implies $p^2 \in \mathbb{R}$ and p must be in the imaginary axis. Hence the zeroes of $F_L(p)$ in $\mathrm{Im}\, p > 0$ satisfy necessarily $\mathrm{Re}\, p = 0$, and their squares provide energies of the L wave bound states for the potential $V(r)$. Conversely, if $E \equiv (\hbar^2/2M)\varepsilon < 0$ is the energy of a L wave bound state,

$p \equiv \mathrm{i}\hbar|\varepsilon|^{1/2}$ will be a zero of F_L: in fact, it is legitimate to call $v_{L,p}$ the reduced radial wave function of this bound state conveniently normalized, and since this function decreases exponentially, (8.244) shows that $F_L(p) = 0$.

We have shown that there is a one-to-one correspondence between the zeroes of F_L in the positive imaginary axis ($\mathrm{Im}\, p > 0$) and the negative energies of L wave bound states. Since these states are non-degenerate, it can be shown that the zeroes are simple.

With respect to the possible zeroes of F_L on the real axis, consider first the case $\mathrm{Re}\, p \neq 0$. This possibility is absurd; from the relations (8.236) and (8.243) we derive $v_{L,p}(r) \equiv 0$, and this contradicts (8.231). Finally suppose $F_L(0) = 0$. In this case the contradiction we just pointed out does not apply anymore, because $v_{L,0}$ and $\tilde{j}_L(0)$ vanish identically. The normalization (8.231) is not convenient to discuss this case and we introduce the redefinitions

$$\bar{v}_{L,p}(r) \equiv \frac{1}{k^{L+1}} v_{L,p}(r) \,, \quad \overline{F}_L(p,r) \equiv k^L F_L(p,r) \tag{8.245}$$

so that

$$\bar{v}_{L,p}(r) \underset{r\to 0}{\sim} \frac{r^{L+1}}{(2L+1)!!} \,, \quad \overline{F}_L(0,r) \underset{r\to 0}{\sim} -\mathrm{i}\frac{(2L-1)!!}{r^L} \,. \tag{8.246}$$

With these new functions, (8.244) can be written as

$$F_L(p) = \mathrm{i}W[\overline{F}_L(p,\,.), \bar{v}_{L,p}(.)] \tag{8.247}$$

and if $F_L(0) = 0$, $\bar{v}_{L,0}$ will be a regular function, not vanishing identically and behaving at large distances as r^{-L}. It will be (up to normalization) the wave function for a L wave bound state at zero energy as long as $L > 0$. Conversely, if in a $L > 0$ wave the potential has a bound state with zero energy, its reduced wave function behaves at the origin as r^{L+1} (and up to a constant it can be taken as $\bar{v}_{L,0}$) and as r^{-L} at infinity, leading in (8.247) to a vanishing wronskian, and therefore $F_L(0) = 0$.

Summarizing, $F_L(p)$ has no zeroes on the real axis unless $p = 0$. In this case if $L > 0$ then $F_L(0) = 0$ if and only if the potential possesses an L wave bound state with zero energy. If $F_0(0) = 0$, this does not imply that the potential has a zero energy s-wave bound state, but as we will show below it has a zero energy resonance. Finally, it can be shown [NE 82] that if $F_L(0) = 0$ and if $\mathrm{Im}\, p \geq 0$, then

$$\begin{aligned} F_0(p) &= \mathrm{i}\gamma_0 k + \mathrm{O}(k^2) \,, \quad & L = 0 \\ F_L(p) &= \beta_L k^2 + \mathrm{O}(k^2) \,, \quad & L > 0 \end{aligned} \tag{8.248}$$

where γ_0 and β_L are real and not equal to zero. Hence, if $p = 0$ is a zero of F_L, this zero is simple if $L = 0$ and double if $L > 0$.

Let us illustrate the previous arguments with the example of the central square well, whose Jost functions are given in (8.239). Furthermore, we only consider s waves. The zeroes of F_0 in the imaginary axis satisfy

$$\mathrm{i}\tan\bar{x} = \frac{\bar{x}}{x}\,, \quad x = \mathrm{i}y\,, \quad y > 0\,. \tag{8.249}$$

It is easy to see that if $U_0 > 0$, there are no solutions to (8.249), while if $U_0 < 0$ there are solutions only if $-y^2 > U_0 R^2$ (i.e. $E > V_0$), and the previous equation leads to (6.60). In a neighborhood of the origin

$$F_0(p) = \cos x_0 + \mathrm{i}\left(\cos x_0 - \frac{\sin x_0}{x_0}\right) x + \mathrm{O}(x^2) \tag{8.250}$$

so that $F_0(0) = 0$ if and only if $\cos x_0 = 0$, namely, $U_0 < 0$ and $x_0 = \pi(n + 1/2)$. In this case the coefficient γ_0 in (8.248) is equal to $\gamma_0 = (-1)^{n+1} R/\pi(n + 1/2)$. The relation $x_0 = \pi(n + 1/2)$ provides the threshold depths of the well for the appearance of a new s-wave bound state [see (6.61)]. Finally, for $L \neq 0$ it is easy to see that $F_L(0) = 0$ if and only if $\tilde{j}_{L-1}(x_0) = 0$ [equivalent to (6.62)] and in a neighborhood of $p = 0$,

$$F_L(p) = \frac{(2L-1)!!}{x_0^L}\tilde{j}_{L-1}(x_0) + \mathrm{O}(k^2)\,. \tag{8.251}$$

c) Levinson's Theorem

There exists an important relation between the number of L wave bound states with energy smaller or equal to zero, and the variation in phase shift δ_L as we change the momentum from ∞ to 0. This relation is known as *Levinson*'s *theorem* [LE 49].

From (8.237) it is possible to estimate that $\exists\varepsilon > 0$ such that

$$|F_L(p) - 1| = \mathrm{O}\left(\frac{1}{|p|^{1/2+\varepsilon}}\right)\,, \quad |p| \to \infty\,, \quad \operatorname{Im} p \geq 0 \tag{8.252}$$

and thus at large values of the momentum we can take the phase of $F_L(p)$ as practically equal to zero. On the other hand (8.235) and (8.236) for p real imply

$$\delta_L(p) = -\arg F_L(p)\ (\mathrm{mod}\ \pi)\,. \tag{8.253}$$

Due to the ambiguity in the definition of phase shifts, we can always choose $\delta_L = -\arg F_L$. This is the determination of the phase shift obtained by continuity (F_L is continuous on the real axis) from $p \to \infty$. Hence $\delta_L(p) \to 0$, $p \to \infty$, a choice already made in Sect. 8.16. Therefore,

$$F_L(p) = |F_L(p)| \exp[-\mathrm{i}\delta_L(p)]\,. \tag{8.254}$$

We next consider the meromorphic function $F_L'(p)/F_L(p)$ in $\operatorname{Im} p > 0$. The relation (8.252) implies that the zeroes of $F_L(p)$ in the closed upper half plane fall within a compact set. Since F_L is analytic in $\operatorname{Im} p > 0$ and since it cannot have zeroes in $\operatorname{Im} p = 0$, $\operatorname{Re} p \neq 0$, the only possible accumulation point for its zeroes is $p = 0$; however, in a neighborhood of this point (8.248) implies that

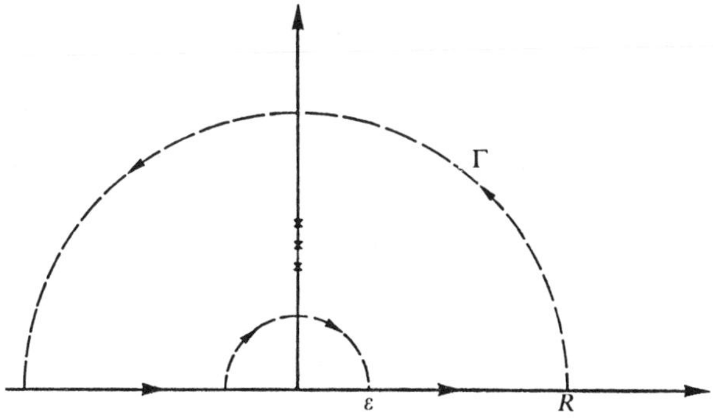

Fig. 8.22. Integration contour in the p plane

$F_L(p)$ depends linearly or quadratically in p, and this excludes the possibility of having $p = 0$ as an accumulation point of zeroes of F_L. Hence, the function F_L has at most a finite number of zeroes in $\operatorname{Im} p \geq 0$, and the potential will bind at most a finite number of L wave states. Integrating $F_L'(p)/F_L(p)$ along the contour Γ represented in Fig. 8.22 where R and ε^{-1} are big enough so that all zeroes of F_L with $\operatorname{Im} p > 0$ fall within Γ, the theorem of residues leads to

$$\int_\Gamma dp\, \frac{F_L'(p)}{F_L(p)} = 2\pi \mathrm{i} \bar{n}_L \tag{8.255}$$

where $\bar{n}_L$ is the number of L wave bound states with energy smaller than zero (notice that the zeroes of F_L with $\operatorname{Im} p > 0$ are simple, and therefore they also provide simple poles in the integrand, with unit residue). Due to (8.252) the integral over the outer semi-circle vanishes as $R \to \infty$, and the integral over the inner semi-circle tends to zero if $\varepsilon \to 0$ when $F_L(0) \neq 0$, but if $F_L(0) = 0$, then this limit is equal to $-\pi\mathrm{i}$ ($L = 0$, simple zero of F_0) or $-2\pi\mathrm{i}$ ($L > 0$, double zero of F_L). Hence

$$\lim_{\substack{\varepsilon \to 0 \\ R \to \infty}} \left(\int_{-R}^{-\varepsilon} + \int_{\varepsilon}^{R} \right) \frac{F_L'(p)}{F_L(p)}\, dp = \begin{cases} 2\pi \mathrm{i} n_L, & (L > 0, \text{ or } L = 0, F_0(0) \neq 0) \\ 2\pi\mathrm{i}(n_0 + 1/2), & (L = 0, F_0(0) = 0) \end{cases} \tag{8.256}$$

where n_L represents the number of L wave bound states with energy ≤ 0 (recall that $\bar{n}_0 = n_0$).

On the other hand, using (8.236) and (8.254) we immediately obtain that the left hand side of (8.256) equals $2\mathrm{i}(\delta_L(0) - \delta_L(\infty))$, therefore

$$\boxed{\frac{1}{\pi}[\delta_L(0) - \delta_L(\infty)] = \begin{cases} n_L & \text{if } L > 0, \text{ or } L = 0 \text{ and } F_0(0) \neq 0 \\ n_0 + 1/2 & \text{if } L = 0, \quad F_0(0) = 0\,. \end{cases}} \tag{8.257}$$

This relation is Levinson's theorem. In particular, taking $\delta_L(\infty) = 0$:

$$n_L = \frac{1}{\pi}\delta_L(0) - \tfrac{1}{2}\sin^2\delta_L(0)\,, \quad L = 0,1,2,\ldots\,. \tag{8.258}$$

This theorem was discussed graphically in Sect. 8.17. In the exceptional case $F_0(0) = 0$ and for a central square well, we must have $\cos x_0 = 0$ and from (8.224) we obtain $\cot\delta_0(0) = 0$ in agreement with (8.257).

d) Low Energy Behavior

The partial amplitude f_L given by (8.195) can be expressed in terms of Jost functions as

$$f_L(p) = \frac{F_L(-p) - F_L(p)}{2\mathrm{i}kF_L(p)}\,. \tag{8.259}$$

If the potential is at least exponentially decreasing, F_L is analytic around $p = 0$ and therefore, if $F_L(0) \neq 0$, f_L will also be analytic around $p = 0$. To determine in this case the dominant behavior of f_L it suffices to rewrite (8.195) as

$$f_L(p) = -\frac{1}{F_L(p)k^2}\int_0^\infty dr\,\tilde{\jmath}_L(kr)U(r)v_{L,p}(r) \tag{8.260}$$

and using estimates on $|\tilde{\jmath}_L(kr)|$ and $|v_{L,p}(r)|$ (the last one is derived from the integral equation (8.232)) it can be shown that

$$|f_L(p)| \leq \frac{\text{const}}{F_L(0)}k^{2L}\int_0^\infty dr\;|U(r)|r^{2L+2} \tag{8.261}$$

hence

$$\boxed{f_L(p) = -a_L k^{2L} + \mathrm{O}(k^{2L+1})} \tag{8.262}$$

justifying (8.217). The quantity a_L, which we call L wave scattering length, is real as (8.259) and (8.236) show, because the dominant part of f_L at $p = 0$ is real if $F_L(0) \neq 0$.

Moreover, starting with the relation

$$\tan\delta_L(p) = \frac{kf_L(p)}{1 + \mathrm{i}kf_L(p)} = \mathrm{i}\frac{F_L(p) - F_L(-p)}{F_L(p) + F_L(-p)} \tag{8.263}$$

and noticing that the right hand side is analytic around $p = 0$ (assuming $F_L(0) \neq 0$ and an exponentially decreasing potential) and odd under $p \to -p$, we obtain for $a_L \neq 0$ and using (8.262) that

$$\boxed{k^{2L+1}\cot\delta_L(p) = -\frac{1}{a_L} + \frac{1}{2}r_L k^2 + \mathrm{O}(k^4)} \tag{8.264}$$

generalizing (8.229). This expansion known as the *effective range expansion* has been justified only for exponentially decreasing potentials. If the potential has a tail of the form $r^{-\alpha}$, $\alpha > 3$, the analytic behavior around $p = 0$ is not valid in general, and the previous expansion fails. Linear or even logarithmic terms may appear.

When $F_L(0) = 0$ we have to modify the previous arguments. For $L = 0$, starting with (8.259) and using the first equation in (8.248) we immediately derive

$$\boxed{f_0(p) = \frac{\mathrm{i}}{k} + \mathrm{O}(1)} \tag{8.265}$$

so that at low energies the s-wave elastic cross section behaves as

$$\sigma_{\mathrm{el}}^{(0)}(p) \underset{k\to 0}{\sim} \frac{4\pi}{k^2} . \tag{8.266}$$

This divergence at low energies justifies the name *zero energy resonance* given to the case $F_0(0) = 0$.

For $L \neq 0$ if $F_L(0) = 0$, using (8.260), the second equation in (8.248) and the estimates leading to (8.261), we find

$$\boxed{f_L(p) = \alpha_L k^{2L-2} + \mathrm{O}(k^{2L-1})} , \quad L \neq 0 \tag{8.267}$$

with α_L real.

These exceptional cases with $F_L(0) = 0$ are rarely found in practice. On the other hand the effective range expansion (8.264) is very useful in nuclear physics in describing for example low energy nucleon-nucleon collisions. In particular, for s-waves, the approximation consisting in taking the first two terms in (8.264) is equivalent to assuming a scattering amplitude of the form

$$f_0(p) = \frac{\mathrm{i}}{k - \mathrm{i}[(1/a_0) - (r_0 k^2/2)]} . \tag{8.268}$$

If $r_0 > 0$, and $2r_0/a_0 < 1$, a very simple Jost function leading to such a f_0 is

$$F_0(p) = \frac{k - \mathrm{i}k_{\mathrm{D}}}{k + \mathrm{i}k_0} \quad \text{with} \tag{8.269}$$

$$\begin{aligned} k_{\mathrm{D}} &= \frac{1}{r_0}\left[1 - \sqrt{1 - (2r_0/a_0)}\right] \\ k_0 &= \frac{1}{r_0}\left[1 + \sqrt{1 - (2r_0/a_0)}\right] . \end{aligned} \tag{8.270}$$

This function is analytic in $\operatorname{Im} p > 0$ and it corresponds to Bargmann potentials which decrease exponentially in this case. (Actually, there exists a family of such potentials leading to the same $\delta_0(p)$ and the same s-wave bound state energy [NE 82].) If $2r_0/a_0 > 0$ the roles of k_0, k_{D} can be exchanged in (8.270) and the choice we have made is in order to have the singularity of $F_0(p)$ as far as possible from the low energy region. For this $F_0(p)$ there is a bound state with energy $-\hbar^2 k_D^2/2M$, assuming $k_D > 0$. This occurs for example in the interaction n-p in the triplet state where $a_0 = 5.39$ fm and $r_0 = 1.703$ fm. With these values $k_D = 0.231\,\text{fm}^{-1}$ leading to a binding energy $B_D = 2.21$ MeV, to be compared with the deuteron binding energy 2.226 MeV.

On the other hand, in the n-p collision in the singlet state, $a_0 = -23.7$ fm, $r_0 = 2.76$ fm and therefore $k_D = -0.0400\,\text{fm}^{-1}$. The function F_0 has a zero in the negative imaginary axis very close to the origin because a_0 is very large and negative. In this case one says that such a zero represents a *virtual state* for it can be shown that if the depth of the well is slightly increased, this zero can move to the upper half plane and become a bound state. The proximity of this virtual state to the low energy region is reflected in the behavior of the cross section in the form

$$\sigma_{\text{el}}^{(0)} \simeq 4\pi a_0^2 \frac{k_D^2}{k^2 + k_D^2} \tag{8.271}$$

up to the factor $k_0^2/(k^2 + k_0^2)$ which in this case is very close to unity because $|a_0| \gg r_0$. Since (8.271) is invariant under $k_D \to -k_D$, the effect of a virtual state is not distinguishable at low energies from a bound state. Finally notice that in this case of n-p scattering $\sigma_{\text{el}}(0) \simeq 70$ barn, very large compared to the geometrical cross section.

The close analogy between the previous discussions and those in Sect. 4.6 is remarkable.

e) Resonances

So far the discussion has concentrated on the analytic properties on the upper half plane, and we have characterized the zeroes of the Jost function as bound states. When the potential is at least exponentially decreasing (something we will assume in this subsection) one can obtain not only effective range formulae as we have seen, but also the zeroes of $F_L(p)$ in $\text{Im}\, p < 0$ may sometimes have an important physical interpretation.

In contrast with the case $\text{Im}\, p > 0$, there can be zeroes of F_L in the lower half plane with non-vanishing real part. The relation (8.238) which would still hold for the analytic continuation of F_L to the region of $\text{Im}\, p < 0$ where the analytic continuation exists, shows that the zeroes of F_L appear symmetrically with respect to the line $\text{Re}\, p = 0$: $F_L(p_R - ip_I) = 0$, $p_I > 0$ implies $F_L(-p_R - ip_I) = 0$. These zeroes in the lower half plane are not necessarily simple.

The zeroes $p = p_R - ip_I$ with $p_R = 0$, $p_I > 0$ and small were discussed in the previous subsection, and they were interpreted as virtual states. When p_I is large, these zeroes do not have a clear physical interpretation. This is to be expected because the analyticity properties of the Jost functions in the lower half plane are very sensitive to modifications of the behavior of the potential at large distances. (The Yukawa cut for example disappears completely if we cut off the potential, independent of the distance where the cut off is performed.) Nevertheless, the behavior close to the real axis are practically insensitive to these modifications,

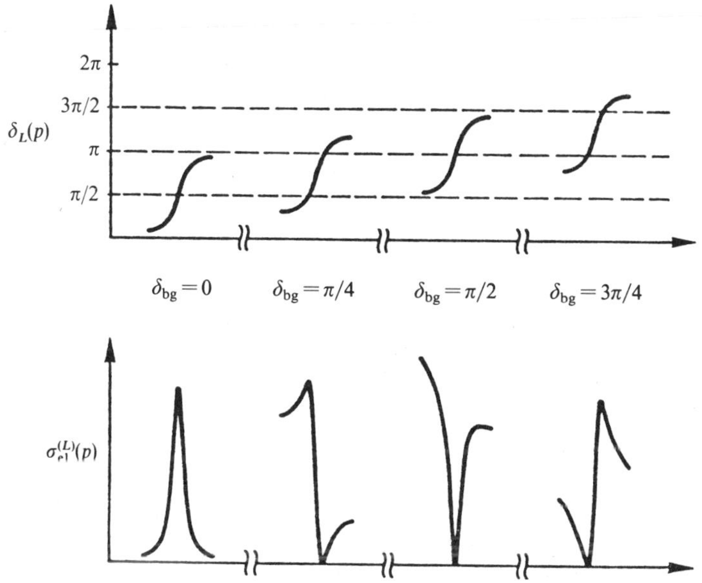

Fig. 8.23. L wave resonant cross section for different values of δ_{bg}; and parameters $p_I : p_R = 0.1 : 5$

and it is reasonable to expect that zeroes with p_I small will possess a physical interpretation.

Suppose that $\bar{p} = p_R - ip_I$, $p_I > 0$, $p_R > 0$ is a simple zero of $F_L(p)$. In a neighborhood of p,

$$F_L(p) \simeq F_L'(\bar{p})(p - \bar{p}) \,. \tag{8.272}$$

If $\bar{p}$ is sufficiently close to the real axis for the approximation (8.272) to be applicable in a real neighborhood of p_R, we will have

$$\delta_L(p) = -\arg F_L(p) \simeq -\arg F_L'(\bar{p}) - \arg(p - \bar{p}) \equiv \delta_{bg} + \delta_{res}(p) \tag{8.273}$$

where δ_{bg} is the background phase shift. As we move the real momentum from the left of p_R to its right, $\delta_{res}(p)$ undergoes an increase very close to π, and this increment is the more sudden the smaller the value of p_I. The same will happen therefore with $\delta_L(p)$. We will say that in this case we have L wave *resonance.* This does not necessarily mean that $\sigma_{el}^{(L)}$ will show a maximum. This depends on the background phase shift δ_{bg}. Indeed,

$$\sigma_{el}^{(L)} \simeq \frac{4\pi}{k^2}(2L+1)\sin^2[\delta_{bg} + \delta_{res}(p)] \tag{8.274}$$

and Fig. 8.23 illustrates the behavior of $\sigma_{el}^{(L)}(p)$ for different values of δ_{bg}.

In terms of energies,

$$\frac{\bar{p}^2}{2M} \equiv \overline{E} = E_{\mathrm{R}} - \mathrm{i}\frac{\Gamma}{2} \quad \text{(in the second Riemann sheet)} \tag{8.275}$$

and assuming $\Gamma \ll E_{\mathrm{R}}$, we have

$$\begin{aligned} &\tan \delta_{\mathrm{res}}(E) \simeq -\frac{\Gamma/2}{E - E_{\mathrm{R}}} \\ &\sigma_{\mathrm{el}}^{(L)}(p) = \frac{4\pi}{k^2}(2L+1)\frac{\Gamma^2/4}{(E-E_{\mathrm{R}})^2 + \Gamma^2/4}\,, \quad \text{if } \delta_{\mathrm{bg}} \simeq 0\,. \end{aligned} \tag{8.276}$$

These formulae reveal the Breit-Wigner form of a resonance in the absence of background. In general the background phase shift will be negligible when the resonance appears at low energy with $L > 0$, for in a neighborhood of $p = 0$ the Jost function behaves as

$$F_L(p) = \alpha_L + \beta_L k^2 + \ldots\,, \qquad L > 0 \tag{8.277}$$

[as deduced from (8.262, 263)] and we will have $\arg F_L'(\bar{p}) \simeq \arg(\beta_L \bar{k}) \simeq 0 \pmod{\pi}$ if $p_{\mathrm{I}} \ll p_{\mathrm{R}}$, hence $\delta_{\mathrm{bg}} \simeq 0 \pmod{\pi}$. This argument does not apply to s-waves.

Finally the presence of a resonance dominating the scattering in a given energy region has manifestations entirely analogous to those discussed in Sect. 4.7: In the time evolution of a scattering wave packet, and with the notation of Sect. 8.13, the scattered wave packet presents a delay given by the *Wigner* formula [WI 55]

$$\tau_L = 2\hbar(d\delta_L(E)/dE)_{E=E_{\mathrm{R}}}\,. \tag{8.278}$$

Causality considerations similar to those in Sect. 4.5 limit the possible decrease of the phase shifts in the form

$$d\delta_L(p)/dk \gtrsim -R \tag{8.279}$$

where R is the range of the potential. Rigorous properties about the delay times in collisions can be found in [JS 72a], [AS 77] and [RS 79].

f) The Poles of the S Matrix

We have seen that for standard and exponentially decaying central potentials, the zeroes of $F_L(p)$ have the following interpretation:

a) $\mathrm{Im}\, p > 0$, $\mathrm{Re}\, p = 0 \leftrightarrow$ bound state.
b) $p = 0$, $L > 0 \leftrightarrow$ zero energy bound state.
   $p = 0$, $L = 0 \rightarrow$ zero energy resonance
c) $\mathrm{Im}\, p < 0$, $\mathrm{Re}\, p = 0 \rightarrow$ virtual state
d) $\mathrm{Im}\, p < 0$, $\mathrm{Re}\, p > 0 \rightarrow$ resonance.

In the complex energy plane these zeroes appear in some of the Riemann sheets if $p \neq 0$, as discussed in Chap. 4.

Since $s_L(p) = F_L(-p)/F_L(p)$, the singularities of $s_L(p)$ in the Bargmann strip come from the zeroes of $F_L(p)$. These zeroes always lead to poles of s_L if $p \neq 0$, for in this case $F_L(-p)$ and $F_L(p)$ cannot vanish simultaneously as a consequence (8.243). Therefore the bound states with non-vanishing energy appear as simple poles of s_L in the positive imaginary axis, while the virtual states and the resonances with energy different from zero appear as poles of s_L in $\operatorname{Im} p < 0$ inside the Bargmann strip. Outside this strip, nothing is known in general with the exception of *Yukawian* potentials (constructed as superpositions of Yukawa potentials), where the strip is replaced by the whole plane minus the Yukawa cuts.

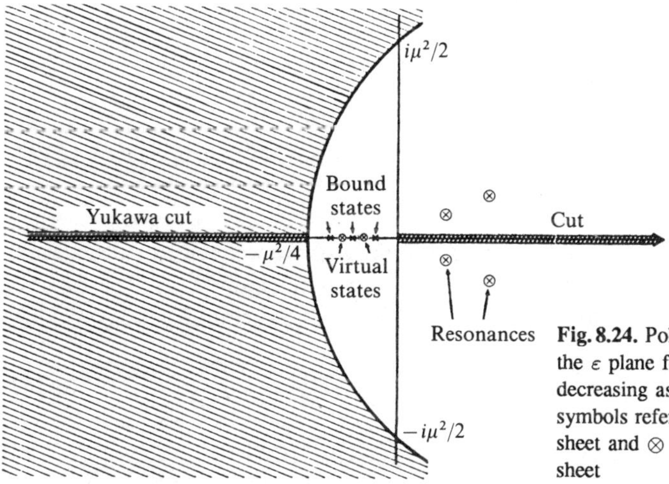

Fig. 8.24. Pole structure of $s_L(\varepsilon)$ in the ε plane for standard potentials decreasing as $\exp(-\mu r)$. The $\times$ symbols refer to the first Riemann sheet and $\otimes$ refer to the second sheet

In the energy plane (or equivalently in the plane of the variable $\varepsilon = k^2$) the analyticity regions of s_L or of the partial amplitude f_L are as indicated in Fig. 8.24. The shadowed region appearing in both sheets is the image under $k \to k^2$ of the region outside the Bargmann strip. Obviously, for Yukawian potentials this region reduces to the cut on the left, and in the phenomenological analysis of s_L for non-potential interaction one usually adopts this Yukawian structure.

8.19 Analyticity Properties of the Scattering Amplitude

The total scattering amplitude $f(p, \cos\theta)$ given by the partial amplitudes $f_L(p)$ through the Legendre series (8.180) also has analyticity properties. We will not discuss in detail how these properties are derived, referring the reader to the bibliography suggested in Sect. 8.18. We simply mention that the amplitude $f(p, \cos\theta)$ as a function of two variables E and q^2 (momentum transfer/$\hbar$)^2 admits an analytic extension except for "poles" to the region $E \in \mathbb{C} - [0, \infty)$, $q^2 \in \mathbb{C} - (-\infty, -\mu^2]$, provided the potential is Yukawian with range μ^{-1}. Subtracting from f the Born amplitude $f_B(p, \cos\theta) \equiv f_B(q^2)$, the difference $f - f_B$ is analytic in E, q^2 except for "poles" in a larger region $(\mathbb{C} - [0,\infty)) \times (\mathbb{C} - (-\infty, -4\mu^2])$. The "poles" are actually the sets $\{E_n\} \times \mathbb{C}$ where $E_1, E_2, \ldots$ are energies of the bound states with negative energies. The joint analyticity properties in E, q^2 are basic in deriving the *Mandelstam representation* which together with unitarity and f_B can replace even the Schrödinger equation for dynamical purposes.

For fixed and phyiscal q^2 i.e. $q^2 \geq 0$, a consequence of the previous arguments is that $f(E, q^2)$ can be extended analytically as a function of E to $\mathbb{C} - [0,\infty)$ except for poles at $E_1, E_2, \ldots$, and it can be shown that the physical amplitude ($E \geq 0$) is obtained as the limit $f(E + \mathrm{i}0, q^2)$. The poles are simple, and the following *dispersion relation* is satisfied:

$$f(E,q^2) = f_B(q^2) + \sum_{n,j} \frac{\Gamma_{n,j} P_{L n,j}(1 - q^2/2\varepsilon_n)}{E - E_n} + \frac{1}{\pi}\int_0^\infty dE' \frac{\operatorname{Im} f(E', q^2)}{E' - (E + \mathrm{i}0)} \tag{8.280}$$

where the sum runs over bound states (which can present accidental degeneracy), and the constants $\Gamma_{j,n}$ are related to the large distance behavior of the normalized bound state wave functions $\phi_{n,j}$ of angular momentum $L_{n,j}$. The function $\operatorname{Im} f(E', q^2)$ is the discontinuity at the cut $[0,\infty)$ of the amplitude

$$\operatorname{Im} f(E', q^2) \equiv \frac{1}{2\mathrm{i}}[f(E' + \mathrm{i}0, q^2) - f(E' - \mathrm{i}0, q^2)] \,. \tag{8.281}$$

This function is real because $f(E, q^2) = (f(E^*, q^{2*}))^*$. It is curious that while the partial amplitudes $f_L(E)$ present two cuts, in the total amplitude for fixed q^2 the cut on the left has disappeared. Its existence was related to the projection required to extract f_L from f. In particular, for forward scattering ($q^2 = 0$) we have

$$\boxed{f(E,0) = f_B(0) + \sum_{n,j} \frac{\Gamma_{n,j}}{E - E_n} + \frac{1}{2\pi^2}\int_0^\infty dk' \frac{k'^2 \sigma_{\text{tot}}(k')}{k'^2 - (k^2 + \mathrm{i}0)}} \tag{8.282}$$

where $\operatorname{Im} f(E', 0)$ has been replaced by the cross section using the optical theorem. Equation (8.282) is very important. In it we find computable quantities such as $\Gamma_{n,j}$ or experimentally measurable ones, E_n, $\sigma_{\text{tot}}(k')$. In the physics of elementary particles the dispersion relations have been used extensively in the description and analysis of the strong interactions. Contrary to (8.282) the general relation (8.280) has the disadvantage that the integration region contains an unphyiscal part due to the fact that the physical energies in the collision process always satisfy $E \geq (\hbar q)^2/8M$. Finally, we can once again derive $\lim_{E\to\infty} f(E, q^2) = f_B(q^2)$, but now using (8.280).

8.20 Coulomb Scattering

The Coulomb potential $V(r) = e_1 e_2/r$ is without a doubt the most important potential in physics. Nevertheless, little of what we have said so far applies to scattering in this potential. As explained in Sect. 8.5 and 7, the reason is that it decreases very slowly at large distances, and its effect persists in the asymptotic region. The study of Coulomb scattering requires therefore special treatment. We noticed in Sect. 8.10 in the study of scattering into cones that the equation (8.80) could be the starting point to deal with this problem. Following *Dollard* [DO 64, 69] we define a modified "free evolution" operator $\exp[-\mathrm{i}H_{0,\mathrm{c}}(t)/\hbar]$:

$$H_{0,\mathrm{c}}(t) \equiv H_0 t + \varepsilon(t)\frac{e_1 e_2}{|\boldsymbol{p}|} M \ln\left(\frac{4H_0|t|}{\hbar}\right) \tag{8.283}$$

with $\varepsilon(t) \equiv \operatorname{sgn} t$. For the modified operator, in contrast with the free evolution operator $\exp(-\mathrm{i}H_0 t/\hbar)$, it can be shown that the Möller operators $\Omega_{\pm,\mathrm{c}}$

$$\Omega_{\pm,\mathrm{c}} \equiv \lim_{t\to\mp\infty} \mathrm{e}^{\mathrm{i}Ht/\hbar}\mathrm{e}^{-\mathrm{i}H_{0,\mathrm{c}}(t)/\hbar}\,, \qquad H = H_0 + V \tag{8.284}$$

exist in the strong sense, they are isometric and asymptotically complete, and the matrix $S_\mathrm{c} \equiv \Omega^\dagger_{-,\mathrm{c}}\Omega_{+,\mathrm{c}}$ is unitary. The change we have made in the evolution operator is exactly the one required for the Coulomb scattering orbits to have asymptotes. These asymptotes although not strictly free, are free in an asymptotic sense:

$$\lim_{t\to\mp\infty}\int_X d^3x\,|(\mathrm{e}^{-\mathrm{i}H_{0,\mathrm{c}}(t)/\hbar}f)(\boldsymbol{x})|^2 = \lim_{t\to\mp\infty}\int_X d^3x\,|(\mathrm{e}^{-\mathrm{i}H_0 t/\hbar}f)(\boldsymbol{x})|^2 \tag{8.285}$$

for all $f \in L^2(\mathbb{R}^3)$. As in the standard case (8.79) it can be shown that for any cone C

$$P(C) \equiv \lim_{t\to\infty}\int_C d^3x\,|(\mathrm{e}^{-2\mathrm{i}tH/\hbar}\mathrm{e}^{\mathrm{i}tH_0/\hbar}f)(\boldsymbol{x})|^2 = \int_C d^3p\,|(S_\mathrm{c}f)\hat{}\,(\boldsymbol{p})|^2 \tag{8.286}$$

and this equality justifies the use of S_c as the scattering operator.

It is convenient to notice that the Hamiltonian operator associated to the modified free evolution $\exp[-\mathrm{i}H_{0,\mathrm{c}}(t)/\hbar]$, namely $H_0 + e_1 e_2/v|t|$, leads classically to modified free trajectories which asymptotically adjust the Coulomb trajectories.

The Möller operators $\Omega_{\pm,\mathrm{c}}$ satisfy as a consequence of their definition the intertwinning property between the operators H_0 and H and therefore S_c commutes with H_0. It can also be shown that given an arbitrary asymptotic state (8.96), the relation (8.97) still holds for the new $\Omega_{\pm,\mathrm{c}}$, and now $|\boldsymbol{p}_\pm\rangle$ are the generalized eigenstates of H corresponding to the asymptotic state $|\boldsymbol{p}\rangle$.

The wave functions $\langle\boldsymbol{x}|\boldsymbol{p}_\pm\rangle = \psi_p^\pm(\boldsymbol{x})$ are regular solutions to the Schrödinger equation

$$\left[-\frac{\hbar^2}{2M}\Delta + \frac{e_1 e_2}{r} - \frac{p^2}{2M}\right]\psi_p^\pm(\boldsymbol{x}) = 0 \tag{8.287}$$

chosen in such a way that the evolution under H of a wave packet constructed with them behaves asymptotically as the packet (8.96) evolving with $\exp[-\mathrm{i}H_{0,c}(t)/\hbar]$.

To determine these functions, and given the rotational invariance of H, it is convenient to choose $\boldsymbol{p}$ in the Oz direction. Then the problem (8.287) will have cylindrical symmetry ($L_z = 0$). Expanding $\psi_p^{\pm}(\boldsymbol{x})$ in Legendre polynomials,

$$\psi_p^{\pm}(\boldsymbol{x}) = (2\pi\hbar)^{-3/2} \sum_{L=0}^{\infty} (2L+1)\mathrm{i}^L \frac{u_{L,p}^{\pm}(r)}{kr} P_L(\cos\theta) \tag{8.288}$$

the radial wave functions must satisfy (8.189) with $U(r)$ defined as

$$U(r) = \frac{2\eta k}{r}, \quad \eta \equiv \frac{Me_1e_2}{\hbar^2 k}. \tag{8.289}$$

To determine $u_{L,p}^{+}$ it suffices to solve (8.189) with the regularity condition. This leads to (see Sect. A.11)

$$u_{L,p}^{+}(r) = C_{L,p}\mathrm{e}^{\mathrm{i}kr}(kr)^{L+1} M(L+1+\mathrm{i}\eta, 2L+2, -2\mathrm{i}kr) \tag{8.290}$$

where the constant $C_{L,p}$ is chosen as

$$\begin{aligned} C_{L,p} &= \frac{2^L \mathrm{e}^{-\pi\eta/2}|\Gamma(L+1+\mathrm{i}\eta)|}{(2L+1)!} \exp[\mathrm{i}\sigma_L(p)] \\ \sigma_L(p) &\equiv \arg\Gamma(L+1+\mathrm{i}\eta) \end{aligned} \tag{8.291}$$

to have the asymptotic behavior

$$u_{L,p}^{+}(r) \underset{r\to\infty}{\sim} \exp[\mathrm{i}\sigma_L(p)] \sin\left(kr - L\tfrac{\pi}{2} + \sigma_L - \eta\ln 2kr\right). \tag{8.292}$$

The Coulomb tail makes it impossible to find the behavior (8.192) of the standard potentials, and there is a progressive change in the phase which ultimately leads to the logarithmic contribution in the right hand side. The condition (8.292) is justified a posteriori. It leads in (8.288) to

$$\psi_p^{+}(\boldsymbol{x}) = (2\pi\hbar)^{-3/2}\mathrm{e}^{-\pi\eta/2}\Gamma(1+\mathrm{i}\eta)\mathrm{e}^{\mathrm{i}kz} M(-\mathrm{i}\eta, 1, \mathrm{i}k(r-z)). \tag{8.293}$$

This also follows directly by solving the problem in parabolic coordinates. It can be shown [MZ 70] that the right hand side in (8.293) has the correct asymptotic behavior for $\psi_p^{+}(\boldsymbol{x})$. Indeed, using (A.144) one easily finds:

$$\begin{aligned} \psi_p^{+}(\boldsymbol{x}) \underset{r-z\to\infty}{\sim} (2\pi\hbar)^{-3/2} \Bigg\{ &\left[1 - \frac{\mathrm{i}\eta^2}{2kr\sin^2(\theta/2)}\right] \exp\{\mathrm{i}[kz + \eta\ln k(r-z)]\} \\ &+ f_c(p,\cos\theta)\frac{\exp[\mathrm{i}(kr - \eta\ln 2kr)]}{r} \Bigg\} \\ f_c(p,\cos\theta) \equiv -\frac{\eta\Gamma(1+\mathrm{i}\eta)}{\Gamma(1-\mathrm{i}\eta)} &\frac{\exp[-2\mathrm{i}\eta\ln\sin(\theta/2)]}{2k\sin^2(\theta/2)}. \end{aligned} \tag{8.294}$$

The asymptotic structure of (8.294) is very similar to (8.151) for standard potentials with the difference that the incident plane wave and the outgoing spherical waves are distorted by logarithmic terms. The *Coulomb scattering amplitude* f_c plays the same role as $f(p, \cos\theta)$. If we proceed formally calculating the incoming and outgoing radial fluxes it is easy to show that $|f_c|^2$ gives the Coulomb differential cross section. However this procedure is not rigorously justified.

Following the rigorous method outlined before, and with the explicit formulae for the states $|\boldsymbol{p}_\pm\rangle$ (take into account that $\langle \boldsymbol{x}|\boldsymbol{p}_-\rangle = \langle \boldsymbol{x}| -\boldsymbol{p}_+\rangle^*$) the scattering amplitude for the process $\boldsymbol{p}_i \to \boldsymbol{p}_f$ is given by $\langle \boldsymbol{p}_{f-}|\boldsymbol{p}_{i+}\rangle = \langle \boldsymbol{p}_f|S_c|\boldsymbol{p}_i\rangle$. The explicit computation of this scalar product [HE 74] leads, for $\boldsymbol{p}_i \neq \boldsymbol{p}_f$, to

$$\langle \boldsymbol{p}_f|S_c|\boldsymbol{p}_i\rangle = -2\pi\mathrm{i}\delta(E_f - E_i)\frac{\eta}{8\pi^2 M p_i}\mathrm{e}^{2\mathrm{i}\sigma_0(p_i)}\left(\frac{1-\hat{\boldsymbol{p}}_i\cdot\hat{\boldsymbol{p}}_f}{2}\right)^{-1-\mathrm{i}\eta} . \tag{8.295}$$

At first sight one might think that the no interaction term $\delta(\boldsymbol{p}_i - \boldsymbol{p}_f)$ does not appear in (8.295) because we assumed $\boldsymbol{p}_i \neq \boldsymbol{p}_f$. However it is another peculiarity of Coulomb potentials that the extension of (8.295) to $\boldsymbol{p}_i = \boldsymbol{p}_f$ is more singular than such a delta function.

The coefficient of $-2\pi\mathrm{i}\,\delta(E_f - E_i)$ in (8.295) is just $T_{E_i,c}(\hat{\boldsymbol{p}}_f, \hat{\boldsymbol{p}}_i)$, and using (8.85) it leads exactly to the amplitude f_c given in (8.294). Using the techniques of Sect. 8.9 it is now straightforward to obtain the Coulomb differential cross section away from the forward direction as

$$\frac{d\sigma_c}{d\Omega} = |f_c(p_i, \cos\theta)|^2 = \left(\frac{e_1 e_2}{2p_i v_i}\right)^2 \frac{1}{\sin^4(\theta/2)} \tag{8.296}$$

which is the famous expression obtained classically by Rutherford [notice that $\hbar$ does not appear in (8.296)]. It is curious that the Born approximation would lead to the same expression for the differential cross section, since it is obvious from (8.294) that

$$f_c^{\mathrm{B}}(p_i, \cos\theta) = -\eta\frac{1}{2k_i\sin^2(\theta/2)} \tag{8.297}$$

taking only terms of order η.

The total amplitude differs from the Born approximation by a phase factor depending on the direction which however cancels out in the computation of the differential cross section.

The singular behavior of $d\sigma_c/d\Omega$ around $\theta = 0$ is a consequence of the long range of the interaction and leads to a divergent elastic cross section. This was predictable according to the discussion of Sect. 8.9.

Finally the phases $\sigma_L(p)$ introduced in (8.291, 292) are the *Coulomb phase shifts*. This can be seen using (8.295) as a tempered distribution which extends

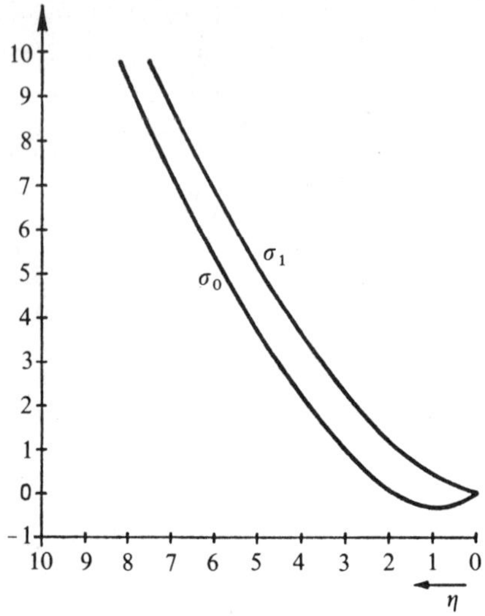

Fig. 8.25. Coulomb phase shifts as functions of η

(uniquely) to the singular case $\boldsymbol{p}_i = \boldsymbol{p}_f$ by the replacement of $-1 - \mathrm{i}\eta$ by $-1 - \mathrm{i}(\eta + \mathrm{i}0)$, and then applying this integral kernel to a square integrable C^1 function of the form $f(\boldsymbol{p}) = g(|\boldsymbol{p}|)Y_L^M(\hat{\boldsymbol{p}})$. We obtain in this way

$$(S_c f)(\boldsymbol{p}) = \exp\left[2\mathrm{i}\sigma_L(|\boldsymbol{p}|)\right] f(\boldsymbol{p}) . \tag{8.298}$$

In Fig. 8.25 we plot σ_0 and σ_1 as functions of η. Notice that σ_0 changes sign at high energies. Furthermore the sign of the phase shifts is opposite to the standard case (8.199). Due to the long range of the potential one should not expect that only a few phase shifts are relevant. Actually all of them are important and one has $\sigma_{L+1}(\eta) \geq \sigma_L(\eta)$ [see (A.161)].

For a general analysis of scattering with long range potentials, see [RS 79], [EN 85].

Occasionally it is interesting to consider the change in the phase shifts when we modify the Coulomb potential by the addition of a central short range potential. In analogy with the ordering of the n, l levels induced by a definite sign of the laplacian of the potential (see Sect. 6.2), something similar happens with the phase shifts. Concretely, it can be shown [BG 85] that if $V(r) = e_1 e_2/r + V_0(r)$, with

$$\lim_{r\to 0} r^2 V(r) > -\frac{1}{4} , \quad \int_{r_0}^{r} dr\, |V_0(r)| < \infty , \ \forall r_0 > 0 , \ V(r) \underset{r\to\infty}{\longrightarrow} 0 \tag{8.299}$$

then the asymptotic behavior of the solutions $u^+_{L,p}$ is of the form (8.291) with σ_L replaced by $\sigma_L + \delta_L$. In addition we have

$$\begin{aligned} \Delta V_0 \geq 0 , \quad \forall r > 0 \Rightarrow \delta_{L+1}(p) \geq \delta_L(p) \\ \Delta V_0 \leq 0 , \quad \forall r > 0 \Rightarrow \delta_{L+1}(p) \leq \delta_L(p) . \end{aligned} \tag{8.300}$$

8.21 Two Body Elastic Scattering

The simple scattering theory presented thus far applies only to the case of two spinless particles one of which is infinitely heavy. In this section we will show (assuming non-relativistic kinematics) that the formalism is easily adapted to the case when the interacting particles have masses of comparable magnitude.

Consider a scattering process involving two different spinless particles $A + B \to A + B$ with momenta $\boldsymbol{p}_1, \boldsymbol{p}_2, \boldsymbol{p}_1', \boldsymbol{p}_2'$, respectively. The interaction is assumed to be mediated by a translationally invariant potential $V(\boldsymbol{x}_1 - \boldsymbol{x}_2)$. The Hamiltonian of the problem decomposes into two parts: one describes the motion of the center of mass, and the other includes the relative kinetic energy and the interaction potential. According to the results of Sect. 6.1:

$$H_0 = H_{0,\text{C.M.}} + \frac{\boldsymbol{p}^2}{2\mu}, \quad H = H_{0,\text{C.M.}} + \frac{\boldsymbol{p}^2}{2\mu} + V \tag{8.301}$$

where μ is the reduced mass and $H_{0,\text{C.M}}$ commutes with the rest of H. The total and free time evolution factorize into one corresponding to the C.M. and another describing the relative motion, and the same happens with the Möller operators and the S matrix. These operators are trivially the identity for the C.M. and the non-trivial dynamical part affects only the relative motion:

$$S = S_{\text{C.M.}} \otimes S_{\text{rel}} = I \otimes S_{\text{rel}} \tag{8.302}$$

where S_{rel} is calculated as with simple scattering where the mass is the reduced mass and the potential is taken to be $V(\boldsymbol{x})$. The subscript "rel" refers to the relative wave function, and therefore the momenta and energies appearing in the matrix elements of S_{rel} are all relative quantities. The following equality is then evident:

$$\begin{aligned}\langle \boldsymbol{p}_1', \boldsymbol{p}_2' | S | \boldsymbol{p}_1, \boldsymbol{p}_2 \rangle &= \delta(\boldsymbol{p}_1' + \boldsymbol{p}_2' - \boldsymbol{p}_1 - \boldsymbol{p}_2) \\ &\times \{\delta(\boldsymbol{p}_{\text{rel}}' - \boldsymbol{p}_{\text{rel}}) - 2\pi \mathrm{i} \delta(E_{\text{rel}}' - E_{\text{rel}}) T_{E_{\text{rel}}}(\hat{\boldsymbol{p}}_{\text{rel}}', \hat{\boldsymbol{p}}_{\text{rel}})\}\end{aligned} \tag{8.303}$$

as long as the potential satisfies the standard conditions. Thus in the collision process the relative energy and the total momentum are conserved. The differential cross section obtained using $T_{E_{\text{rel}}}$ is evidently the one corresponding to the C.M. and the cross section in any other reference system is obtained using the results of Sect. 8.4.

If the particles A, B participating in the collision have spins S_A, S_B it is necessary to make slight modifications in the formalism. Mostly notational changes. It is not enough to specify the relative momentum in the C.M. to characterize the initial or final states; we also need for example the projection of the third component of spin for each of the particles. Omitting for simplicity the subscript "rel", we obtain in this case

$$\begin{aligned}\langle \boldsymbol{p}' m_1' m_2' | S | \boldsymbol{p} m_1 m_2 \rangle &= \delta(\boldsymbol{p}' - \boldsymbol{p}) \delta_{m_1' m_1} \delta_{m_2' m_2} \\ &- 2\pi \mathrm{i} \delta(E' - E) T_E(\hat{\boldsymbol{p}}' m_1' m_2'; \hat{\boldsymbol{p}} m_1 m_2)\end{aligned} \tag{8.304}$$

where m_1 is the third component of spin for the particle incoming in the direction $\hat{\boldsymbol{p}}$, m_2 is the projection for the particle incoming in the direction $-\hat{\boldsymbol{p}}$, and similarly for m_1', m_2' with respect to $\hat{\boldsymbol{p}}'$. The computation of T_E is carried out as in the case of spinless simple scattering, but taking into account that the interaction may depend on the spin orientation: the potential is a $[(2S_A + 1)(2S_B + 1)] \times [(2S_A + 1)(2S_B + 1)]$ matrix in spin space. We assume each component of this matrix to satisfy the standard conditions, thus ensuring the existence and unitarity of S. In the particular case where the potential does not depend on the spin orientation, it is clear that

$$T_E(\hat{\boldsymbol{p}}' m_1' m_2'; \hat{\boldsymbol{p}} m_1 m_2) = \delta_{m_1' m_1} \delta_{m_2' m_2} T_E(\hat{\boldsymbol{p}}', \hat{\boldsymbol{p}}) \tag{8.305}$$

and we can neglect completely the presence of spins.

Often the initial spin state is not well defined, and it is only given in terms of a $[(2S_A + 1)(2S_B + 1)] \times [(2S_A + 1)(2S_B + 1)]$ density matrix ϱ. Even though in the basis $|m_1 m_2\rangle$ the matrix ϱ is not necessarily diagonal, there always exists a basis in spin space $|\xi_j\rangle$ such that (Sect. 2.7)

$$\varrho = \sum_j \lambda_j |\xi_j\rangle\langle\xi_j| \tag{8.306}$$

and the differential cross section in the C.M. to produce a final state with polarizations m_1', m_2' is simply a statistical average with weights λ_j of the result corresponding to initial spins described by $|\xi_j\rangle$:

$$\frac{d\sigma}{d\Omega} = \sum_j \lambda_j \frac{d\sigma(\boldsymbol{p}\xi_j \to \boldsymbol{p}' m_1' m_2')}{d\Omega} = (2\pi\hbar)^4 \left(\frac{k}{v}\right)^2 (T_E \varrho T_E^\dagger)_{m_1' m_2', m_1' m_2'} \tag{8.307}$$

and in this equation we consider T_E as a matrix in spin space. If we are not interested in the final polarizations, we have to sum (8.307) over the values of m_1', m_2':

$$\boxed{\frac{d\sigma(\boldsymbol{p}\varrho \to \boldsymbol{p}', \text{all})}{d\Omega} = (2\pi\hbar)^4 \left(\frac{k}{v}\right)^2 \operatorname{Tr}\{T_E \varrho T_E^\dagger\}} \tag{8.308}$$

which in the case of unpolarized incident beams becomes

$$\frac{d\sigma(\boldsymbol{p} \to \boldsymbol{p}')}{d\Omega} = (2\pi\hbar)^4 \left(\frac{k}{v}\right)^2 \frac{1}{(2S_A + 1)(2S_B + 1)} \operatorname{Tr}\{T_E T_E^\dagger\} . \tag{8.309}$$

So far we have assumed that the particles A, B are distinguishable. In other words we can devise an experimental procedure to differentiate them. Classically it is in principle possible to follow their individual trajectories even if the particles are of the same type, for example electrons. Quantum mechanically, when the particles collide, the finite size of their wave packets makes it impossible to discern which of the particles came from which incident beam. This indistinguishability translates (as we will discuss in detail in Chap. 13) in the fact that the initial wave function of two identical particles must be symmetric if their spin is integral, and antisymmetric otherwise. This forces us to modify the previous formulae in order to account for indistinguishability.

The initial state in the C.M. will now take the form

$$|i\rangle \equiv \frac{1}{\sqrt{2}}[|\boldsymbol{p}m_1m_2\rangle + (-1)^{2S}|-\boldsymbol{p}m_2m_1\rangle] \tag{8.310}$$

where S is the spin of each particle. The transition amplitude of this state to a similar final state with momentum $\boldsymbol{p}'$ and spin projections m_1', m_2' becomes

$$\boxed{T_E(f,i) = T_E(\hat{\boldsymbol{p}}'m_1'm_2'; \hat{\boldsymbol{p}}m_1m_2) + (-1)^{2S}T_E(-\hat{\boldsymbol{p}}'m_2'm_1'; \hat{\boldsymbol{p}}m_1m_2)} \tag{8.311}$$

derived from the property

$$T_E(\hat{\boldsymbol{p}}'m_1'm_2'; \hat{\boldsymbol{p}}m_1m_2) = T_E(-\hat{\boldsymbol{p}}', m_2'm_1'; -\hat{\boldsymbol{p}}m_2m_1)$$

which in turn is due to the fact that $H_0, H, \Omega_\pm$ and S all commute with the operator permuting the two particles as a consequence of their indistinguishability.

Hence, in contrast with the classical treatment where the detection probability of a particle independently of the beam where it came from is the sum over the probabilities of the two possibilities, in quantum mechanics it is the amplitudes which get superposed. This originates interferences between the two possibilities: constructive if the spin is an integer, and destructive if the spin is a half odd integer. Defining the scattering amplitude according to (8.85), the C.M. differential cross section becomes

$$\begin{aligned}\frac{d\sigma}{d\Omega}(\boldsymbol{p}m_1m_2 \to \boldsymbol{p}'m_1'm_2') = |&f(\boldsymbol{p}m_1m_2 \to \boldsymbol{p}'m_1'm_2') \\ &+ (-1)^{2S}f(\boldsymbol{p}m_2m_1 \to \boldsymbol{p}'m_2'm_1')|^2 .\end{aligned} \tag{8.312}$$

In the special case of a spin independent central interaction with unpolarized initial beams, and if we do not measure the final polarizations, the differential cross section is

$$\begin{aligned}\frac{d\sigma}{d\Omega}(\boldsymbol{p} \to \boldsymbol{p}') &= \frac{1}{(2S+1)^2}\sum_{\substack{m_1m_2\\ m_1'm_2'}} |f(p,\cos\theta)\delta_{m_1m_1'}\delta_{m_2m_2'} \\ &\quad + (-1)^{2S}f(p,-\cos\theta)\delta_{m_1m_2'}\delta_{m_2m_1'}|^2 \\ &= |f(p,\cos\theta)|^2 + |f(p,-\cos\theta)|^2 \\ &\quad + 2\frac{(-1)^{2S}}{2S+1}\mathrm{Re}\,\{f^*(p,\cos\theta)f(p,-\cos\theta)\} .\end{aligned} \tag{8.313}$$

The last term has a purely quantum mechanical origin, and it reflects the interference due to indistinguishability. For $\theta = \pi/2$

$$\frac{d\sigma}{d\Omega} = 2\left[1 + \frac{(-1)^{2S}}{2S+1}\right] |f(p,0)|^2 = \begin{cases} 4|f(p,0)^2 & \text{if } S = 0 \\ |f(p,0)|^2 & \text{if } S = 1/2 \end{cases} \tag{8.314}$$

while for distinguishable particles its value would have been $2|f(p,0)|^2$.

The specialization of (8.313) to the Coulomb case is known as Mott's formula, and the interference effect has been observed experimentally. (See for example [BK 61] for the case ^{12}C–^{12}C.)

Finally, the cross section integrated over all angles and multiplied by the incident flux measures the total rate of scattered particles, irrespective of the beam where they came from. If the total elastic cross section times the incident flux has to measure the rate of particles lost by the incident beam, it is clear that we have to divide the integrated differential cross section by two for identical particles:

$$\sigma_{\text{el}}(p) = \frac{1}{2} \int d\Omega \frac{d\sigma}{d\Omega} \ . \tag{8.315}$$

8.22 Multichannel Scattering

The problems pointed out in Sect. 6.1 can be translated to the case of N bodies $A_1, \ldots, A_N$, which for simplicity will be assumed to be spinless and distinguishable. If we only include interactions of the form

$$V(\boldsymbol{x}_1, \ldots \boldsymbol{x}_N) = \sum_{1 \le i < j \le N} V_{ij}(\boldsymbol{x}_i - \boldsymbol{x}_j) \tag{8.316}$$

some of the results quoted there can be generalized [RS 79], [SI 82], [SI 84]. Thus for example, if $\tilde{H}(A_1, \ldots, A_N)$ is the Hamiltonian after the trivial center of mass motion has been subtracted, the following properties hold:

i) $\tilde{H}(A_1, \ldots, A_N)$ is essentially self-adjoint in $C_0^\infty(\mathbf{R}^{3(N-1)})$ if for each pair i, j:

$$V_{ij} \in L_u^2 \equiv \left\{ f : \sup_x \int_{|\boldsymbol{y}-\boldsymbol{x}| \le 1} d^3y \, |f(\boldsymbol{y})|^2 < \infty \right\} . \tag{8.317}$$

Assuming furthermore that

$$\lim_{|\boldsymbol{x}| \to \infty} \int_{|\boldsymbol{y}-\boldsymbol{x}| \le 1} d^3y \, |V_{ij}(\boldsymbol{y})|^2 \to 0 \tag{8.318}$$

one can prove an important theorem due to Hunziker-Van Winter-Zhislin:

$$\begin{aligned} &\sigma_{\text{es}}(\tilde{H}(A_1, \ldots, A_N)) = [\Sigma, \infty) \\ &\Sigma = \min_{1 \le n < N} [\inf \sigma(\tilde{H}(A_{i_1}, \ldots, A_{i_n}))] \\ &\qquad + \inf \sigma(\tilde{H}(A_{i_{n+1}}, \ldots, A_{i_N}))] \ . \end{aligned} \tag{8.319}$$

The minimum is computed over all the permutations $i_1, \ldots, i_N$ of $1, \ldots, N$, and groupings into two clusters.

ii) With regard to the singularly continuous part of the spectrum, it is known for example that if the potentials are of Coulomb or Yukawa type, $\sigma_{\text{s.c.}} = \emptyset$. The same applies to potentials leading to asymptotic completeness (see below).

iii) The finiteness of σ_{disc} holds if the functions $V_{ij}(\boldsymbol{r})$ are bounded and decrease faster than $1/r^2$ at large distances, and if the energy spectrum of any grouping of the N particles into three or more decoupled clusters is strictly larger than the threshold Σ previously introduced.

(iv) Finally, if $V_{ij} \in L^2 + L^\infty_\varepsilon$ (or more generally $R + L^\infty_\varepsilon$), the ground state, if it exists (i.e. if $\inf \sigma \in \sigma_{\text{p}}$), is non-degenerate and its wave function is strictly positive.

The study of multichannel scattering is necessarily more complicated, and as in the analysis of bound states for multiparticle systems, the known rigorous results are scarcer than for two particle systems. To be concrete, consider three spinless particles A_1, A_2, A_3, with interactions mediated by the potentials V_{ij} between the particles A_i and A_j. There may be configurations where A_i, A_j are bound forming a fragment $(A_iA_j)_n$ where n enumerates the different types of fragments, each characterized by its internal energy and angular momentum. In the collision $A_1 + (A_2A_3)_n$ several things can happen:

a)	$A_1 + (A_2A_3)_n \to A_1 + (A_2A_3)_n$	(elastic channel)
b)	$\to A_1 + (A_2A_3)_m$	$m \neq n$ ((de)-excitation channel)
c)	$\to A_1 + A_2 + A_3$	(breaking channel)
d)	$\to (A_1A_2)_m + A_3$ $\to (A_1A_3)_m + A_2$	(pick up channel) .

The possibility of three body bound states as an outgoing channel is forbidden by energy conservation, at least for reasonable potentials. We will not consider the possibility of outgoing particles different from those coming in.

The clusters in the final state will move freely with a free Hamiltonian depending on the channel excited. Thus for the different cases considered previously we have the Hamiltonians

$$\begin{aligned}
&\text{a, b)}\ H_{1(23)} \equiv \frac{\boldsymbol{p}_1^2}{2M_1} + \left(\frac{\boldsymbol{p}_2^2}{2M_2} + \frac{\boldsymbol{p}_3^2}{2M_3} + V_{23}\right)\\
&\text{c)}\ H_{123} \equiv \frac{\boldsymbol{p}_1^2}{2M_1} + \frac{\boldsymbol{p}_2^2}{2M_2} + \frac{\boldsymbol{p}_3^2}{2M_3}\\
&\text{d)}\ H_{(12)3} \equiv \left(\frac{\boldsymbol{p}_1^2}{2M_1} + \frac{\boldsymbol{p}_2^2}{2M_2} + V_{12}\right) + \frac{\boldsymbol{p}_3^2}{2M_3}\\
&\quad\ H_{(13)2} \equiv \left(\frac{\boldsymbol{p}_1^2}{2M_1} + \frac{\boldsymbol{p}_3^2}{2M_2} + V_{13}\right) + \frac{\boldsymbol{p}_2^2}{2M_2} .
\end{aligned} \tag{8.320}$$

The asymptotic states of each channel are characterized by the C.M. momentum and the internal wave function of each fragment. Such states span a closed subspace D_α of $L^2(\mathbb{R}^9) \equiv \mathcal{H}$, and it can be shown [HU 68] that if each $V_{ij}(\boldsymbol{x}_i - \boldsymbol{x}_j) \equiv V_{ij}(\boldsymbol{x}_{ij}) \in L^2(\mathbb{R}^3) + L^p(\mathbb{R}^3)$, $2 \le p < 3$ (for example standard potentials) then there exist Möller operators $\Omega^\alpha_\pm$ defined as strong limits in D_α

$$\Omega^\alpha_\pm = \lim_{t \to \mp\infty} \exp[itH/\hbar] \exp[-itH_\alpha/\hbar] \tag{8.321}$$

where H_α is the restriction of the appropriate free Hamiltonian (8.320) to D_α. For instance, in the channel, $(A_1A_2)_m + A_3$, H_α is simply

$$\frac{(\boldsymbol{p}_1 + \boldsymbol{p}_2)^2}{2(M_1 + M_2)} + E_{(12),m} + \frac{\boldsymbol{p}_3^2}{2M_3} \tag{8.322}$$

where $E_{(12),m}$ is the internal energy of the cluster $(A_1A_2)_m$. The operators $\Omega^\alpha_\pm$ extend to $\mathcal{H}$, taking them to vanish in $\mathcal{H} \ominus D_\alpha$. Hence they are partial isometries.

Define $R^\alpha_\pm$ as the ranges of $\Omega^\alpha_\pm$: $R^\alpha_\pm = \Omega^\alpha_\pm D_\alpha$. An important property of these ranges it their orthogonality

$$R^\alpha_+ \perp R^\beta_+ \,, \quad R^\alpha_- \perp R^\beta_- \quad \text{if} \quad \alpha \neq \beta \tag{8.323}$$

which can be seen through the following argument: given for example $|f\rangle \in R^\alpha_+, |g\rangle \in R^\beta_+$, there exist $|f_\alpha\rangle \in D_\alpha, |g_\beta\rangle \in D_\beta$ such that $|f\rangle = \Omega^\alpha_+|f_\alpha\rangle, |g\rangle = \Omega^\beta_+|g_\beta\rangle$, and therefore

$$\begin{aligned} \langle g|f\rangle &= \lim_{t \to -\infty} \langle g_\beta| \exp[itH_\beta/\hbar] \exp[-itH_\alpha/\hbar]|f_\alpha\rangle \\ &= \lim_{t \to -\infty} \langle g_\beta| \exp[it(H_\beta - H_\alpha)/\hbar]|f_\alpha\rangle = 0 \,. \end{aligned} \tag{8.324}$$

In (8.324) we have used: a) $[H_\beta, H_\alpha] = 0$ (both are functions of the momenta); b) the spectrum of $H_\beta - H_\alpha$ is absolutely continuous if β, α are channels with different groupings of particles (e.g. $A_1 + (A_2A_3)_n$ and $(A_1A_2)_m + A_3$, or $A_1 + A_2 + A_3$ and $(A_1A_2)_m + A_3$), for then $H_\beta - H_\alpha$ is a quadratic function of the momenta, and we can apply the Riemann-Lebesgue theorem; c) if the channels β and α have the particles with the same grouping (e.g. $(A_1A_2)_m + A_3$ and $(A_1A_2)_n + A_3, m \neq n$), then $H_\beta - H_\alpha = \text{const.} \times I$, and $\langle g|f\rangle = 0$ since $\langle g_\beta|f_\alpha\rangle = 0$ because the bound states $(A_1A_2)_m$ and $(A_1A_2)_n$ differ in some internal quantum number.

The previous arguments do not apply for the domains D_α. For instance, in channel c), $D_\alpha = \mathcal{H}$. This apparent paradox disappears when we observe as an example that even though the states $A_1 + A_2 + A_3$ and $A_1 + (A_2A_3)_n$ are not orthogonal, they will be so asymptotically, since they evolve with different Hamiltonians, and the overlap between their wave functions vanishes when $t \to \pm\infty$.

As in the case of simple scattering, it is easy to show that $R^\alpha_\pm \perp \mathcal{H}_{\text{bound}}$, where $\mathcal{H}_{\text{bound}}$ is the space of three particle bound states $(A_1A_2A_3)$.

We say that the system is *asymptotically complete* if

$$\mathcal{H} = \mathcal{H}_{\text{bound}} \oplus [\oplus_\alpha R^\alpha_+] = \mathcal{H}_{\text{bound}} \oplus [\oplus_\alpha R^\alpha_-] \,. \tag{8.325}$$

Notice that we require now $\oplus_\alpha R_+^\alpha = \oplus_\alpha R_-^\alpha = \mathcal{H} \ominus \mathcal{H}_{\text{bound}}$. The condition $R_+^\alpha = R_-^\alpha$ implying asymptotic completeness for single channel scattering cannot be required anymore, for it would imply that the incoming channel α cannot give rise to a different outgoing channel. The question of verifying explicitly (8.325) is very difficult. For the case of arbitrary N, asymptotic completeness has been shown only in the case that either all the potentials V_{ij} are repulsive, or they are sufficiently weak that no bound states of two or more particles are possible (in other words, the only channel is $A_1 + \ldots + A_N \to A_1 + \ldots + A_N$) [AM 74], [RS 79] or all V_{ij} are short range [SS 87]. For $N = 3$ however, there are sufficiently general results about completeness including very singular potentials, or even long range ones [EN 84], [SI 84], [EN 85]. Here we will assume (8.325) to hold.

To define the S operator, it is convenient to introduce an auxiliary Hilbert space $\mathcal{H}'$ incorporating directly the asymptotic orthogonality of the different asymptotic subspaces

$$\mathcal{H}' \equiv \oplus_\alpha D_\alpha \tag{8.326}$$

and to translate to $\mathcal{H}'$ the Möller operators in the sense

$$\begin{aligned} &\Omega_\pm : \mathcal{H}' \to \mathcal{H} \\ &\Omega_\pm : \{|f_{\alpha_1}\rangle, |f_{\alpha_2}\rangle, |f_{\alpha_3}\rangle, \ldots\} \to \sum \Omega_\pm^{\alpha_i} |f_{\alpha_i}\rangle \,. \end{aligned} \tag{8.327}$$

Defining finally the adjoints of the isometric operators $\Omega_\pm$,

$$\Omega_\pm^\dagger : |f\rangle \to \begin{cases} \Omega_\mp^{-1}|f\rangle, & \text{if } |f\rangle \in \mathcal{H} \ominus \mathcal{H}_{\text{bound}} \\ 0, & \text{if } |f\rangle \in \mathcal{H}_{\text{bound}} \end{cases} \tag{8.328}$$

the scattering operator becomes

$$S = \Omega_-^\dagger \Omega_+ \tag{8.329}$$

mapping isometrically $\mathcal{H}'$ onto $\mathcal{H}'$; in other words, it is a *unitary operator* in the auxiliary Hilbert space generated by the asymptotes defined as orthogonal for different channels.

To compute the transition probability from an initial to a final state with asymptotes $|\phi_i^\alpha\rangle \in D_\alpha \subset \mathcal{H}'$, $|\phi_f^\beta\rangle \in D_\beta \subset \mathcal{H}'$ respectively, it suffices to evaluate the matrix element

$$\langle \phi_f^\beta | S | \phi_i^\alpha \rangle = \langle \phi_f^\beta | \Omega_-^\dagger \Omega_+ | \phi_i^\alpha \rangle = \langle \phi_f^\beta | \Omega_-^{\beta\dagger} \Omega_+^\alpha | \phi_i^\alpha \rangle \,. \tag{8.330}$$

For brevity we are using the same notation for $|\phi_i^\alpha\rangle$ as a state in $\mathcal{H}'$ and as an element of $\mathcal{H}$; the index α labeling the channel removes the possible ambiguity.

Apart from its unitarity, the S operator conserves the free energies of the channels. Indeed, as in single channel scattering we have the intertwinning property

$$H \Omega_\pm^\alpha = \Omega_\pm^\alpha H_\alpha \tag{8.331}$$

and from it we easily obtain (D_α is stable under H_α)

$$H_0'S = SH_0' \quad (H_0' \equiv \oplus_\alpha (H_\alpha \upharpoonright D_\alpha)) \tag{8.332}$$

expressing the conservation law just mentioned.

From the assumed translational invariance of the interaction potentials we derive as in Sect. 8.21 that the S matrix factorizes into a trivial part accounting for the motion of the C.M. and the non-trivial relative S matrix. Separating the trivial part, i.e. working in the C.M. in the following, most of the results for the single channel case can be translated to multichannel scattering with a slight complication in notation. For simplicity we discuss only the case of scattering with two spinless particles or fragments in the final state. From (8.332) we immediately derive that if $|\boldsymbol{p}\alpha\rangle$, $|\boldsymbol{p}'\alpha'\rangle$ are the asymptote states in the channels α and α' respectively with (relative) momenta $\boldsymbol{p}$ and $\boldsymbol{p}'$, the S matrix elements between these two states can be written as

$$\langle \boldsymbol{p}'\alpha'|S|\boldsymbol{p}\alpha\rangle = \delta(\boldsymbol{p}' - \boldsymbol{p})\delta_{\alpha'\alpha} - 2\pi \mathrm{i}\delta(E_{p',\alpha'} - E_{p,\alpha})T_{E_{p,\alpha}}(\hat{\boldsymbol{p}}'\alpha', \hat{\boldsymbol{p}}\alpha) \tag{8.333}$$

analogous to (8.55). In this relation $E_{p,\alpha}$ and $E_{p',\alpha'}$, are respectively the free initial and final energies. Thus, if the α channel is $A_1 + (A_2A_3)_n$,

$$\begin{aligned} E_{p,\alpha} &= \frac{p^2}{2\mu_{1(23)}} + E_{(23),n} \\ \frac{1}{\mu_{1(23)}} &\equiv \frac{1}{M_1} + \frac{1}{M_2 + M_3} \,. \end{aligned} \tag{8.334}$$

Repeating the reasoning in Sect. 8.9 to derive the differential cross section, we obtain (on the energy shell $E_{p',\alpha'} = E_{p,\alpha}$):

$$\boxed{\frac{d\sigma}{d\Omega}(\boldsymbol{p}\alpha \to \boldsymbol{p}'\alpha') = (2\pi\hbar)^4 \frac{k'^2}{vv'}|T_{E_{p,\alpha}}(\hat{\boldsymbol{p}}'\alpha', \hat{\boldsymbol{p}}\alpha)|^2} \tag{8.335}$$

and defining the scattering amplitude

$$\boxed{f(\boldsymbol{p}\alpha \to \boldsymbol{p}'\alpha') \equiv -(2\pi\hbar)^2 \sqrt{\frac{kk'}{vv'}} T_{E_{p,\alpha}}(\hat{\boldsymbol{p}}'\alpha', \hat{\boldsymbol{p}}\alpha)} \tag{8.336}$$

which coincides with (8.85) for elastic scattering, (8.335) becomes

$$\boxed{\frac{d\sigma}{d\Omega}(\boldsymbol{p}\alpha \to \boldsymbol{p}'\alpha') = \frac{k'}{k}|f(\boldsymbol{p}\alpha \to \boldsymbol{p}'\alpha')|^2} \,. \tag{8.337}$$

To calculate the quantity $T_{E_{p,\alpha}}(\hat{\boldsymbol{p}}'\alpha', \hat{\boldsymbol{p}}\alpha)$ we formally copy all the steps in Sect. 8.12 with the Lippmann-Schwinger equations:

$$\begin{aligned} T_{E_{p,\alpha}}(\hat{\boldsymbol{p}}'\alpha', \hat{\boldsymbol{p}}\alpha) &= \langle \boldsymbol{p}'\alpha | V_{\alpha'} | \boldsymbol{p}\alpha_+ \rangle = \langle \boldsymbol{p}'\alpha'_- | V_\alpha | \boldsymbol{p}\alpha \rangle \\ &= \langle \boldsymbol{p}'\alpha' | T_{\alpha'\alpha}(E_{p,\alpha} + \mathrm{i}0) | \boldsymbol{p}\alpha \rangle \\ &= \langle \boldsymbol{p}'\alpha' | \overline{T}_{\alpha'\alpha}(E_{p,\alpha} + \mathrm{i}0) | \boldsymbol{p}\alpha \rangle \end{aligned} \tag{8.338}$$

where the operators V_α, $V_{\alpha'}$ represent the interactions between different fragments in the channels α, α'

$$V_\alpha \equiv H - H_\alpha \,, \quad V_{\alpha'} \equiv H - H_{\alpha'} \,. \tag{8.339}$$

The states $\pm$ are defined by application of the corresponding Möller operators, and the operators $T_{\alpha'\alpha}(z)$, $\overline{T}_{\alpha'\alpha}(z)$ are

$$\begin{aligned} T_{\alpha'\alpha}(z) &\equiv V_\alpha + V_{\alpha'} G(z) V_\alpha \\ \overline{T}_{\alpha'\alpha}(z) &\equiv V_{\alpha'} + V_{\alpha'} G(z) V_\alpha \,. \end{aligned} \tag{8.340}$$

Although there are many representations for V_α, all of them are equivalent in the computation of (8.338), in other words, they are equivalent in D_α. One of these representations is simply the difference, as expressed by (8.339), between the total and the α channel Hamiltonians (8.322). This expression is complicated because it depends on the momenta as well as on all the potentials. A simpler expression can be obtained when we replace H_α in (8.339) by one of the fragmentation Hamiltonians (8.320) which is equivalent to H_α on D_α. This has the advantage that V_α appears simply as the sum of the potentials mediating the interaction between two particles belonging to different fragments. Thus, for the channel $A_1 + (A_2A_3)_n$, $V_\alpha = V_{12} + V_{13}$. In what follows H_α will denote the fragmentation Hamiltonian (8.320) corresponding to the α channel.

It is also necessary to remark that even though $T_{\alpha'\alpha}$ and $\overline{T}_{\alpha'\alpha}$ are very different operators, both lead to the same matrix element on the energy shell. This is because $T_{\alpha'\alpha}(z) - \overline{T}_{\alpha'\alpha}(z) = V_\alpha - V_{\alpha'} = H_{\alpha'} - H_\alpha$ and the last operator has a vanishing matrix element between the states $|\boldsymbol{p}\alpha\rangle$ and $|\boldsymbol{p}'\alpha'\rangle$ if $E_{p,\alpha} = E_{p',\alpha'}$.

Unfortunately, the explicit formulae (8.340) for $T_{\alpha'\alpha}$, $\overline{T}_{\alpha'\alpha}$ are not very useful because they involve $G(z)$. Using equations between resolvents it is easy to obtain integral equations similar to (8.105). For example,

$$T_{\alpha'\alpha}(z) = V_\alpha + V_{\alpha'} G_{\alpha'}(z) T_{\alpha'\alpha}(z) \tag{8.341}$$

where $G_{\alpha'}(z) \equiv (z - H_{\alpha'})^{-1}$. These equations are solved formally by iteration. The Born approximation

$$\boxed{T_{\alpha'\alpha}(z) \simeq V_\alpha} \tag{8.342}$$

often provides reasonable results in atomic problems, and one expects this approximation to be good for weak potentials or high energies. There is however little known about the convergence criteria for the iterative series solving (8.341). We should point out that even though $G_{\alpha'}(z)$ is simpler than $G(z)$, it is certainly not simpler than $G_0(z)$ in simple scattering (except in the breaking channel), due to the presence in H_α of the interactions between particles inside each fragment. Therefore, in the Born approximation:

$$\boxed{T^{\mathrm{B}}_{E_{p,\alpha}}(\hat{\boldsymbol{p}}'\alpha', \hat{\boldsymbol{p}}\alpha) = \langle \boldsymbol{p}'\alpha'|V_\alpha|\boldsymbol{p}\alpha\rangle = \langle \boldsymbol{p}'\alpha'|V_{\alpha'}|\boldsymbol{p}\alpha\rangle} \; . \tag{8.343}$$

Finally we would like to add that the theory of multichannel scattering in its time independent formulation via Lippmann-Schwinger integral equations is most difficult due to its very singular nature. Nevertheless there has been some important progress in this direction initiated by the work of *Faddeev* [FA 65], [HE 69].

8.23 General Form of the Optical Theorem

We will now complete the discussion in Sect. 8.13 by proving the optical theorem with more generality. For this purpose it is necessary to have an expression analogous to (8.151) in the multichannel formalism. For convenience we once again restrict our arguments to spinless fragments.

Consider the asymptote state composed of two fragments with momentum (relative to the C.M.) $\boldsymbol{p}$ in the α channel. The corresponding scattering state satisfies the Lippmann-Schwinger equation

$$|\boldsymbol{p}\alpha_+\rangle = |\boldsymbol{p}\alpha\rangle + \frac{1}{E_{\boldsymbol{p},\alpha} + \mathrm{i}0 - H_\alpha} V_\alpha |\boldsymbol{p}\alpha_+\rangle \; . \tag{8.344}$$

The wave function of $|\boldsymbol{p}\alpha_+\rangle$ will in general be very complicated due to the possibility of having excitations, breaking, etc. However if we are only interested in the projection onto the same channel α, it is enough to consider $\langle \boldsymbol{x}\alpha|\boldsymbol{p}\alpha_+\rangle$ where $\boldsymbol{x}$ is the relative position vector of the two fragments. (We do not take into account the internal structure of each fragment because it is unchanged in the elastic channel.) The computation is now straightforward along the lines used in the derivation of (8.149), and hence (8.151), because the resolvent appearing in (8.344) has on D_α an expression analogous to the kernel of $G_0(z)$. This yields

$$\langle \boldsymbol{x}\alpha|\boldsymbol{p}\alpha_+\rangle \underset{|\boldsymbol{x}|\to\infty}{\sim} (2\pi\hbar)^{-3/2}\left[\mathrm{e}^{\mathrm{i}\boldsymbol{k}\cdot\boldsymbol{x}} + f(\boldsymbol{p}\alpha \to p\hat{\boldsymbol{x}}, \alpha)\frac{\mathrm{e}^{\mathrm{i}k|\boldsymbol{x}|}}{|\boldsymbol{x}|}\right] \tag{8.345}$$

completely identical to (8.151), and justifying the reasoning in Sect. 8.13 leading to the optical theorem

$$\boxed{\mathrm{Im}\, f(\boldsymbol{p}\alpha \to \boldsymbol{p}\alpha) = \frac{k}{4\pi}\sigma_{\mathrm{tot},\alpha}(p)} \tag{8.346}$$

where $\sigma_{\mathrm{tot},\alpha}$ is the total cross section in the incoming channel α.

8.24 Symmetries in Multichannel Scattering

It is clear that the symmetries of H and of the H_α translate into invariance properties for the Möller operators and the S matrix. For simplicity we treat the case of a scattering problem with two fragments in the "in" and "out" states and we analyze some of the invariances in the C.M. system. The action of these symmetries U leaving the D_α invariant carries over naturally to $\mathcal{H}'$ in the form $\oplus_\alpha(U \restriction D_\alpha)$.

a) Rotational Invariance

If for example all the potentials are central, H and H_α commute with the operator $U(\mathcal{R})$ implementing a rotation $\mathcal{R}$ of the whole system around its C.M., and restricting to spinless fragments,

$$f(\boldsymbol{p}\alpha \to \boldsymbol{p}'\alpha') = f(\mathcal{R}\boldsymbol{p}, \alpha \to \mathcal{R}\boldsymbol{p}', \alpha') \tag{8.347}$$

(obviously, the index labeling the channel does not change in this case because D_α is stable under rotations). Similarly, the total angular momentum (about the C.M.) and its third component commute with S, and therefore the subspace $\mathcal{H}'_{LM}$ with well defined values L, M of these observables will be stable under S. Denoting by S_{LM} the restriction of S to $\mathcal{H}'_{LM}$, it is clear that S_{LM} and $S_{LM\pm 1}$ will be unitarily equivalent and we can define an operator S_L. Such S_L is unitary, but its components $S_L^{\alpha'\alpha} \equiv P'_{\alpha'} S_L P'_\alpha$ (P'_α is the orthogonal projection of $\mathcal{H}'$ onto D_α) which appear in the transition $\alpha(LM) \to \alpha'(LM)$ are not necessarily unitary. Their norm can decrease even if $\alpha = \alpha'$ due to the possibility of changing channels. In other words, the eigenvalues of $S_L^{\alpha\alpha}$ have always absolute values ≤ 1:

$$\begin{aligned} &S_L^{\alpha\alpha}|pLM, \alpha\rangle = \exp[2\mathrm{i}\delta_{L,\alpha}(p)]|pLM, \alpha\rangle \\ &\operatorname{Im} \delta_{L,\alpha}(p) \geq 0 \end{aligned} \tag{8.348}$$

and all these values have absolute value equal to one if and only if all the channels except for α are closed at the energies considered. The complex numbers $\delta_{L,\alpha}(p)$ are the new phase shifts, and the moduli $|\exp[2\mathrm{i}\delta_{L,\alpha}(p)]|$ are called the *inelasticity* factors.

Using (8.347) and (8.348), the elastic scattering amplitude in the α channel depends only on $|\boldsymbol{p}| = p$ and $\hat{\boldsymbol{p}}' \cdot \hat{\boldsymbol{p}} = \cos\theta$, and in complete analogy with (8.180) we can write:

$$f_\alpha(p, \cos\theta) = \frac{\mathrm{i}}{2k} \sum_{L=0}^{\infty} (2L+1)(1 - \exp[2\mathrm{i}\delta_{L,\alpha}(p)])P_L(\cos\theta) . \tag{8.349}$$

From the orthogonality of the Legendre polynomials we immediately derive the elastic cross section:

$$\boxed{\sigma_{\mathrm{el},\alpha}(p) = \frac{\pi}{k^2}\sum_{L=0}^{\infty}(2L+1)|1-\exp[2\mathrm{i}\delta_{L,\alpha}(p)]|^2} \ . \tag{8.350}$$

On the other hand, (8.349) and the optical theorem (8.346) imply

$$\boxed{\sigma_{\mathrm{tot},\alpha}(p) = \frac{2\pi}{k^2}\sum_{L=0}^{\infty}(2L+1)\left(1-\mathrm{Re}\ \exp[2\mathrm{i}\delta_{L,\alpha}(p)]\right)} \tag{8.351}$$

hence, their difference provides the *absorption cross section*

$$\boxed{\sigma_{\mathrm{abs},\alpha}(p) = \frac{\pi}{k^2}\sum_{L=0}^{\infty}(2L+1)\left(1-|\exp[2\mathrm{i}\delta_{L,\alpha}(p)]|^2\right)} \ . \tag{8.352}$$

It is evident from these formulae that $\sigma_{\mathrm{abs},\alpha} \neq 0$ requires $\sigma_{\mathrm{el},\alpha} \neq 0$. This is expected physically. If the absorption cross section in the α channel is different from zero, some particles will disappear from the incident beam, and its attenuation requires a non-vanishing interference with the elastically scattered wave in this channel in order to account for the depletion of the transmitted flux in the incident direction. In particular the elastically scattered wave must not vanish, and $\sigma_{\mathrm{el},\alpha} \neq 0$.

A typical example is provided by the extreme case of complete absorption when the impact parameter ϱ is smaller than R, and complete transparency for $\varrho > R$. Omitting the index α of the channel for simplicity, and assuming $kR \gg 1$, we have $\exp(2\mathrm{i}\delta_L(p)) = 0$ or 1, depending on whether $kR > L$ or $< L$; and therefore

$$\sigma_{\mathrm{el}}(p) = \frac{\pi}{k^2}\sum_{L=0}^{kR}(2L+1) \simeq \pi R^2 \quad (kR \gg 1) \ . \tag{8.353}$$

Similarly $\sigma_{\mathrm{abs}}(p) \simeq \pi R^2$, $\sigma_{\mathrm{tot}}(p) \simeq 2\pi R^2$.

b) Time Reversal Invariance

If H and H_α are invariant under time reversal (for instance, if the fragments have no spin and the potentials are real), the scattering operator S satisfies $U_T S = S^\dagger U_T$, and in particular, for the scattering amplitudes into two fragment channels, with spin, we have

$$\begin{aligned} &f(\boldsymbol{p} m_1 m_2 \alpha \to \boldsymbol{p}' m_1' m_2' \alpha') \\ &\quad = f(-\boldsymbol{p}', -m_1', -m_2', \alpha' \to -\boldsymbol{p}, -m_1, -m_2, \alpha) \end{aligned} \tag{8.354}$$

a reciprocity relation analogous to (8.170). (The inclusion of spins avoided previously for simplicity does not present any major complications in multichannel scattering as in the elastic scattering treated in Sect. 8.21.)

An important consequence of (8.354) is the *detailed balance* between a given reaction and its inverse. Consider the collision $A + B \to C + D$, in the C.M. where the incoming beams are unpolarized, and the final polarizations are not observed. The particles A, B, C, D are fragments as in previous discussions. If $\boldsymbol{p}, \boldsymbol{p}'$ are the relative momenta in the channels $A + B \equiv \alpha$, and $C + D \equiv \alpha'$, respectively, the differential cross section for the reaction is

$$\frac{d\sigma}{d\Omega}(\boldsymbol{p}\alpha \to \boldsymbol{p}'\alpha') = \frac{1}{(2S_A+1)(2S_B+1)}\frac{k'}{k} \times \sum_{m_1 m_2 m_1' m_2'} |f(\boldsymbol{p}m_1m_2\alpha \to \boldsymbol{p}'m_1'm_2'\alpha')|^2 . \tag{8.355}$$

In the same way

$$\frac{d\sigma}{d\Omega}(-\boldsymbol{p}'\alpha' \to -\boldsymbol{p}\alpha) = \frac{1}{(2S_C+1)(2S_D+1)}\frac{k}{k'} \times \sum_{m_1 m_2 m_1' m_2'} |f(-\boldsymbol{p}'m_1'm_2'\alpha' \to -\boldsymbol{p}m_1m_2\alpha)|^2 . \tag{8.356}$$

The equality (8.354) implies that

$$\boxed{\begin{aligned}&(2S_A+1)(2S_B+1)k^2\frac{d\sigma}{d\Omega}(\boldsymbol{p}\alpha \to \boldsymbol{p}'\alpha')\\ &= (2S_C+1)(2S_D+1)k'^2\frac{d\sigma}{d\Omega}(-\boldsymbol{p}'\alpha' \to -\boldsymbol{p}\alpha) .\end{aligned}} \tag{8.357}$$

This relation expresses the detailed balance, and its experimental verification was discussed in Sect. 7.13. If, as is usually the case, the dynamics is rotationally invariant, we can eliminate the minus signs in the last factor in (8.357). It is interesting to realize that (8.357) with the minus signs suppressed, is also valid in the Born approximation, even if there is no time reversal symmetry. This is because

$$f_{\mathrm{B}}(\boldsymbol{p}m_1m_2\alpha \to \boldsymbol{p}'m_1'm_2'\alpha') = f_{\mathrm{B}}^*(\boldsymbol{p}'m_1'm_2'\alpha' \to \boldsymbol{p}m_1m_2\alpha) . \tag{8.358}$$

8.25 The Optical Potential

In nuclear physics collisions are almost always accompanied by absorption, and we have to use the multichannel formalism. Since this formalism is quite complicated in practice, in the study of the elastic channel one often replaces all other channels by an appropriate modification of the interaction potential thus transforming the problem into a single channel scattering problem. This new potential is not easily calculable, but it is amenable to very interesting phenomenological approximations. An essential feature of the modified potential, known as the *optical potential* is its non-hermiticity. This is to be expected, because it has to contain the probability of absorption. Among its major drawbacks are its non-locality and its energy dependence.

As an example of the techniques of the multichannel formalism, we will show how to obtain the optical potential. Consider a channel α with two spinless fragments. If the asymptote state (in the C.M.) is $|\boldsymbol{p}\alpha\rangle$, its scattering state will be

$$|\boldsymbol{p}\alpha_+\rangle = |\boldsymbol{p}\alpha\rangle + \frac{1}{E_{\boldsymbol{p},\alpha} + \mathrm{i}0 - H_\alpha} V_\alpha |\boldsymbol{p}\alpha_+\rangle \,. \tag{8.359}$$

The wave function in the elastic channel is $\langle \boldsymbol{x}\alpha|\boldsymbol{p}\alpha_+\rangle$, where the internal structure of the fragments is not considered for the reasons pointed out in Sect. 8.23. If we knew a Schrödinger equation for this wave function, the asymptotic behavior (8.345) of its solution would provide the elastic scattering amplitude. Since one certainly has

$$\langle \boldsymbol{x}\alpha|H|\boldsymbol{p}\alpha_+\rangle = E_{\boldsymbol{p},\alpha}\langle \boldsymbol{x}\alpha|\boldsymbol{p}\alpha_+\rangle \tag{8.360}$$

writing $H = H_\alpha + V_\alpha$ we obtain

$$H_\alpha(-\mathrm{i}\hbar\boldsymbol{\nabla})\langle \boldsymbol{x}\alpha|\boldsymbol{p}\alpha_+\rangle + \langle \boldsymbol{x}\alpha|V_\alpha|\boldsymbol{p}\alpha_+\rangle = E_{\boldsymbol{p},\alpha}\langle \boldsymbol{x}\alpha|\boldsymbol{p}\alpha_+\rangle \tag{8.361}$$

where $H_\alpha(-\mathrm{i}\hbar\boldsymbol{\nabla})$ is the differential operator representing the free energy in the α channel in the position representation. The equation (8.361) is the Schrödinger equation we are looking for if the potential term can be written as

$$\langle \boldsymbol{x}\alpha|V_\alpha|\boldsymbol{p}\alpha_+\rangle = \int d^3y \; \mathcal{V}_{\mathrm{opt},\alpha}(\boldsymbol{x},\boldsymbol{y})\langle \boldsymbol{y}\alpha|\boldsymbol{p}\alpha_+\rangle \,. \tag{8.362}$$

Since the optical potential $\mathcal{V}_{\mathrm{opt},\alpha}$ depends only on the relative coordinates between the fragments in the α channel, equation (8.362) imposes on it the condition

$$\langle P_\alpha V_\alpha \Omega_+^\alpha P_\alpha\rangle = \mathcal{V}_{\mathrm{opt},\alpha}\langle P_\alpha \Omega_+^\alpha P_\alpha\rangle \tag{8.363}$$

where P_α is the orthogonal projection in $\mathcal{H}$ onto D_α, and $\langle A\rangle$ represents the average of A over the internal structure of the fragments since we are making $\mathcal{V}_{\mathrm{opt},\alpha}$ depend only on the relative coordinates. To arrive at an expression of $\mathcal{V}_{\mathrm{opt},\alpha}$ satisfying (8.363), we take into account the identity

$$P_\alpha V_\alpha \Omega_+^\alpha P_\alpha = P_\alpha V_\alpha P_\alpha \Omega_+^\alpha P_\alpha + P_\alpha V_\alpha Q_\alpha \Omega_+^\alpha P_\alpha \tag{8.364}$$

with $Q_\alpha \equiv I - P_\alpha$, and we also use the operator expression following from (8.359):

$$Q_\alpha \Omega_+^\alpha P_\alpha = \frac{1}{E_{\boldsymbol{p},\alpha} + \mathrm{i}0 - H_\alpha} Q_\alpha V_\alpha (P_\alpha + Q_\alpha)\Omega_+^\alpha P_\alpha \tag{8.365}$$

which formally yields

$$Q_\alpha \Omega_+^\alpha P_\alpha = \frac{1}{E_{p,\alpha} + \mathrm{i}0 - H_\alpha - Q_\alpha V_\alpha Q_\alpha} Q_\alpha V_\alpha P_\alpha \Omega_+^\alpha P_\alpha$$
$$= \frac{1}{E_{p,\alpha} + \mathrm{i}0 - Q_\alpha H Q_\alpha} Q_\alpha V_\alpha P_\alpha \Omega_+^\alpha P_\alpha \,. \tag{8.366}$$

Substituting this expression in the second term on the right hand side of (8.364) we obtain an equation of the form (8.363) but without an average. Averaging we derive

$$\boxed{\mathcal{V}_{\mathrm{opt},\alpha} = \langle P_\alpha \left[V_\alpha + V_\alpha \frac{Q_\alpha}{E_{p,\alpha} + \mathrm{i}0 - Q_\alpha H Q_\alpha} V_\alpha \right] P_\alpha \rangle} \,. \tag{8.367}$$

The first term $\langle P_\alpha V_\alpha P_\alpha \rangle$ is nothing but the static hermitian potential felt by the incoming particle if there were no other channels. In other words it is the average of V_α over the internal states of the fragments. The second term is due to the presence of other open channels, it vanishes when the other channels are not available ($Q_\alpha = 0$), and it is hermitian when the energy $E_{p,\alpha}$ is small enough to be unable to excite other channels i.e. $E_{p,\alpha} < Q_\alpha H Q_\alpha$. However, if the energy is big enough to excite other channels, this term will not be hermitian and it has a negative definitive antihermitian part. Indeed, writing $\mathcal{V}_{\mathrm{opt},\alpha} = \mathrm{Re}\, \mathcal{V}_{\mathrm{opt},\alpha} + \mathrm{i}\, \mathrm{Im} \mathcal{V}_{\mathrm{opt},\alpha}$, we have:

$$\mathrm{Im}\, \mathcal{V}_{\mathrm{opt},\alpha} = -\pi \langle P_\alpha V_\alpha Q_\alpha \delta(E_{p;\alpha} - Q_\alpha H Q_\alpha) Q_\alpha V_\alpha P_\alpha \rangle \leq 0 \,. \tag{8.368}$$

The imaginary part of the optical potential is negative due to the absorption process as can be seen by rederiving the continuity equation for complex potentials. Hence, it is not surprising that the phase shifts $\delta_{L,\alpha}(p)$ obtained like in the previous section when the optical potential commutes with rotations have non-negative imaginary parts.

Obviously, dealing with these non-local and energy dependent optical potentials is very difficult. In nuclear physics it is common to replace them by approximate local optical potentials which are nonetheless energy dependent [FE 58].

9. The W.B.K. Method

9.1 Introduction

It is interesting to realize that the old Sommerfeld-Wilson-Ishiwara quantization method (Sect. 1.7) gives in many cases of physical interest values for the energy levels very close to those one would obtain after solving the Schrödinger equation exactly.This fact was explained after *Wentzel* [WE 26], and almost simultaneously and independently, *Brillouin* [BR 26] proposed an approximation method to solve the Schrödinger equation based on mathematical techniques developed by *Jeffreys* [JE 24]. This method was later improved by *Kramers* [KR 26], and it is commonly known as the W.B.K. method or semiclassical approximation. It provides a procedure to compute approximately energy levels and eigenfunctions which complements, sometimes with great efficiency, more general computational methods that will be introduced in later chapters.

We first apply the W.B.K. method to the Schrödinger equation for one-dimensional problems. Next we discuss the three-dimensional case with central potentials. A rigorous study of approximations of the type treated in this chapter and their generalization to several degrees of freedom can be found in [MA 72].

9.2 The W.B.K. Method in One-Dimensional Problems

Consider the one-dimensional Schrödinger equation

$$-\frac{\hbar^2}{2M}\psi''(x) + V(x)\psi(x) = E\psi(x) \,. \tag{9.1}$$

Its solutions are implicit functions of $\hbar$. Since we are interested in their behavior as $\hbar \to 0$ (namely when the natural actions in the problem are very large relative to $\hbar$), the change of variable

$$\psi(x) = \left(\frac{dS(x;\hbar)}{dx}\right)^{-1/2} \exp\left\{\frac{\mathrm{i}}{\hbar}S(x;\hbar)\right\} , \tag{9.2}$$

where the explicit dependence in $\hbar$ is manifest, transforms (9.1) into

$$\left(S'(x;\hbar)\right)^2 - \hbar^2 \left(S'(x;\hbar)\right)^{1/2} \frac{d^2}{dx^2} \left(S'(x;\hbar)\right)^{-1/2} = 2M(E - V(x)) \,, \tag{9.3}$$

where the primes indicate derivatives with respect to x. This equation is easier to study than (9.1) in terms of a formal series expansion in powers of $\hbar$:

$$S(x;\hbar) = \sum_{n=0}^{\infty} \hbar^{2n} S_n(x) . \tag{9.4}$$

Substituting (9.4) into (9.3), and setting to zero terms with the same power of $\hbar$ leads to a system of first-order differential equations for $S_n(x)$, which can be solved by elementary integrations after $S_0, \ldots, S_{n-1}$ are known. In particular, the first two of these equations are

$$\begin{aligned} &(S_0')^2 = 2M(E - V(x)) \equiv p^2(x) , \\ &2S_0'S_1' - (S_0')^{1/2} \frac{d^2}{dx^2} (S_0')^{-1/2} = 0 , \end{aligned} \tag{9.5}$$

with solutions

$$\begin{aligned} &S_0(x) = \pm \int^x dx' p(x') , \\ &S_1(x) = \pm \frac{1}{2} \int^x dx' (p(x'))^{-1/2} \frac{d^2}{dx^2} (p(x'))^{-1/2} . \end{aligned} \tag{9.6}$$

The structure of (9.3) guarantees that $S_1' \equiv 0$ implies $S_m' = 0$, $m \geq 1$.

In the limit of $\hbar \to 0$ one should expect the first term in (9.4) to dominate, so that a good approximation to $\psi(x)$ is given by a linear combination of the functions

$$\psi_\pm(x) = \frac{1}{\sqrt{p(x)}} \exp\left\{ \pm \frac{i}{\hbar} \int^x dx' p(x') \right\} , \tag{9.7}$$

known as the *basic W.B.K. solutions.*

The two terms in the right-hand-side have a reasonable interpretation in the "classical" regions $E > V(x)$: $p(x')$ defined in (9.5) plays the role of a "local" particle momentum, so that the exponentials are essentially $\exp(\pm i p_{\text{average}} x/\hbar)$; on the other hand, $|\psi_\pm(x)|^2 = |p(x)|^{-1}$ should represent the probability density to find the particle at the point x, and it is therefore reasonable that it should be proportional to its inverse local velocity at that point in the limit $\hbar \to 0$ where the classical results should be recovered. This reasoning suggests that the approximation to $\psi(x)$ given by (9.7) is acceptable so long as the associated wavelength $\lambda\!\!\!^{-}(x) \equiv \hbar/p(x)$ changes very little as x varies; namely, when

$$\left| \frac{d\lambda\!\!\!^{-}(x)}{dx} \right| \ll 1 . \tag{9.8}$$

Similarly, the equations (9.6) imply that $|\hbar^2 S_1'(x)| \ll |S_0'(x)|$ if the wavelength varies slowly and smoothly with x.

Since the functions $\psi_\pm(x)$ diverge at the *turning points* (the points where $p(x) = 0$ and which separate the allowed and forbidden regions for the motion of a classical particle with energy E), it is clear that the classical W.B.K. solutions are useful only far from the turning points.

In the following we will always assume that far away from the turning points, the applicability conditions of the *W.B.K. approximation* $S(x;\hbar) \simeq S_0(x)$ are satisfied, so that in the region between two adjacent turning points, the general solution to the Schrödinger equation will in this approximation be given by

1)
$$\boxed{\begin{aligned} &\text{if} \quad V(x) > E \quad \text{and} \quad p_1(x) \equiv \{2M[V(x)-E]\}^{1/2} \\ &\psi(x) = \frac{A}{\sqrt{p_1(x)}} \exp\left[\frac{1}{\hbar}\int_{x_0}^{x} dx' p_1(x')\right] \\ &\qquad + \frac{B}{\sqrt{p_1(x)}} \exp\left[-\frac{1}{\hbar}\int_{x_0}^{x} dx' p_1(x')\right] , \end{aligned}} \tag{9.9}$$

2)
$$\boxed{\begin{aligned} &\text{if} \quad V(x) < E \quad \text{and} \quad p_2(x) \equiv \{2M[E-V(x)]\}^{1/2} \\ &\psi(x) = \frac{A}{\sqrt{p_2(x)}} \exp\left[\frac{\mathrm{i}}{\hbar}\int_{x_0}^{x} dx' p_2(x')\right] \\ &\qquad + \frac{B}{\sqrt{p_2(x)}} \exp\left[-\frac{\mathrm{i}}{\hbar}\int_{x_0}^{x} dx' p_2(x')\right] . \end{aligned}} \tag{9.10}$$

In a region between two adjacent turning points, one usually takes x_0 inside that region, and incidentally it could coincide with either of its extremes. A and B are arbitrary constants.

9.3 Connection Formulae

Even though we have obtained a good approximation to the solution of the Schrödinger equation in the region between two adjacent turning points, it is not possible with the machinery developed so far to relate the pairs of constants A, B appearing in different regions. The usefulness of this method depends on the possibility of finding a procedure relating the constants corresponding to two adjacent zones. Let us for simplicity assume the situation depicted in Fig. 9.1, where $x = 0$ is a turning point. Taking the distance a sufficiently large, the W.B.K. approximation will be acceptable in regions I and IV. There we can write

$$\begin{aligned} \psi_{\mathrm{I}}(x) = &\frac{A_{\mathrm{I}}}{\sqrt{p_1(x)}} \exp\left[\frac{1}{\hbar}\int_0^x dx' p_1(x')\right] \\ &+ \frac{B_{\mathrm{I}}}{\sqrt{p_1(x)}} \exp\left[-\frac{1}{\hbar}\int_0^x dx' p_1(x')\right] , \end{aligned} \tag{9.11}$$

$$\begin{aligned} \psi_{\mathrm{IV}}(x) = &\frac{A_{\mathrm{IV}}}{\sqrt{p_2(x)}} \sin\left(\frac{1}{\hbar}\int_0^x dx' p_2(x')\right) \\ &+ \frac{B_{\mathrm{IV}}}{\sqrt{p_2(x)}} \cos\left(\frac{1}{\hbar}\int_0^x dx' p_2(x')\right) . \end{aligned} \tag{9.12}$$

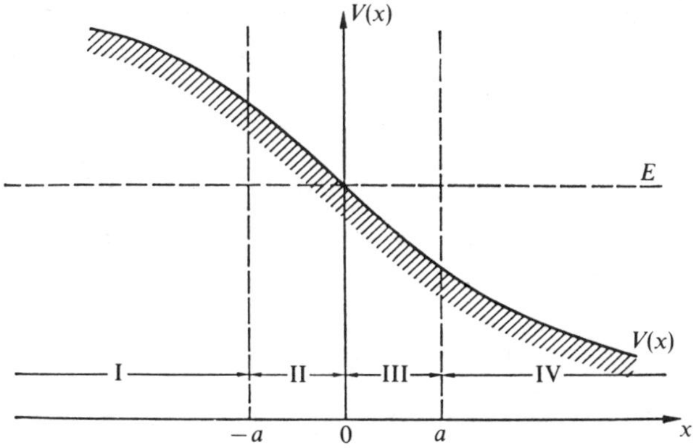

Fig. 9.1. Regions to be considered near a turning point $x = 0$ with $V(x) > E$ for $x < 0$

Our problem is reduced to finding two of the arbitrary constants in (9.11) and (9.12) in terms of the other two.

This is a difficult question in the analysis of asymptotic expansions, and it has generated a considerable controversy [FF 65], [BM 72]. Even today it receives important contributions [SI 85]. Its study through the formal series (9.4) is possible using the complex x-plane to go from one allowed region to an adjacent one without crossing the turning point separating them. We follow here an alternative method due to *Langer, Cherry* and *Silverstone* [LA 37], [CH 50], [SI 85] that is more transparent and powerful from a mathematical point of view. It is based on the comparison between (9.1) and another explicitly solvable Schrödinger equation which (9.1) approaches as $\hbar \to 0$, and the effect of the non-zero value of $\hbar$ is incorporated via a small distortion of the independent variable, which can be controlled rigorously.

More concretely, write $\psi(x)$ as

$$\psi(x) = (\phi'(x;\hbar))^{-1/2}\left[a\,\mathrm{Ai}(-\hbar^{-3/2}\phi(x;\hbar)) + b\,\mathrm{Bi}(-\hbar^{-3/2}\phi(x;\hbar))\right], \quad (9.13)$$

where a, b are constants, and Ai, Bi are the Airy functions [AS 64]. Such ψ satisfies (9.1) if and only if $\phi(x;\hbar)$ is a solution of

$$\phi(x;\hbar)(\phi'(x;\hbar))^2 - \hbar^2(\phi'(x;\hbar))^{1/2}\frac{d^2}{dx^2}(\phi'(x;\hbar))^{-1/2} = 2M(E - V(x)) \quad (9.14)$$

whose similarity with (9.3) is quite apparent.

If the potential V has a turning point at x_0 for the energy E, and it is strictly linear, $V(x) = E - \alpha(x - x_0)$, $\alpha \neq 0$, a solution of (9.14) is $\phi(x;\hbar) = (2M\alpha)^{1/3}(x - x_0)$, and (9.13) in this case provides the explicit solution to the Schrödinger equation, because the Airy functions are solutions to the equation $w''(z) - zw(z) = 0$. Close to a turning point $x_0(= 0)$ as in Fig. 9.1 the potential does not necessarily coincide with its tangent, and $\phi(x;\hbar)$ describes the change of parameter x, in general non-linear, required to linearize the potential in this region.

As in (9.3) we can write

$$\phi(x;\hbar) = \sum_{n=0}^{\infty} \hbar^{2n}\phi_n(x) \tag{9.15}$$

and as in the previous case, the coefficients $\phi_n(x)$ of the expansion can be evaluated by elementary integration. In particular

$$\begin{aligned} &\phi_0(x)(\phi_0'(x))^2 = 2M(E - V(x)) , \\ &\phi_1(x)(\phi_0'(x))^2 + 2\phi_0(x)\phi_0'(x)\phi_1'(x) - (\phi_0'(x))^{1/2}\frac{d^2}{dx^2}(\phi_0'(x)^{-1/2} = 0 . \end{aligned} \tag{9.16}$$

Notice that $\phi_0 > 0\,(\phi_0 < 0)$ in the classically allowed (forbidden) regions. The solutions to (9.16) appropriate for a turning point x_0 are

$$\begin{aligned} &\phi_0(x) = \left\{\frac{3}{2}\int_{x_0}^{x} dx' p(x')\right\}^{2/3} , \\ &\phi_1(x) = (\phi_0(x))^{-1/2}\frac{1}{2}\int_{x_0}^{x} dx'(\phi_0(x')\phi_0'(x'))^{-1/2}\frac{d^2}{dx'^2}(\phi_0'(x'))^{-1/2} . \end{aligned} \tag{9.17}$$

The integration constants have been chosen so that both functions vanish at x_0 and are analytic in a neighborhood of the turning point, assuming that V is analytic and does not have a vanishing slope at x_0. It can be shown that similar results are obtained for the other ϕ_n, $n \geq 2$.

It is possible to prove that the expansion (9.15) is asymptotic (for $\hbar \to 0$), and uniform in a small neighborhood including the turning point x_0 [CH 50]. The next step consists of using the asymptotic behavior of the Airy functions appearing in (9.13). It is known [AS 64], [SN 85] that for $|z| \to \infty$:

$$\begin{aligned} &\mathrm{Ai}(-z) \sim \tfrac{1}{2}\pi^{-1/2}z^{-1/4}\left[\mathrm{e}^{-\mathrm{i}\pi/4}\beta(\mathrm{i}\xi) + \mathrm{e}^{\mathrm{i}\pi/4}\beta(-\mathrm{i}\xi)\right] , \quad |\arg z| < \pi/3 \\ &\mathrm{Ai}(z) \sim \tfrac{1}{2}\pi^{-1/2}z^{-1/4}\beta(-\xi), \quad |\arg z| < 2\pi/3 \\ &\mathrm{Bi}(-z) \sim \tfrac{1}{2}\pi^{-1/2}z^{-1/4}\left[\mathrm{e}^{\mathrm{i}\pi/4}\beta(\mathrm{i}\xi) + \mathrm{e}^{-\mathrm{i}\pi/4}\beta(-\mathrm{i}\xi)\right] , \quad |\arg z| < \pi/3 \\ &\mathrm{Bi}(z) \sim \tfrac{1}{2}\pi^{-1/2}z^{-1/4}[2\beta(\xi) \pm \mathrm{i}\beta(-\xi)] \quad 0 < \pm\arg z < 2\pi/3 , \end{aligned} \tag{9.18}$$

where

$$\xi \equiv \tfrac{2}{3}z^{3/2}, \quad \beta(\xi) \equiv \mathrm{e}^{\xi}\sum_{k=0}^{\infty} c_k\xi^{-k} \quad c_k \equiv \frac{\Gamma(k+5/6)\Gamma(k+1/6)}{\Gamma(5/6)\Gamma(1/6)2^k k!} . \tag{9.19}$$

The formal series $\beta(\xi)$ is divergent, although Borel summable in the complex ξ-plane cut along $[0,\infty)$. With this Borel sum, the equations (9.18) are valid in the given domains.

While the first three results are well known, the last is rather surprising, for it includes an imaginary coefficient, which would apparently imply that $\mathrm{Bi}(x)$ is not real for $x > 0$. Note however that the positive real axis is not contained in its domain of validity; on the other hand, this axis represents a cut in the Borel sum of $\beta(\xi)$, hence the right-hand side of the expression for $\mathrm{Bi}(z)$ in (9.18) has to be computed on the real axis by analytic continuation, and it is precisely through this analytic continuation that the function $\beta(\xi)$ acquires its limit values $\beta(x \pm \mathrm{i}0)$, $x > 0$, which are *complex* in spite of the reality of the coefficients c_k; the term $\pm\mathrm{i}\beta(-x)$ exactly cancels the imaginary part of $\beta(x \pm \mathrm{i}0)$, thus giving a real value for $\mathrm{Bi}(x)$.

The substitution of (9.18) in (9.13) leads immediately to the asymptotic expansion of $\psi(x)$ to the left and right of the turning point in the limit as $\hbar \to 0$, in terms of $\phi(x;\hbar)$ [in this limit, this quantity is positive (negative) in the classically allowed (forbidden) region]. It now suffices to write these results in terms of the function $S(x;\hbar)$ defined in (9.2) to obtain the desired connection formulae. To achieve this we define now [SI 85]:

$$S(x;\hbar) \equiv -\mathrm{i}\frac{\hbar}{2}\ln\frac{\beta\left((2\mathrm{i}/3)\phi^{3/2}/\hbar\right)}{\beta\left(-(2\mathrm{i}/3)\phi^{3/2}/\hbar\right)} . \tag{9.20}$$

It is easy to verify that such $S(x;\hbar)$ satisfies (9.3), and that both S and S' are positive in the classically allowed regions. Furthermore, using the expression for the wronskian of Airy functions,

$$\begin{aligned}(S'(x;\hbar))^{-1/2}&\exp\left\{\pm\frac{\mathrm{i}}{\hbar}S(x;\hbar)\right\}\\ &= (\phi(x;\hbar))^{-1/4}(\phi'(x;\hbar))^{-1/2}\beta((\pm 2\mathrm{i}/3)\phi^{3/2}(x;\hbar)/\hbar) .\end{aligned} \tag{9.21}$$

Once ϕ is known the equation (9.20) provides a concrete solution for (9.3) without any ambiguities in the integration constants or signs appearing in the computation of $S_n(x)$. Specifically, (9.20) leads to the asymptotic behavior $S_n(x) \sim k_n(x - x_0)^{3/2-3n}$, $x \to x_0$, with k_n constant.

Combining (9.13), (9.18) and (9.21) leads straightforwardly for ψ real to

$$\psi(x) \sim A[S'(x;\hbar)]^{-1/2}\cos\left\{\frac{1}{\hbar}S(x;\hbar) - \frac{\pi}{4} + \eta\right\} \tag{9.22}$$

in the classically allowed region, with $\tan\eta \equiv b/a$.

An entirely analogous procedure allows us to write $\psi(x)$ in the classically forbidden region as

$$\begin{aligned}\psi(x) \sim A[\mp Q'(x \pm \mathrm{i}0;\hbar)]^{-1/2}&\left\{\sin\eta\exp\left[\pm\frac{1}{\hbar}Q(x \pm \mathrm{i}0;\hbar)\right]\right.\\ &\left.+\frac{1}{2}\exp(\mp\mathrm{i}\varepsilon\eta)\exp\left[\mp\frac{1}{\hbar}Q(x \pm \mathrm{i}0;\hbar)\right]\right\} ,\end{aligned} \tag{9.23}$$

where $Q(z;\hbar) \equiv \mathrm{i}S(z;\hbar)$, with $S(z;\hbar)$ computed via analytic continuation from (9.20). This analytic continuation presents a cut precisely at the values of interest in (9.23), and it is therefore necessary to specify the path followed in the complex plane to reach these points. According to whether the continuation is made maintaining $\mathrm{Im}\, z > 0\, (< 0)$, we should take the $-$ (+) sign in (9.23) to ensure the reality of $\psi(x)$. The coefficient $\varepsilon = \pm 1$ depends on whether the allowed region is to the right or to the left of the turning point.

From (9.22) and (9.23) we immediately derive the exact connection formula valid to all orders in $\hbar$:

$$[S'(x;\hbar)]^{-1/2}\cos\left\{\frac{1}{\hbar}S(x;\hbar)-\frac{\pi}{4}+\eta\right\}\longleftrightarrow$$

$$[\mp Q'(x\pm \mathrm{i}0,\hbar)]^{-1/2}\left[\sin\eta\exp\left\{\pm\frac{1}{\hbar}Q(x+\mathrm{i}0,\hbar)\right\}\right.$$

$$\left.+\frac{1}{2}\exp(\mp\mathrm{i}\varepsilon\eta)\exp\left\{\mp\frac{1}{\hbar}Q(x\pm \mathrm{i}0,\hbar]\right\}\right] . \tag{9.24}$$

Since both members are nothing but the expression of $\psi(x)$ in the different regions, the bidirectional character of (9.24) is quite clear.

In practice, the exact computation of $S(x;\hbar)$ is not generally feasible. The W.B.K. approximation consists of taking $S(x;\hbar)\simeq S_0(x;\hbar)=\frac{2}{3}(\phi_0(z;\hbar))^{3/2}$. Hence on the real axis:

$$\begin{aligned} S_0(x;\hbar) &= \left|\int_{x_0}^{x}dx'p_2(x')\right| , \qquad x\in \text{allowed region} ,\\ Q_0(x\pm \mathrm{i}0,\hbar) &= \pm\left|\int_{x_0}^{x}dx'p_1(x')\right| , \qquad x\in \text{forbidden region} , \end{aligned} \tag{9.25}$$

with the notation used in Sect. 9.2. We now derive explicitly the form of (9.24) in this approximation. For concreteness we only treat the case depicted in Fig. 9.1.

Suppose $\psi(x)$ in the classically allowed region has the leading asymptotic form (for $\hbar\to 0$):

$$\psi(x)\sim\frac{A}{\sqrt{p_2(x)}}\cos\left\{\frac{1}{\hbar}\int_{x_0}^{x}dx'p_2(x')-\frac{\pi}{4}+\eta\right\} . \tag{9.26}$$

If $\sin\eta\neq 0$, from (9.24) we obtain

$$\psi(x)\sim\frac{A\sin\eta}{\sqrt{p_1(x)}}\exp\left\{\frac{1}{\hbar}\int_{x}^{x_0}dx'p_1(x')\right\} , \tag{9.27}$$

where we have neglected the second term of (9.24), not only because it does not appear to leading order, but also because it is unnecessary when Q is approximated by a real function. This justifies the second equation in Table 9.1; its one-directional character is due to the fact that with a single growing exponential in the forbidden region it is not possible to determine the angle η. In this case it can be included in a redefinition of A. On the other hand, if $\sin\eta = 0$, we have in the forbidden region

$$\psi(x)\sim\frac{A}{\sqrt{p_1(x)}}\exp\left\{-\frac{1}{\hbar}\int_{x}^{x_0}dx'p_1(x')\right\} . \tag{9.28}$$

Table 9.1. Connection formulae in the W.B.K. method, with turning point $x = x_0$. Their validity requires $\sin\eta \neq 0$

$E > V(x)$ if $x > x_0$	$\psi(x) = \frac{A}{\sqrt{p_1(x)}} e^{-\frac{1}{\hbar}\int_x^{x_0} dx'\, p_1(x')}$	$\longrightarrow$	$\psi(x) = \frac{2A}{\sqrt{p_2(x)}} \cos\left(\frac{1}{\hbar}\int_{x_0}^{x} dx'\, p_2(x') - \frac{\pi}{4}\right)$
	$\psi(x) = \frac{A \sin\eta}{\sqrt{p_1(x)}} e^{+\frac{1}{\hbar}\int_x^{x_0} dx'\, p_1(x')}$	$\longleftarrow$	$\psi(x) = \frac{A}{\sqrt{p_2(x)}} \cos\left(\frac{1}{\hbar}\int_{x_0}^{x} dx'\, p_2(x') - \frac{\pi}{4} + \eta\right)$
$E < V(x)$ if $x > x_0$	$\psi(x) = \frac{2A}{\sqrt{p_2(x)}} \cos\left(\frac{1}{\hbar}\int_x^{x_0} dx'\, p_2(x') - \frac{\pi}{4}\right)$	$\longleftarrow$	$\psi(x) = \frac{A}{\sqrt{p_1(x)}} e^{-\frac{1}{\hbar}\int_{x_0}^{x} dx'\, p_1(x')}$
	$\psi(x) = \frac{A}{\sqrt{p_2(x)}} \cos\left(\frac{1}{\hbar}\int_x^{x_0} dx'\, p_2(x') - \frac{\pi}{4} + \eta\right)$	$\longrightarrow$	$\psi(x) = \frac{A \sin\eta}{\sqrt{p_1(x)}} e^{+\frac{1}{\hbar}\int_{x_0}^{x} dx'\, p_1(x')}$

From (9.24) we obtain for the allowed region:

$$\psi(x) \sim \frac{2A}{\sqrt{p_2(x)}} \cos\left\{\frac{1}{\hbar}\int_{x_0}^{x} dx' p_2(x') - \frac{\pi}{4}\right\}, \tag{9.29}$$

justifying the first equation in Table 9.1. Again its one-directional character is a consequence of the fact that the terms ignored in (9.29) could generate an angle $\eta = O(\hbar)$. This would automatically give rise to a growing exponential in the forbidden region which would end up by dominating over the behavior (9.28).

In Table 9.1 we have summarized the connecting formulae when the turning point is $x = x_0$, whether $V(x) > E$ for $x < x_0$ or not. One often finds in the literature an indiscriminate use of these relations without paying due attention to the sense of the arrows in such formulae. This may in principle lead to wrong results. We will return to this point later.

The reader interested in the study of turning points which are zeros of $V(x) - E$ of order higher than one, is referred to the work of *Fröman* and *Fröman* [FF 65].

9.4 Bound State Energies

In the first application of the W.B.K. method we will obtain an approximate formula for the bound state energy levels of a one-dimensional potential well $V(x)$. In order to avoid complications with the directionality of the connecting formulae we assume that $V(x)$ has only two turning points for the energies E to be considered.

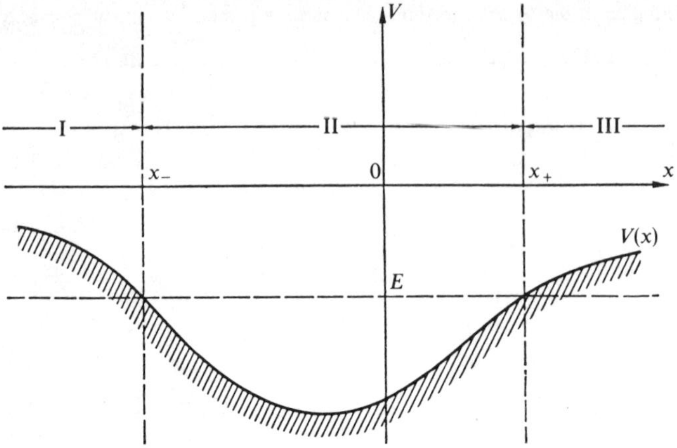

Fig. 9.2. Potential $V(x)$ with two turning points $x = x_-$, $x = x_+$ for an energy E

A typical situation appears in Fig. 9.2. In region I and far from the turning point x_- the bound wave function of energy E is, in the W.B.K. approximation, given by

$$\psi_{\mathrm{I}}(x) = \frac{A}{\sqrt{p_1(x)}} \exp\left\{-\frac{1}{\hbar}\int_x^{x_-} dx' p_1(x')\right\} , \tag{9.30}$$

since the normalizability of bound states forbids the presence of growing exponentials. Using the connecting formulae, we obtain

$$\begin{aligned} \psi_{\mathrm{II}}(x) &= \frac{2A}{\sqrt{p_2(x)}} \cos\left(\frac{1}{\hbar}\int_{x_-}^{x} dx' p_2(x') - \frac{\pi}{4}\right) \\ &= \frac{2A}{\sqrt{p_2(x)}} \cos\left(\frac{1}{\hbar}\int_{x}^{x_+} dx' p_2(x') - \frac{\pi}{4} + \eta\right) , \end{aligned} \tag{9.31}$$

where

$$\eta = \frac{\pi}{2} - \frac{1}{\hbar}\int_{x_-}^{x_+} dx' p_2(x') . \tag{9.32}$$

If $\sin\eta \neq 0$, the connecting formulae given in Table 9.1 would imply the existence of a growing exponential in region III. The bound state condition thus requires

$$\sin\left(\frac{1}{\hbar}\int_{x_-}^{x_+} dx\, p_2(x) - \frac{\pi}{2}\right) = 0 , \tag{9.33}$$

namely

$$\boxed{\int_{x_-}^{x_+} dx \sqrt{2M[E - V(x)]} = \hbar\pi\left(n + \frac{1}{2}\right) , \quad n \text{ integer} \geq 0} . \tag{9.34}$$

Conversely, if (9.34) is satisfied [i.e., (9.33)], the exponentially decreasing W.B.K. solutions in regions I and III lead to a unique W.B.K. solution in region II up to an overall normalization constant, as a straightforward consequence of the results listed in Table 9.1.

The reader should be aware that in the derivation of (9.34) we have implicitly assumed the generally unjustified hypothesis that the regions around x_- and x_+, where the asymptotic expansions are valid, always overlap in the classically allowed region.

Equation (9.34) gives an approximation to the energy levels E_n for bound states in energy regions with only two turning points. We will denote as $E_{n,\mathrm{WBK}}$ the energy levels obtained using (9.34).

It would be useful to have an estimate of the error in the approximation of E_n by $E_{n,\mathrm{WBK}}$. This is a difficult problem. *Fröman* and *Fröman* [FF 65] have analyzed the corrections to the approximation given by (9.34). Their results are however too complicated to give a useful bound on the error made using it. A careful analysis of these corrections led to the characterization of families of potentials whose exact energy levels are given by an equation similar to (9.34) with $V(x)$ replaced by a suitable function [RK 68]. This procedure, whose basic idea goes back to *Langer* [LA 37], will be useful when we discuss central potentials.

From the numerical analysis of many potentials stems a practical rule which to the best of our knowledge has never been proved rigorously: The error in the W.B.K. approximation (9.34) is small if the following two characteristic actions in the problem (measured in units of $\hbar$)

$$
\begin{aligned}
A_1 &\equiv \frac{1}{\hbar}\int_{x_-}^{x_+} dx\sqrt{2M[E_n - V(x)]} \simeq n\pi\ , \\
A_2 &\equiv \frac{1}{\hbar}\int_{x_-}^{x_+} dx\sqrt{2M[V_0 - E_n]}
\end{aligned}
\tag{9.35}
$$

are large. In (9.35), $V_0 = \min\{V(+\infty), V(-\infty)\}$. For potentials with an infinite number of bound states we can say that the W.B.K. approximation becomes better and better as n grows. For potentials with a finite number of bound states, one should expect that the largest errors appear in the computations of the energies of those states at the beginning and the end points of the bound state spectrum.

As a last remark, notice that replacing n by $(n+1/2)$ in (9.34) reproduces the SWI quantization rule. Thus the partial agreement of the energy levels computed with the old quantum theory with those obtained using the Schrödinger equation is not surprising anymore.

The equation (9.34) is the leading approximation to an exact quantization procedure due to *Dunham* [DU 32]. Suppose E_n is an exact energy level for the analytic potential $V(x)$, and let $\psi_n(x)$ be the associated analytic eigenfunction with n nodes. Let x_-, x_+ be the turning points which we assume to be unique. If γ is a simple contour surrounding the interval $[x_-, x_+]$ in the positive sense in the complex x-plane, the argument principle in complex analysis implies that

$$\frac{1}{2\pi \mathrm{i}} \int_\gamma dz \, \frac{\psi_n'(z)}{\psi_n(z)} = n \, . \tag{9.36}$$

Writing $\psi_n(x) = \exp[\sigma(x;\hbar)/\hbar]$, (9.36) becomes

$$\int_\gamma dz \, \sigma'(z;\hbar) = 2\pi \mathrm{i} n \hbar \, . \tag{9.37}$$

The Schrödinger equation for ψ_n leads to

$$\hbar\sigma''(x;\hbar) + [\sigma'(x;\hbar)]^2 = 2M(V(x) - E_n) \equiv -p^2(x) \, , \tag{9.38}$$

and writing $\sigma(x;\hbar) = \sum_0^\infty \hbar^n \sigma_n(x)$, we have (taking $\mathrm{Im}\, p \geq 0$):

$$\sigma_0'(x) = \mathrm{i}p(x) \, , \quad \sigma_1' = -\frac{1}{2}\frac{p'(x)}{p(x)} \, , \quad \sigma_2' = -\frac{\mathrm{i}}{8}\frac{p'^2}{p^3} - \frac{\mathrm{i}}{4}\left(\frac{p'}{p}\right)' \, , \ldots \, . \tag{9.39}$$

Take into account that for $x > x_+$, $\psi_n'(x)/\psi_n(x) < 0$, and this fixes the sign of $\sigma_0'(x)$. It is easy to see that the approximation $\sigma \simeq \sigma_0 + \hbar\sigma_1$ leads to (9.34), where σ_0 is responsible for the term $n\pi\hbar$ and $\hbar\sigma_1$ for the extra contribution $\pi\hbar/2$. The corrections $\hbar^2\sigma_2 \ldots$ modify (9.34) giving rise through (9.37) to a presumably asymptotic representation as $\hbar \to 0$ of the exact levels. The application of this technique to the computation of the energy spectrum in several families of simple potentials can be found in [BO 77]. The possibility of having to add subleading terms in the asymptotic expansion of Dunham's formula is studied in [BP 79].

For supersymmetric potentials [WI 81] an alternative formula to (9.34) has been proposed to calculate approximately the energy levels [CB 85].

Next we apply what we have learned about the W.B.K. method to estimate the bound state energy levels in a variety of potentials.

a) *The Harmonic Oscillator*. The potential is $V(x) = kx^2/2$, and for an energy $E\,(> 0)$, the turning points are $x_\pm = \pm[2E/k]^{1/2}$. Applying (9.34) we obtain

$$E_{n,\mathrm{WBK}} = \hbar\omega(n + 1/2) \, , \quad n = 0, 1, 2 \ldots , \quad \omega \equiv \sqrt{k/M} \, , \tag{9.40}$$

which coincides for all n with the exact result E_n.

b) $V(x) = a|x|$, $a > 0$. In this case the turning points for an energy $E(> 0)$ are $x_\pm = \pm E/a$, and we obtain from (9.34)

$$E_{n,\mathrm{WBK}} = \alpha^{-2/3}\left[\frac{3\pi(2n+1)}{8}\right]^{2/3} \equiv \alpha^{-2/3} z_n^{2/3} \, , \tag{9.41}$$

where $\alpha \equiv \sqrt{2M}/\hbar a$. For this potential the Schrödinger equation can be solved explicitly in terms of the Airy functions [AS 64]. The bound states with negative parity are determined by the equation $\mathrm{Ai}(-\alpha^{2/3}E) = 0$, where $\mathrm{Ai}(x)$ is the Airy function, whereas the states of positive parity are given by $\mathrm{Ai}'(-\alpha^{2/3}E) = 0$. The zeroes of these functions are well known. The exact and WBK results are given in Table 9.2 for the first few bound states. For values of z_n large enough we obtain

$$\alpha^{2/3}E_n = z_n^{2/3}\left[1 - \frac{7}{48}z_n^{-2} + \frac{35}{288}z_n^{-4} - \dots\right], \quad n \text{ even},$$
$$\alpha^{2/3}E_n = z_n^{2/3}\left[1 + \frac{5}{48}z_n^{-2} - \frac{5}{36}z_n^{-4} + \dots\right], \quad n \text{ odd}, \qquad (9.42)$$

and therefore the error in using the W.B.K. approximation decreases as $1/n^2$.

Table 9.2. Comparison between the W.B.K. approximation and the exact results for the first ten bound states for the potential well $V(x) = a|x|$

n	$\alpha^{2/3}E_n$	$\alpha^{2/3}E_{n,\mathrm{WBK}}$	$\dfrac{E_{n,\mathrm{WBK}} - E_n}{E_n}$
0	1.018 793	1.115 460	$+9.49 \times 10^{-2}$
1	2.338 107	2.320 251	-7.64×10^{-3}
2	3.248 198	3.261 626	$+4.13 \times 10^{-3}$
3	4.087 949	4.081 810	-1.50×10^{-3}
4	4.820 099	4.826 316	$+1.29 \times 10^{-3}$
5	5.520 560	5.517 164	-6.15×10^{-4}
6	6.163 307	6.167 128	$+6.20 \times 10^{-4}$
7	6.786 708	6.784 454	-3.32×10^{-4}
8	7.372 177	7.374 853	$+3.63 \times 10^{-4}$
9	7.944 134	7.942 487	-2.07×10^{-4}

c) We now analyze the eigenvalues of the Schrödinger equation:

$$\frac{d^2\psi(x)}{dx^2} + [\varepsilon - U(x)\psi(x) = 0$$
$$\varepsilon \equiv \frac{2M}{\hbar^2}E, \quad U(x) \equiv \frac{2M}{\hbar^2}V(x) = -A\frac{e^x}{1+e^x} - B\frac{e^x}{(1+e^x)^2}. \qquad (9.43)$$

These potentials are such that $U(x) \to -A$ as $x \to +\infty$, $U(x) \to 0$ for $x \to -\infty$, and they have a minimum at $x = \ln[(B+A)/(B-A)]$, with $U_{\min} = -(A+B)^2/4B$, if $B > |A|$ (only cases we will consider). Typical shapes of these potentials are displayed in Fig. 9.3. The discrete spectrum of energies ε for this potential lies in the interval $(-(A+B)^2/4B, \min(0, -A))$.

For energies in the interval mentioned, the turning points are

$$x_\pm = \ln\left\{\frac{1}{2(|\varepsilon| - A)}\left[(B + A - 2|\varepsilon|) \pm \sqrt{(A+B)^2 - 4B|\varepsilon|}\right]\right\}. \qquad (9.44)$$

Changing variables to $t = e^x$, and defining $t_\pm = \exp x_\pm$, equation (9.34) becomes

$$\int_{t_-}^{t_+} dt\frac{1}{t(1+t)}\{-(|\varepsilon| - A)t^2 + (A + B - 2|\varepsilon|)t - |\varepsilon|\}^{1/2} = \pi\left(n + \frac{1}{2}\right). \qquad (9.45)$$

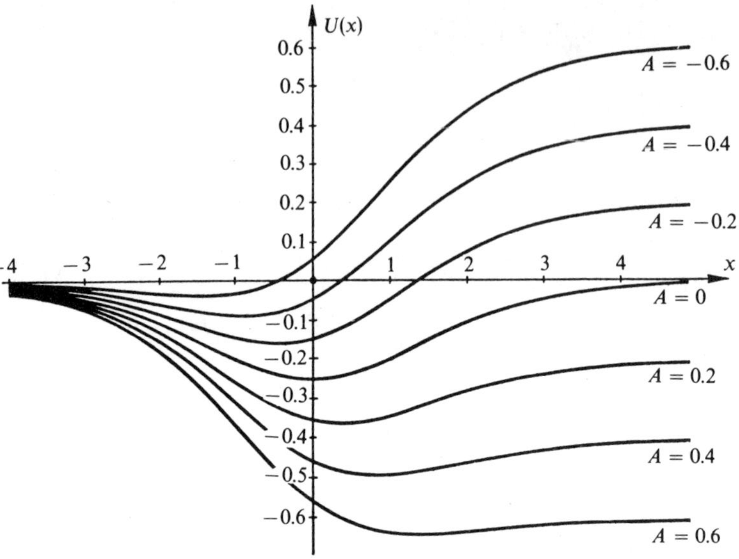

Fig. 9.3. Graphical representation of the potential $V(x)$ in (9.43) for $B = 1$ and different values of A

This integral can be evaluated in closed form, and it gives for the W.B.K. estimate of bound state energies:

$$\sqrt{|\varepsilon|} + \sqrt{|\varepsilon| - A} = -n + \tfrac{1}{2}\left(\sqrt{4B} - 1\right), \quad n = 0, 1, 2, \ldots . \tag{9.46}$$

The eigenvalue problem (9.43) can be solved in terms of hypergeometric functions (Sect. 6.9), and the exact values of the energy levels satify

$$\sqrt{|\varepsilon|} + \sqrt{|\varepsilon| - A} = -n + \tfrac{1}{2}\left(\sqrt{4B + 1} - 1\right), \quad n = 0, 1, 2, \ldots . \tag{9.47}$$

Comparing (9.46) and (9.47), we immediately see that the largest discrepancies between the exact results and those obtained with the W.B.K. approximation occur for the smallest possible values of $B(> |A|)$. In Fig. 9.4 we give for $A = 0$ the bound state energies as functions of B, for small values of this parameter.

As expected, the largest errors appear for the ground state, and they decrease as n increases. Notice, however, that the nth excited state appears when B crosses the value $B = n(n+1)$, whereas in the W.B.K. approximation, the threshold for the appearence of the nth excited state occurs at $B = n(n+1) + 1/4$, hence when this state is weakly bound, the error of the approximation can be large, independently of the value of n. This is because, even though the quantity A_1 introduced previously may be large, A_2 is small, and the W.B.K. method is not accurate.

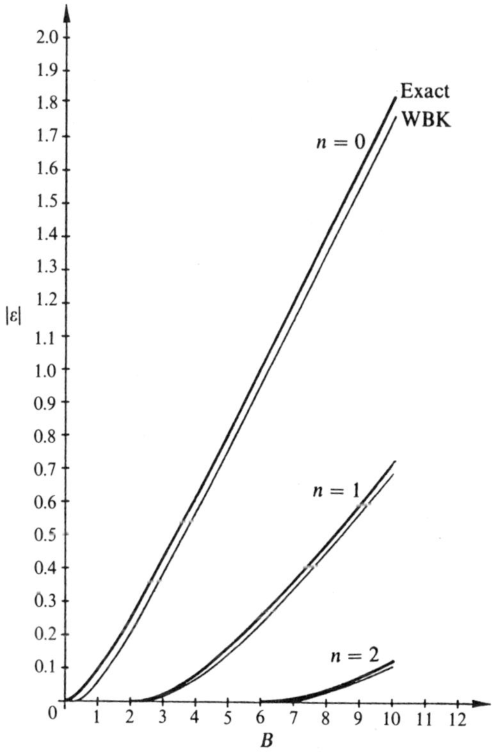

Fig. 9.4. Bound state energies for the potential in (9.43) for $A = 0$ and different values of B

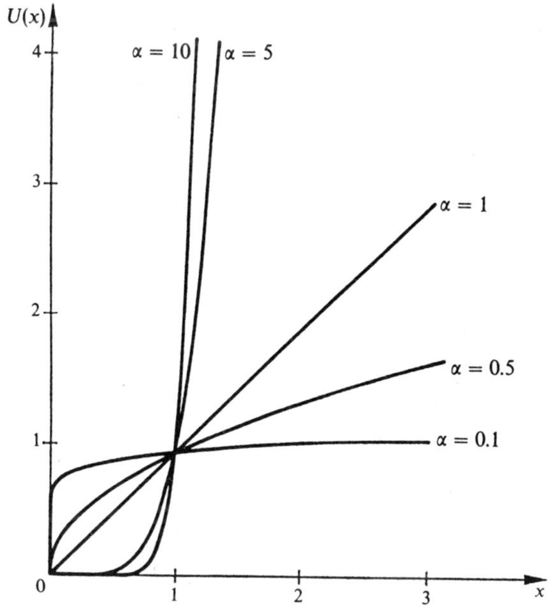

Fig. 9.5. Potential $U(x) = |x|^\alpha$ for $x > 0$ and different values of α

d) Finally we consider the family of potentials $U(x) = 2MV(x)/\hbar^2 = |x|^\alpha$ (in appropriate units), where α is any positive number. Some representative cases appear in Fig. 9.5. For $\alpha = 2$ this is the potential for a harmonic oscillator, and for $\alpha = 1$ we obtain the example studied in (b). When $\alpha \to \infty$, this potential becomes a square well, and, finally, when $\alpha \to 0$, $V(x) = 1$, with a continuous energy spectrum filling the interval $[1, \infty)$.

For an energy $\varepsilon = 2ME/\hbar^2 > 0$ the turning points are $x_\pm = \pm\varepsilon^{1/\alpha}$, and (9.34) becomes

$$2\int_0^{\varepsilon^{1/\alpha}} dx \sqrt{\varepsilon - x^\alpha} = \pi\left(n + \frac{1}{2}\right) . \tag{9.48}$$

Making the change of variables $y = x^\alpha/\varepsilon$ and taking into account

$$\int_0^1 dy\, y^{1/\alpha - 1}(1-y)^{1/2} = \frac{\sqrt{\pi}}{2}\frac{\Gamma(1/\alpha)}{\Gamma(1/\alpha + 3/2)} , \tag{9.49}$$

we immediately obtain

$$\varepsilon_{n,\mathrm{WBK}}(\alpha) = \left\{\sqrt{\pi}\left(n + \frac{1}{2}\right)\alpha\frac{\Gamma(3/2 + 1/\alpha)}{\Gamma(1/\alpha)}\right\}^{2\alpha/(2+\alpha)} , \tag{9.50}$$

$$n = 0, 1, 2, \ldots .$$

Since for $\alpha \to \infty$, $\Gamma(1/\alpha) \sim \alpha$, in this approximation the energy levels of the square well become

$$\varepsilon_{n,\mathrm{WBK}}(\infty) = \frac{\pi^2}{4}\left(n + \frac{1}{2}\right)^2 , \quad n = 0, 1, 2, \ldots \tag{9.51}$$

to be compared with the exact result

$$\varepsilon_n(\infty) = \frac{\pi^2}{4}(n+1)^2 , \quad n = 0, 1, 2, \ldots . \tag{9.52}$$

The relative error decreases therefore as $1/n$. For low values of n the error is particularly large, because $V(x)$ changes discontinuously at the turning points.

Table 9.3. Comparison of the energy levels in the family of potentials $|x|^\alpha$

α	$\varepsilon_{0,\text{WBK}}$	ε_0	$\varepsilon_{1,\text{WBK}}$	ε_1
10	0.737	1.299	4.596	5.098
9	0.748	1.263	4.515	4.935
8	0.762	1.226	4.419	4.756
7	0.779	1.186	4.303	4.559
6	0.801	1.145	4.161	4.339
5	0.829	1.102	3.983	4.089
4	0.867	1.060	3.752	3.800
3	0.921	1.023	3.441	3.451
2	1.000	1.000	3.000	3.000
1	1.115	1.019	2.320	2.338
0.9	1.128	1.025	2.231	2.251
0.8	1.140	1.032	2.136	2.157
0.7	1.151	1.040	2.035	2.058
0.6	1.161	1.049	1.927	1.950
0.5	1.167	1.058	1.812	1.833
0.4	1.170	1.066	1.687	1.707
0.3	1.164	1.073	1.551	1.568
0.2	1.147	1.074	1.401	1.413
0.1	1.107	1.062	1.229	1.236

In those limiting cases where the potential may possess an infinite barrier at some point, say at $x = a$, it seems reasonable to impose in the W.B.K. approximation that the wave function vanish at this point. Once we do this for the infinite square well, the W.B.K. approximation leads naturally to the exact result, and not to the one obtained by direct application of (9.34) (an equation not justifiable in such cases.)

For $\alpha \to 0$ and n fixed, it is easy to show that

$$\varepsilon_{n,\text{WBK}}(\alpha) \sim \alpha^\alpha \to 1\ , \tag{9.53}$$

namely, the energies of the lowest bound states tend to accumulate in a neighborhood of $\varepsilon = 1$ as expected. Letting n increase when α decreases we can obtain a limit anywhere in the interval $[1, \infty)$.

In Table 9.3 the results (9.50) are compared with those obtained by solving the Schrödinger equation numerically for the fundamental and first excited states. As expected, the energy ε_1 of the first excited state is better approximated in the W.B.K. method than the energy ε_0 of the ground state. On the other hand the approximation becomes worse when α becomes large. With the exact Dunham quantization formula (9.36), we can calculate systematically the successive corrections to (9.50) when α is even [BO 77]. Including the first correction explicitly, we obtain

$$\varepsilon_n(\alpha) = \varepsilon_{n,\text{WBK}}(\alpha)\left[1 + \frac{\alpha(\alpha-1)}{3\pi(\alpha+2)^2}\left(\cot\frac{\pi}{\alpha}\right)\frac{1}{(n+1/2)^2} + \mathrm{O}((n+1/2)^{-4})\right]. \tag{9.54}$$

This expansion is presumably asymptotic, and it provides a good number of correct significant digits for n sufficiently large. For instance, if $\alpha = 4$ and $n = 10$, the two terms included in (9.54) give $\varepsilon_{10}(4)$ with a precision of three parts in 10^7 while taking the next terms up to $(n+1/2)^{-12}$ we obtain a precision of 1 part in 10^{15}.

9.5 The Potential Barrier

The problem we want to address now with the W.B.K. method is the approximate computation of the transmission and reflection coefficients T, R, respectively of a particle of energy E in a potential barrier $V(x)$. We envisage the situation represented in Fig. 9.6, where the particles are incident from the left. Let x_- and x_+ be the turning points. In region III there is only a transmitted wave, and therefore in the W.B.K. approximation it may be expressed as

$$\psi_{\rm III}(x) = \frac{A}{\sqrt{p_2(x)}} \exp\left(\frac{\rm i}{\hbar}\int_{x_+}^{x} dx' p_2(x')\right) , \tag{9.55}$$

or equivalently

$$\begin{aligned}\psi_{\rm III}(x) &= \frac{A}{\sqrt{p_2(x)}} \cos\left(\frac{1}{\hbar}\int_{x_+}^{x} dx' p_2(x') - \frac{\pi}{4} + \frac{\pi}{4}\right) \\ &+ \frac{{\rm i}A}{\sqrt{p_2(x)}} \cos\left(\frac{1}{\hbar}\int_{x_+}^{x} dx' p_2(x') - \frac{\pi}{4} - \frac{\pi}{4}\right) .\end{aligned} \tag{9.56}$$

Using the connection formulae in Table 9.1 we obtain

$$\begin{aligned}\psi_{\rm II}(x) &= \frac{A{\rm e}^{-{\rm i}\pi/4}}{\sqrt{p_1(x)}} \exp\left[\frac{1}{\hbar}\int_{x}^{x_+} dx' p_1(x')\right] \\ &= \frac{A{\rm e}^{-{\rm i}\pi/4}}{\sqrt{p_1(x)}} \exp\left[\frac{1}{\hbar}\int_{x_-}^{x_+} dx' p_1(x')\right] \exp\left[-\frac{1}{\hbar}\int_{x_-}^{x} dx' p_1(x')\right] ,\end{aligned} \tag{9.57}$$

and applying these formulae once again,

$$\begin{aligned}\psi_{\rm I}(x) &= \frac{2A{\rm e}^{-{\rm i}\pi/4}}{\sqrt{p_2(x)}} \exp\left[\frac{1}{\hbar}\int_{x_-}^{x_+} dx' p_1(x')\right] \\ &\times \cos\left(\frac{1}{\hbar}\int_{x}^{x_-} dx' p_2(x') - \frac{\pi}{4}\right) .\end{aligned} \tag{9.58}$$

A direct computation allows us to rewrite the previous equation in the form

$$\begin{aligned}\psi_{\rm I}(x) &= \frac{A}{\sqrt{p_2(x)}}{\rm e}^{-{\rm i}\pi/2} \exp\left[\frac{1}{\hbar}\int_{x_-}^{x_+} dx' p_1(x')\right] \exp\left[-\frac{\rm i}{\hbar}\int_{x_-}^{x} dx' p_2(x')\right] \\ &+ \frac{A}{\sqrt{p_2(x)}} \exp\left[\frac{1}{\hbar}\int_{x_-}^{x_+} dx' p_1(x')\right] \exp\left[\frac{\rm i}{\hbar}\int_{x_-}^{x} dx' p_2(x')\right] ,\end{aligned} \tag{9.59}$$

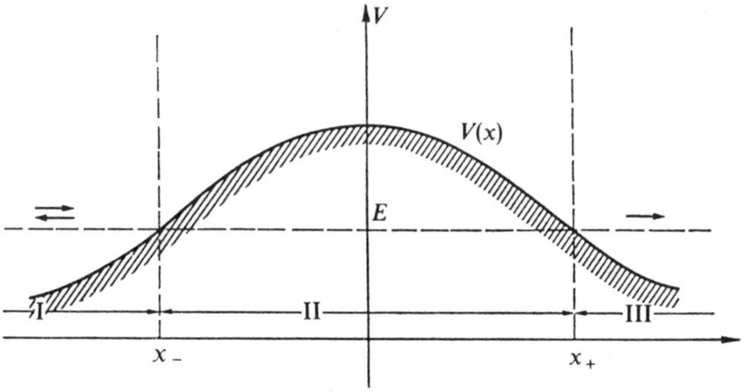

Fig. 9.6. Potential barrier $V(x)$ with a beam of particles of energy E approaching it from $-\infty$

where the second term represents a wave travelling along the positive Ox-axis and the first term is another travelling wave in the opposite direction. With the simplifying hypothesis $p_2(\infty) = p_2(-\infty)$ the comparison between (9.58) and (9.59) gives for the transmission and reflection coefficients the approximate expressions

$$\boxed{\begin{aligned} T_{\mathrm{WBK}} &= \exp\left[-\frac{2}{\hbar}\int_{x_-}^{x_+} dx\, p_1(x)\right] \equiv \mathrm{e}^{-2K} \\ R_{\mathrm{WBK}} &= 1\,. \end{aligned}} \tag{9.60}$$

It is well known (4.97) that unitarity requires $T + R = 1$. This relation is approximately satisfied in the W.B.K. method for its results are only acceptable if the characteristic actions in the problem, measured in units of $\hbar$, are large, and therefore, $T_{\mathrm{WBK}} \ll 1$.

Some authors prefer for the transmission and reflection coefficients the expressions

$$\boxed{T'_{\mathrm{WBK}} = (1 + \mathrm{e}^{2K})^{-1}\,, \quad R'_{\mathrm{WBK}} = (1 + \mathrm{e}^{2K})^{-1}\mathrm{e}^{2K}}\,, \tag{9.61}$$

satisfying $T'_{\mathrm{WBK}} + R'_{\mathrm{WBK}} = 1$. They can be obtained [KE 35], [FF 65], [BM 72], using more sophisticated connecting formulae than those given here. [In the formal derivation of (9.57) we have neglected an exponentially decreasing contribution, which after changing the reference point ($x_+ \to x_-$) would become exponentially increasing. This is the reason why we have obtained an essentially real function ψ_{III}. Had we not neglected this contribution, we would have obtained results similar to (9.61).] It often happens that the equations (9.61) have a range of validity larger than that of (9.60) because they can be used for those cases where the energy E is slightly smaller than the maximum of the barrier, namely when K is not very large. Obviously, for $K \gg 1$, the approximations (9.60) and (9.61) coincide for all practical purposes.

Before exhibiting some illustrative examples of the results obtained, we would like to add that the W.B.K. method has also been applied to situations where E is bigger than the barrier's maximum and thus, the turning points are complex [FF 65].

a) In this first example (purely academic, and devoid of any physical interest) we consider the barrier $V(x) = A^2 - B^2x^2$, and we assume that the energy E of the incident particle is smaller than the maximum A^2 of the barrier. The turning points are $x_\pm = \pm(A^2 - E)^{1/2}B^{-1}$ and, therefore $K = \pi\alpha/2$, where $\alpha = \sqrt{2M}(A^2 - E)/\hbar B$. The problem posed now admits an exact solution. If $b = \sqrt{2MB/\hbar}$, the Schrödinger equation is

$$\frac{d^2\psi(x)}{dx^2} + b^2\left(x^2 - \frac{\alpha}{b}\right)\psi(x) = 0\,. \tag{9.62}$$

Introducing the change of variables $y = -\mathrm{i}bx^2$ and defining the quantity $a = (1 + \mathrm{i}\alpha)/4$ we find that $\phi(y) \equiv \mathrm{e}^{y/2}\psi(x)$ must satisfy the equation

$$y\frac{d^2\phi(y)}{dy^2} + \left(\frac{1}{2} - y\right)\frac{d\phi(y)}{dy} - a\phi(y) = 0\,, \tag{9.63}$$

which is a confluent hypergeometric equation. The general solution of (9.62) turns out to be

$$\begin{aligned}\psi(x) = {} & A_1\mathrm{e}^{\mathrm{i}bx^2/2}M(a, 1/2, -\mathrm{i}bx^2) \\ & + A_2\mathrm{e}^{\mathrm{i}bx^2/2}\sqrt{b}x\mathrm{e}^{-\mathrm{i}\pi/4}M(a + 1/2, 3/2, -\mathrm{i}bx^2)\,.\end{aligned} \tag{9.64}$$

Taking into account the asymptotic behavior of confluent hypergeometric functions (Appendix A):

$$\begin{aligned}\psi(x) \underset{x\to\infty}{\sim} & \left\{A_1\frac{\Gamma(1/2)}{\Gamma(a)} + A_2\frac{\Gamma(3/2)}{\Gamma(1/2 + a)}\right\}\mathrm{e}^{-\mathrm{i}\pi(a-1/2)/2}(bx^2)^{a-1/2}\mathrm{e}^{-\mathrm{i}bx^2/2} \\ & + \left\{A_1\frac{\Gamma(1/2)}{\Gamma(1/2 - a)} - \mathrm{i}A_2\frac{\Gamma(3/2)}{\Gamma(1 - a)}\right\} \\ & \times \mathrm{e}^{-\mathrm{i}\pi a/2}(bx^2)^{-a}\mathrm{e}^{\mathrm{i}bx^2/2}\,.\end{aligned} \tag{9.65}$$

The first term could be considered as a travelling wave moving along the negative x-axis and therefore it should not appear in our problem. Consequently

$$A_1 = \frac{\Gamma(3/2)}{\Gamma(1/2 + a)}A\,, \quad A_2 = -\frac{\Gamma(1/2)}{\Gamma(a)}A\,, \tag{9.66}$$

where A is a normalization constant. We thus obtain

$$\begin{aligned}\psi(x) \underset{x\to\infty}{\sim} & \frac{A\Gamma(1/2)\Gamma(3/2)}{\Gamma(a)\Gamma(1/2 + a)}\left\{\frac{\Gamma(a)}{\Gamma(1/2 - a)} + \mathrm{i}\frac{\Gamma(1/2 + a)}{\Gamma(1 - a)}\right\} \\ & \times \mathrm{e}^{-\mathrm{i}\pi a/2}(bx^2)^{-a}\mathrm{e}^{\mathrm{i}bx^2/2}\,.\end{aligned} \tag{9.67}$$

Similarly we obtain the behavior of $\psi(x)$ when $x \to -\infty$:

$$\psi(x) \underset{x\to-\infty}{\sim} \frac{A\Gamma(1/2)\Gamma(3/2)}{\Gamma(a)\Gamma(1/2+a)}$$
$$\times \left\{ 2b^{a-1/2}|x|^{2a-1}\mathrm{e}^{-\mathrm{i}\pi a/2}\mathrm{e}^{\mathrm{i}\pi/4}\exp(-\mathrm{i}b|x|^2/2) \right.$$
$$+ \left[\frac{\Gamma(a)}{\Gamma(1/2-a)} - \mathrm{i}\frac{\Gamma(1/2+a)}{\Gamma(1-a)} \right]$$
$$\left. \times\, b^{-a}\mathrm{e}^{-\mathrm{i}\pi a/2}|x|^{-2a}\exp(\mathrm{i}b|x|^2/2) \right\} . \qquad (9.68)$$

Taking into account the asymptotic expressions (9.67) and (9.68) and that $2a = 1 - 2a^*$ we obtain, in terms of the quantity α, that the transmission coefficient computed as a ratio of fluxes at points located symmetrically with respect to the barrier is

$$T = \frac{1}{4}\left| \frac{\Gamma(1/4+\mathrm{i}\alpha/4)}{\Gamma(1/4-\mathrm{i}\alpha/4)} + \mathrm{i}\frac{\Gamma(3/4+\mathrm{i}\alpha/4)}{\Gamma(3/4-\mathrm{i}\alpha/4)} \right|^2 . \qquad (9.69)$$

Since $\Gamma^*(z) = \Gamma(z^*)$ and moreover [AS 64],

$$\Gamma(1/4+\mathrm{i}y)\Gamma(3/4-\mathrm{i}y) = \pi\sqrt{2}[\cosh\pi y + \mathrm{i}\sinh\pi y]^{-1} \qquad (9.70)$$

a direct computation yields

$$T = (1+\mathrm{e}^{\pi\alpha})^{-1} , \qquad (9.71)$$

which agrees for all values of $E \le A^2$ with T'_{WBK}, whereas it agrees with T_{WBK} only in the limit of large values of α.

b) In this second example (as academic as the previous one) we consider the potential barrier $V(x) = V_0(1-|x|/a)$, $V_0 > 0$, $a > 0$. When $E < V_0$, the turning points are $x_\pm = \pm a(1 - E/V_0)$. Defining $\alpha = (2MV_0a^2/\hbar^2)^{1/3}(1 - E/V_0)$ we find after an elementary integration, $K = (4/3)\alpha^{3/2}$. This problem can be solved exactly using the Airy functions [AS 64] and their derivatives. The transmission coefficient is

$$T = \{1 + \pi^2[\mathrm{Ai}(\alpha)\mathrm{Ai}'(\alpha) + \mathrm{Bi}(\alpha)\mathrm{Bi}'(\alpha)]^2\}^{-1} . \qquad (9.72)$$

Table 9.4. Comparison of T, T_{WBK} and T'_{WBK} for different values of the energy

α	T	T_{WBK}	T'_{WBK}
0	7.500×10^{-1}	1.000	5.000×10^{-1}
0.2	6.202×10^{-1}	7.878×10^{-1}	4.407×10^{-1}
0.4	4.575×10^{-1}	5.094×10^{-1}	3.375×10^{-1}
0.6	2.919×10^{-1}	2.896×10^{-1}	2.245×10^{-1}
0.8	1.600×10^{-1}	1.484×10^{-1}	1.292×10^{-1}
1.0	7.671×10^{-2}	6.948×10^{-2}	6.497×10^{-2}
1.5	8.034×10^{-3}	7.454×10^{-3}	7.399×10^{-3}
2.0	5.539×10^{-4}	5.301×10^{-4}	5.298×10^{-4}
2.5	2.717×10^{-5}	2.643×10^{-5}	2.643×10^{-5}
3.0	9.787×10^{-7}	9.599×10^{-7}	9.599×10^{-7}

In Table 9.4 we compare the exact result T with the approximations T_{WBK} and T'_{WBK}. We see clearly that in this case T'_{WBK} is not a better approximation to the exact result T that T_{WBK}.

c) In the two previous examples we have elaborated on simple but unrealistic cases, because for these barriers the particle velocity tends to infinity when $x \to \pm\infty$. This does not happen with the potential

$$V(x) = V_0 \cosh^{-2} ax\,, \quad V_0 > 0\,, \quad a > 0\,. \tag{9.73}$$

If E is the particle's energy and $0 < E < V_0$, the turnig points $x_\pm$ are given by $\cosh ax_\pm = \pm\sqrt{V_0/E}$. Defining $U_0 \equiv 2MV_0/\hbar^2$ and $k = \sqrt{2ME}/\hbar$ one obtains

$$K = 2\int_0^{x_+} dx\,[-k^2 + U_0 \cosh^{-2} ax]^{1/2}\,, \tag{9.74}$$

which after a change of variables $y = \cosh^2 ax$ becomes

$$K = \frac{1}{a}\int_1^{\sqrt{U_0/k^2}} dy\,\frac{1}{y}\left[\frac{U_0 - k^2 y}{y-1}\right]^{1/2}\,, \tag{9.75}$$

so that

$$\begin{aligned} T_{\text{WBK}} &= \exp\left(-\frac{2\pi}{a}[\sqrt{U_0} - k]\right)\,, \\ T'_{\text{WBK}} &= \left[1 + \exp\left(\frac{2\pi}{a}[\sqrt{U_0} - k]\right)\right]^{-1}\,. \end{aligned} \tag{9.76}$$

On the other hand the Schrödinger equation admits an exact solution in this case. Indeed, if in the equation

$$\frac{d^2\psi(x)}{dx^2} + [k^2 - U_0 \cosh^{-2} ax]\psi(x) = 0 \tag{9.77}$$

we implement the change $\xi = \tanh ax$ the function $\phi(\xi) \equiv \psi(x)$ satisfies

$$(1-\xi^2)\frac{d^2\phi(\xi)}{d\xi^2} - 2\xi\frac{d\phi(\xi)}{d\xi} - \frac{U_0}{a^2}\phi(\xi) + \frac{k^2}{a^2(1-\xi^2)}\phi(\xi) = 0\,. \tag{9.78}$$

If $4U_0 \le a^2$ it is convenient to introduce

$$s = \frac{1}{2}\left[-1 + \sqrt{1 - \frac{4U_0}{a^2}}\right]\,, \tag{9.79}$$

so that $s(s+1) = -U_0/a^2$. With the change $\phi(\xi) = (1-\xi^2)^{-ik/2a}\varphi(\xi)$, (9.78) becomes

$$\begin{aligned} &(1-\xi^2)\frac{d^2\varphi(\xi)}{d\xi^2} - 2\xi\left(1 - \frac{ik}{a}\right)\frac{d\varphi(\xi)}{d\xi} \\ &\quad + \left[\frac{ik}{a} + s(s+1) + \frac{k^2}{a^2}\right]\varphi(\xi) = 0\,. \end{aligned} \tag{9.80}$$

If, finally we write $\xi = 1 - 2t$, then $y(t) \equiv \varphi(\xi)$ must be a solution to

$$t(1-t)\frac{d^2y(t)}{dt^2} + \left(1 - \frac{ik}{a}\right)(1-2t)\frac{dy(t)}{dt} + \left[\frac{k^2}{a^2} + s(s+1) + \frac{ik}{a}\right] y(t) = 0 , \tag{9.81}$$

which is a hypergeometric equation. Taking the solution of (9.81) with correct behavior of $\psi(x)$ when $x \to +\infty$ we obtain

$$\psi(x) = (1-\xi^2)^{-i\bar{k}/2} F(-i\bar{k} - s, -i\bar{k} + s + 1, -i\bar{k}; (1-\xi)/2) \tag{9.82}$$

with $\bar{k} \equiv k/a$. Indeed, for $x \to +\infty$ we have ξ behaving like $1 - 2\exp(-2ax)$ and, therefore $(1-\xi^2) \sim 4\exp(-2ax)$ implying that $\psi(x) \sim 4^{-i\bar{k}/2}\exp(ikx)$, which is the desired behavior. Using the properties of hypergeometric functions, we can write (9.82) as

$$\begin{aligned}\psi(x) = (1-\xi^2)^{-i\bar{k}/2} \Big\{ &\frac{\Gamma(1-i\bar{k})\Gamma(-i\bar{k})}{\Gamma(1+s)\Gamma(-s)} \\ &\times F(-i\bar{k} - s, -i\bar{k} + s + 1, 1 - i\bar{k}; (1+\xi)/2) \\ &+ [(1+\xi)/2]^{i\bar{k}} \frac{\Gamma(1-i\bar{k})\Gamma(-i\bar{k})}{\Gamma(-i\bar{k}-s)\Gamma(-i\bar{k}+s+1)} \\ &\times F(1+s, -s, i\bar{k}; (1+\xi)/2) \Big\} .\end{aligned} \tag{9.83}$$

When $x \to -\infty$ one has $\xi + 1 \sim 2\exp(2ax)$, and in this limit

$$\begin{aligned}\psi(x) \sim 4^{-i\bar{k}/2} \Big\{ &\frac{\Gamma(1-i\bar{k})\Gamma(-i\bar{k})}{\Gamma(1+s)\Gamma(-s)} e^{-ikx} \\ &+ \frac{\Gamma(1-i\bar{k})\Gamma(-i\bar{k})}{\Gamma(-i\bar{k}-s)\Gamma(-i\bar{k}+s+1)} e^{ikx} \Big\} .\end{aligned} \tag{9.84}$$

The exact transmission amplitude is therefore

$$\sigma = \frac{\Gamma(i\bar{k}-s)\Gamma(-i\bar{k}+s+1)}{\Gamma(1-i\bar{k})\Gamma(-i\bar{k})} . \tag{9.85}$$

From the analogous expression for the reflection coefficient derivable from (9.84) we easily see that the reflection coefficient vanishes whenever s is an integer. In particular, when $s = 1$, the associated attractive potential coincides with the soliton potential discussed in Sect. 4.9, completely transparent for all energies. Using (9.85) together with $\Gamma^*(z) = \Gamma(z^*)$, $\Gamma(-iy)\Gamma(+iy) = \pi/y\sinh(\pi y)$ if $y \neq 0$, and $\Gamma(z)\Gamma(1-z) = \pi\csc\pi z$ if $0 < \mathrm{Re}\, z < 1$ a straightforward calculation yields the exact transmisssion coefficient

$$T = \frac{\sinh^2(\pi k/a)}{\sinh^2(\pi k/a) + \cos^2\left[(\pi/2)\sqrt{1-(4U_0/a^2)}\right]} , \quad 4U_0 \leq a^2 . \tag{9.86}$$

Similarly one obtains:

$$T = \frac{\sinh^2(\pi k/a)}{\sinh^2(\pi k/a) + \cosh^2\left[(\pi/2)\sqrt{(4U_0/a^2) - 1}\right]}\,, \quad 4U_0 \geq a^2\,. \tag{9.87}$$

For $k \gg a$ and $\sqrt{U_0} \gg a$, T approaches T'_{WBK}, while for T_{WBK} to be a good approximation requires $\sqrt{U_0} - k \gg a$ as well; i.e., that the particle energy should be small compared to the height of the barrier.

9.6 The Miller-Good Method

The W.B.K. method explained so far allowed us to successfully understand the Sommerfeld-Wilson-Ishiwara quantization rules and it also gave us some approximate formulae for the bound state energies of a well with two turning points as well as the transmission coefficient of a potential barrier. The approximation considered however, has the disadvantage that the wave functions its produces become infinite (and therefore meaningless) at the turning points.

Langer [LA 37] tried to remedy this difficulty, and later, *Miller* and *Good* [MG 53] developed a method which not only reproduces the results of (9.34) and (9.61), but also gives approximate wave functions for all values of x. Following this work, we will derive now the energies and eigenfunctions of the bound states in a potential $V(x)$ with only two turning points. The reader interested in the study of transmission by potential barriers using this method should look up the original article or [BM 72]. We only mention that with this procedure one obtains T'_{WBK} for the transmission coefficient.

The Schrödinger equation for the posed problem is

$$\begin{aligned} &\frac{d^2\psi(x)}{dx^2} + \frac{1}{\hbar^2}p^2(x)\psi(x) = 0 \\ &p^2(x) \equiv 2M[E - V(x)]\,. \end{aligned} \tag{9.88}$$

Notice that $p^2(x) \geq 0$ for $x_- \leq x \leq x_+$ where $x_\pm$ are the turning points, and $p^2(x) < 0$ elsewhere. Define a new variable $\phi \equiv \phi(x;\hbar)$ and a function $P^2(\phi)$ such that the equation

$$\frac{d^2 f(\phi)}{d\phi^2} + \frac{1}{\hbar^2}P^2(\phi)f(\phi) = 0 \tag{9.89}$$

can be solved exactly. A simple computation shows that

$$\psi(x) = (\phi')^{-1/2} f(\phi(x;\hbar)) \tag{9.90}$$

(where the prime indicates differentiation with respect to x) will be a solution to (9.88) if ϕ satisfies

$$P^2(\phi)(\phi')^2 - \hbar^2(\phi')^{1/2}\frac{d^2}{dx^2}(\phi')^{-1/2} = p^2 . \tag{9.91}$$

The similarity between (9.91) and (9.14) is obvious, the latter corresponding to the choice $P^2 \equiv \phi$ appropriate for the asymptotic analysis around some given turning point. Analogously, the W.B.K. approximation to the wave function is obtained after we neglect the order $\hbar^2$ term in (9.91) and taking $P^2 \equiv 1$. The Miller-Good method for attractive potentials with only two simple turning points consists of replacing (9.91) by the approximate expression $P(\phi)\phi' \simeq p$, and to take for $P(\phi)$ the value for the momentum in a harmonic potential: $P^2(\phi) = \hbar^2(\alpha - \phi^2)$, with α chosen conveniently. In the Miller-Good approximation to (9.91) we have

$$\int_{\phi_0}^{\phi(x)} d\sigma P(\sigma) = \int_{x_0}^{x} d\xi\, p(\xi) . \tag{9.92}$$

The constants ϕ_0 and x_0 are determined so that the wave function (9.90) in this approximation $\psi_{\rm MG}$ is finite at every point. The accuracy of the approximation relies on the smallness of the terms neglected in (9.91) with respect to p^2.

According to the results obtained in the study of the harmonic oscillator, the most general solution to (9.89) for harmonic P^2 will be

$$f(\phi) = A_1 \mathrm{e}^{-\phi^2/2} M(a, \tfrac{1}{2}, \phi^2) + A_2 \phi \mathrm{e}^{-\phi^2/2} M(a + \tfrac{1}{2}, \tfrac{3}{2}, \phi^2) , \tag{9.93}$$

with $a = (1 - \alpha)/4$.

Let us now proceed to choose the constants x_0, ϕ_0, and α in such a way that the wave function $\psi_{\rm MG}(x)$ remains continuous at the turning points. The continuity at $x = x_-$ is achieved if $\phi_0 = -\sqrt{\alpha}$ and $x_0 = x_-$ since then $\phi'(x)$ is regular at $x = x_-$. The same will occur at $x = x_+$ if we choose α so that for $x = x_+$, $\phi = \sqrt{\alpha}$; explicitly

$$\int_{x_-}^{x_+} d\xi\, p(\xi) = \hbar \int_{-\sqrt{\alpha}}^{+\sqrt{\alpha}} d\sigma\, (\alpha - \sigma^2)^{1/2} = \tfrac{\pi}{2}\hbar\alpha . \tag{9.94}$$

If we impose the condition that the wave function corresponding to a bound state should be normalizable we obtain as with the harmonic oscillator that $\alpha = 2n+1$ with $n = 0, 1, 2, \ldots$. Hence the Miller-Good method gives the following bound state energies:

$$\int_{x_-}^{x_+} dx \sqrt{2M[E - V(x)]} = \left(n + \tfrac{1}{2}\right)\pi\hbar , \quad n = 0, 1, 2, \ldots \tag{9.95}$$

that is to say, they agree with those obtained in the W.B.K. approximation. The associated wave functions are

$$\psi_{n,\rm MG}(x) = \frac{A_n}{\sqrt{\phi'(x)}} \mathrm{e}^{-\phi^2(x)/2} H_n(\phi(x)) , \tag{9.96}$$

where A_n is a normalization constant, $H_n(z)$ are the Hermite polynomials and $\phi = \phi(x)$ is the real function (depending of the level considered) defined by

$$\hbar \int_{-\sqrt{2n+1}}^{\phi(x)} d\sigma \, (2n+1-\sigma^2)^{1/2} = \int_{x_-}^{x} d\xi \, p(\xi) \,, \tag{9.97}$$

and from this

$$\phi'(x) = \frac{p(x)}{\hbar[2n+1-\phi^2(x)]^{1/2}} \,. \tag{9.98}$$

Once $\psi_{\text{MG}}(x)$ is determined, it is easy to compute the value of A_n by numerical integration. *Miller* [MI 54] has shown that to a very good approximation

$$|A_n| \simeq \pi^{1/4} \left\{ 2^n n! \hbar \int_{x_-}^{x_+} \frac{d\xi}{p(\xi)} \right\}^{-1/2} \,. \tag{9.99}$$

Finally it is not surprising that for the harmonic oscillator the W.B.K. method yields the exact values for the energy levels: the Miller-Good approximation is exact in this case.

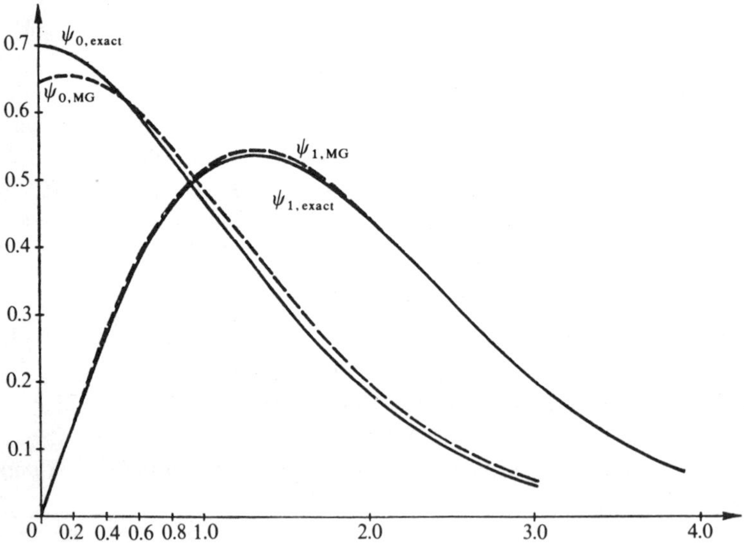

Fig. 9.7. Comparison between the exact and Miller-Good wave functions for the potential $V = |x|$

As an application of this method, let us consider the potential $V(x) = a|x|$, already discussed as an example of the applications of the W.B.K. method. Due to the symmetry in the problem, it suffices to study the wave function for $x \geq 0$. The bound state energies, which as mentioned, coincide with those of the W.B.K. approximation, are determined by (9.95). From (9.97), a straightforward computation yields

$$\begin{aligned}
&\phi(2n+1-\phi^2)^{1/2}+(2n+1)\arcsin\frac{\phi}{\sqrt{2n+1}}\\
&\quad=\frac{\pi}{2}(2n+1)-\frac{4\alpha}{3}(E_{n,\mathrm{WBK}}-ax)^{3/2}\ ,\quad 0\le x\le E_{n,\mathrm{WBK}}/a\ ,\\
&-\phi(\phi^2-2n-1)^{1/2}+(2n+1)\ln\frac{\phi+\sqrt{\phi^2-2n-1}}{\sqrt{2n+1}}\\
&\quad=-\frac{4\alpha}{3}(ax-E_{n,\mathrm{WBK}})^{3/2}\ ,\quad x\ge E_{n,\mathrm{WBK}}/a\ ,
\end{aligned}\tag{9.100}$$

where $\alpha\equiv\sqrt{2M}/\hbar a$. From this we extract $\phi(x)$ and (9.96) provides the wave function $\psi_{n,\mathrm{MG}}(x)$. Finally, it can be shown that the exact wave functions are

$$\psi_n=N_n\mathrm{Ai}[\alpha^{2/3}(ax-E_n)]\ ,\tag{9.101}$$

where N_n is a normalization constant and E_n is the exact energy of the nth bound state. In Fig. 9.7 the Miller-Good and exact wave functions for the ground state ψ_0 and first excited state ψ_1 are compared in units $\alpha=a=1$. Both wave functions have been normalized to one numerically.

9.7 Transmission by Double Potential Barriers

The transmission problem can also be treated in the case of a finite but arbitrary number of turning points, using the connection formulae derived before. As an example we study the potential represented in Fig. 9.8. Assume that in the region V the wave function is purely outgoing:

$$\psi_{\mathrm{V}}(x)=\frac{A}{\sqrt{p_2(x)}}\exp\left[\frac{\mathrm{i}}{\hbar}\int_d^x dx'p_2(x')\right]\ .\tag{9.102}$$

Using the techniques developed in the study of the simple barrier we immediately obtain

$$\begin{aligned}
\psi_{\mathrm{III}}(x)=&\frac{A}{\sqrt{p_2(x)}}\mathrm{e}^{-\mathrm{i}\pi/2}\exp\left[\frac{1}{\hbar}\int_c^d dx'p_1(x')\right]\exp\left[-\frac{\mathrm{i}}{\hbar}\int_c^x dx'p_2(x')\right]\\
&+\frac{A}{\sqrt{p_2(x)}}\exp\left[\frac{1}{\hbar}\int_c^d dx'p_1(x')\right]\exp\left[\frac{\mathrm{i}}{\hbar}\int_c^x dx'p_2(x')\right]\ .
\end{aligned}\tag{9.103}$$

If the integrals over the interval (x,c) are written in terms of integrals extended over (b,x) and the analogous expressions to (9.102, 103) are used to go from region III to region I as well as their complex conjugates, we obtain

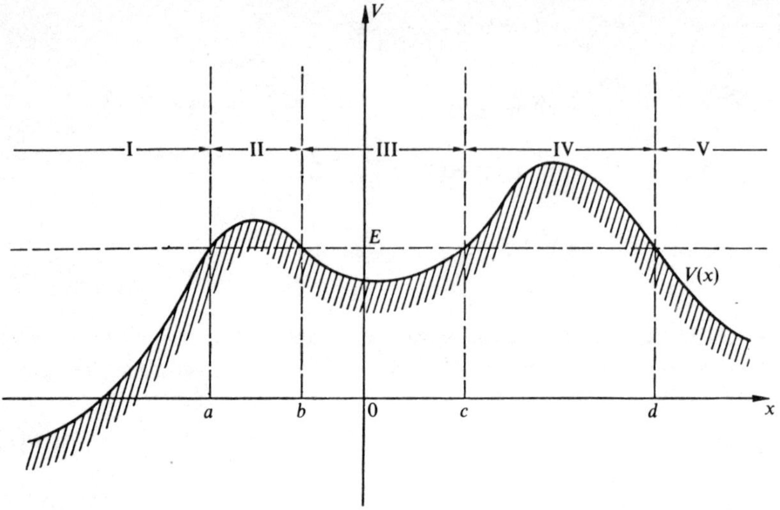

Fig. 9.8. Double potential barrier, and an energy E with four turning points

$$\psi_{\mathrm{I}}(x) = -\frac{2\mathrm{i}A}{\sqrt{p_2(x)}}\mathrm{e}^{K_{\mathrm{II}}+K_{\mathrm{IV}}}\cos L_{\mathrm{III}}\exp\left[-\frac{\mathrm{i}}{\hbar}\int_a^x dx' p_2(x')\right] + \frac{2A}{\sqrt{p_2(x)}}\mathrm{e}^{K_{\mathrm{II}}+K_{\mathrm{IV}}}\cos L_{\mathrm{III}}\exp\left[\frac{\mathrm{i}}{\hbar}\int_a^x dx' p_2(x')\right], \tag{9.104}$$

where

$$K_{\mathrm{II}} \equiv \frac{1}{\hbar}\int_a^b dx' p_1(x'), \quad K_{\mathrm{IV}} \equiv \frac{1}{\hbar}\int_c^d dx' p_1(x'), \quad L_{\mathrm{III}} \equiv \frac{1}{\hbar}\int_b^c dx' p_2(x'). \tag{9.105}$$

Thus, the transmission coefficient, in the W.B.K. approximation becomes

$$T_{\mathrm{WBK}} = \frac{\exp[-2(K_{\mathrm{II}}+K_{\mathrm{IV}})]}{4\cos^2 L_{\mathrm{III}}} \tag{9.106}$$

It is obvious that (9.106) does not hold if $\cos L_{\mathrm{III}} \approx 0$. This condition corresponds to bound states in the W.B.K. approximation, for the well between the two barriers. At these special energies it is to be expected that the particle spends a large amount of time (when compared with the collision time) captured by the well, thus producing a resonance phenomenon. A detailed discussion of this situation as well as the derivation of formulae more general than (9.106), can be found in [FD 70].

9.8 Potential Wells: Several Turning Points

To illustrate what happens when one tries to compute in the W.B.K. approximation the bound state energies in a one-dimensional potential well with more than two turning points, let us consider the situation represented in Fig. 9.9. In region I and far for the turning point the W.B.K. wave function is

$$\psi_{\mathrm{I}}(x) = \frac{A}{\sqrt{p_1(x)}} \exp\left[-\frac{1}{\hbar}\int_x^a dx' p_1(x')\right] , \tag{9.107}$$

and using the connection formulae

$$\psi_{\mathrm{II}}(x) = \frac{2A}{\sqrt{p_2(x)}} \cos\left(\frac{1}{\hbar}\int_a^x dx' p_2(x') - \frac{\pi}{4}\right) , \tag{9.108}$$

which can be written as

$$\begin{aligned}\psi_{\mathrm{II}}(x) = \frac{2A}{\sqrt{p_2(x)}} &\left\{\sin\left(\frac{1}{\hbar}\int_a^b dx\, p_2(x)\right) \cos\left(\frac{1}{\hbar}\int_x^b dx' p_2(x') - \frac{\pi}{4}\right)\right.\\ &\left. - \cos\left(\frac{1}{\hbar}\int_a^b dx\, p_2(x)\right) \cos\left(\frac{1}{\hbar}\int_x^b dx' p_2(x') - \frac{\pi}{4} - \frac{\pi}{2}\right)\right\} .\end{aligned} \tag{9.109}$$

Furthermore, in region V

$$\psi_{\mathrm{V}}(x) = \frac{B}{\sqrt{p_1(x)}} \exp\left[-\frac{1}{\hbar}\int_d^x dx' p_1(x')\right] , \tag{9.110}$$

and using the connection formulae once again

$$\begin{aligned}\psi_{\mathrm{IV}}(x) = \frac{2B}{\sqrt{p_2(x)}} &\left\{\sin\left(\frac{1}{\hbar}\int_c^d dx\, p_2(x)\right) \cos\left(\frac{1}{\hbar}\int_c^x dx' p_2(x') - \frac{\pi}{4}\right)\right.\\ &\left. - \cos\left(\frac{1}{\hbar}\int_c^d dx\, p_2(x)\right) \cos\left(\frac{1}{\hbar}\int_c^x dx' p_2(x') - \frac{\pi}{4} - \frac{\pi}{2}\right)\right\} .\end{aligned} \tag{9.111}$$

The difficulty arises when we try to continue $\psi_{\mathrm{II}}(x)$ and $\psi_{\mathrm{IV}}(x)$ to region III, because the connection formulae are mono-directional. Neglecting this restriction, we easily obtain

$$\begin{aligned}\psi_{\mathrm{III}}(x) = \frac{A}{\sqrt{p_1(x)}} &\left[\sin\left(\frac{1}{\hbar}\int_a^b dx\, p_2(x)\right) \exp\left[-\frac{1}{\hbar}\int_b^x dx' p_1(x')\right]\right.\\ &\left. +2\cos\left(\frac{1}{\hbar}\int_a^b dx\, p_2(x)\right) \exp\left[\frac{1}{\hbar}\int_b^x dx' p_1(x')\right]\right] ,\end{aligned} \tag{9.112}$$

as well as

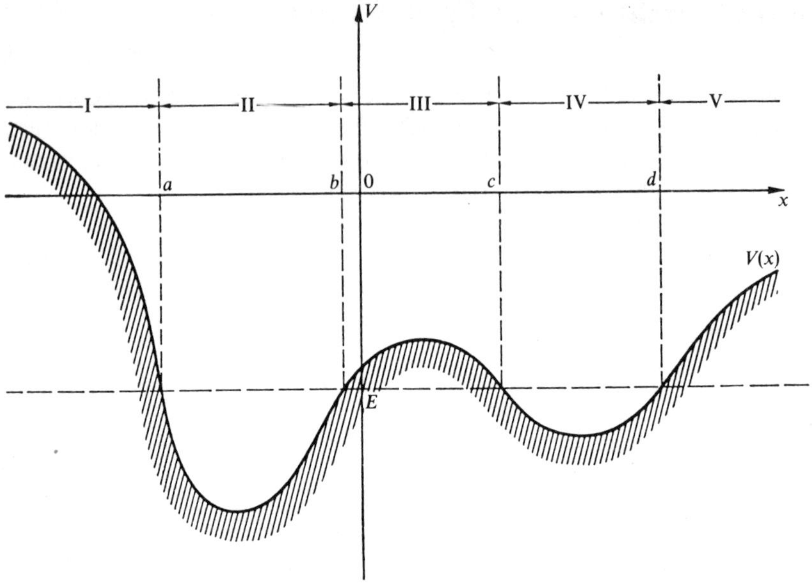

Fig. 9.9. Potential well $V(x)$ and energy E with four turning points

$$\psi_{\rm III}(x) = \frac{B}{\sqrt{p_1(x)}}\left[\sin\left(\frac{1}{\hbar}\int_c^d dx\, p_2(x)\right)\exp\left[-\frac{1}{\hbar}\int_x^c dx' p_1(x')\right]\right.$$
$$\left. +2\cos\left(\frac{1}{\hbar}\int_c^d dx\, p_2(x)\right)\exp\left[\frac{1}{\hbar}\int_x^c dx' p_1(x')\right]\right] , \qquad (9.113)$$

The equality of both expressions implies that the bound state energies are solutions to the equation

$$\cot\left(\frac{1}{\hbar}\int_a^b dx\sqrt{2M[E-V(x)]}\right)\cot\left(\frac{1}{\hbar}\int_c^d dx\sqrt{2M[E-V(x)]}\right)$$
$$= \frac{1}{4}\exp\left(-\frac{2}{\hbar}\int_b^c dx\sqrt{2M[V(x)-E]}\right) . \qquad (9.114)$$

Notice that when the potential barrier between both wells is very large, the right-hand side is practically zero and, therefore, the bound states of the system coincide with those of each well. This is very reasonable physically, since the barrier penetration probability will be very small.

For symmetric potentials $a = -d$, $b = -c$ and the previous equation reduces to

$$\cot\left(\frac{1}{\hbar}\int_c^d dx\sqrt{2M[E-V(x)]}\right)$$
$$= \pm\frac{1}{2}\exp\left(-\frac{1}{\hbar}\int_{-c}^c dx\sqrt{2M[V(x)-E]}\right) . \qquad (9.115)$$

We point out again that (9.114) and (9.115) have been obtained applying the connection formulae in an unjustified sense. However the analysis of many practical cases assures us that their numerical accuracy is similar to the one obtained when there are only two turning points, in other words they give a good approximation to the exact results whenever the actions typical in the problem are large in units of $\hbar$.

In the limit of an infinitely high or wide barrier, the second term in (9.115) would vanish, and this formula would give rise to a spectrum of energies $E_{n,\mathrm{WKB}}$ doubly degenerate. The presence of the second term produces in general the lifting of this degeneracy, and around the original level, we obtain two simple levels with energy splitting ΔE_n exponentially small in (natural action/$\hbar$). A naive use of (9.115) gives

$$\Delta E_{n,\mathrm{WBK}} = \frac{\hbar}{M}\left[\int_c^d dx\, p^{-1}(x)\right]^{-1} \exp\left(-\frac{1}{\hbar}\int_{-c}^{c} dx\, |p(x)|\right) , \tag{9.116}$$

with $p(x) \equiv (2M(E_{n,\mathrm{WKB}} - V(x)))^{1/2}$.

This splitting, which originated by tunneling through the intermediate barrier in double wells, has received considerable attention in recent years both in the functional integral formulation (*instanton* contributions [CO 77]) as well as with more rigorous methods [HA 80], [SI 84a]. A remarkable result due to *Simon* [SI 84a] establishes that if $V(x)$ is a symmetric double potential well with vanishing minima at x_-, x_+, then, under very general conditions, the exact splitting ΔE_0 for the fundamental level satisfies

$$\ln \Delta E_0 \underset{\hbar\downarrow 0}{\sim} -\frac{1}{\hbar}\int_{x_-}^{x_+} dx\sqrt{2MV(x)} , \tag{9.117}$$

and it coincides with the leading term of (9.116).

A typical, and particularly interesting example is the double anharmonic well:

$$H = -\frac{d^2}{dx^2} - x^2 + \lambda x^4 , \quad \lambda > 0 \tag{9.118}$$

with $\lambda \ll 1$ (corresponding to $\hbar \to 0$). The application of (9.117) to this potential yields

$$\ln \Delta E_0 \underset{\lambda\downarrow 0}{\sim} -(\sqrt{2}/3\lambda) . \tag{9.119}$$

Rigorous results for any level and more detailed than (9.119) for the ground state can be found in [HA 80]. For a recent discussion of this anharmonic oscillator in the spirit of the W.B.K. and Miller-Good approximations see [BH 85].

9.9 Central Potentials

For central potentials $V(r)$ the reduced radial wave function $u(r)$ satisfies the equation

$$\frac{d^2u(r)}{dr^2} + \left\{\frac{2M}{\hbar^2}[E - V(r)] - \frac{l(l+1)}{r^2}\right\} u(r) = 0\,, \quad 0 < r < \infty\,. \tag{9.120}$$

In principle we can think of (9.120) as a one-dimensional problem with an effective potential

$$V_{\text{eff}}(r) = \begin{cases} +\infty\,, & r < 0 \\ V(r) + \dfrac{\hbar^2}{2M}\dfrac{l(l+1)}{r^2}\,, & 0 < r < \infty\,. \end{cases} \tag{9.121}$$

If $V_{\text{eff}}(r)$ has two turning points r_0 and r_1 for a given energy E, we might estimate the bound states energies using (9.34) so that in the W.B.K. approximation they would be solutions to

$$\int_{r_0}^{r_1} dr \left[\frac{2ME}{\hbar^2} - \frac{2M}{\hbar^2}V(r) - \frac{l(l+1)}{r^2}\right]^{1/2} = \left(n + \frac{1}{2}\right)\pi\,, \tag{9.122}$$
$$n = 0, 1, 2, \ldots\,.$$

For s waves, if $r_0 = 0$, and following the arguments used for the square well, it would be more natural to impose the condition $u_{\text{WBK}}(0) = 0$ rather than considering the origin as a turning point. Nevertheless, we will not insist upon this point due to the modification suggested by *Kramers* [KR 26], and elaborated by *Kemble* [KE 37] and *Langer* [LA 37].

Langer implemented in (9.120) the change of independent and dependent variables $r = \mathrm{e}^x$, $u(r) = \mathrm{e}^{x/2}\varphi(x)$. The new function $\varphi(x)$ satisfies

$$\frac{d^2\varphi(x)}{dx^2} + \left\{\frac{2M}{\hbar^2}[E - V(\mathrm{e}^x)]\mathrm{e}^{2x} - (l + 1/2)^2\right\}\varphi(x) = 0\,, \tag{9.123}$$
$$-\infty < x < +\infty\,.$$

He then applied the W.B.K. approximation to this equation, and obtained for the bound state energies:

$$\boxed{\begin{aligned} &\int_{r_0'}^{r_1'} dr \left[\frac{2ME}{\hbar^2} - \frac{2M}{\hbar^2}V(r) - \frac{(l+1/2)^2}{r^2}\right]^{1/2} \\ &= \left(n + \frac{1}{2}\right)\pi\,, \quad n = 0, 1, 2, \ldots\,, \end{aligned}} \tag{9.124}$$

where r_0' and r_1' are the two roots of the integrand (assuming that there are only two). We will next show that in general (9.124) produces better results than (9.122) even when this might not be true for some very particular cases (see [FF 65]). For these radial problems, it is also possible to find an analog of Dunham's exact quantization formula. The correction $l(l+1) \to (l+1/2)^2$ useful in the W.B.K. approximation must be modified in order to maintain the regularity condition at the origin when one wishes to include higher order W.B.K. corrections [SV 84].

We now consider some illustrative examples of the previous equations.

1) Coulomb Potential. Let $V(r) = -Ze^2/r\,(Z > 0)$, and to simplify the notation we define the quantities $a \equiv -2M|E|/\hbar^2$, $2b \equiv 2MZe^2/\hbar^2$, $c \equiv -l(l+1)$ [in (9.122)], $c \equiv -(l+1/2)^2$ [in (9.124)]; then the W.B.K. bound state energies are determined by

$$\begin{aligned} &\int_{r_0}^{r_1} dr\,\frac{1}{r}(ar^2+2br+c)^{1/2} = \left(n+\frac{1}{2}\right)\pi \\ &r_0 \equiv -\frac{1}{a}\left(b-\sqrt{b^2-ac}\right)\,, \quad r_1 \equiv -\frac{1}{a}\left(b+\sqrt{b^2-ac}\right)\,, \end{aligned} \tag{9.125}$$

and from this we obtain

$$E = -\frac{MZ^2e^4}{2\hbar^2(n+1/2+\sqrt{-c})^2}\,, \tag{9.126}$$

which coincides with the exact result if $c = -(l+1/2)^2$, namely, if we apply the prescription given in (9.124).

2) Isotropic Harmonic Oscillator. The potential is now $V(r) = M\omega^2 r^2/2$, and defining $a = -M^2\omega^2/\hbar^2$, $2b = 2ME/\hbar^2$ and c defined as before, the change of variables $x = r^2$ turns the equation for the bound state energies into

$$\begin{aligned} &\frac{1}{2}\int_{x_0}^{x_1} dx\,\frac{1}{x}(ax^2+2bx+c)^{1/2} = \left(n+\frac{1}{2}\right)\pi \\ &x_0 \equiv -\frac{1}{a}\left(b-\sqrt{b^2-ac}\right)\,, \quad x_1 \equiv -\frac{1}{a}\left(b+\sqrt{b^2-ac}\right)\,. \end{aligned} \tag{9.127}$$

Hence

$$E = \hbar\omega\left(2n+1+\sqrt{-c}\right)\,, \tag{9.128}$$

which agrees with the exact result if $c = -(l+1/2)^2$, namely, if we use (9.124).

3) Hylleraas Potential. In this case $V(r) = V_0\cosh^{-2}(r/R)$ with $V_0 < 0$. For $l = 0$ waves the equation (9.122) has as integration limits $r_0 = 0$ and the root r_1 of $\cosh^2(r_1/R) = V_0/E$. The change of variables $x = \cosh^2(r/R)$ transforms the integral into an easily computable one, and we obtain

$$E_{n,\mathrm{WBK}} = -\frac{\hbar^2}{2MR^2}\left[\frac{\sqrt{2M|V_0|R^2}}{\hbar} - 2n - 1\right]^2\,, \tag{9.129}$$

where the only allowed values of n are those non-negative integers for which the quantity between brackets is positive. In the modified W.B.K. method (9.124), the energies $E'_{n,\mathrm{WBK}}$ are determined by

$$\int_{r'_0}^{r'_1} dr \left[-\frac{2M|E|}{\hbar^2} + \frac{2M|V_0|}{\hbar^2} \cosh^{-2}(r/R) - \frac{1}{4r^2} \right]^{1/2} = \left(n + \frac{1}{2} \right) \pi \,. \tag{9.130}$$

Table 9.5. Bound state energies for the Hylleraas potential with $V_0 = -51\,\mathrm{MeV}$, $M = 938.2\,\mathrm{MeV}/c^2$, $R = 5.8\,\mathrm{fm}$. We list the exact answer, the W.B.K. values $E_{n,\mathrm{WBK}}$ and the modified W.B.K. results $E'_{n,\mathrm{WBK}}$

n	$-E_n$(MeV)	$-E_{n,\mathrm{WBK}}$(MeV)	$-E'_{n,\mathrm{WBK}}$(MeV)
0	35.69	40.40	35.51
1	19.39	22.90	19.24
2	8.02	10.33	7.92
3	1.59	2.70	1.53
4	—	0.0053	—

In Table 9.5 we compare in a particular case the exact results (Sect. 5.9) with the two approximations considered. It is obvious that (9.129) produces worse results than the modified W.B.K. method and it also gives a fifth bound state, which does not exist in this case.

The Hylleraas potential is one of those types of potentials for which the quantization formula (9.134) is exact, if the potential is corrected appropriately [RK 68].

Finally, the W.B.K. method also provides approximate information about the phase shifts $\delta_L(p)$ (where p is the free momentum) produced by the central potential $V(r)$. Suppose that for a given orbital angular momentum L, there is a single turning point r'_0 (for the effective potential with Langer's substitution $L(L+1) \to (L+1/2)^2$). In the forbidden region near the origin, the reduced radial wave function satisfying the regularity condition becomes in the W.B.K. approximation:

$$u(r) = \frac{A}{\sqrt{p_1(r)}} \exp\left[-\frac{1}{\hbar} \int_r^{r'_0} dr' p_1(r') \right] \,, \quad r < r'_0 \,. \tag{9.131}$$

One easily verifies that this function vanishes at the origin as r^{L+1}. If instead of Langer's substitution we had used (9.121), we would not have been able to obtain this behavior. For the allowed region, the connection formulae yield

$$u(r) = \frac{2A}{\sqrt{p_2(r)}} \sin\left[\frac{1}{\hbar} \int_{r'_0}^{r} dr' p_2(r') + \frac{\pi}{4} \right] \,, \quad r > r'_0 \,. \tag{9.132}$$

Since in the absence of interaction, the wave function behaves asymptotically like $\sin(pr/\hbar - L\pi/2)$, the following phase shifts result in the W.B.K. approximation

$$\delta_L^{\mathrm{WBK}}(p) = \left(L+\frac{1}{2}\right)\frac{\pi}{2} + \left\{\int_{r_0'}^{\infty} dr\left[\sqrt{k^2 - U(r) - \frac{(L+1/2)^2}{r^2}} - k\right] - kr_0'\right\}, \tag{9.133}$$

where $k \equiv p/\hbar$. Notice that the integral converges if $U(r) = \mathrm{O}(1/r^{1+\varepsilon})$, $\varepsilon > 0$, $r \to \infty$ and that for $U(r) \equiv 0$, $\delta_L^{\mathrm{WBK}}(p) = 0$. It can be shown [NE 82] that the W.B.K. approximation to the phase shifts is exact in the limit of very intense potentials, for fixed values of the momentum p and the angular momentum L. This indicates that only when the phase shifts are large, should one expect the W.B.K. approximation to be acceptable. This again confirms the naive intuition that this method works whenever the characteristic actions are large, compared with $\hbar$.

It is however remarkable that even for small phase shifts where the approximation is not reliable, we can nonetheless predict a decrease of δ_L when L increases (a qualitatively correct behavior). To exhibit this, we write (9.133) in the equivalent form

$$\delta_L^{\mathrm{WBK}}(p) = \int_{r_0'}^{\infty} dr\sqrt{k^2 - U(r) - \frac{(L+1/2)^2}{r^2}} - \int_{(L+1/2)/k}^{\infty} dr\sqrt{k^2 - \frac{(L+1/2)^2}{r^2}}\,. \tag{9.134}$$

For large L, r_0' is very close to $(L+1/2)/k$, assuming $U(r) = \mathrm{O}(r^{-\alpha})$, $\alpha > 1$. Expanding the first integrand in powers of U, we obtain in this limit an asymptotic expression for δ_L^{WBK}:

$$\delta_L^{\mathrm{WBK}}(p) \underset{L\to\infty}{\sim} -\frac{1}{2}\int_{(L+1/2)/k}^{\infty} dr\,\frac{U(r)}{\sqrt{k^2 - [(L+1/2)^2/r^2]}}\,. \tag{9.135}$$

If $U(r) \sim a/r^{\alpha}$, $\alpha > 1$, $r \to \infty$, then

$$\delta_L^{\mathrm{WBK}}(p) \underset{L\to\infty}{\sim} -\frac{a}{2}\frac{k^{\alpha-2}}{L^{\alpha-1}}\int_1^{\infty}\frac{dx}{x^{\alpha-1}\sqrt{x^2-1}}, \tag{9.136}$$

in agreement with the rigorous behavior $|f_L(p)| = \mathrm{O}(L^{-\alpha+1})$, quoted in Sect. 8.15.

10. Time-Independent Perturbation Theory and Variational Method

10.1 Introduction

It is often the case in the applications of QM that the eigenstates and eigenvalues corresponding to the bound states of a Hamiltonian H are too complicated to be obtained exactly. We will discuss two different procedures to obtain approximate solutions to these problems: *Time-independent perturbation theory*, and *variational method.*

The first method is useful when H can be decomposed in the form $H = H_0 + \lambda H_1$, where H_0 is simple enough so that the Schrödinger equation can be solved exactly, and H_1 is small in a sense that will be made precise later. We first give a simplified treatment of this method and later on we present the general formalism due to *Kato* [KA 49, 80], [RS 78].

The variational method is particularly useful in those cases where one has some idea about the dependence of the eigenfunctions of H on the coordinates appearing in the problem. This method gives in principle any discrete eigenvalue and the corresponding eigenfunction, although in practice it can be used successfully only in the determination of the lowest energy states, after fixing the quantum numbers corresponding to the symmetries of the problem, if there are any.

We will discuss throughout this chapter many examples of physical interest which help clarify the general methods presented.

10.2 Time-Independent Perturbations. The Non-Degenerate Case

Let H be the Hamiltonian of interest, and let us assume that it can be decomposed in the form

$$H = H_0 + \lambda H_1 \ . \tag{10.1}$$

The Hamiltonian H_0 describes the unperturbed physical system, whose eigenvalues $E_i^{(0)}$ and normalized eigenfunctions $|\psi_i^{(0)}\rangle$ are known. H_0 may have both a point and continuous spectrum. The term λH_1, where λ is real, is called the perturbation.

Let $E_i^{(0)}$ be a non-degenerate discrete (i.e. isolated from the rest of the spectrum) eigenvalue of the unperturbed Hamiltonian H_0 and let $|\psi_i^{(0)}\rangle$ be the associated eigenfunction. We assume the existence of a neighborhood Λ of $\lambda = 0$ such that for $\lambda \in \Lambda$, $H_0 + \lambda H_1$ has a single non-degenerate eigenvalue $E_i(\lambda)$ with eigenfunction $|\psi_i(\lambda)\rangle$; and both of them with an analytic dependence on λ in Λ. To determine $E_i(\lambda)$ and $|\psi_i(\lambda)\rangle$ we must solve the equation

$$H|\psi_i(\lambda)\rangle = E_i(\lambda)|\psi_i(\lambda)\rangle \; . \tag{10.2}$$

Using the freedom in the normalization of $|\psi_i(\lambda)\rangle$, we may require for sufficiently small λ that

$$\langle\psi_i^{(0)}|\psi_i(\lambda)\rangle = \langle\psi^{(0)}|\psi_i^{(0)}\rangle = 1 \; . \tag{10.3}$$

If $\lambda \in \Lambda$, we can expand in a power series of λ:

$$E_i(\lambda) = \sum_{n=0}^{\infty} \lambda^n E_i^{(n)} \; , \qquad |\psi_i(\lambda)\rangle = \sum_{n=0}^{\infty} \lambda^n |\psi_i^{(n)}\rangle \; . \tag{10.4}$$

The first series is called the *Rayleigh-Schrödinger series*. Replacing (10.1) and (10.4) in (10.2), and reordering the power series, one formally obtains

$$(H_0 - E_i^{(0)})|\psi_i^{(0)}\rangle + \sum_{n=1}^{\infty} \lambda^n \times \left[H_0|\psi_i^{(n)}\rangle + H_1|\psi_i^{(n-1)}\rangle - \sum_{k=0}^{n} E_i^{(k)}|\psi_i^{(n-k)}\rangle \right] = 0 \; . \tag{10.5}$$

This equation is satisfied for all λ only if

$$\boxed{\begin{aligned} &(H_0 - E_i^{(0)})|\psi_i^{(0)}\rangle = 0 \\ &(H_0 - E_i^{(0)})|\psi_i^{(n)}\rangle = (E_i^{(1)} - H_1)|\psi_i^{(n-1)}\rangle \\ &\qquad + \sum_{k=2}^{n} E_i^{(k)}|\psi_i^{(n-k)}\rangle \; , \quad n \geq 1 \; . \end{aligned}} \tag{10.6}$$

Similarly, the substitution of (10.4) in the normalization condition (10.3) yields

$$\boxed{\langle\psi_i^{(0)}|\psi_i^{(n)}\rangle = 0 \; , \quad n \geq 1} \; . \tag{10.7}$$

Equations (10.6) and (10.7) give a complete solution to the problem because they give $E_i^{(n)}$ and $|\psi_i^{(n)}\rangle$ for every value of n; indeed, the scalar product of the second equation in (10.6) with $|\psi_i^{(0)}\rangle$ together with (10.7) gives

$$\boxed{E_i^{(n)} = \langle\psi_i^{(0)}|H_1|\psi_i^{(n-1)}\rangle} \; . \tag{10.8}$$

Assuming for simplicity in this preliminary formal treatment of perturbation theory that H_0 has only a simple discrete spectrum, we can compute the scalar product of (10.6) with $|\psi_j^{(0)}\rangle$ for $j \neq i$. Using $H_0|\psi_j^{(0)}\rangle = E_j^{(0)}|\psi_j^{(0)}\rangle$, we obtain

$$\langle\psi_j^{(0)}|\psi_i^{(n)}\rangle = \frac{1}{E_i^{(0)} - E_j^{(0)}}\left[\langle\psi_j^{(0)}|H_1|\psi_i^{(n-1)}\rangle - \sum_{k=1}^{n-1} E_i^{(k)}\langle\psi_j^{(0)}|\psi_i^{(n-k)}\rangle\right]. \quad (10.9)$$

Since the eigenfunctions of H_0 form a complete set, from (10.7) and (10.9) we have

$$|\psi_i^{(n)}\rangle = \sum_{j\neq i}\frac{1}{E_i^{(0)} - E_j^{(0)}} \times\left[\langle\psi_j^{(0)}|H_1|\psi_i^{(n-1)}\rangle - \sum_{k=1}^{n-1} E_i^{(k)}\langle\psi_j^{(0)}|\psi_i^{(n-k)}\rangle\right]|\psi_j^{(0)}\rangle\,. \quad (10.10)$$

In order to work with a more compact notation, it is convenient to define the operator

$$S_i \equiv \frac{Q_i^{(0)}}{E_i^{(0)} - H_0} = \sum_{j\neq i}\frac{|\psi_j^{(0)}\rangle\langle\psi_j^{(0)}|}{E_i^{(0)} - E_j^{(0)}}\,, \quad (10.11)$$

so that

$$\boxed{|\psi_i^{(n)}\rangle = S_i\left[H_1|\psi_i^{(n-1)}\rangle - \sum_{k=1}^{n-1} E_i^{(k)}|\psi_i^{(n-k)}\rangle\right]}\,. \quad (10.12)$$

With (10.8) and (10.12), we can iteratively determine $E_i^{(1)}$, $|\psi_i^{(1)}\rangle$, $E_i^{(2)}$, $|\psi_i^{(2)}\rangle,\ldots$ and the problem is solved in principle. To lowest orders in this perturbation theory we explicitly have

$$\boxed{\begin{aligned} E_i^{(1)} &= \langle\psi_i^{(0)}|H_1|\psi_i^{(0)}\rangle\,, \\ |\psi_i^{(1)}\rangle &= S_iH_1|\psi_i^{(0)}\rangle = \sum_{j\neq i}\frac{\langle\psi_j^{(0)}|H_1|\psi_i^{(0)}\rangle}{E_i^{(0)} - E_j^{(0)}}|\psi_j^{(0)}\rangle\,, \\ E_i^{(2)} &= \langle\psi_i^{(0)}|H_1S_iH_1|\psi_i^{(0)}\rangle = \sum_{j\neq i}\frac{|\langle\psi_i^{(0)}|H_1|\psi_j^{(0)}\rangle|^2}{E_i^{(0)} - E_j^{(0)}}\,. \end{aligned}} \quad (10.13)$$

When we consider the ground state energy, the correction $E_i^{(2)}$ is always non-positive as a simple consequence of the last equation in (10.13). The series (with an infinite number of terms generically) that gives $|\psi_i^{(1)}\rangle$ cannot be handled easily, and it is often convenient to resolve directly the inhomogeneous equation (10.6) with the condition (10.7). The situation becomes more complicated whenever H_0 contains continuous eigenvalues, for in this case, although (10.13) formally holds, the sums therein will also include integration over the continuous eigenvalues together with the sum over the discrete part of the spectrum.

From (10.3), the eigenvector $|\bar{\psi}_i(\lambda)\rangle$ with unit norm will be of the form $|\bar{\psi}_i(\lambda)\rangle = Z_i^{1/2}(\lambda)|\psi_i(\lambda)\rangle$, where the renormalization constant $Z_i(\lambda)$ is determined by the equation

$$Z_i(\lambda)\langle\psi_i(\lambda)|\psi_i(\lambda)\rangle = 1 \, . \tag{10.14}$$

Writing $Z_i(\lambda)$ in the form

$$Z_i(\lambda) = \sum_{n=0}^{\infty} \lambda^n Z_i^{(n)} \tag{10.15}$$

with $Z_i^{(0)} = 1$, and replacing (10.4) and (10.15) in (10.14), we obtain

$$\sum_{k=0}^{n}\sum_{r=0}^{n-k} Z_i^{(k)}\langle\psi_i^{(r)}|\psi_i^{(n-k-r)}\rangle = 0 \, , \quad n \geq 1 \, , \tag{10.16}$$

which once again gives a recursive procedure to determine $Z_i^{(n)}$. In particular

$$Z_i^{(1)} = 0 \, , \quad Z_i^{(2)} = -\langle\psi_i^{(1)}|\psi_i^{(1)}\rangle \, . \tag{10.17}$$

We would like to make two observations: First, $Z_i(\lambda) = |\langle\psi_i^{(0)}|\bar{\psi}_i(\lambda)\rangle|^2$, and hence $Z_i(\lambda)$ measures the probability to observe the unperturbed state in the perturbed one, and second, we have

$$Z_i(\lambda) = 1 - \lambda^2 \sum_{j \neq i} \frac{|\langle\psi_j^{(0)}|H_1|\psi_i^{(0)}\rangle|^2}{(E_i^{(0)} - E_j^{(0)})^2} + \mathrm{O}(\lambda^3) \, , \tag{10.18}$$

so that to second order in λ

$$Z_i(\lambda) = \frac{\partial E_i(\lambda)}{\partial E_i^{(0)}} \, . \tag{10.19}$$

It can be shown that this result holds true to all orders of perturbation theory, [use of (10.71) and (10.78) is needed in the proof], and thus $Z_i(\lambda)$ is given by the derivative of the perturbed energy eigenvalue with respect to the unperturbed value of the energy.

According to (10.8), the computation of the nth order correction $E_i^{(n)}$ to the energy requires knowledge of the $(n-1)$th order correction to the wave function. It can be shown however [DA 61], that knowing $|\psi_i^{(r)}\rangle$, $r = 0, 1, \ldots, n$ is enough to compute $E_i^{(s)}$ for $s = 1, \ldots, 2n+1$. We can illustrate this fact in the case of $n = 1$. By hypothesis we know $|\psi_i^{(0)}\rangle$, $|\psi_i^{(1)}\rangle$; then (10.8) immediately yields $E_i^{(1)}$ and $E_i^{(2)}$. Using now (10.12) and (10.13), we have

$$\boxed{\begin{aligned} E_i^{(3)} &= \langle\psi_i^{(0)}|H_1|\psi_i^{(2)}\rangle = \langle\psi_i^{(0)}|H_1 S_i(H_1 - E_i^{(1)})|\psi_i^{(1)}\rangle \\ &= \langle\psi_i^{(1)}|(H_1 - E_i^{(1)})|\psi_i^{(1)}\rangle \, , \end{aligned}} \tag{10.20}$$

expressing $E_i^{(3)}$ in terms of $|\psi_i^{(0)}\rangle$, $|\psi_i^{(1)}\rangle$.

Similarly one easily obtains:

$$\boxed{\begin{aligned} E_i^{(4)} &= \langle\psi_i^{(1)}|(H_1 - E_i^{(1)})|\psi_i^{(2)}\rangle - E_i^{(2)}\langle\psi_i^{(1)}|\psi_i^{(1)}\rangle \\ E_i^{(5)} &= \langle\psi_i^{(2)}|(H_1 - E_i^{(1)}|\psi_i^{(2)}\rangle - 2E_i^{(2)}\mathrm{Re}\langle\psi_i^{(1)}|\psi_i^{(2)}\rangle \\ &\quad - E_i^{(3)}\langle\psi_i^{(1)}|\psi_i^{(1)}\rangle \,. \end{aligned}} \tag{10.21}$$

It is generally not easy to give sufficient conditions for the convergence of the series in (10.4). These expansions are useful only if the first few terms give a good approximation to the exact result. A physically plausible criterion, although mathematically neither necessary nor sufficient, is that the approximation is good if

$$|\langle\psi_j^{(0)}|\lambda H_1|\psi_i^{(0)}\rangle| \ll |E_j^{(0)} - E_i^{(0)}| \,, \quad j \neq i \,. \tag{10.22}$$

An interesting case when condition (10.22) does not hold appears when the Hamiltonian H_0 has an energy level $E_j^{(0)}$ very close to $E_i^{(0)}$, so that the perturbation, though small, violates (10.22). Under these circumstances, the method explained will not be valid in general. What is often done, is to decompose the original Hamiltonian as

$$H = H_0 + \lambda H_1' + \lambda H_1'' \,, \tag{10.23}$$

where

$$H_1' = \sum_{k=i,j}\sum_{l=i,j} |\psi_k^{(0)}\rangle\langle\psi_k^{(0)}|H_1|\psi_l^{(0)}\rangle\langle\psi_l^{(0)}| \,, \quad H_1'' = H_1 - H_1' \,. \tag{10.24}$$

The idea is now to exactly solve $H_0' = H_0 + \lambda H_1'$, and then treat $\lambda H_1''$ as a perturbation. The diagonalization of H_0' is trivial, because the states $|\psi_k^{(0)}\rangle$, $k \neq i, j$, are still eigenvectors of H_0' with eigenvalues $E_k^{(0)}$, because $H_1'|\psi_k^{(0)}\rangle = 0$. The states $|\psi_i^{(0)}\rangle$ and $|\psi_j^{(0)}\rangle$ are not eigenstates of H_0', but one can find linear combinations of them satisfying the eigenvalue equations. If $|\varphi\rangle = [\alpha|\psi_i^{(0)}\rangle + \beta|\psi_j^{(0)}\rangle]/(|\alpha|^2 + |\beta|^2)^{1/2}$, the equation $H_0'|\varphi\rangle = \varepsilon|\varphi\rangle$ becomes

$$\begin{aligned} &[E_i^{(0)} + \lambda E_i^{(1)} - \varepsilon]\alpha + \lambda\langle\psi_i^{(0)}|H_1|\psi_j^{(0)}\rangle\beta = 0 \\ &\lambda\langle\psi_j^{(0)}|H_1|\psi_i^{(0)}\rangle\alpha + [E_j^{(0)} + \lambda E_j^{(1)} - \varepsilon]\beta = 0 \,, \end{aligned} \tag{10.25}$$

with solutions

$$\begin{aligned} \varepsilon_\pm &= \tfrac{1}{2}(E_i^{(0)} + \lambda E_i^{(1)} + E_j^{(0)} + \lambda E_j^{(1)}) \\ &\quad \pm [\tfrac{1}{4}(E_i^{(0)} + \lambda E_i^{(1)} - E_j^{(0)} - \lambda E_j^{(1)})^2 + \lambda^2|\langle\psi_i^{(0)}|H_1|\psi_j^{(0)}\rangle|^2]^{1/2} \\ \alpha &= \lambda\langle\psi_i^{(0)}|H_1|\psi_j^{(0)}\rangle \,, \quad \beta_\pm = \varepsilon_\pm - E_i^{(0)} - \lambda E_i^{(1)} \,. \end{aligned} \tag{10.26}$$

Thus the original difficulty disappears if we start with the states $|\varphi_+\rangle$, $|\varphi_-\rangle$ and $|\psi_k^{(0)}\rangle$ with $k \neq i, j$, and using $\lambda H_1''$ as the perturbation. Now $\langle\varphi_\pm|H_1''|\varphi_\pm\rangle =$

$\langle\varphi_\pm|H_1''|\varphi_\mp\rangle = 0$, so that for this new problem in perturbation theory the criterion (10.22) is satisfied. It is obvious that the previous arguments can be extended to situations when there are n closely spaced levels.

The situation just described is rather common in atomic, molecular and nuclear physics. There are eigenstates of H_0 with very similar energies, and the presence of some other interaction produces very strong mixtures between these states. This situation is known as a mixing of levels or configurations.

To conclude this section, we would like to add that (10.13) is a particular case of the *Hellmann-Feynman* [HE 35], [FE 39] formula. If $E_i(\lambda)$ is a simple, discrete eigenvalue of a Hamiltonian $H(\lambda)$ with a possibly non-linear dependence on a parameter λ, and $|\phi_i(\lambda)\rangle$ is the corresponding normalized eigenstate, then under appropriate conditions of regularity and differentiability for $H(\lambda)$, $E_i(\lambda)$ and $|\phi_i(\lambda)\rangle$ we have

$$\frac{dE_i(\lambda)}{d\lambda} = \langle\frac{d}{d\lambda}H(\lambda)\rangle_{\phi_i(\lambda)}\,, \tag{10.27}$$

as follows immediately by differentiating $E_i(\lambda) = \langle\phi_i(\lambda)|H(\lambda)|\phi_i(\lambda)\rangle$ and $\|\phi_i(\lambda)\|^2 = 1$. Similarly, it should be pointed out that the inequality $E_0^{(2)} \leq 0$ for the second correction to the ground state energy follows from the *concavity* of E_0 as a function of λ (and this in turn is derived from the Min-Max principle to be discussed in Sect. 10.9).

10.3 The Harmonic Oscillator with λx^4 Perturbation

As a first application of the methods introduced in the previous section, let us consider the problem of an anharmonic oscillator with Hamiltonian

$$\begin{aligned} H &= H_0 + \lambda H_1 \\ H_0 &= -\frac{\hbar^2}{2M}\frac{d^2}{dx^2} + \frac{1}{2}kx^2\,, \quad H_1 = x^4\,. \end{aligned} \tag{10.28}$$

In order to make the notation as simple as possible, let us take $\hbar = 2M = k/2 = 1$. The spectrum of H_0 is simple and discrete, with eigenvalues and eigenfunctions

$$E_n^{(0)} = 2n+1\,, \quad \psi_n^{(0)} = [\pi^{1/2}2^n n!]^{-1/2}H_n(x)\mathrm{e}^{-x^2/2}\,, \tag{10.29}$$

where $H_n(x)$ are the Hermite polynomials. Using (A.55, 56), we can evaluate the matrix element $\langle\psi_n^{(0)}|x^4|\psi_m^{(0)}\rangle = I^4_{nm}$. Direct application of (10.13) gives, for the energy of the perturbed ground state, the expression

$$E_0(\lambda) = 1 + \tfrac{3}{4}\lambda - \tfrac{21}{16}\lambda^2 + \mathrm{O}(\lambda^3)\,. \tag{10.30}$$

Using Kato's techniques, to be explained below, it is easy to compute a few more terms in this expansion:

$$E_0(\lambda) = 1 + \frac{3}{4}\lambda - \frac{21}{16}\lambda^2 + \frac{333}{64}\lambda^3 - \frac{30885}{1024}\lambda^4 + O(\lambda^5) . \tag{10.31}$$

Writing this series as $E_0(\lambda) = \sum_{n=0}^{\infty} B_n \lambda^n$, $B_0 \equiv 1$, *Bender* and *Wu* [BW 69] computed the first 75 coefficients B_n and inferred numerically the following behavior for large values of n:

$$B_n \underset{n\to\infty}{\sim} (-1)^n 2\sqrt{6}\pi^{-3/2}(3/2)^n \Gamma(n+1/2) . \tag{10.32}$$

A powerful method due to *Lipatov* [LI 77] and based on the path integral techniques gives asymptotic expansions similar to the previous one for many potentials [BL 77], and even $1/n$ corrections to them. (For an update on this subject, see [SI 82a], where references to the literature are given and rigorous derivations of (10.32) discussed.)

The equation (10.32) reveals that the formal perturbative series for $E_0(\lambda)$ is not convergent in a neighborhood of $\lambda = 0$ [this is not very surprising at first sight since for $\lambda < 0$ the defect indices of H, initially defined in $C_0^\infty(\mathbb{R})$, are (2,2)]. It can be shown nevertheless that the function $E_0(\lambda)$, $\lambda > 0$ can be analytically continued to the complex λ-plane cut along the negative real axis [LM 70], and the perturbative series $\sum_0^\infty B_n \lambda^n$ is a strong asymptotic expansion of $E_0(\lambda)$ around $\lambda = 0$ in that sector [GG 70]: given $\varepsilon > 0$, there exist $A_\varepsilon > 0$, $B_\varepsilon > 0$ such that

$$\left| E_0(\lambda) - \sum_{n=0}^{N} B_n \lambda^n \right| \leq (N+1)! A_\varepsilon^{N+1} |\lambda|^{N+1} ,$$
$$|\arg \lambda| \leq \pi - \varepsilon , \quad |\lambda| < B_\varepsilon . \tag{10.33}$$

This important property guarantees the Borel summability of the perturbative series, with Borel sum precisely $E_0(\lambda)$ for all λ in the plane cut along $(-\infty, 0]$ [HI 82]: the series $\sum_0^\infty B_n \lambda^n / n!$, known as the Borel transform of $\sum_0^\infty B_n \lambda^n$, has a convergence radius equal to 2/3 as a consequence of (10.32), and thus, for $|\lambda| < 2/3$, it defines some analytic function $e_0(\lambda)$. This function has an analytic extension to a neighborhood of the positive real axis, so that the integral

$$\int_0^\infty dt\, e^{-t} e_0(t\lambda) \tag{10.34}$$

is absolutely convergent $\forall$ Re$\lambda > 0$, and its value, known as the Borel sum of the series $\sum_0^\infty B_n \lambda^n$, coincides with $E_0(\lambda)$ in this region.

It is important to stress that the strong asymptotic nature of the perturbative expansion for $E_0(\lambda)$ which implies the Borel summability, is more restrictive than the usual notion of asymptotic series in the sense of Poincaré. The latter only requires that if $E_0(n, \lambda) \equiv \sum_{j=0}^{n} B_j \lambda^j$, then $R(n, \lambda) \equiv \lambda^{-n}[E_0(\lambda) - E_0(n, \lambda)]$ should satisfy

$$\lim_{\lambda \downarrow 0} R(n, \lambda) = 0 , \quad \forall n \geq 1 . \tag{10.35}$$

Table 10.1. Values of B_n and $E_0(n, 0.1)$ for $n = 1, 2, \ldots, 20$. A numerical evaluation gives $E_0(0.1) = 1.065285$. The numbers in parentheses refer to powers of ten

n	B_n	$E_0(n,\lambda) \equiv 1 + \sum_{k=1}^{n} \lambda^k B_k$ $\lambda = 0.1$
1	+0.750 000 000 (0)	1.075 000
2	−1.312 500 000 (0)	1.061 875
3	+5.203 125 000 (0)	1.067 078
4	−3.016 113 281 (1)	1.064 062
5	+2.238 112 792 (2)	1.066 300
6	−1.999 462 921 (3)	1.064 301
7	+2.077 708 948 (4)	1.066 378
8	−2.456 891 772 (5)	1.063 921
9	+3.256 021 887 (6)	1.067 178
10	−4.781 043 106 (7)	1.062 396
11	+7.708 333 164 (8)	1.070 105
12	−1.354 432 468 (10)	1.056 560
13	+2.577 262 349 (11)	1.082 333
14	−5.281 751 322 (12)	1.029 516
15	+1.160 166 746 (14)	1.145 532
16	−2.719 757 615 (15)	0.873 556
17	+6.778 794 692 (16)	1.551 436
18	−1.790 210 195 (18)	−0.238 774
19	+4.994 011 921 (19)	+4.755 238
20	−1.467 514 010 (21)	−9.919 902

Thus we can achieve

$$|\lambda^{-n}[E_0(\lambda) - E_0(n, \lambda)]| < \varepsilon \tag{10.36}$$

for arbitrary $\varepsilon > 0$, taking $\lambda > 0$ sufficiently small. A typical behavior of $E_0(n, \lambda)$ appears in Table 10.1, for $\lambda = 0.1$. The values of $E_0(n, 0.1)$ approach the correct result when n increases from 1 to 6 and $E_0(6, 0.1)$ differs in 0.92 per 1000 from the exact result. If n keeps on increasing, $E_0(n, 0.1)$ begins to differ from the exact value, and for $n \geq 19$, the difference between them is substantial. For λ smaller than 0.1 one obtains better approximations to $E_0(\lambda)$ than in the previous case, whereas if λ is sufficiently large, the values of $E_0(n, \lambda)$ differ considerably from $E_0(\lambda)$ for all values of n.

Before closing this section we would like to add that a useful expression for $E_0(\lambda)$ is obtained using Padé approximants. Given the power series $f(z) = \sum_0^\infty a_n z^n$ its Padé approximant of order $[N, M]$ is defined by

$$f^{[N,M]}(z) = \frac{P^{[N,M]}(z)}{Q^{[N,M]}(z)}, \tag{10.37}$$

where $P^{[N,M]}$, $Q^{[N,M]}(\neq 0)$ are polynomials in z of orders at most M and N, respectively, and such that

$$f(z) - f^{[N,M]}(z) = O(z^{N+M+1}). \tag{10.38}$$

These conditions determine the rational function $f^{(N,M)}(z)$ uniquely, whenever it exists. Its explicit form is [BA 75]:

$$f^{[N,M]}(z) = \frac{\begin{vmatrix} a_{M-N+1} & a_{M-N+2} & \cdots\cdots & a_{M+1} \\ \vdots & \vdots & & \vdots \\ a_M & a_{M+1} & \cdots\cdots & a_{M+N} \\ \sum\limits_{k=N}^{M} a_{k-N} z^k & \sum\limits_{k=N-1}^{M} a_{k-N+1} z^k & \cdots\cdots & \sum\limits_{k=0}^{M} a_k z^k \end{vmatrix}}{\begin{vmatrix} a_{M-N+1} & a_{M-N+2} & \cdots\cdots & a_{M+1} \\ \vdots & \vdots & & \vdots \\ a_M & a_{M+1} & \cdots\cdots & a_{M+N} \\ z^N & z^{N-1} & \cdots\cdots & 1 \end{vmatrix}} \tag{10.39}$$

with the understanding that $a_j \equiv 0$ if $j < 0$ and provided the denominator in (10.39) is not identically zero. Obviously, $f^{[0,M]}(z)$ gives the truncated Taylor series expansion.

It has been shown that in this case [LM 69], [SI 70a] any sequence of Padé approximants $E_0^{[N,N+j]}(\lambda)$ (with variable N and j fixed and non-negative) associated to the series $E_0(\lambda) = \sum_{n=0}^{\infty} B_n \lambda^n$ converges uniformly towards $E_0(\lambda)$ on compact subsets of the complex plane cut along $(-\infty, 0]$.

The use of approximants notably improves the numerical estimates. For example, the series (10.31) allows us to obtain $E_0^{[2,2]}(\lambda)$:

$$E_0^{[2,2]}(\lambda) = \frac{3968 + 31724\lambda + 39134\lambda^2}{3968 + 28748\lambda + 22781\lambda^2} \,. \tag{10.40}$$

In Table 10.2 we compare $E_0(4, \lambda)$, $E_0^{[2,2]}(\lambda)$ with the exact result $E_0(\lambda)$ obtained numerically for different values of λ, making manifest the power of this method. The computation of higher order Padé approximants can be found in [SI 70a]. Since for $\lambda \gg 1$, $H \sim p^2 + \lambda x^4$, and $p^2 + \lambda x^4$ is unitarily equivalent to $\lambda^{1/3}(p^2 + x^4)$ under the dilatation $x \to \lambda^{-1/6} x$, it is clear that $E_0(\lambda) \sim c_0 \lambda^{1/3}$, $\lambda \to \infty$, where c_0 is the ground state energy of $p^2 + x^4$ ($c_0 = 1.0603621$). Therefore, no Padé $E_0^{[N,M]}(\lambda)$ with N, M fixed can simulate this behavior, and the approximation $E_0(\lambda) \simeq E_0^{[N,M]}(\lambda)$ will fail necessarily for λ sufficiently large. The combination of perturbative and variational methods allows us to capture the correct behavior for large λ, and to obtain algebraic approximants with uniform precision for all $\lambda > 0$ [PA 79].

Table 10.2. Comparison between $E_0(4,\lambda)$, $E_0^{[2,2]}(\lambda)$ and the exact result $E_0(\lambda)$ obtained by numerically integrating the Schrödinger equation

λ	$E_0(4,\lambda)$	$E_0^{[2,2]}(\lambda)$	$E_0(\lambda)$
0.1	1.064 06	1.065 22	1.065 29
0.2	1.090 87	1.117 54	1.118 29
0.3	1.003 05	1.161 48	1.164 05
0.4	0.650 88	1.199 19	1.204 81
0.5	− 0.187 81	1.231 98	1.241 85
0.6	− 1.807 51	1.260 82	1.275 98
0.7	− 4.575 14	1.286 38	1.307 75
0.8	− 8.930 00	1.309 21	1.337 55
0.9	−15.383 77	1.329 74	1.365 67
1.0	−24.520 51	1.348 29	1.392 35

The anharmonic oscillator has become a test laboratory for new approximation techniques in the computation of energy levels. There is an extensive bibliography in this subject. Besides the references already quoted, we find very interesting: [BD 71], [BA 76], [BB 78] and [SZ 79], not only because they offer alternative computational methods, but also because of the great precision of their analysis. Other extensively studied models are $H = p^2 - x^2 + \lambda x^4$, $\lambda > 0$ (anharmonic double well) [SZ 79], [BP 83], [DP 84] and $H = p^2 + x^2 + \lambda x^{2n}$, $n \geq 2$, $\lambda > 0$ [GG 70, 78].

10.4 The Harmonic Oscillator with λx^3 Perturbation

We next study the linear harmonic oscillator with a λx^3 perturbation, namely:

$$H = H_0 + \lambda H_1 \ ,$$
$$H_0 = -\frac{d^2}{dx^2} + x^2 \ , \quad H_1 = x^3 \ , \tag{10.41}$$

where as in the previous case we take $\hbar = 2M = k/2 = 1$. In Fig. 10.1 we plot the potentials $x^2 + \lambda x^3$ and x^2. The energy corresponding to the perturbed "ground state" can be computed using the techniques developed in Sect. 10.2, or more conveniently using Kato's theory. We obtain

$$E_0(\lambda) = 1 - \frac{11}{16}\lambda^2 - \frac{465}{256}\lambda^4 - \frac{39709}{4096}\lambda^6 + O(\lambda^8) \tag{10.42}$$

and, as before, the coefficients in the expansion in powers of λ are all finite but grow very fast with n.

From a mathematical point of view, the situation here is far more complex than in the previous section. With the results of Sect. 4.10 it is possible to show that the defect indices of H are (1,1), and therefore H, originally defined on $C_0^\infty(\mathbb{R})$-functions, admits infinitely many self-adjoint extensions, all of them with discrete spectra, and there are no physical criteria allowing us to choose

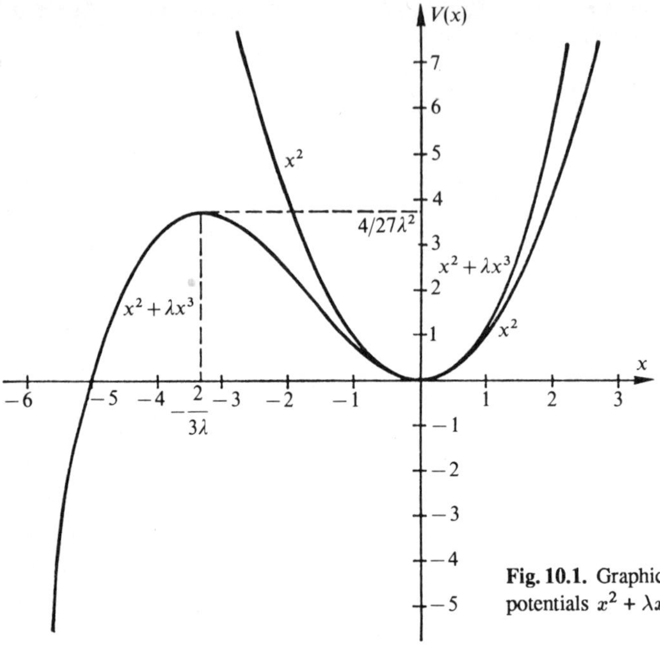

Fig. 10.1. Graphical representation of the potentials $x^2 + \lambda x^3$ and x^2 with $\lambda = 0.2$

one among them. (This may not be surprising, for even classically it takes only a finite amount of time for a particle moving towards the left region to reach $x = -\infty$, and it is thus necessary to instruct the particle on how to proceed after it gets there, although it is certainly not clear how to do that. These situations are called *incomplete* [RR 73]). It is however amusing to observe that if we consider the differential equation

$$\left[-\frac{d^2}{dx^2} + x^2 + \lambda x^3\right]\psi(x) = E\psi(x) , \tag{10.43}$$

the bounded asymptotic solutions for values of E and $\lambda > 0$ fixed are given by

$$\begin{aligned} \psi(x) &\sim A(E,\lambda)x^{-3/4}\exp\left[-\tfrac{2}{5}\lambda^{1/2}x^{5/2}\right] , \quad x \to \infty , \\ \psi(x) &\sim B(E,\lambda)|x|^{-3/4}\cos\left[\tfrac{2}{5}\lambda^{1/2}|x|^{5/2} + \alpha(E,\lambda)\right] , \quad x \to -\infty , \end{aligned} \tag{10.44}$$

and a similar result is obtained if $\lambda < 0$. Due to the term $|x|^{-3/4}$ there is a square integrable solution of (10.43) for any real λ and complex E. In other words, the point spectrum of the maximal differential operator associated to H fills the whole complex plane. The standard criterion for selecting eigenvalues by imposing the quadratic integrability of the solutions to the Schrödinger equation is insufficient in this case.

The question now arises whether the energy computed by means of the truncated series (10.42) for $\lambda \ll 1$ has any special significance. Numerical calculations with the solutions of (10.43) indicate that the energies given by the truncated series (10.42) are characterized by the fact that the corresponding wave functions are large in the region near the origin and very small elsewhere. If we take for

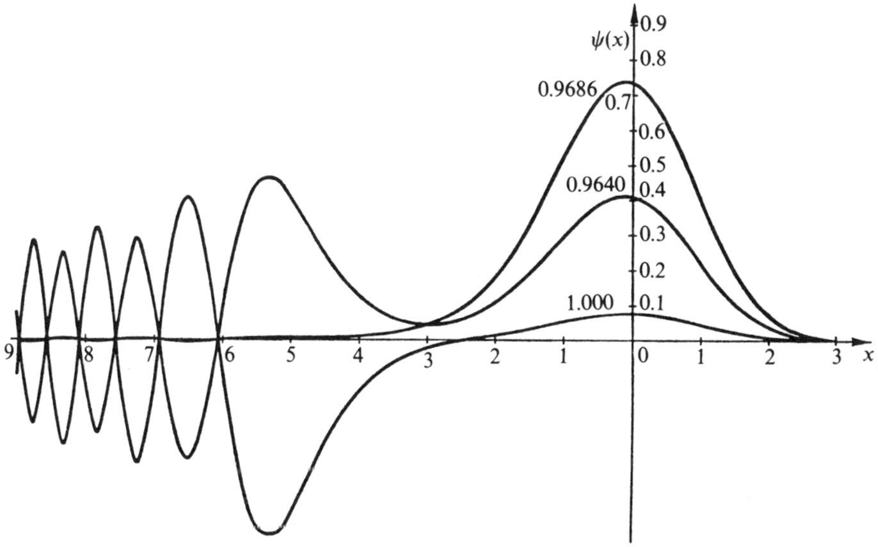

Fig. 10.2. Numerical solution of (10.43) for those values of E indicated in the curves. $\psi(x)$ has been normalized (to one) in all cases

example $\lambda = 0.2$, using (10.42) we obtain $E_0(0.2) \simeq 0.9690$ and therefore the phenomenon alluded to should arise at energies very close to this value. The results obtained numerically are given in Fig. 10.2, in complete agreement with the arguments presented. The energy corresponding to a maximal concentration in a neighborhood of the origin is $E = 0.9686 \pm 0.0001$.

Physically we should expect that if the particles are described by an eigenstate of $p^2 + x^2$, and then we introduce the perturbation λx^3, the particle would tend to escape through tunneling to the region of negative potential energy, the amount of time it takes to tunnel through the barrier is longer the higher the barrier that has to be penetrated. In other words it will be the longer, the lower the energy level is with respect to the height of the barrier. Using the time-energy uncertainty relation, the state, which is not stationary anymore, will not have a well-defined energy. It will present an uncertainty of $\Delta E \sim \hbar/\tau$, where $\tau \sim$ lifetime of the state or escape time. Therefore none of the possible extensions of (10.41) has a direct physical meaning, for they yield a discrete spectrum, and from the physical considerations just described, we expect the energies to be complex, with $\operatorname{Re} E$ being the mean position of the energy level, and $\operatorname{Im} E = -\Gamma/2$, where Γ is its width (see in this regard Sect. 4.7).

There are rigorous studies [CG 80] of these questions based on techniques of complex scaling: If in the operator H we implement the scale change $x \to e^{\theta} x$, with $0 < \operatorname{Im} \theta < \pi/8$, we obtain the new operator

$$H_\theta = e^{-2\theta} p^2 + e^{2\theta} x^2 + \lambda e^{3\theta} x^3 , \tag{10.45}$$

closed in $D(p^2) \cap D(x^3)$ with a spectrum independent of θ and purely discrete if $\lambda (> 0)$ is sufficiently small. The eigenvalues $E_n(\lambda)$ of H_θ, $n = 1, 2, \ldots$ converge to the energy level $2n + 1$ of $p^2 + x^2$ when $\lambda \to 0$, and they are the quasi-bound levels mentioned previously. The perturbative series for the nth level is Borel summable, and its Borel sum is Re $E_n(\lambda)$, and the widths $\Gamma_n(\lambda) \equiv -2\text{Im}\, E_n(\lambda)$ are positive for almost all values of λ. For the numerical computations of these levels, see [YB 78], [DR 81], [AL 88]. In particular, for the example discussed previously, $E_0(0.2) = 0.968631994 - i1.61204190 \times 10^{-5}$.

10.5 Two-Electron Atoms (I)

Consider now a nucleus of charge $Z|e|$ with two bound electrons. After neglecting the Hughes-Eckart terms (see Sect. 14.1) the Hamiltonian can be decomposed in the form

$$H = H_0 + H_1$$

$$H_0 = \sum_{i=1}^{2} \left[-\frac{\hbar^2}{2\mu}\Delta_i - \frac{Ze^2}{r_i} \right] \equiv \sum_i H_{0i} \, , \quad H_1 = \frac{e^2}{r_{12}} \, , \tag{10.46}$$

where $\boldsymbol{r}_1$ and $\boldsymbol{r}_2$ are the position vectors of the two electrons, $r_{12} = |\boldsymbol{r}_1 - \boldsymbol{r}_2|$ and μ is the reduced mass of an electron with respect to the nucleus. In the ground state of H_0 the two electrons occupy $1s$ states (with antiparallel spins to ensure the total antisymmetry of the combined states for two identical fermions, Sect. 13.2), and therefore the unperturbed ground state is non-degenerate and has energy

$$E_0^{(0)} = -\mu(Z\alpha c)^2 \, . \tag{10.47}$$

Its normalized wave function is

$$\psi_0^{(0)}(\boldsymbol{r}_1, \boldsymbol{r}_2) = (\pi a^3)^{-1} \mathrm{e}^{-(r_1+r_2)/a} \, , \tag{10.48}$$

where $a = \hbar/\mu(Z\alpha c)$. The first correction to the energy in perturbation theory is

$$E_0^{(1)} = \frac{e^2}{\pi^2 a^6} \int d^3r_1 d^3r_2 \frac{1}{r_{12}} \mathrm{e}^{-2(r_1+r_2)/a} \, . \tag{10.49}$$

Taking into account (A.135) and (A.41) the angular integrations can be performed immediately yielding

$$E_0^{(1)} = \frac{16e^2}{a^6} \int_0^\infty dr_1 \, r_1^2 \times \left[\int_0^{r_1} dr_2 \frac{r_2^2}{r_1} \mathrm{e}^{-2(r_1+r_2)/a} + \int_{r_1}^\infty dr_2 \, r_2 \mathrm{e}^{-2(r_1+r_2)/a} \right] . \tag{10.50}$$

A further elementary integration gives $E_0^{(1)} = 5\mu(Z\alpha c)^2/8Z$; thus, to first order in perturbation theory the energy of the ground state is given by

$$E_{\text{pert}} = -\mu(Z\alpha c)^2 \left(1 - \frac{5}{8Z}\right) . \tag{10.51}$$

Table 10.3 contains the values of the energies (10.51) and those computed using the variational method (see Sect. 10.10) for various atoms. The experimental data are extracted from [KU 62]. It is to be expected that as Z increases, the Coulomb repulsion energy between the two electrons should become relatively smaller than the interaction energy between the electrons and the nucleus, the perturbative corrections should be less important, and the agreement between theory and experiment should improve. This is clearly the case in Table 10.3.

Table 10.3. Comparison between E_{pert} and the experimental value. For atoms we use the spectroscopic notation. The roman numeral indicates the number of electrons removed, plus one. All values are given in eV

Atom	$-E_0^{(0)}$	$E_0^{(1)}$	$-E_{\text{pert}}$	$-E_{\text{var}}$	$E_{0,\text{exp}}$
He	108.831 68 (36)	34.01	74.82	77.48	79.003 49 (58)
Li II	244.885 72 (82)	51.02	193.87	196.52	198.083 8 (31)
Be III	435.359 9 (14)	68.02	367.33	369.99	371.574 (12)
B IV	680.257 4 (22)	85.03	595.22	597.88	599.502 (25)
C V	979.574 6 (32)	102.04	877.19	880.19	881.885 (37)
N VI	1 333.318 8 (44)	119.05	1 214.27	1 216.93	1 218.741 (62)
O VII	1 741.486 0 (58)	136.05	1 605.09	1 608.09	1 610.068 (74)
F VIII	2 204.080 2 (72)	153.06	2 051.02	2 053.67	2 055.909 (99)

For the computation of higher order perturbative corrections, there are mixed techniques incorporating both perturbation theory and variational methods initially proposed by *Hylleraas* [HY 30]. In [MI 65] one can find the coefficients e_n of the perturbative expansion

$$E_0 = -\mu(Z\alpha c)^2 \sum_{n=0}^{\infty} e_n Z^{-n} \tag{10.52}$$

for $n \leq 21$. For instance

$$\begin{aligned} e_0 &= 1 , \quad e_1 = -\tfrac{5}{8} \\ e_2 &= 0.157666428 , \\ e_3 &= -0.008699029 , \\ e_4 &= 0.000888705 , \\ e_5 &= 0.001036374 . \end{aligned} \tag{10.53}$$

The sum of this perturbative series truncated at $n = 21$ reproduces the experimental results with a precision of the order of 10^{-4}, and one should not expect an improvement of this result, because the terms neglected in the Hamiltonian (Hughes-Eckart, relativistic corrections etc.) start making contributions at this level of accuracy.

Kato's theory, to be presented in Sect. 10.7 guarantees the convergence of (10.53) for $Z > 7.7$ [KA 80], namely, starting with O VII. On the other hand, the numerical analysis of the series leads us to believe that it converges for $Z > 0.894$ [ST 66], [SI 81] which includes not only the important case of non-negative ions with two electrons, but also the ion $H^-(Z = 1)$. For this last case, the perturbative series to order $n = 21$ gives $E_0 = -0.527751\mu(\alpha c)^2$, to be compared with the variational result of *Pekeris* [PE 62] $E_0 = -0.5277510062954\mu(\alpha c)^2$. An important difference between this case and other ions with two electrons and $Z > 1$ is that while the latter always present an infinite number of bound states [RS 78], [TH 81], the ion H^- has a unique discrete level [HI 77, 77a].

10.6 Van der Waals Forces (I)

Another important application of perturbation theory is the derivation of the interaction energy between two atoms at large distances. To understand the origin of these forces, consider the case of two hydrogen atoms whose nuclei are separated by a distance R, large compared with the Bohr radius. To label the positions of the particles participating in the problem we follow the notation defined in Fig. 10.3.

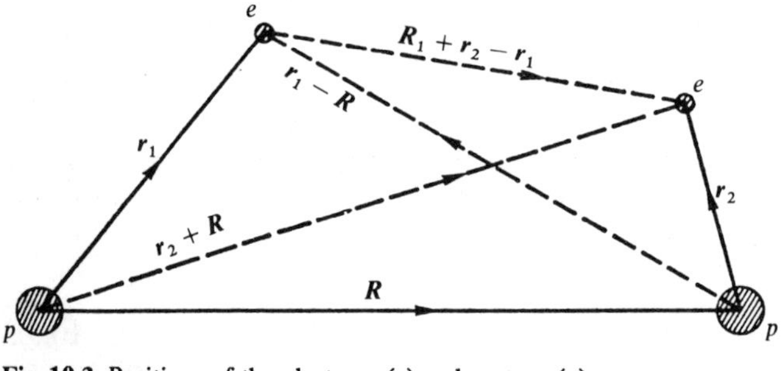

Fig. 10.3. Positions of the electrons (e) and protons (p)

Assuming the positions of the nuclei as fixed, the Hamiltonian of the system can be written to a good approximation as

$$\begin{aligned} H &= H_0 + H_1 \,, \\ H_0 &= -\frac{\hbar^2}{2\mu}(\Delta_1 + \Delta_2) - \frac{e^2}{r_1} - \frac{e^2}{r_2} \,, \\ H_1 &= \frac{e^2}{R} + \frac{e^2}{|\boldsymbol{R}+\boldsymbol{r}_2-\boldsymbol{r}_1|} - \frac{e^2}{|\boldsymbol{R}+\boldsymbol{r}_2|} - \frac{e^2}{|\boldsymbol{R}-\boldsymbol{r}_1|} \,. \end{aligned} \tag{10.54}$$

Assuming that in the absence of H_1 both atoms are in their ground states, the unperturbed energy and wave functions are

$$E_0^{(0)} = -\mu(\alpha c)^2 , \quad \psi_0^{(0)}(\boldsymbol{r}_1, \boldsymbol{r}_2) = (\pi a_0^3)^{-1} e^{-(r_1+r_2)/a_0} , \tag{10.55}$$

where $a_0 = \hbar/\mu(\alpha c)$ is the Bohr radius. Ignoring the spin variable (whose effect we neglect in this problem), the state considered is non-degenerate, and we can apply the perturbation theory developed so far. In the matrix elements of H_1 the only important contributions come from those values of r_1 and r_2 which are at most of the order of a_0. Since $R \gg a_0$ we can replace H_1 by the expression obtained after expanding in powers of r_i/R, keeping only the lowest order terms. Using (A.135) or by a direct computation we obtain

$$H_1 = e^2 \left[\frac{\boldsymbol{r}_1 \cdot \boldsymbol{r}_2}{R^3} - 3\frac{(\boldsymbol{r}_1 \cdot \boldsymbol{R})(\boldsymbol{r}_2 \cdot \boldsymbol{R})}{R^5} \right] , \tag{10.56}$$

where we recognize the interaction energy between two electric dipoles associated to the instantaneous configuration of both atoms. The terms neglected will represent dipole-quadrupole, quadrupole-quadrupole interactions, etc. ....

It is evident that $E_0^{(1)}$ is equal to zero because in its calculation we find integrals vanishing by parity arguments. Furthermore the contributions to this order of the terms neglected is equal to zero. [Actually, the expectation value of H_1 given in (10.54) in the unperturbed ground state decreases exponentially with R at large distances, and therefore its expansion in a power series of $1/R$ vanishes identically.] Finally we conclude that up to second order in perturbation theory, the energy of both atoms is given by

$$\begin{gathered} E_0 = -\mu(\alpha c)^2 \left[1 + \left(\frac{a_0}{R} \right)^6 \xi \right] \\ \xi \equiv \frac{\hbar^2}{\mu a_0^6} \sum_{j \neq 0} \frac{|\langle \psi_0^{(0)} | S | \psi_j^{(0)} \rangle|^2}{E_j^{(0)} - E_0^{(0)}} , \qquad S \equiv \boldsymbol{r}_1 \cdot \boldsymbol{r}_2 - 3(\boldsymbol{r}_1 \cdot \hat{\boldsymbol{R}})(\boldsymbol{r}_2 \cdot \hat{\boldsymbol{R}}) , \end{gathered} \tag{10.57}$$

where the sum also includes integration over the continuous spectrum of H_0. Notice that ξ is a dimensionless positive constant. We can say that the second-order correction to the energy acts as an attractive effective potential energy proportional to R^{-6}, and this is the origin of the Van der Waals forces. Its origin is purely quantum mechanical, and it is due to the mutual polarization of the atoms. The simple calculation presented here goes back to *London* [LO 30].

Even though the precise computation of ξ is very complicated, it is not difficult to find a bound for it. The state $|\psi_j^{(0)}\rangle$ of lowest energy appearing in the equation defining ξ is the one where both electrons are excited to $n = 2$ states with energy $E_j^{(0)} = -\mu(\alpha c)^2/4$. States with one $n = 1$ electron and another with $n \geq 2$ do not contribute to the sum, because the relevant matrix elements vanish by parity invariance. Hence all the denominators in the sum are large than or equal to $3\mu(\alpha c)^2/4$ and we can write the inequality

$$\xi \leq \frac{4}{3a_0^4} \sum_{j \neq 0} |\langle \psi_0^{(0)} | S | \psi_j^{(0)} \rangle|^2 . \tag{10.58}$$

Using the closure relation and $\langle\psi_0^{(0)}|S|\psi_0^{(0)}\rangle = 0$ we obtain

$$\xi \leq \frac{4}{3a_0^4}\langle\psi_0^{(0)}|S^2|\psi_0^{(0)}\rangle. \tag{10.59}$$

Taking into account that for a spherically symmetric function $f(\boldsymbol{x})$

$$\int d^3x\, f(\boldsymbol{x}) x_i x_j = \frac{\delta_{ij}}{3}\int d^3x\, f(\boldsymbol{x})\boldsymbol{x}^2\,, \tag{10.60}$$

we immediately obtain

$$\xi \leq \frac{8}{9a_0^4}\langle\psi_0^{(0)}|r_1^2 r_2^2|\psi_0^{(0)}\rangle\,, \tag{10.61}$$

and using the results derived in the study of the hydrogen atom, $\xi \leq 8$, while the precise value of ξ is 6.499027 [KO 67].

The approximations leading to (10.57) are only valid if $R \gg a_0$, so that the effects due to the overlap between the wave functions of the two electrons are negligible. If R is very large, the time delay in the propagation of the interactions has to be taken into consideration, and interactions of the form $1/R^7$ seem to be more appropriate, rather than the one derived here [CP 48], [FS 70]. Finally, the effect of nuclear recoil herein ignored can be important to account for the Van der Waals forces between light systems like positronium-hydrogen atom, positronium-positronium at not very large distances [MR 85].

10.7 Kato's Theory

For the general treatment of perturbation theory, applicable to degenerate levels, there is an elegant method due to Kato [KA 49, 80], [RS 78] based on the study of the resolvent of H. We will only give a formal presentation of this method.

The resolvent of H is defined to be the operator

$$G(z) = \frac{1}{zI - H} \tag{10.62}$$

as a function of the complex variable z. (In the notation of Appendix C, $G(z) = R_z(H)$.) The spectrum of the self-adjoint operator H is given by those values of z where the operator $G(z)$ either does not exist or it is not bounded. For simplicity we assume the spectrum of H to be discrete, although the results obtained can be trivially extended to the case where part of the spectrum is continuous. (See Appendix C for more details.) Denoting by $E_0, E_1, E_2, \ldots, E_i, \ldots$ the eigenvalues of H and by P_i the orthogonal projection on the subspace generated by the eigenvectors with eigenvalue E_i we have

$$G(z)P_i = \frac{P_i}{z - E_i} \tag{10.63}$$

and therefore

$$G(z) = \sum_i \frac{P_i}{z - E_i} . \tag{10.64}$$

If Γ_i is a simple closed contour in the complex z plane, traversed in the positive sense, without crossing the spectrum of H and containing in its interior only the eigenvalue E_i, Cauchy's theorem implies

$$P_i = \frac{1}{2\pi \mathrm{i}} \int_{\Gamma_i} G(z)\, dz . \tag{10.65}$$

If the contour contains in its interior n eigenvalues, the result of the integral is the sum of the n projections associated to these eigenvalues. On the other hand, taking into account $HG(z) = zG(z) - I$, we obtain

$$HP_i = \frac{1}{2\pi \mathrm{i}} \int_{\Gamma_i} zG(z)\, dz . \tag{10.66}$$

Consider now the Hamiltonian $H = H_0 + \lambda H_1$. The resolvents of H_0 and H are

$$G_0(z) = \frac{1}{zI - H_0} , \quad G(z) = \frac{1}{zI - H} . \tag{10.67}$$

It is easy to verify that $G(z)$ satisfies the equation

$$G(z) = G_0(z) + \lambda G_0(z) H_1 G(z) . \tag{10.68}$$

Solving it by iteration, we can write $G(z)$ as a power series expansion in λ

$$G(z) = \sum_{n=0}^{\infty} \lambda^n G_0(z)[H_1 G_0(z)]^n . \tag{10.69}$$

Let $\{|\psi_{i,\alpha}^{(0)}\rangle\}$ be an orthonormal set of eigenvectors of H_0 with eigenvalues $E_i^{(0)}$. The index $\alpha = 1, 2, \dots, g_i$ counts the possible degeneracy of the level.

For sufficiently regular perturbations, in a sense that will be specified later, and if λ is small enough, we will find a set of eigenvalues of $H, E_{i,\alpha}(\lambda)$ in a neighborhood of $E_i^{(0)}$. They may not be all different, but their multiplicities should add up to g_i. Let $|\psi_{i,\alpha}(\lambda)\rangle$ be a set of orthonormalizable eigenvectors of H with eigenvalues $E_{i,\alpha}(\lambda)$.

For sufficiently small λ we assume that it is possible to find a contour Γ_i enclosing $E_i^{(0)}$ and all the $E_{i,\alpha}(\lambda)$, and excluding any other point in the spectra of H_0 and H. Then the projection P_i on the subspace generated by $|\psi_{i,\alpha}(\lambda)\rangle$ satisfies

$$P_i = \frac{1}{2\pi \mathrm{i}} \int_{\Gamma_i} G(z)\, dz . \tag{10.70}$$

We now expand P_i in a power series of λ. To achieve this we substitute (10.69) in (10.70) and exchange the order of summation and integration:

$$\boxed{P_i = P_i^{(0)} + \sum_{n=1}^{\infty} \lambda^n A_i^{(n)}} \,, \qquad A_i^{(n)} \equiv \frac{1}{2\pi \mathrm{i}} \int_{\Gamma_i} G_0(z)[H_1 G_0(z)]^n dz \,, \tag{10.71}$$

where $P_i^{(0)}$ is the projection onto the subspace generated by $|\psi_{i,\alpha}^{(0)}\rangle$, $\alpha = 1,2,\ldots,g_i$. If $Q_i^{(0)} \equiv I - P_i^{(0)}$, the resolvent $G_0(z)$ can be written as

$$G_0(z) = \frac{P_i^{(0)}}{z - E_i^{(0)}} + \frac{Q_i^{(0)}}{(z - E_i^{(0)})I + E_i^{(0)}I - H_0} \tag{10.72}$$

and expanding in a power series of $(z - E_i^{(0)})$, we obtain:

$$G_0(z) = \sum_{k=0}^{\infty} (-1)^{k-1} (z - E_i^{(0)})^{k-1} S_i^{(k)} \,, \qquad S_i^{(0)} \equiv -P_i^{(0)} \,, \quad S_i^{(k)} \equiv (S_i)^k \,, \quad k \geq 1 \,, \quad S_i \equiv \frac{Q_i^{(0)}}{E_i^{(0)} I - H_0} \,. \tag{10.73}$$

As a consequence of (10.71), the only singularity appearing in the expression for $A_i^{(n)}$ is a pole of order $n+1$ at $z = E_i^{(0)}$. Taking into account (10.73) and choosing the contour Γ_i so that the series (10.73) converges in norm for any point in it we easily obtain

$$\boxed{A_i^{(n)} = -\sum_{(n)} S_i^{(k_1)} H_1 S_i^{(k_2)} H_1 \ldots H_1 S_i^{(k_{n+1})}} \,, \tag{10.74}$$

where $\sum_{(p)}$ is a sum over all combinations satisfying $k_1 + k_2 + \ldots + k_{n+1} = p$, with $k_1, k_2, \ldots, k_{n+1} \geq 0$. Similarly, starting with (10.66) we can derive:

$$\boxed{\begin{aligned} HP_i &= E_i^{(0)} P_i + \sum_{n=1}^{\infty} \lambda^n B_i^{(n)} \,, \\ B_i^{(n)} &= \sum_{(n-1)} S_i^{(k_1)} H_1 S_i^{(k_2)} H_1 \ldots H_1 S_i^{(k_{n+1})} \,. \end{aligned}} \tag{10.75}$$

The equations (10.71, 74) and (10.75) are the basic formulae needed to solve the problem. Indeed, the eigenvalues and eigenvectors of H, that we wish to find, are the eigenvalues and eigenvectors of HP_i in the subspace with projection P_i, and the equations derived allow us to construct this operators to any order in λ. Hence the problem is reduced to diagonalizing a matrix in a space of dimension g_i.

The explicit expansions of P_i and HP_i to second order in λ are

$$\begin{aligned} P_i = P_i^{(0)} &+ \lambda[P_i^{(0)} H_1 S_i + S_i H_1 P_i^{(0)}] + \lambda^2 [P_i^{(0)} H_1 S_1 H_1 S_i \\ &+ S_i H_1 P_i^{(0)} H_1 S_i + S_i H_1 S_i H_1 P_i^{(0)} - P_i^{(0)} H_1 P_i^{(0)} H_1 S_i^{(2)} \\ &- P_i^{(0)} H_1 S_i^{(2)} H_1 P_i^{(0)} - S_i^{(2)} H_1 P_i^{(0)} H_1 P_i^{(0)}] + \mathrm{O}(\lambda^3) , \end{aligned} \tag{10.76}$$

$$\begin{aligned} (H - E_i^{(0)}) P_i = \lambda P_i^{(0)} H_1 P_i^{(0)} &+ \lambda^2 [P_i^{(0)} H_1 P_i^{(0)} H_1 S_i \\ &+ P_i^{(0)} H_1 S_i H_1 P_i^{(0)} + S_i H_1 P_i^{(0)} H_1 P_i^{(0)}] + \mathrm{O}(\lambda^3) . \end{aligned} \tag{10.77}$$

1) Non-degenerate Level: $g_i = 1$. In this case $E_i = \mathrm{Tr}\,(HP_i)$, and taking (10.75) into account

$$E_i = E_i^{(0)} + \sum_{n=1}^{\infty} \lambda^n \mathrm{Tr}\,\{B_i^{(n)}\} , \tag{10.78}$$

which should be equivalent to the expansion given in (10.40). This is straightforward to verify, since

$$\begin{aligned} \mathrm{Tr}\,\{B_i^{(1)}\} &= \mathrm{Tr}\,\{P_i^{(0)} H_1 P_i^{(0)}\} = \langle \psi_i^{(0)} | H_1 | \psi_i^{(0)} \rangle = E_i^{(1)} \\ \mathrm{Tr}\,\{B_i^{(2)}\} &= \mathrm{Tr}\,\{P_i^{(0)} H_1 P_i^{(0)} H_1 S_i + P_i^{(0)} H_1 S_i H_1 P_i^{(0)} + S_i H_1 P_i^{(0)} H_1 P_i^{(0)}\} \\ &= \mathrm{Tr}\,\{P_i^{(0)} H_1 S_i H_1 P_i^{(0)}\} = \sum_{j \neq i} \frac{|\langle \psi_i^{(0)} | H_1 | \psi_j^{(0)} \rangle|^2}{E_i^{(0)} - E_j^{(0)}} = E_i^{(2)} \end{aligned} \tag{10.79}$$

and so on. We have used that the cyclic property of the trace implies $\mathrm{Tr}\,\{S_i M P_i^{(0)}\} = 0$ for any operator M. The third-order correction to the energy becomes

$$E_i^{(3)} = \langle \psi_i^{(0)} | H_1 S_i H_1 S_i H_1 | \psi_i^{(0)} \rangle - \langle \psi_i^{(0)} | H_1 | \psi_i^{(0)} \rangle \langle \psi_i^{(0)} | H_1 S_i^{(2)} H_1 | \psi_i^{(0)} \rangle , \tag{10.80}$$

which agrees with (10.20). The eigenvector of H corresponding to the eigenvalue $E_i(\lambda)$ is obtained immediately from $|\psi_i\rangle = P_i|\psi_i^{(0)}\rangle$ and using the expansion (10.71), whose coefficients appear in (10.74).

We should point out however that $P_i|\psi_i^{(0)}\rangle$ does not coincide in general with the vector $|\psi_i(\lambda)\rangle$ introduced in Sect. 10.2. They are related by

$$|\psi_i(\lambda)\rangle = \frac{1}{\langle \psi_i^{(0)} | P_i | \psi_i^{(0)} \rangle} P_i |\psi_i^{(0)}\rangle \tag{10.81}$$

and neither of them is necessarily normalized.

2) Degenerate Level: $g_i > 1$. Let $\mathcal{E}_i^{(0)}$ and $\mathcal{E}_i$ be the g_i-dimensional spaces determined by the projections $P_i^{(0)}$ and P_i respectively. For λ sufficiently small the norm $\|P_i^{(0)} - P_i\|$ will be small , and we will have two bijective maps $P_i : \mathcal{E}_i^{(0)} \to \mathcal{E}_i$ and $P_i^{(0)} : \mathcal{E}_i \to \mathcal{E}_i^{(0)}$. On the other hand, the operators

$$H_i \equiv P_i^{(0)} H P_i P_i^{(0)} , \quad K_i \equiv P_i^{(0)} P_i P_i^{(0)} \tag{10.82}$$

are hermitian in $\mathcal{E}_i^{(0)}$, and since according to the previous remarks any eigenvector $|\psi_{i,\alpha}\rangle$ of H in $\mathcal{E}_i$ with eigenvalue $E_{i,\alpha}$ can be written as $|\psi_{i,\alpha}\rangle = P_i|\psi_{i,\alpha}^{(0)}\rangle$, with $|\psi_{i,\alpha}^{(0)}\rangle \in \mathcal{E}_i^{(0)}$, we have

$$HP_i|\psi_{i,\alpha}^{(0)}\rangle = E_{i,\alpha}P_i|\psi_{i,\alpha}^{(0)}\rangle \,. \tag{10.83}$$

The necessary and sufficient condition for this relation in $\mathcal{E}_i$ to be satisfied, is that its projection to $\mathcal{E}_i^{(0)}$ holds:

$$P_i^{(0)}HP_i|\psi_{i,\alpha}^{(0)}\rangle = E_{i,\alpha}P_i^{(0)}P_i|\psi_{i,\alpha}^{(0)}\rangle \,. \tag{10.84}$$

Using the definitions (10.82), the previous equation can be written as

$$H_i|\psi_{i,\alpha}^{(0)}\rangle = E_{i,\alpha}K_i|\psi_{i,\alpha}^{(0)}\rangle \,, \tag{10.85}$$

which is the desired equation. We have therefore found that the eigenvalues of H approaching $E_i^{(0)}$ as $\lambda \to 0$ are given by the g_i roots of the equation

$$\det|H_i - EK_i| = 0 \,. \tag{10.86}$$

To obtain the corresponding eigenvectors of H, one simply acts with the projection operator P_i on the eigenvectors of (10.85) which belong to $\mathcal{E}_i^{(0)}$ to obtain vectors in $\mathcal{E}_i$.

Taking into account the expansions obtained for P_i and HP_i, it is straightforward to show that

$$\begin{aligned} K_i &= P_i^{(0)} - \lambda^2 P_i^{(0)} H_1 S_i^{(2)} H_1 P_i^{(0)} + \ldots \\ H_i &= E_i^{(0)} K_i + \lambda P_i^{(0)} H_1 P_i^{(0)} + \lambda^2 P_i^{(0)} H_1 S_i H_1 P_i^{(0)} + \ldots \,. \end{aligned} \tag{10.87}$$

For its particular interest, we explicitly write down (10.86) to second order of perturbation theory. If $|\varphi_{i,\alpha}\rangle$ is a set of g_i orthonormal eigenvectors in $\mathcal{E}_i^{(0)}$, then (10.86) becomes

$$\det|\lambda\langle\varphi_{i,\beta}|H_1|\varphi_{i,\alpha}\rangle + \lambda^2\langle\varphi_{i,\beta}|H_1S_iH_1|\varphi_{i,\alpha}\rangle - (E - E_i^{(0)})\delta_{\beta\alpha}| = 0 \,, \tag{10.88}$$

where we have neglected a term $\lambda^2(E - E_i^{(0)})$ which is of higher order, because $E - E_i^{(0)} = \mathrm{O}(\lambda)$. It may happen that $\langle\varphi_{i,\beta}|H_1|\varphi_{i,\alpha}\rangle = \delta_{\beta\alpha}E_i^{(0)}$. In this case which appears often in physical applications, the degeneracy is not lifted to first order in perturbation theory, and (10.88) reduces to

$$\det|\lambda^2\langle\varphi_{i,\beta}|H_1S_iH_1|\varphi_{i,\alpha}\rangle - (E - E_i^{(0)} - \lambda E_i^{(1)})\delta_{\beta\alpha}| = 0 \,. \tag{10.89}$$

This concludes our formal exposition of Kato's method for stationary perturbations to any order in the interaction.

There are mathematical conditions on H_0 and H_1 which justify the formal methods presented. Unfortunately, these conditions are very restrictive and difficult to verify in practical cases. To be concrete [KA 80], if there are two non-negative constants a and b such that

$$\|H_1|\psi\rangle\| \le a\|H_0|\psi\rangle\| + b\||\psi\rangle\| \tag{10.90}$$

for all $|\psi\rangle \in D(H_0)$, then P_i defined by (10.71) is an absolutely convergent power series for all λ satisfying

$$|\lambda| < r_0 \equiv d/2[b + a(d + |E_i^{(0)}|)] , \tag{10.91}$$

where d is the distance from $E_i^{(0)}$ to the rest of the spectrum of H_0. If under the perturbation the degeneracy of $E_i^{(0)}$ is not lifted, the eigenvalue $E_i(\lambda)$ has the same analyticity domain as P_i; however, when the initial degeneracy is partially or totally lifted, the new eigenvalues and projection operators are analytic around $\lambda = 0$, but their radii of convergence can be smaller than that of P_i. The condition (10.91) takes a particularly simple form when the operator H_1 is bounded:

$$|\lambda| < r_0 = \frac{d}{2\|H_1\|} . \tag{10.92}$$

As remarked before, the *regularity* condition (10.90) can be too restrictive in cases of interest. For example, it is not satisfied for perturbations of the type λx^4 to the harmonic oscillator ($\lambda > 0$) or for physical phenomena as important as the Stark or Zeeman effects. Nevertheless, in these and many other cases all the terms of the formal perturbative series are finite (this is not the case for a harmonic oscillator with perturbation $\lambda \exp(x^4)$, $\lambda > 0$ [DK 75]), and even though it is divergent (λx^4, Stark effect), or possibly divergent (Zeeman effect), they are Borel summable asymptotic series whose sum gives the eigenvalue of $H_0 + \lambda H_1$ corresponding to $E_i^{(0)}$ for the cases of λx^4 and the Zeeman effect (the Stark effect is discussed in the next section). When the perturbations are not regular, there are no general results about the behavior of the perturbative series, and all we can do is to study many different particular cases. For example, if $H_0 = -\Delta - 1/r$, $H_1 = 1/r$, the perturbative expansion for the ground state is convergent with sum $-(1-\lambda)^2/4$ whatever λ, whereas for $\lambda \ge 1$, $H_0 + \lambda H_1$ has no eigenvalues [SI 82a]. A less trivial case is provided by [HS 78]:

$$H(\lambda) = a^\dagger(\lambda)a(\lambda) , \quad a(\lambda) \equiv \frac{d}{dx} + x + \lambda x^2 , \tag{10.93}$$

giving the harmonic oscillator $-d^2/dx^2 + x^2 - 1$ perturbed by $-2\lambda x + 2\lambda x^3 + \lambda^2 x^4$. With this perturbation, the perturbative series in powers of λ for the ground state vanishes identically, while the true ground state energy $E_0(\lambda)$ of $H(\lambda)$ satisfies $E_0(\lambda) > 0$, $\forall \lambda \neq 0$. In spite of this, this series is asymptotic, for $\lambda \to 0$, to $E_0(\lambda)$ because $E_0(\lambda) < c\exp(-d/\lambda^2)$, $c, d > 0$. Situations like this appear quite often in supersymmetric models [WI 81].

10.8 The Stark Effect

Stark [ST 13] discovered that in the presence of electric fields each line in the Balmer series of hydrogen splits into a certain number of components. This effect as noted already by *Stark*, is typical of all atoms and is today known as the Stark effect. Using the Bohr model of the atom *Schwarzschild* [SC 14] gave a theoretical explanation of it. The quantitative agreement between theory and experiment was excellent, and this provided considerable evidence which helped consolidate Bohr's model.

We now proceed to calculate the Stark effect for hydrogen-like atoms and for electric fields small enough so that the numerical results of the truncated asymptotic perturbative expansion make sense. Let $\mathcal{E}$ be the electric field strength, directed along the positive z-axis. Omitting spins and the energy of the nucleus in the field (whose contributions are negligible) the Hamiltonian of the system is

$$H = H_0 + H_1$$
$$H_0 = -\frac{1}{2\mu}\Delta - \frac{Ze^2}{r} , \quad H_1 = -e\mathcal{E}z . \tag{10.94}$$

If $\mathcal{E} = 0$ the energy levels $E_n^{(0)}$ of one-electron atoms are n^2-times degenerate, and it is convenient to take the n^2 orthonormal functions necessary to carry out the perturbative treatment in parabolic coordinates $\psi_{n_1 n_2 m}(\xi, \eta, \phi)$. We need to evaluate the matrix elements $\langle n_1' n_2' m' | H_1 | n_1 n_2 m \rangle$ and the advantage of using these coordinates is that, as we will see, this matrix is already diagonal. This does not happen if we compute $\langle n' l' m' | H_1 | n l m \rangle$ in spherical coordinates. In parabolic coordinates the interaction is

$$H_1 = -\frac{e\mathcal{E}}{2}(\xi - \eta) . \tag{10.95}$$

Combining (6.134, 135, 137) we easily find that, if $n_1 + n_2 + |m| + 1 = n = n_1' + n_2' + |m'| + 1$, then

$$\langle n_1' n_2' m' | H_1 | n_1 n_2 m \rangle = -\delta_{m'm}\delta_{n_1' n_1}\delta_{n_2' n_2}$$
$$\times \frac{e\mathcal{E}a}{4}\left[\int_0^\infty dx\, x^2 f_{n_1 m}^2(x) - \int_0^\infty dx\, x^2 f_{n_2 m}^2(x)\right] \tag{10.96}$$

and using (A.15):

$$\int_0^\infty dx\, x^2 f_{pm}^2(x) = 6[p^2 + p|m| + p] + m^2 + 3|m| + 2 , \tag{10.97}$$

so that

$$\langle n_1' n_2' m' | H_1 | n_1 n_2 m \rangle = -\delta_{m'm}\delta_{n_1' n_1}\delta_{n_2' n_2}\frac{3e\mathcal{E}a}{2}n(n_1 - n_2) . \tag{10.98}$$

Hence, the energy to first order of perturbation theory is

$$E_{n,k} = \frac{1}{2}\mu(Z\alpha c)^2\frac{1}{n^2} - \frac{3}{2}e\mathcal{E}\frac{\hbar}{\mu(Z\alpha c)}nk$$
$$k = n_1 - n_2 = 0, \pm 1, \ldots, \pm(n-1) \tag{10.99}$$

and the multiplicity of the sublevel k is $n-|k|$. Approximating μ in the first-order correction by the mass of the electron and with V/cm units for $\mathcal{E}$ we obtain

$$E_{n,k} = -R\frac{Z^2}{n^2} - 7.94 \times 10^{-9}\frac{nk\mathcal{E}}{Z}\text{eV}\,, \tag{10.100}$$

where R is Rydberg's constant for the one-electron atom considered. This equation gives a good explanation for the Stark effect in hydrogen, up to electric fields of 10^5 V/cm. The ground state ($n = 1$) does not change to first order while all others produce $(2n-1)$ levels, one of which coincides with the unperturbed one and the reamining $2(n-1)$ are symmetrically distributed around it. These sublevels are equidistant in energy, with an energy difference proportional to $n\mathcal{E}$.

The quadratic Stark effect has been studied by *Epstein* [EP 16], *Wentzel* [WE 26], *Waller* [WA 26] and *Van Vleck* [VL 26], and the cubic Stark effect is treated by *Ishida* and *Hiyama* [IH 28]. The quadratic correction to (10.100) is

$$E^{(2)}_{n,k,m} = -\frac{1}{16}\frac{\hbar^2}{\mu^3(Z\alpha c)^4}(e\mathcal{E})^2 n^4[17n^2 - 3k^2 - 9m^2 + 19]$$
$$= -6.43 \times 10^{-22}\frac{\mathcal{E}^2}{Z^4}n^4[17n^2 - 3k^2 - 9m^2 + 19]\,\text{eV}\,. \tag{10.101}$$

In hydrogen, and for electric fields up to 10^6 V/cm equation (10.101) has been confirmed experimentally.

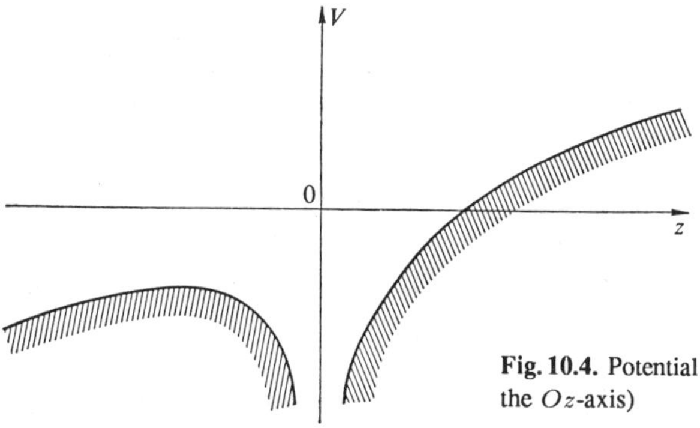

Fig. 10.4. Potential energy in the Stark effect (along the Oz-axis)

The potential $V(\boldsymbol{r}) = -(Ze^2/r) - e\mathcal{E}z$ is plotted in Fig. 10.4 for $x = y = 0$. We see that the ionization energy is decreased by an amount $\Delta E = 2|e|\sqrt{|e|\mathcal{E}Z}$ and therefore all the bound states in hydrogen-like atoms with binding energy smaller than ΔE belong now to the continuous spectrum. Furthermore, those electrons in bound states with binding energy larger than ΔE will acquire a probability of penetrating the potential barrier in the presence of an electric field, and become free electrons. This produces a natural broadening of the spectral lines, according to the uncertainty principle. This effect was studied theoretically by *Oppenheimer* [OP 28] and verified experimentally in hydrogen by *Traubenberg* and *Gebauer* [TG 29], [GT 30] and by these authors together with *Lewin* [TG 30].

The rigorous analysis of the perturbative series for the Stark effect was initiated by *Conley* and *Rejto* [CR 66], who proved the spectral concentration property, and revived by *Graffi* and *Grecchi* [GG 78a], who proved the Borel summability. Today we know, among other things, that $\sigma(H) = \sigma_{\mathrm{a.c.}}(H) = \mathbb{R}$, for any real value of $\mathcal{E} \neq 0$, and that the perturbative series [PR 80]

$$E_0(\mathcal{E}) = -\frac{1}{2} - \frac{9}{4}\mathcal{E}^2 - \frac{3555}{64}\mathcal{E}^4 - \frac{2512779}{512}\mathcal{E}^6 - \cdots \tag{10.102}$$

for the ground state (in atomic units) is divergent, although Borel summable, whose sum [also denoted by $E_0(\mathcal{E})$] is a complex number! The real part Re $E_0(\mathcal{E})$ gives the mean position of the energy band originated by the ground state due to the interaction with the electric field, and the imaginary part Im $E_0(\mathcal{E})$ is negative (if $\mathcal{E} \neq 0$) and provides the width $\Gamma_0(\mathcal{E}) \equiv -2\mathrm{Im}\, E_0(\mathcal{E})$ of the energy band. Asymptotically the width behaves as

$$\Gamma_0(\mathcal{E}) \underset{\mathcal{E}\to 0}{\sim} \frac{4}{|\mathcal{E}|} \exp\left(-\frac{2}{3|\mathcal{E}|}\right) . \tag{10.103}$$

This result is due to Oppenheimer (but for corrections). The exponentially decreasing behavior of $\Gamma_0(\mathcal{E})$ for $\mathcal{E} \to 0$ suggests that the spectral projection of H in an interval of width $\mathrm{O}(|\mathcal{E}|^n)$ around Re $E_0(\mathcal{E})$ converges to the projection of H_0 onto the state $|1s\rangle$ in this limit, for all n (this is known as the phenomenon of *spectral concentration*). More information and details can be found in [HU 79], [TH 81], [SI 82a].

10.9 The Variational Method

The method presented in this section is particularly useful in those cases where the Hamiltonian of the problem is not easily decomposable in the form $H = H_0 + \lambda H_1$, with H_0 exactly soluble and λH_1 treatable as a small perturbation. The variational method yields good numerical estimates whenever some physical or mathematical reasoning gives an approximate idea of the behavior of the wave functions. Let H be the Hamiltonian of the problem. Suppose for simplicity the spectrum to be discrete with energies $E_0, E_1, E_2, \ldots, E_i, \ldots$, in increasing order. Let $|\psi_{i,\alpha}\rangle$ be the corresponding orthonormal eigenstates where the index $\alpha = 1, 2, \ldots, g_i$ counts the possible degeneracy of the level E_i. Any state $|\psi\rangle$ can be written as

$$|\psi\rangle = \sum_i \sum_{\alpha=1}^{g_i} c_{i,\alpha} |\psi_{i,\alpha}\rangle , \tag{10.104}$$

so that

$$\langle\psi|H|\psi\rangle = \sum_i \sum_{\alpha=1}^{g_i} E_i|c_{i,\alpha}|^2 \geq E_0 \sum_i \sum_{\alpha=1}^{g_i} |c_{i,\alpha}|^2 = E_0\langle\psi|\psi\rangle \tag{10.105}$$

and therefore

$$\boxed{E_0 \leq \frac{\langle\psi|H|\psi\rangle}{\langle\psi|\psi\rangle}} \; . \tag{10.106}$$

In other words, for any state $|\psi\rangle$ the expectation value of the energy gives an upper bound on the ground state energy. If $\mathcal{E}_i$ is the subspace spanned by $|\psi_{i,\alpha}\rangle$, $\alpha = 1, 2, \dots, g_i$ and if $|\psi\rangle$ belongs to $\mathcal{E}_0$ then $\langle\psi|H|\psi\rangle/\langle\psi|\psi\rangle = E_0$.

In the applications of this method, it is customary to begin with a family of states, known as the *trial functions* or *states*, depending on n parameters $\lambda_0, \lambda_1, \dots, \lambda_n$. With them we compute the expectation value of H and we obtain an upper bound on E_0 depending on these parameters. The result obtained is then minimized with respect to the λ's and in this way we obtain the best bound of the whole family.

It is convenient to realize that any normalized state $|\psi\rangle$ can be written in the form $|\psi\rangle = P_{\mathcal{E}_0}|\psi\rangle + \varepsilon|\varphi\rangle$, $|\varphi\rangle \perp \mathcal{E}_0$, where $P_{\mathcal{E}_0}$ is the orthogonal projection onto $\mathcal{E}_0$ and $|\varphi\rangle$ is normalized. The inequality (10.106), together with $\|P_{\mathcal{E}_0}|\psi\rangle\|^2 + |\varepsilon|^2 = 1$ yields

$$E_0 \leq \langle H\rangle_\psi = E_0 + |\varepsilon|^2[\langle\varphi|H|\varphi\rangle - E_0] \; . \tag{10.107}$$

Hence if the state $|\psi\rangle$ differs from one in $\mathcal{E}_0$ by terms of order ε, the bound obtained for the energy coincides with E_0 up to terms of order $|\varepsilon|^2$. One should keep in mind however that in the bound (10.107) the coefficient of $|\varepsilon|^2$ can sometimes be very large.

If the space $\mathcal{E}_0$ is known, we can choose $|\psi\rangle$ so that $|\psi\rangle \perp \mathcal{E}_0$ and then repeat the previous argument. Since now $c_{0,\alpha} = 0$ for all α, we have

$$E_1 \leq \frac{\langle\psi|H|\psi\rangle}{\langle\psi|\psi\rangle} \quad \text{if} \quad |\psi\rangle \perp \mathcal{E}_0 \tag{10.108}$$

and by the same token we obtain an upper bound on E_i if $|\psi\rangle$ is orthogonal to the subspace $\mathcal{E}_0 \oplus \mathcal{E}_1 \oplus \dots \oplus \mathcal{E}_{i-1}$.

In practice, the subspaces $\mathcal{E}_0, \dots, \mathcal{E}_{i-1}$ are generally unknown. Sometimes, there are some quantum numbers associated to constants of motion, whose wave functions have a known structure. This happens for example with angular momentum. Under these circumstances, it suffices to choose trial functions with this quantum number fixed to be sure that the expectation values of the energy for the states are not lower than the energy of the levels with this quantum number. If for instance the problem has spherical symmetry and we take trial functions all with the same angular momentum J and third component M, we obtain an upper bound on the energy of the ground level corresponding to a J-wave.

If the ground state is non-degenerate and we have some approximate knowledge of the normalized eigenvector corresponding to E_0, say $|\bar{\psi}_0\rangle$ to be distinguished from the actual (normalized) state $|\psi_0\rangle$, we might take a trial state $|\psi\rangle$ orthogonal to $|\bar{\psi}_0\rangle$ and normalized. We can write $|\psi\rangle = \varepsilon|\psi_0\rangle + \sqrt{1-|\varepsilon|^2}|\psi_\perp\rangle$ with $|\psi_\perp\rangle$ orthogonal to $|\psi_0\rangle$. Then

$$\langle H\rangle_\psi = |\varepsilon|^2 E_0 + (1-|\varepsilon|^2)\langle\psi_\perp|H|\psi_\perp\rangle \tag{10.109}$$

and since $\langle\psi_\perp|H|\psi_\perp\rangle \geq E_1$:

$$E_1 + |\varepsilon|^2(E_0 - E_1) \leq \langle H\rangle_\psi\,. \tag{10.110}$$

Therefore, $\langle H\rangle_\psi$ gives a good bound on E_1 if $|\varepsilon|^2(E_0 - E_1)$ is negligible.

Since $|\psi\rangle \perp |\bar{\psi}_0\rangle$, ε will be smaller, the closer $|\psi_0\rangle$ is to $|\bar{\psi}_0\rangle$. Indeed

$$|\varepsilon|^2 = |\langle\psi_0|\psi\rangle|^2 = \langle\psi_0|I - P_{\bar{\psi}_0}|\psi\rangle^2 \leq \|(I - P_{\bar{\psi}_0})\psi_0\|^2\,, \tag{10.111}$$

where $P_{\bar{\psi}_0} = |\bar{\psi}_0\rangle\langle\bar{\psi}_0|$. Hence,

$$|\varepsilon|^2 \leq 1 - |\langle\bar{\psi}_0|\psi_0\rangle|^2\,. \tag{10.112}$$

Since we also have

$$\begin{aligned}\overline{E}_0 \equiv \langle\bar{\psi}_0|H|\bar{\psi}_0\rangle &= \langle P_{\psi_0}\bar{\psi}_0|H|P_{\psi_0}\bar{\psi}_0\rangle + \langle(I-P_{\psi_0})\bar{\psi}_0|H|(I-P_{\psi_0})\bar{\psi}_0\rangle \\ &\geq E_0\|P_{\psi_0}\bar{\psi}_0\|^2 + E_1\|(I-P_{\psi_0})\bar{\psi}_0\|^2\,,\end{aligned} \tag{10.113}$$

then

$$\overline{E}_0 - E_0 \geq (1 - |\langle\bar{\psi}_0|\psi_0\rangle|^2)(E_1 - E_0)\,. \tag{10.114}$$

Combining (10.110, 112) and (10.114) the final result is

$$E_1 \leq \langle H\rangle_\psi + [\overline{E}_0 - E_0]\,, \tag{10.115}$$

an inequality generalizing (10.108).

The inequalities (10.106, 108) and (10.115) are valid even if $\sigma(H)$ is not purely discrete. In fact, the first two are a direct consequence of the *Min-Max Principle* (Appendix C), valid for any self-adjoint operator. Due to its importance, we reproduce it here, in a slightly more general version, and only for semi-bounded operators:

"If H is the semi-bounded Hamiltonian of a system, and

$$E_0 \leq E_1 \leq \ldots [< E_* \equiv \inf\sigma_{\mathrm{es}}(H)] \tag{10.116}$$

are the eigenvalues of H to the left of its essential spectrum, each one repeated as many times as indicated by its multiplicity, then it can be shown [TH 81] that

$$\begin{aligned}E_i + E_{i+1} + \ldots + E_{i+j-1} &= \inf_{D_{i+j}\subset D(H)}\ \sup_{D_j\subset D_{i+j}} [\mathrm{Tr}_{D_j} H] \\ &= \sup_{D_i\subset D(H)}\ \inf_{\substack{D_j\perp D_i \\ D_j\subset D(H)}} [\mathrm{Tr}_{D_j} H]\end{aligned} \tag{10.117}$$

where $\mathrm{Tr}_{D_j} H$ is the trace of the matrix representing H in any orthonormal basis of D_j. If the number of eigenvalues $< E_*$ were finite, say N, counting multiplicities, (10.117) should be interpreted by taking $E_k \equiv E_*$, $\forall k \geq N$. When $i+j-1 < N$, the family of subspaces D_{i+j}, D_i appearing in the inf and sup of (10.117) can be restricted to subspaces spanned by eigenvectors of H, without altering its validity."

The inequality (10.115) is now easy to prove. Take in (10.117) $i = 0$, $j = 2$:

$$E_0 + E_1 = \inf_{D_2 \subset D(H)} [\mathrm{Tr}_{D_2} H] \tag{10.118}$$

hence, if $D_2 = \mathrm{span}\{\bar{\psi}_0, \psi\}$, we have

$$E_0 + E_1 \leq \mathrm{Tr}_{D_2} H = \langle H \rangle_{\bar{\psi}_0} + \langle H \rangle_\psi = \overline{E}_0 + \langle H \rangle_\psi \tag{10.119}$$

equivalent to (10.115). Similarly, (10.106) and (10.108) are consequences of (10.117).

A very important result following from the Min-Max principle is the *Rayleigh-Ritz method*: If $D_n \subset D(H)$, then the eigenvalues $E_0(D_n) \leq E_1(D_n) \leq \ldots \leq E_n(D_n)$ of the matrix representing H in D_n provide upper bounds $E_0(D_n) \geq E_0$, $E_1(D_n) \geq E_1, \ldots, E_n(D_n) \geq E_n$. To understand this, it suffices to notice that according to (10.117)

$$E_i(D_n) = \inf_{D_{i+1} \subset D_n} \sup_{\substack{\phi \in D_{i+1} \\ (\phi \neq 0)}} \langle H \rangle_\phi \geq \inf_{D_{i+1} \subset D(H)} \sup_{\substack{\phi \in D_{i+1} \\ (\phi \neq 0)}} \langle H \rangle_\phi = E_i \,. \tag{10.120}$$

As the dimension of the test space D_n increases (i.e. considering a sequence $D_1 \subset D_2 \subset \ldots \subset D_n \subset \ldots$) the levels $E_i(D_n)$ will decrease, and it is reasonable to expect that under certain conditions they approximate E_i as closely as one wishes [MI 64]. For example, the basis of the harmonic oscillator is appropriate to analyze in the Rayleigh-Ritz method the effect on it of an interaction described by an arbitrary semi-bounded polynomial $P(x)$ [RS 78].

The variational method provides upper bounds. Its ideal complement would be a procedure giving lower bounds, so that with both methods combined, we would narrow the energy eigenvalues to certain intervals. There are several techniques to obtain lower bounds. Among them, *Temple*'s *inequality* [TE 28] stands out for its simplicity: If $E_0 \equiv \inf \sigma(H) \in \sigma_{\mathrm{dis}}(H)$, and $\overline{E} \equiv \inf[\sigma(H) \backslash \{E_0\}]$, then the inequality $(H - E_0)(H - \overline{E}) \geq 0$ is obvious, and taking expectation values in a normalized state ϕ, we obtain:

$$\langle H^2 \rangle_\phi - (E_0 + \overline{E})\langle H \rangle_\phi + E_0 \overline{E} \geq 0 \tag{10.121}$$

hence, if $\langle H \rangle_\phi < \overline{E}$, we have

$$\langle H \rangle_\phi - \frac{\Delta_\phi^2 H}{\overline{E} - \langle H \rangle} \leq E_0 \,, \tag{10.122}$$

which is the inequality mentioned. In its applications it suffices to have a lower bound estimate of $\overline{E}$.

For example, let H be the Hamiltonian (10.28) for the anharmonic oscillator in units $2M = \hbar = k/2 = 1$. Taking for ϕ the ground state $\psi_0^{(0)}$ of the harmonic oscillator (10.29), a simple computation leads to

$$\langle H \rangle_\phi = 1 + \tfrac{3}{4}\lambda\,, \quad \Delta_\phi^2 H = 6\lambda^2 \tag{10.123}$$

and therefore, Temple's inequality becomes

$$1 + \frac{3}{4}\lambda - \frac{1}{\overline{E}(\lambda) - (1 + 3\lambda/4)} \leq E_0(\lambda)\,, \quad \text{if} \quad 1 + \frac{3}{4}\lambda < \overline{E}(\lambda)\,. \tag{10.124}$$

Since the perturbation λx^4 is repulsive ($\lambda \geq 0$) and even, restricting ourselves to the even parity subspace we can take $\overline{E}(\lambda) \geq E_2(0) = 5$, so that

$$1 + \frac{3}{4}\lambda - \frac{6\lambda^2}{4 - 3\lambda/4} \leq E_0(\lambda)\,, \quad 0 \leq \lambda \leq \frac{16}{3}\,. \tag{10.125}$$

The upper bound $\langle H \rangle_\phi = 1 + 3\lambda/4 \geq E_0(\lambda)$ can easily be improved with little extra effort if we apply the Rayleigh-Ritz method with $D_2 = \{\psi_0^{(0)}, \psi_2^{(0)}\}$, i.e., the space spanned by the first two even eigenvectors of the harmonic oscillator. The matrix representing H in D_2 is

$$H_{D_2} = \begin{pmatrix} 1 + \frac{3}{4}\lambda & \frac{3}{\sqrt{2}}\lambda \\ \frac{3}{\sqrt{2}}\lambda & 5 + \frac{39}{4}\lambda \end{pmatrix}, \tag{10.126}$$

whose lowest eigenvalue gives an upper bound on $E_0(\lambda)$ better than $1 + 3\lambda/4$. Thus, for $\lambda = 0.2$ we obtain, using (10.125) and (10.126): $1.08766 \leq E_0(0.2) \leq 1.11913$, to be compared with the exact value $E_0(0.2) = 1.11829$.

For more sophisticated methods (intermediate Hamiltonians, projection techniques, etc.), of great utility in atomic and molecular physics, see [WS 72], [GH 78], [TH 81].

10.10 Two-Electron Atoms (II)

The problem of computing the ground state energy for a nucleus of charge $Z|e|$ with two bound electrons, treated in Sect. 10.5 from the point of view of perturbation theory, will be analyzed here using the variational method. The Hamiltonian appears in (10.46), and if the interaction terms were absent, the space part of the wave function for the ground state 1^1S_0 (in the notation $n\,^{2S+1}L_J, n =$ order of the level) is

$$\psi(\boldsymbol{r}_1, \boldsymbol{r}_2) = (\pi a^3)^{-1} e^{-(r_1 + r_2)/a}\,, \tag{10.127}$$

where $a = \hbar/\mu(Z\alpha c)$. We can take (10.127) as a family of trial functions, leaving a as an arbitrary parameter. It is straightforward to obtain

$$\langle\psi| - \frac{1}{2\mu}\Delta_1|\psi\rangle = \langle\psi| - \frac{1}{2\mu}\Delta_2\psi\rangle = \frac{1}{2\mu a^2} ,$$
$$\langle\psi| - \frac{Ze^2}{r_1}|\psi\rangle = \langle\psi| - \frac{Ze^2}{r_2}|\psi\rangle = -\frac{Ze^2}{a} . \tag{10.128}$$

On the other hand we know from Sect. 10.5 that

$$\langle\psi|\frac{1}{r_{12}}|\psi\rangle = \frac{5}{8a} , \tag{10.129}$$

Since $|\psi\rangle$ is normalized, the expectation value of the energy is

$$\langle\psi|H|\psi\rangle = \mu(\alpha c)^2 Z' \left[Z' - 2Z + \tfrac{5}{8}\right] . \tag{10.130}$$

with $Z' \equiv \hbar/\mu a(\alpha c)$. Expression (10.130) has a minimum at $Z' = Z - 5/16$, indicating the extent to which one electron sees the nucleus screened by the other. For the ground state energy we obtain the inequality

$$E_0 \leq E_{\text{var}} = -\mu(Z\alpha c)^2 \left(1 - \frac{5}{16Z}\right)^2 \tag{10.131}$$

and comparing this result with (10.51),

$$E_{\text{pert}} = E_{\text{var}} + \frac{25}{256}\mu(\alpha c)^2 = E_{\text{var}} + 2.6573879(88)\frac{\mu}{m_e}\,\text{eV} . \tag{10.132}$$

It is therefore clear, in this case, that the variational computation gives a better approximation to the energy than the first order of perturbation theory (see Table 10.3).

To improve the agreement between theory and experiment it suffices in principle to choose a wider family of trial functions [HY 30]. Notice first of all that in the sector ${}^1S_0$ the Hamiltonian H has some level with an energy smaller than E_{var} given in (10.131). Moreover, in any other sector ${}^{2S+1}L_J$ the positivity of the perturbation implies that the energy is larger than $-(5/8)\mu(Z\alpha c)^2$. Hence, the ground state will be in the sector ${}^1S_0$ as long as $Z > 1.50$, and to compute it, we can restrict its variational search to functions symmetric under the exchange $\boldsymbol{r}_1 \leftrightarrow \boldsymbol{r}_2$ and spin singlet. The electrostatic repulsion between the electrons induces a correlation in their positions which is not incorporated in the choice (10.127). Functions often used including these features are

$$\psi(\boldsymbol{r}_1, \boldsymbol{r}_2) = (\pi a^3)^{-1} e^{-(r_1+r_2)/a}[1 + Af(r_1, r_2, r_{12})] . \tag{10.133}$$

Sometimes these trial functions depend on hundreds of parameters [KI 57], [PE 62]. With them it has been possible to obtain excellent variational estimates and rather sharp lower bounds. For example, it is known that for He in atomic units $-2.9037474 \leq E_0 \leq -2.9037237$ [BA 60]. And there are computations indicating that $E_0 = -2.9037244$ [FP 66]. For a rigorous discussion of this problem see [GH 78], [TH 81].

Before considering other examples it is instructive to study the energies for the levels 2^1P and 2^3P in the helium atom ($Z = 2$). The variational method is particularly useful in this case because it automatically provides upper bounds to the lowest levels with $L = 1$ and $S = 0$ and 1. In this discussion $\psi^{(i)}_{nlm}(\boldsymbol{r})$ will stand for the normalized position space wave function of an electron in a one-electron atom with charge $Z_i|e|$ and quantum numbers (n, l, m), and χ^σ_s will be the normalized spin wave function for a two-electron system with spin s and third component σ. The dependence of $\chi^\sigma_s(m_1 m_2)$ on the third components of the individual spins is such that $\sum_{m_1 m_2} \chi^\sigma_s(m_1 m_2)|m_1 m_2\rangle$ is the state with total spin s and third component σ. It is constructed using the Clebsch-Gordan coefficients. It would seem reasonable to begin with trial functions:

$$\begin{aligned} 2^3P &: \psi(\xi_1, \xi_2) = \psi^{(1)}_{100}(\boldsymbol{r}_1)\psi^{(2)}_{21m}(\boldsymbol{r}_2)\chi^\sigma_1 \,, \\ 2^1P &: \psi(\xi_1, \xi_2) = \psi^{(1)}_{100}(\boldsymbol{r}_1)\psi^{(2)}_{21m}(\boldsymbol{r}_2)\chi^0_0 \,, \end{aligned} \tag{10.134}$$

where ξ_i are the space and spin coordinates of electron i, and taking Z_1 and Z_2 as variational parameters. These trial functions are inadequate for we are not explicitly including the antisymmetry of the total wave function for a two-electron system with respect to the exchange of ξ_1 and ξ_2, as indicated in Chap. 8 and as will be discussed extensively in Sect. 13.2. [The wave function (10.127) is multiplied by χ^0_0 and it is therefore antisymmetric as expected.] Taking this into account the correct normalized trial functions are

$$\begin{aligned} 2^3P &: \psi_-(\xi_1, \xi_2) = \frac{1}{\sqrt{2}}[\psi^{(1)}_{100}(\boldsymbol{r}_1)\psi^{(2)}_{210}(\boldsymbol{r}_2) - \psi^{(1)}_{100}(\boldsymbol{r}_2)\psi^{(2)}_{210}(\boldsymbol{r}_1)]\chi^0_1 \,, \\ 2^1P &: \psi_+(\xi_1, \xi_2) = \frac{1}{\sqrt{2}}[\psi^{(1)}_{100}(\boldsymbol{r}_1)\psi^{(2)}_{210}(\boldsymbol{r}_2) + \psi^{(1)}_{100}(\boldsymbol{r}_2)\psi^{(2)}_{210}(\boldsymbol{r}_1)]\chi^0_0 \,, \end{aligned} \tag{10.135}$$

where we have chosen $m = \sigma = 0$, because the rotational symmetry and conservation of spin by H implies that its expectation values are independent of this choice. One immediately obtains

$$\begin{aligned} \langle\psi_\pm|H|\psi_\pm\rangle &= \langle\psi^{(1)}_{100}(\boldsymbol{r}_1)|H_{01}|\psi^{(1)}_{100}(\boldsymbol{r}_1)\rangle + \langle\psi^{(2)}_{210}(\boldsymbol{r}_1)|H_{01}|\psi^{(2)}_{210}(\boldsymbol{r}_1)\rangle \\ &\quad + I(Z_1, Z_2) \pm J(Z_1, Z_2) \\ I(Z_1, Z_2) &\equiv e^2\langle\psi^{(1)}_{100}(\boldsymbol{r}_1)\psi^{(2)}_{210}(\boldsymbol{r}_2)|r_{12}^{-1}|\psi^{(1)}_{100}(\boldsymbol{r}_1)\psi^{(2)}_{210}(\boldsymbol{r}_2)\rangle \\ J(Z_1, Z_2) &\equiv e^2\langle\psi^{(1)}_{100}(\boldsymbol{r}_1)\psi^{(2)}_{210}(\boldsymbol{r}_2)|r_{12}^{-1}|\psi^{(1)}_{100}(\boldsymbol{r}_2)\psi^{(2)}_{210}(\boldsymbol{r}_1)\rangle \,. \end{aligned} \tag{10.136}$$

The quantity $I(Z_1, Z_2)$ is known as the *direct integral* and $J(Z_1, Z_2)$ is the *exchange integral*. The exchange integral is a direct consequence of the antisymmetry of the wave function and it would not have appeared had we used (10.134) as trial functions; this would have resulted in $E(2^1P) = E(2^3P)$, contradicting experimental evidence. Using the explicit form of the wave functions for the hydrogen-like atom given in Sect. 6.5 yields

$$\begin{aligned}\langle\psi^{(1)}_{100}(\boldsymbol{r}_1)|H_{01}|\psi^{(1)}_{100}(\boldsymbol{r}_1)\rangle &= \mu(\alpha c)^2\left(\tfrac{1}{2}Z_1^2 - ZZ_1\right) ,\\ \langle\psi^{(2)}_{210}(\boldsymbol{r}_1)|H_{01}|\psi^{(2)}_{210}(\boldsymbol{r}_1)\rangle &= \mu(\alpha c)^2\left(\tfrac{1}{8}Z_2^2 - \tfrac{1}{4}ZZ_2\right) .\end{aligned} \tag{10.137}$$

The evaluation of the I and J integrals is carried out by means of (A.135):

$$\begin{aligned}I(Z_1, Z_2) &= \frac{e^2}{6a_1^3a_2^5}\int_0^\infty dr_1\int_0^\infty dr_2\,\frac{r_1^2r_2^4}{r_>}\,\mathrm{e}^{-2r_1/a_1}\mathrm{e}^{-r_2/a_2}\\ J(Z_1, Z_2) &= \frac{e^2}{18a_1^3a_2^5}\int_0^\infty dr_1\int_0^\infty dr_2\,\frac{r_<r_1^3r_2^3}{r_>^2}\\ &\quad\times\exp\left[-\left(\frac{1}{a_1}+\frac{1}{2a_2}\right)(r_1+r_2)\right] ,\end{aligned} \tag{10.138}$$

where $a_i \equiv \hbar/\mu(Z_i\alpha c)$. A straightforward integration leads to

$$\begin{aligned}I(Z_1, Z_2) &= \mu(\alpha c)^2\left[\frac{Z_2}{4} - \frac{Z_1Z_2^5}{(2Z_1+Z_2)^5} - \frac{Z_2^5}{4(2Z_1+Z_2)^4}\right] ,\\ J(Z_1, Z_2) &= \mu(\alpha c)^2\frac{112Z_1^3Z_2^5}{3(Z_1+Z_2)^7} .\end{aligned} \tag{10.139}$$

All we have to do now is to minimize the expectation value of the Hamiltonian with respect to the parameters Z_1 and Z_2. A numerical computation leads to the results given in Table 10.4. Notice first of all that the exact results are reproduced with an error of about 2 per 100. The energy splitting between the levels 2^3P and 2^1P is 0.253922 eV, to be compared with the value 0.225858 eV due entirely to the exchange integral. The values of Z_1 and Z_2 are reasonable because the $1s$ electron feels a nuclear charge $\simeq 2|e|$, while the $2p$ electron feels a "nuclear" charge $\simeq |e|$ due to the screening produced by the $1s$ electron. It is important to emphasize that we can predict why the state 2^3P would be the one most tightly bound. For this state the spatial wave function is antisymmetric and thus the probability for both electrons to be close is very small, and this reduces the repulsive Coulomb force acting between them. As will be seen later on, this is a general effect known as Hund's rule.

Table 10.4. Results in eV of the variational computation of the energies of the 2^3P and 2^1P states in the helium atom. In the next to last column, we give the ionization energy computed according to the techniques discussed in the text. In the last column we give a very accurate value of this quantity

State	Z_1	Z_2	$<\psi\|H\|\psi>$	$-(\langle\psi\|H\|\psi\rangle - 2\mu(\alpha c)^2$	Ionization energy
2^3P	1.991 2	1.089 2	−57.971 678	3.555 838	3.624 094
2^1P	2.003 0	0.964 7	−57.745 820	3.329 980	3.370 172

10.11 Van der Waals Forces (II)

In Sect. 10.6 the perturbative method allowed us to estimate a lower bound for the ground state energy of two hydrogen atoms far apart. We now proceed to find an upper bound using the variational method.

The first question to address is a suitable choice for the family of trial functions. We could take those given in (10.55) leaving a_0 as an arbitrary parameter, but since $H_1 \propto R^{-3}$, this would give a bound on the energy varying as R^{-3}. To achieve the expected dependence R^{-6} we choose as trial functions:

$$\psi(\boldsymbol{r}_1, \boldsymbol{r}_2) = (\pi a_0)^{-3} \mathrm{e}^{-(r_1+r_2)/a_0}[1 + AH_1] , \tag{10.140}$$

where A is a real variational parameter and H_1 is the function given in (10.56). The average value of the energy is

$$\langle H\rangle_\psi \equiv \frac{\langle\psi|H_0 + H_1|\psi\rangle}{\langle\psi|\psi\rangle} = \frac{\langle 0|(1 + AH_1)(H_0 + H_1)(1 + AH_1)|0\rangle}{\langle 0|(1 + AH_1)^2|0\rangle} , \tag{10.141}$$

where the wave function corresponding to $|0\rangle$ is (10.140) with $A = 0$. Parity arguments imply:

$$\langle H\rangle_\psi = \frac{E_0 + 2A\langle 0|H_1 H_1|0\rangle + A^2\langle 0|H_1 H_0 H_1|0\rangle}{1 + A^2\langle 0|H_1 H_1|0\rangle} . \tag{10.142}$$

Using the results of Sect. 10.6 we have $E_0 = -\mu(\alpha c)^2$, $\langle 0|H_1 H_1|0\rangle = 6(ea_0)^4/R^6$. To evaluate the remaining term we proceed as follows:

$$\langle 0|H_1 H_0 H_1|0\rangle = \langle 0|H_1[H_0, H_1]|0\rangle - \frac{6\mu(\alpha c)^2(ea_0)^4}{R^6} . \tag{10.143}$$

Taking into account the expressions for H_0 and H_1 given respectively in (10.54) and (10.56),

$$\langle 0|H_1[H_0, H_1]|0\rangle = -\frac{\hbar^2 e^4}{\mu R^6}\langle 0|[(\boldsymbol{r}_1\cdot\boldsymbol{r}_2)-3(\boldsymbol{r}_1\cdot\hat{\boldsymbol{R}})(\boldsymbol{r}_2\cdot\hat{\boldsymbol{R}})]^2\frac{1}{r_1}\frac{\partial}{\partial r_1}|0\rangle \tag{10.144}$$

and using (10.60):

$$\langle 0|H_1[H_0, H_1]|0\rangle = -\frac{4\hbar^2 e^4}{3\mu R^6}\langle 0|r_1^2 r_2^2\frac{1}{r_1}\frac{\partial}{\partial r_1}|0\rangle = \frac{6\mu(\alpha c)^2(ea_0)^4}{R^6} , \tag{10.145}$$

so that $\langle 0|H_1 H_0 H_1|0\rangle = 0$. We finally obtain

$$\begin{aligned}\langle H\rangle_\psi &= \frac{-\mu(\alpha c)^2 + [12(ea_0)^4 A]/R^6}{1 + [6(ea_0)^4 A^2]/R^6} \\ &\simeq -\mu(ac)^2 + \frac{6(ea_0)^4}{R^6}[2A + \mu(\alpha c)^2 A^2] ,\end{aligned} \tag{10.146}$$

where we have used $R \gg a_0$ and that we are only interested in terms up to order R^{-6}. The minimum for $\langle H\rangle_\psi$ is found at $A = -1/\mu(\alpha c)^2$, and the best bound on the energy obtained with our family of trial functions is

$$E_{\text{var}}(R) = -\mu(\alpha c)^2 \left[1 + 6\left(\frac{a_0}{R}\right)^6\right] . \quad (10.147)$$

Since according to perturbation theory $\xi \leq 8$, from (10.57), we conclude:

$$6 \leq \xi \leq 8 , \quad (10.148)$$

to be compared with the more accurate value $\xi = 6.499027$ [KO 67].

The attraction between neutral Coulomb systems is not only true for hydrogen atoms. It is a universal fact. Recently, *Lieb* and *Thirring* [LT 85] have shown, using variational techniques, that the ground state energy $E_0(R)$ of a system composed of two neutral components with a distance R between their centers of mass always satisfies an inequality of the form

$$E_0(R) \leq E_0^{(1)} + E_0^{(2)} - cR^{-6} , \quad c > 0 , \quad (10.149)$$

where $E_0^{(i)}$ is the ground state energy of the ith component, and c is a constant depending only on their intrinsic properties. This inequality always holds when R is larger than the sum of the diameters of the two components. We point out that this result does not rely on the dipole-dipole approximation to describe the interactions.

10.12 One-Electron Atoms

The Hamiltonian describing these atoms is

$$H = -\frac{\hbar^2}{2\mu}\Delta - \frac{Ze^2}{r} . \quad (10.150)$$

The exact solutions for its bound states were given in Chap. 6, but it is educationally interesting to study this problem via the variational method. In the ground state the wave function is spherically symmetric. Let us consider here three families of normalized trial functions:

$$\psi_1(r) = \sqrt{\frac{b^3}{\pi}} e^{-br} , \quad \psi_2(r) = \sqrt{\frac{b^3}{\pi}} (b^2 r^2 + 1)^{-1} , \quad \psi_3(r) = \sqrt{\frac{b^5}{3\pi}} r e^{-br} \quad (10.151)$$

with $b > 0$. A trivial computation shows that

$$\langle\psi_1|H|\psi_1\rangle = \frac{\hbar^2 b^2}{2\mu} - Ze^2 b , \quad \langle\psi_2|H|\psi_2\rangle = \frac{\hbar^2 b^2}{4\mu} - \frac{2Ze^2 b}{\pi} , \quad \langle\psi_3|H|\psi_3\rangle = \frac{\hbar^2 b^2}{6\mu} - \frac{Ze^2 b}{2} \quad (10.152)$$

and, therefore, the values of b corresponding to the lowest energy are

$$b_1 = \frac{1}{a}, \quad b_2 = \frac{4}{\pi a}, \quad b_3 = \frac{3}{2a}, \quad a \equiv \frac{\hbar}{\mu(Z\alpha c)}, \tag{10.153}$$

leading to the variational energies

$$E_1 = -\frac{1}{2}\mu(Z\alpha c)^2, \quad E_2 = -\frac{4}{\pi^2}\mu(Z\alpha c)^2, \quad E_3 = -\frac{3}{8}\mu(Z\alpha c)^2. \tag{10.154}$$

E_1, not very surprisingly, coincides with the exact energy, since the exact wave function belongs to the family ψ_1. In the other cases we find $E_2 = (8/\pi^2)E_1 \simeq 0.8106E_1$ and $E_3 = 0.75E_1$, i.e., the family ψ_2 gives a better bound to the energy than the family ψ_3. The functions ψ_1, ψ_2, ψ_3 for the parameters (10.153) are represented in Fig. 10.5.

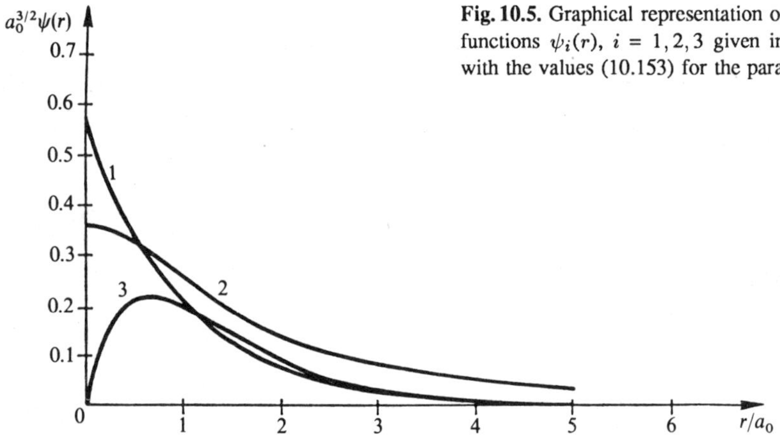

Fig. 10.5. Graphical representation of the wave functions $\psi_i(r)$, $i = 1, 2, 3$ given in (10.151), with the values (10.153) for the parameters

Let ψ_{0i} be the wave function in the family ψ_i giving the best bound on the energy; the previously introduced quantity

$$|\varepsilon_i|^2 \equiv 1 - |\langle\psi_0|\psi_{0i}\rangle|^2 \tag{10.155}$$

gives an idea of how close ψ_{0i} is to the actual wave function ψ_0. We find that

$$\begin{aligned} \varepsilon_1 &= 0 \\ |\varepsilon_2|^2 &= 1 - 4\left\{1 - \frac{\pi}{4\sqrt{2}}\left[\frac{\pi}{2} + \mathrm{Ci}\left(\frac{\pi}{4}\right) - \mathrm{Si}\left(\frac{\pi}{4}\right)\right]\right\}^2 \simeq 0.2035, \\ |\varepsilon_3|^2 &= 1 - 2^9 3^6 5^{-8} \simeq 0.0445, \end{aligned} \tag{10.156}$$

where $\mathrm{Ci}(x)$ and $\mathrm{Si}(x)$ are the so-called sine and cosine integrals [AS 64]. Hence ψ_{03} gives a better approximation to the wave function than ψ_{02}, even though it gave a worse bound on the energy. This is because its behavior at the origin is $\propto r$, making the Coulomb attraction in this region less relevant. The fact that $|\varepsilon_3|$ is small even if E_3 is not a good estimate for the energy, is a good example of the warnings made concerning the inequality (10.107). Similarly if $\langle r\rangle_i = \langle\psi_{0i}|r|\psi_{0i}\rangle$, we find,

$$\langle r\rangle_1 = \tfrac{3}{2}a\,, \quad \langle r\rangle_2 = \infty\,, \quad \langle r\rangle_3 = \tfrac{5}{3}a\,. \tag{10.157}$$

These results clearly indicate that the accuracy in the estimate of the energy given by a wave function does not necessarily imply that it will also give a good approximation to other physical observables, and vice versa.

If we want to determine the binding energy of the first bound state with a given angular momentum l, we may use the family of normalized trial functions

$$\psi_{lm} = \left[\frac{2^{2l+3}b^{2l+3}}{(2l+2)!}\right]^{1/2} r^l e^{-br} Y_l^m(\hat{\boldsymbol{r}}) \tag{10.158}$$

orthogonal to any other wave function with angular momentum $l' \neq l$ and with the correct behavior as $r \to 0$. A simple calculation gives

$$\langle\psi_{lm}|H|\psi_{lm}\rangle = \frac{\hbar^2 b^2}{2\mu} - \frac{Ze^2 b}{l+1}\,, \tag{10.159}$$

with a minimum at $b = 1/(l+1)a$. Consequently the variational estimate of the energy is $E_l = -\mu(Z\alpha c)^2/2(l+1)^2$, in agreement with the exact result, once again because the familiy considered contains the exact wave function.

10.13 Eigenvalues for Large Coupling Constants

Central potentials behaving like $1/r$ at the origin are often found in atomic and nuclear physics. With the exception of the Coulomb potential, none of these can be solved in terms of elementary functions for all orbital angular momentum. A typical example is the Yukawa potential, for which there are good numerical estimates for the lower eigenvalues in $l = 0$ waves, obtained with variational methods [HL 51].

We want to present here a procedure [GP 75] which calculates $E_{nl}(g)$ for a central potential $V(r)$ such that $V(r) \sim -2g/r$, when $r \to 0$, as a power series in $1/g$. The basic idea is very simple: for given quantum numbers n, l, the coupling constant g can be made sufficiently large so that the difference between the given potential and the Coulomb potential of the same strength can be considered relatively small and therefore treated as a regular perturbation. The perturbation theory developed in Sects. 10.2 and 10.7 gives a method to evaluate $E_{nl}(g)$ as a convergent series essentially in powers of $1/g$. If the coefficients of this series, which in general depend on g, are expanded in powers of $1/g$, $E_{nl}(g)$ will become a power series in $1/g$ with constant coefficients. The series thus obtained may have lost its convergence, but it may still maintain its asymptotic character. When we treat this series using Padé approximants in $1/g$, the expression obtained is generally valid for those values of g for which the (n, l) wave is bound. We will compute $E_{nl}(g)$ only up to terms of order $O(1/g^2)$, and we refer the reader interested in more details to the original paper [GP 75].

The reduced radial Schrödinger equation ($\hbar = 2M = 1$) is

$$\left[-\frac{d^2}{dr^2} + \frac{l(l+1)}{r^2} - \frac{2g}{r}v(r)\right] u_{nl}(r) = E_{nl}u_{nl}(r) \, . \tag{10.160}$$

We assume $g > 0$, and the expansion

$$v(r) = 1 + v_0 r + v_1 r^2 + v_2 r^3 + \dots \tag{10.161}$$

and we split the Hamiltonian (10.160) in the form:

$$\begin{aligned} H &= H_0 + V(r) \\ H_0 &= -\frac{d^2}{dr^2} + \frac{l(l+1)}{r^2} - \frac{2g}{r} \\ V(r) &= -\frac{2g}{r}[v(r) - 1] = -2g[v_0 + v_1 r + v_2 r^2 + \dots] \, . \end{aligned} \tag{10.162}$$

According to the arguments in Sect. 10.2 we can write

$$\begin{aligned} E_{nl} &= E_{nl}^{(0)} + E_{nl}^{(1)} + E_{nl}^{(2)} + \dots \, , \\ E_{nl}^{(0)} &= -g^2/n^2 \, , \\ E_{nl}^{(1)} &= \langle nl|V(r)|nl\rangle \, , \\ E_{nl}^{(2)} &= \langle nl|V(r)|nl^{(1)}\rangle \, , \end{aligned} \tag{10.163}$$

where $|nl\rangle$ is the exact solution to the Coulomb problem with Hamiltonian H_0 (in this section n is the principal quantum number). The next terms in this expansion are calculable by formulae analogous to (10.20) and (10.21). It is easy to realize that $\langle nl|r^p|nl\rangle = \langle p\rangle g^{-p}$, $p \geq 0$, where $\langle p\rangle$ is a pure number; in particular

$$\langle 1\rangle = \frac{1}{2}[3n^2 - l(l+1)] \, , \quad \langle 2\rangle = \frac{n^2}{2}[5n^2 + 1 - 3l(l+1)] \, . \tag{10.164}$$

We thus obtain

$$E_{nl}^{(1)} = -2g\left[v_0 + v_1\langle 1\rangle\frac{1}{g} + v_2\langle 2\rangle\frac{1}{g^2}\right] + \mathrm{O}(1/g^2) \, . \tag{10.165}$$

It is also possible to prove that in $E_{nl}^{(k)} (k \geq 2)$ the dominant term is $1/g^{2(k-1)}$, and therefore

$$E_{nl} = -\frac{g^2}{n^2} - 2v_0 g - 2v_1\langle 1\rangle - 2v_2\langle 2\rangle\frac{1}{g} + \mathrm{O}(1/g^2) \, , \tag{10.166}$$

which is expected to give a good asymptotic expansion for large values of g.

If g_{nl} is the threshold value of g; i.e., the state (n, l) is bound only if $g \geq g_{nl}$, then [TA 72]

$$E_{n0} \sim -\text{const.} \times (g - g_{n0})^2 \,, \quad g \downarrow g_{n0} \,, \qquad E_{nl} \sim -\text{const.} \times (g - g_{nl}) \,, \quad g \downarrow g_{nl} \,, \quad l \neq 0 \,. \tag{10.167}$$

This suggests that the quantity to be approximated via Padé's is $(n/g)\sqrt{-E_{n0}}$ for $l = 0$ waves and $-(n/g)^2 E_{nl}$ for $l \neq 0$ waves. The quasi-diagonal approximants of highest order, which can be computed from (10.166) are [1,2] and [2,1].

Restricting our attention to a Yukawa potential, $v(r) = \mathrm{e}^{-r}$, and therefore $v_0 = -1$, $v_1 = 1/2$, $v_2 = -1/6, \ldots$, we obtain, for instance

$$\sqrt{-E_{n0}}^{[1,2]} = \frac{g}{n} \frac{1 + (1 - n^2)/3g - n^2(4 + 5n^2)/12g^2}{1 + (1 + 2n^2)/3g}$$

$$-E_{nl}^{[1,2]} = \frac{g^2}{n^2} \frac{1 + (-2n^2 + \langle 2\rangle/3\langle 1\rangle)/g + n^2(\langle 1\rangle - 2\langle 2\rangle/3\langle 1\rangle)/g^2}{1 + \langle 2\rangle/3\langle 1\rangle g} \,, \tag{10.168}$$

$l \neq 0$.

The threshold couplings in this approximation, $g_{nl}^{(1,2)}$, are given by the highest root of the polynomial in the numerator. We obtain, for example,

$$g_{n0}^{[1,2]} = \tfrac{1}{6}[n^2 - 1 + \sqrt{16n^4 + 10n^2 + 1}] \,, \tag{10.169}$$

giving $g_{10}^{(1,2)} = 0.8660$, $g_{20}^{(1,2)} = 3.372$, $g_{30}^{(1,2)} = 7.540, \ldots$ to be compared with the variational values [RG 70] $g_{10} = 0.8399$, $g_{20} = 3.223$, $g_{30} = 7.171, \ldots$.

Table 10.5. Values of g necessary to produce a given binding energy

	[1, 2]	[2, 3]	[3, 4]	[4, 5]	[5, 6]	Numerical
g_{10}	0.866 025 4	0.844 547 9	0.840736 8	0.840 046 7	0.839 923 7	0.839 9
g_{20}	3.372 281	3.261 391	3.234 803	3.226 999	3.224 536	3.223
g_{21}	5.236 068	4.788 544	4.680 991	4.628 623	4.595 768	4.541
g_{30}	7.540 408	7.277 918	7.210 490	7.186 518	7.176 660	7.171
g_{31}	10.778 52	9.428 641	9.150 784	9.068 761	9.007 249	8.872
g_{32}	12.147 81	11.331 69	11.142 05	11.050 25	11.004 51	10.947

In [GP 75] the general theory is explained and detailed computations for the Yukawa potential have been carried out up to terms of order $O(1/g^9)$ for the energies, which allows us to compute $E_{nl}^{[5,6]}$. In Table 10.5 we give the values of the coupling constant needed to produce a given energy E_{10}, using Padé approximants; and the comparison of these results with the variational calculations [HL 51]. We find a very good agreement, which improves as $-E_{10}$ increases.

In Table 10.6 we compute g_{nl} for different Padé approximants and they are compared with existing computations [RG 70]. Finally in Fig. 10.6 we plot $-E_{nl}^{[5,6]}(g)$ as a function of g for the lowest bound states.

Table 10.6. Threshold values g_{nl} for different Padé approximants, and their known numerical values [RG 70]

$-E_{10}$	$g^{[1,2]}$	$g^{[3,4]}$	$g^{[5,6]}$	g variational
0	0.866 025 40	0.840 736 76	0.839 923 68	0.839 909 75
$(0.05)^2$	0.919 776 51	0.896 909 61	0.896 250 89	0.896 240 5
$(0.15)^2$	1.026 643 3	1.007 691 1	1.007 245 6	1.007 240
$(0.35)^2$	1.238 308 5	1.224 723 7	1.224 500 4	1.224 498 65
1^2	1.914 213 6	1.908 349 3	1.908 310 4	1.908 310 35
2^2	2.936 491 7	2.934 147 4	2.934 141 8	2.934 141 8
4^2	4.958 039 9	4.957 358 8	4.957 358 4	4.957 358 3
9^2	9.977 225 6	9.977 113 9	9.977 113 9	9.977 113 9

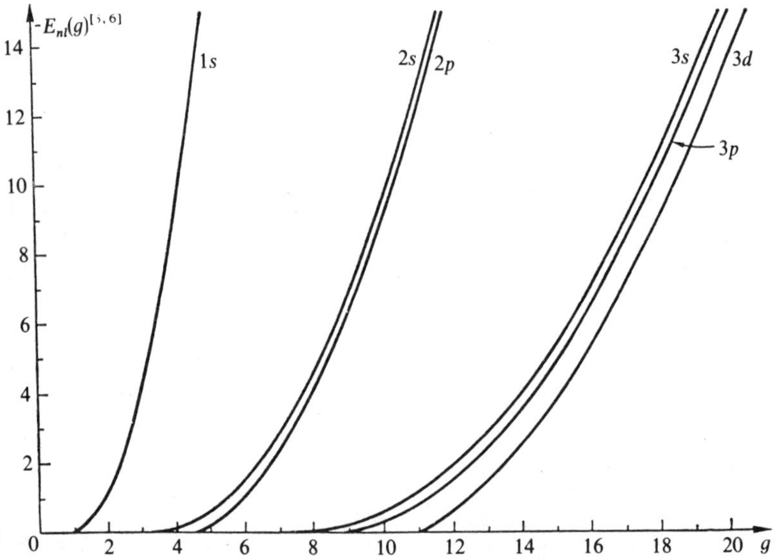

Fig. 10.6. Binding energies as functions of the coupling constant for the Yukawa potential

11. Time-Dependent Perturbation Theory

11.1 Introduction

In previous chapters we have learned how to solve either exactly or approximately the time-independent Schrödinger equation. This is however not enough in general, for there are many instances where one is required to understand the time evolution of a quantum system. If the Hamiltonian in the Schrödinger picture is $H(t)$, the equation we want to integrate is

$$i\hbar\frac{d|\Psi(t)\rangle}{dt} = H(t)|\Psi(t)\rangle\ , \tag{11.1}$$

assuming $|\Psi(t_0)\rangle$ is known. This is equivalent to the computation of the operator $U(t, t_0)$ introduced in Chap. 2. In most cases the mathematical structure of $H(t)$ is so complicated that no practical methods of computation are available.

Sometimes, the Hamiltonian can be decomposed in the form

$$H(t) = H_0 + H_1(t)\ , \tag{11.2}$$

where H_0 has no explicit time dependence, and it determines the dynamics of an unperturbed conservative system, whose time evolution is described by the equation

$$i\hbar\frac{d|\Phi(t)\rangle}{dt} = H_0|\Phi(t)\rangle\ , \tag{11.3}$$

which will be assumed manageable. The explicit time dependence is entirely contained in the term $H_1(t)$ which will be treated as a small perturbation in a sense to be detailed later. Its smallness depends not only on the mathematical structure of $H_1(t)$, but also on the time elapsed since the perturbation was turned on.

Before attempting to describe time-dependent perturbation theory in detail, it is convenient to analyze some simple exactly solvable examples which illustrate the difficulties encountered in these type of problems.

11.2 Nuclear Spin Resonance

In this problem one is interested in solving the Schrödinger equation for a spin 1/2 particle in the presence of a uniform magnetic field $\boldsymbol{B}(t)$. The equation of motion is [RA 37]

$$\mathrm{i}\hbar\frac{d\Psi(t)}{dt} = -\gamma\boldsymbol{S}\cdot\boldsymbol{B}(t)\Psi(t)\,, \tag{11.4}$$

where $\Psi(t)$ is a spinor (two component object necessary to describe spin 1/2 particles), the spin operator is $\boldsymbol{S} = \hbar\boldsymbol{\sigma}/2$, with $\boldsymbol{\sigma}$ the Pauli matrices, and γ is a constant with units ($\mathrm{s}^{-1}\mathrm{G}^{-1}$). The magnetic field $\boldsymbol{B}(t)$ has the following components

$$\boldsymbol{B}(t) = (B_1\cos\omega t, B_1\sin\omega t, B_0) \tag{11.5}$$

with constant B_0 and B_1. Obviously the Hamiltonian (11.4) can be written in the form (11.2) with

$$\begin{aligned} H_0 &= -\gamma\boldsymbol{S}\cdot\boldsymbol{B_0}\,, & \boldsymbol{B_0} &\equiv (0,0,B_0)\,, \\ H_1(t) &= -\gamma\boldsymbol{S}\cdot\boldsymbol{B_1}(t)\,, & \boldsymbol{B_1}(t) &\equiv (B_1\cos\omega t, B_1\sin\omega t, 0)\,. \end{aligned} \tag{11.6}$$

If we denote by $\Psi_+(t)$ and $\Psi_-(t)$ the upper and lower components respectively of the spinor $\Psi(t)$ and introduce the constants

$$\omega_0 \equiv -\gamma B_0\,, \quad \omega_1 \equiv -\gamma B_1 \tag{11.7}$$

which will be taken to be positive, the equations of the motion become

$$\begin{aligned} \mathrm{i}\frac{d\Psi_+(t)}{dt} &= \frac{\omega_0}{2}\Psi_+(t) + \frac{\omega_1}{2}\mathrm{e}^{-\mathrm{i}\omega t}\Psi_-(t)\,, \\ \mathrm{i}\frac{d\Psi_-(t)}{dt} &= -\frac{\omega_0}{2}\Psi_-(t) + \frac{\omega_1}{2}\mathrm{e}^{\mathrm{i}\omega t}\Psi_+(t)\,. \end{aligned} \tag{11.8}$$

To solve this system we make the change of variables (equivalent to going into the interaction picture to be discussed later)

$$\Psi_\pm(t) = \mathrm{e}^{\mp\mathrm{i}\omega_0 t/2}\phi_\pm(t)\,. \tag{11.9}$$

The new $\phi_\pm(t)$ variables satisfy

$$\mathrm{i}\frac{d\phi_\pm(t)}{dt} = \frac{\omega_1}{2}\mathrm{e}^{\pm\mathrm{i}\varepsilon t}\phi_\mp(t)\,, \quad \varepsilon \equiv \omega_0 - \omega\,. \tag{11.10}$$

In the absence of interactions $\omega_1 = 0$, and $\phi_\pm$ are time-independent and equal to their initial values. Introducing

$$\xi_-(t) \equiv \mathrm{e}^{\mathrm{i}\varepsilon t}\phi_-(t)\,, \tag{11.11}$$

the previous system becomes

$$\begin{aligned} & \mathrm{i}\frac{d\phi_+(t)}{dt} = \frac{\omega_1}{2}\xi_-(t) , \\ & \mathrm{i}\frac{d\xi_-(t)}{dt} + \varepsilon\xi_-(t) = \frac{\omega_1}{2}\phi_+(t) . \end{aligned} \tag{11.12}$$

Differentiating the last equation with respect to t, and using the first, we obtain

$$\frac{d^2\xi_-(t)}{dt^2} - \mathrm{i}\varepsilon\frac{d\xi_-(t)}{dt} + \frac{\omega_1^2}{4}\xi_-(t) = 0 . \tag{11.13}$$

Thus

$$\xi_-(t) = A\mathrm{e}^{\mathrm{i}(\varepsilon+\Omega)t/2} + B\mathrm{e}^{\mathrm{i}(\varepsilon-\Omega)t/2} , \quad \Omega \equiv \sqrt{\varepsilon^2+\omega_1^2} \tag{11.14}$$

and

$$\begin{aligned} & \Psi_+(t) = -\frac{A\omega_1}{\Omega+\varepsilon}\mathrm{e}^{-\mathrm{i}(\omega-\Omega)t/2} + \frac{B\omega_1}{\Omega-\varepsilon}\mathrm{e}^{-\mathrm{i}(\omega+\Omega)t/2} , \\ & \Psi_-(t) = A\mathrm{e}^{\mathrm{i}(\omega+\Omega)t/2} + B\mathrm{e}^{\mathrm{i}(\omega-\Omega)t/2} . \end{aligned} \tag{11.15}$$

This explicit solution to the equations of motion can be summarized in the following evolution operator $U(t,0)$:

$$U(t,0) = \exp\{-\mathrm{i}\omega t\sigma_3/2\} \exp\{-\mathrm{i}[(\omega_0-\omega)\sigma_3 + \omega_1\sigma_1]t/2\} . \tag{11.16}$$

Notice that the first term corresponds to changing coordinates to a moving frame where the magnetic field $B(t)$ is constant, and lies in the $y=0$ plane.

If at time $t=0$, $\Psi_+(0)=1$ and $\Psi_-(0)=0$, the integration constants turn out to be $A=-B=-\omega_1/2\Omega$, and

$$\begin{aligned} & \Psi_+(t) = \left[\cos\frac{\Omega t}{2} - \mathrm{i}\frac{\varepsilon}{\Omega}\sin\frac{\Omega t}{2}\right]\mathrm{e}^{-\mathrm{i}\omega t/2} , \\ & \Psi_-(t) = -\mathrm{i}\frac{\omega_1}{\Omega}\sin\frac{\Omega t}{2}\mathrm{e}^{\mathrm{i}\omega t/2} . \end{aligned} \tag{11.17}$$

Hence the probability that at a time t the third component of spin which at $t=0$ was $\hbar/2$, be $-\hbar/2$ is given by

$$P(t) = |\Psi_-(t)|^2 = \frac{\omega_1^2}{(\omega_0-\omega)^2+\omega_1^2}\sin^2\left[\frac{t}{2}\sqrt{(\omega_0-\omega)^2+\omega_1^2}\right] . \tag{11.18}$$

It is obvious that when $\omega_1=0$, this transition probability vanishes. For any other value of $\omega_1 \neq 0$, the particle oscillates between the two possible spin eigenstates. The probability $P(t)$ that the third component of the spin operator be $-\hbar/2$ is largest for times t_n:

$$t_n = \frac{(2n+1)\pi}{\sqrt{(\omega_0-\omega)^2+\omega_1^2}} , \quad n=0,1,\ldots \tag{11.19}$$

and obviously

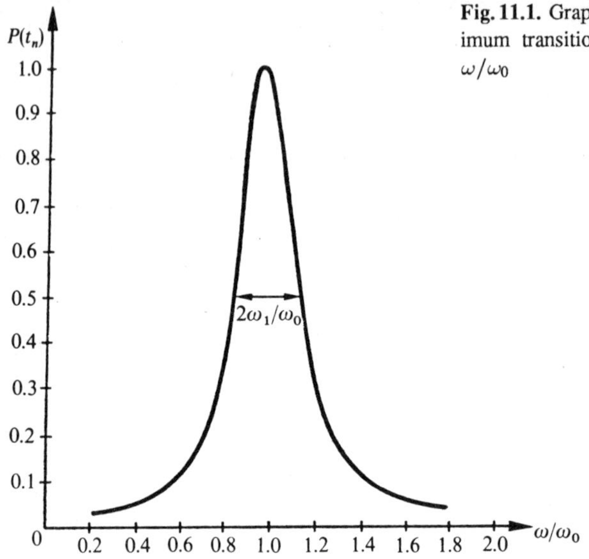

Fig. 11.1. Graphical representation of the maximum transition probability as a function of ω/ω_0

$$P(t_n) = \frac{\omega_1^2}{(\omega_0 - \omega)^2 + \omega_1^2} . \tag{11.20}$$

Notice that for given ω_0 and any $\omega_1 \neq 0$, there exists a frequency $\omega = \omega_0$ such that $P(t_n) = 1$ i.e. we obtain a resonance phenomenon. This is most easily seen by the Breit-Wigner type dependence of $P(t_n)$ in ω illustrated in Fig. 11.1. The transition probabilities are important only for frequencies ω satisfying $|\omega - \omega_0| \lesssim \omega_1$, and they attain their maximum for $\omega = \omega_0$. This highest value of the resonance happens exactly when $\hbar\omega$ equals the energy difference between the two states of the system before the interacton was turned on.

11.3 The Forced Harmonic Oscillator

The next problem we wish to consider is a forced harmonic oscillator, i.e. a system described by the Hamiltonian

$$H(t) = -\frac{\hbar^2}{2M}\frac{d^2}{dx^2} + \frac{1}{2}kx^2 - f(t)x , \tag{11.21}$$

where $f(t)$ is a real locally integrable function. The problem is quite interesting in various branches of physics such as quantum field theory, the Mössbauer effect, diffusion of neutrons and X-rays by phonons, vibrational excitations of molecules in collisions, etc. Taking into account the results obtained in Sect. 4.4, using the formalism of creation and annihilation operators allows us to express the Hamiltonian (11.21) as

$$H(t) = \hbar\omega\left[a^\dagger a + \frac{1}{2}\right] - \sqrt{\frac{\hbar}{2M\omega}} f(t)(a^\dagger + a) \equiv H_0 + H_1(t) . \tag{11.22}$$

In the integration of (11.1) we will follow the method explained in [GG 66]. Removing from $|\Psi(t)\rangle$ the time dependence introduced by the unperturbed Hamiltonian

$$|\Psi_I(t)\rangle \equiv e^{iH_0 t/\hbar}|\Psi(t)\rangle \ , \tag{11.23}$$

we obtain for $|\Psi_I(t)\rangle$ the equation

$$i\hbar\frac{d|\Psi_I(t)\rangle}{dt} = e^{iH_0 t/\hbar} H_1(t) e^{-iH_0 t/\hbar}|\Psi_I(t)\rangle \ . \tag{11.24}$$

Using (4.68)

$$i\hbar\frac{d|\Psi_I(t)\rangle}{dt} = -\sqrt{\frac{\hbar}{2M\omega}} f(t)[a e^{-i\omega t} + a^\dagger e^{i\omega t}]|\Psi_I\rangle \ . \tag{11.25}$$

Defining now the function

$$K(t) \equiv \frac{1}{\sqrt{2M\omega\hbar}}\int_0^t ds\, f(s) e^{i\omega s} \ , \tag{11.26}$$

the previous equation may be written as

$$\frac{d|\Psi_I(t)\rangle}{dt} = \left[i\frac{dK^*(t)}{dt}a + i\frac{dK(t)}{dt}a^\dagger\right]|\Psi_I(t)\rangle \ . \tag{11.27}$$

This equation is not yet easily integrable because a and $a^\dagger$ do not commute. Thus we try to eliminate one of these operators, say $a^\dagger$, by introducing a new evolution picture

$$|\Psi_R(t)\rangle \equiv e^{-iK(t)a^\dagger}|\Psi_I(t)\rangle \ . \tag{11.28}$$

Now

$$\frac{d|\Psi_R(t)\rangle}{dt} = i e^{-iK(t)a^\dagger}\frac{dK^*(t)}{dt} a e^{iK(t)a^\dagger}|\Psi_R(t)\rangle \ , \tag{11.29}$$

and since

$$e^{-iK(t)a^\dagger} a e^{iK(t)a^\dagger} = a + iK(t) \ , \tag{11.30}$$

we finally obtain

$$\frac{d|\Psi_R(t)\rangle}{dt} = i\frac{dK^*(t)}{dt}[a + iK(t)]|\Psi_R(t)\rangle \ , \tag{11.31}$$

whose integration is trivial, thus yielding

$$|\Psi(t)\rangle = e^{-iH_0 t/\hbar} e^{iK(t)a^\dagger} e^{iK^*(t)a} \exp\left[-\int_0^t ds\, K(s)\frac{dK^*(s)}{ds}\right]|\Psi(0)\rangle. \tag{11.32}$$

Hence, the transition probability between two normalized eigenstates $|\psi_n\rangle$ and $|\psi_m\rangle$ of H_0 is

$$
\begin{aligned}
P_{n\to m}(t) &= |\langle\psi_m|e^{iK(t)a^\dagger}e^{iK^*(t)a}\exp\left[-\int_0^t ds\, K(s)\frac{dK^*(s)}{ds}\right]|\psi_n\rangle|^2 \\
&= e^{-|K(t)|^2}|\langle\psi_m|e^{iK(t)a^\dagger}e^{iK^*(t)a}|\psi_n\rangle|^2 . \qquad (11.33)
\end{aligned}
$$

Taking into account that

$$
e^{iK^*(t)a}|\psi_n\rangle = \sqrt{n!}\sum_{r=0}^{n}\frac{i^{n-r}K^*(t)^{n-r}}{(n-r)!\sqrt{r!}}|\psi_r\rangle , \qquad (11.34)
$$

the amplitude (11.33) is easily seen to be

$$
\begin{aligned}
&\langle\psi_m|e^{iK(t)a^\dagger}e^{iK^*(t)a}|\psi_n\rangle \\
&= \sqrt{n!m!}\, i^{n+m}K^*(t)^n K(t)^m \sum_{s=0}^{(m,n)}\frac{(-1)^s|K(t)|^{-2s}}{(n-s)!(m-s)!s!} . \qquad (11.35)
\end{aligned}
$$

where $(m,n) \equiv \min(m,n)$. Consequently, the square root of the transition probability is, up to a sign,

$$
P^{1/2}_{n\to m}(t) = e^{-|K(t)|^2/2}\sqrt{n!m!}\sum_{s=0}^{(m,n)}\frac{(-1)^{m-s}|K(t)|^{m+n-2s}}{s!(m-s)!(n-s)!} . \qquad (11.36)
$$

Using [A. 137] it becomes

$$
\begin{aligned}
P^{1/2}_{n\to m}(t) &= \sqrt{\frac{n!}{m!}}\frac{|K(t)|^{n-m}}{(n-m)!}e^{-|K(t)|^2/2}M(-m, n-m+1, |K(t)|^2), n \geq m, \\
&= (-1)^{m-n}\sqrt{\frac{m!}{n!}}\frac{|K(t)|^{m-n}}{(m-n)!}e^{-|K(t)|^2/2} \\
&\quad \times M(-n, m-n+1, |K(t)|^2), \quad n \leq m . \qquad (11.37)
\end{aligned}
$$

Notice that $P_{n\to m}(t) = P_{m\to n}(t)$ and that the Kummer functions involved are all polynomials.

Using the recurrence relations for these functions we get

$$
P^{1/2}_{n\to m+1}(t) = \frac{n-m-|K(t)|^2}{\sqrt{m+1}|K(t)|}P^{1/2}_{n\to m}(t) - \sqrt{\frac{m}{m+1}}P^{1/2}_{n\to m-1}(t) \qquad (11.38)
$$

valid for any values of n and m. In the computation of transition probabilities it is convenient to use (11.38) together with

$$
\begin{aligned}
P^{1/2}_{n\to 0}(t) &= \frac{|K(t)|^n}{\sqrt{n!}}e^{-|K(t)|^2/2} , \\
P^{1/2}_{n\to 1}(t) &= \frac{|K(t)|^{n-1}}{\sqrt{n!}}[n-|K(t)|^2]e^{-|K(t)|^2/2} .
\end{aligned} \qquad (11.39)
$$

From (11.37) we easily find that if $|K(t)| \ll 1$ and we keep only terms up to order $|K(t)|^2$, the nonvanishing transition probabilities are those satisfying $|\Delta n| \leq 1$:

$$
\begin{aligned}
&P_{n\to n-1}(t) \simeq n|K(t)|^2 , \\
&P_{n\to n}(t) \simeq 1-(2n+1)|K(t)|^2 , \qquad (11.40)\\
&P_{n\to n+1}(t) \simeq (n+1)|K(t)|^2 .
\end{aligned}
$$

This result is obtained also using the first order perturbation theory that will be presented in the next section. As $|K(t)|$ increases, the number of final states for which the transition probability is not negligible, increases. A typical situation is plotted in Fig. 11.2.

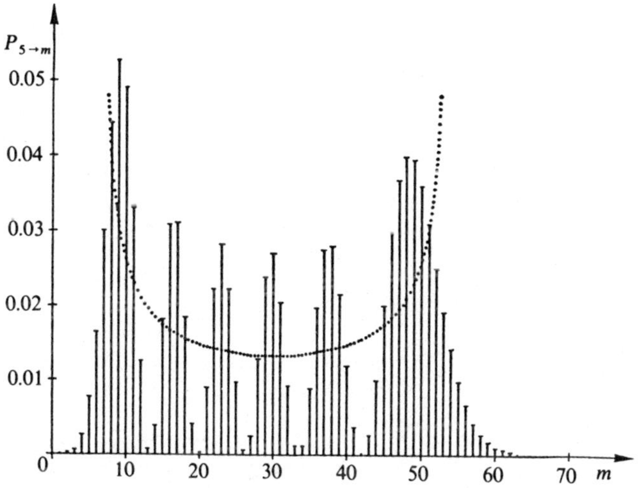

Fig. 11.2. Transition probability from the initial state $n = 5$ to a final state m for $|K(t)| = 5$. The dotted line represents the classical probability distribution for the energy when $|K(t)| = 5$

In particular the transition probability from the ground state to any state m is

$$
P_{0\to m}(t) = \frac{|K(t)|^{2m}}{m!} e^{-|K(t)|^2} , \qquad (11.41)
$$

whose behavior as a function of $|K(t)|$ appears in Fig. 11.3 for small values of m.

Obviously when $|K(t)|$ increases there are more transition probabilities to be taken into account. (See for example Fig. 11.4).

Taking logarithms in (11.41) and approximating $\ln m!$ by Stirling's formula, $\ln m! \simeq m \ln m$, it is easy to show that if $|K(t)|^2 \gg 1$ the transition probability has a maximum for m close to $|K(t)|^2$, as shown in Fig. 11.4.

In the classical study of the forced harmonic oscillator (11.21), if $z(t) = p - \mathrm{i}M\omega x$, where p is the momentum, the equation of motion is

$$
\dot{z}(t) = f(t) - \mathrm{i}\omega z(t) , \qquad (11.42)
$$

with general solution

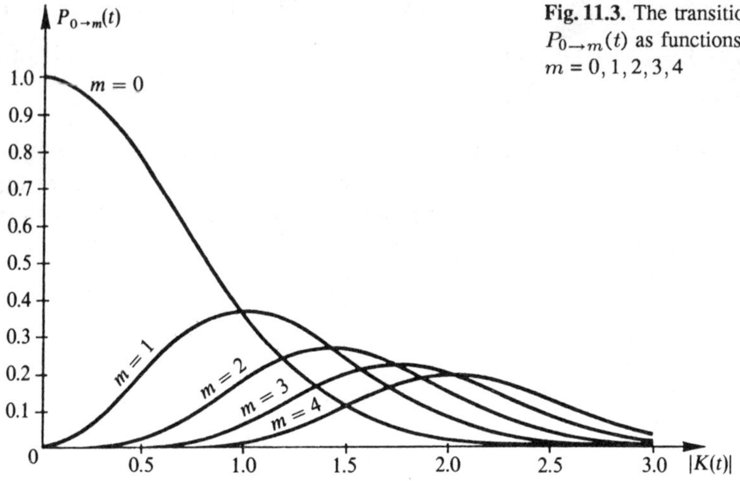

Fig. 11.3. The transition probabilities $P_{0\to m}(t)$ as functions of $|K(t)|$ for $m = 0, 1, 2, 3, 4$

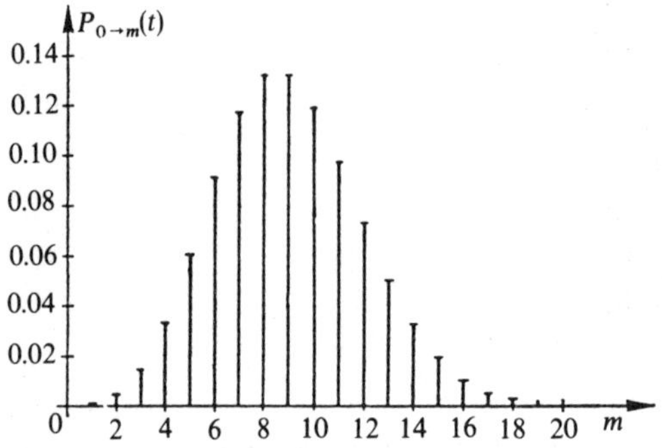

Fig. 11.4. Transition probabilities from the ground state to a state m, for $|K(t)| = 3$

$$z(t)e^{i\omega t} = z(0) + \sqrt{2M\hbar\omega}\, K(t) . \tag{11.43}$$

The energy of the harmonic oscillator at time t is $E(t) = |z(t)|^2/2M$. If $\varepsilon(t)$ is the energy measured in units of $\hbar\omega$:

$$\varepsilon(t) = \varepsilon(0) + |K(t)|^2 + 2\sqrt{\varepsilon(0)}\,|K(t)| \cos\theta(t) , \tag{11.44}$$

where $\theta(t)$ is the phase angle between $K(t)$ and $z(0)$. Thus

$$\left[\sqrt{\varepsilon(0)} - |K(t)|\right]^2 \le \varepsilon(t) \le \left[\sqrt{\varepsilon(0)} + |K(t)|\right]^2 . \tag{11.45}$$

Assuming an ensemble of oscillators with an equiprobable distribution in θ, the probability to find the energy in the interval $[\varepsilon - d\varepsilon/2, \varepsilon + d\varepsilon/2]$ is given by

$$P(\varepsilon) = \frac{1}{\pi}\left|\frac{d\theta}{d\varepsilon}\right| = \frac{1}{\pi}[4|K(t)|^2\varepsilon(0) - \{\varepsilon(0) + |K(t)|^2 - \varepsilon\}^2]^{-1/2} , \tag{11.46}$$

whose local average agrees with the quantum results if $\varepsilon(0) \gg 1$, $|K(t)| \gg 1$, according to the correspondence principle (see Fig. 11.2).

We now interpret the results obtained in some special cases. Suppose first that

$$f(t) = [2M\hbar\omega^3]^{1/2}\left[\theta(t)\theta(T-t)\frac{t}{T} + \theta(t-T)\right] , \tag{11.47}$$

where $\theta(t)$ is the step function. Namely, at $t = 0$ we turn on an anharmonic term which grows linearly with time reaching at $t = T$ the value $[2M\hbar\omega^3]^{1/2}$ and remaining constant afterwards. Then

$$|K(T)|^2 = \frac{1}{\omega^2 T^2}\left(2 + \omega^2 T^2 - 2\cos\omega T - 2\omega T \sin\omega T\right) . \tag{11.48}$$

If the time T which takes the interaction to reach its maximum value is much smaller than the characteristic time of the system ($T\omega \ll 1$) we say that the interaction has been introduced *suddenly*, and in this case $|K(T)| \simeq (\omega T)/2 \ll 1$. Now the relevant transition probabilities are given in (11.40), and in particular we see that in the limiting case $\omega T \to 0$ the state of the system does not change when we introduce the interaction: due to the sudden appearence of the interaction, the wave function cannot change during such a small amount of time.

Consider now the opposite situation, when the time T is very large with respect to the characteristic time of the system ($\omega T \gg 1$). In the limit $\omega T \to \infty$ (the interaction is turned on *adiabatically*) it is clear that $|K(T)| \to 1$ and, therefore,

$$P_{0\to m} \sim \frac{1}{m!}\mathrm{e}^{-1} , \tag{11.49}$$

where for simplicity we only exhibit transitions $0 \to m$.

The previous relation is not sufficient to fix the asymptotic behaviour in T of the wave function. In the adiabatic limit and taking into account (11.32), we find for the evolution of the ground state

$$\begin{aligned} |\Psi_0(T)\rangle &= \exp\left[-\int_0^T ds\, K(s)\frac{dK^*(s)}{ds}\right] \mathrm{e}^{-\mathrm{i}H_0 T/\hbar}\mathrm{e}^{\mathrm{i}K(T)a^\dagger}|\psi_0\rangle \\ &= \exp\left[-\int_0^T ds\, K(s)\frac{dK^*(s)}{ds}\right] \mathrm{e}^{-\mathrm{i}H_0 T/\hbar}\mathrm{e}^{|K(T)|^2/2}|\underline{\mathrm{i}K(T)}\rangle \\ &= \exp\left[-\int_0^T ds\, K(s)\frac{dK^*(s)}{ds}\right] \mathrm{e}^{|K(T)|^2/2}\mathrm{e}^{-\mathrm{i}\omega T/2}|\underline{\mathrm{i}K(T)\mathrm{e}^{-\mathrm{i}\omega T}}\rangle \\ &\sim \mathrm{e}^{-\mathrm{i}\omega T/6}|\underline{1}\rangle , \end{aligned} \tag{11.50}$$

where we have used the notation and properties of coherent states given in Sect. 4.4. We point out that the state $|\underline{1}\rangle$ is the ground state of the system for $t \geq T$. Finally, it is easy to verify that the phase $-\omega T/6$ would disappear had we shifted for every t the origin of energies, so that the energy of the ground state would always vanish. There is a very important consequence of (11.50):

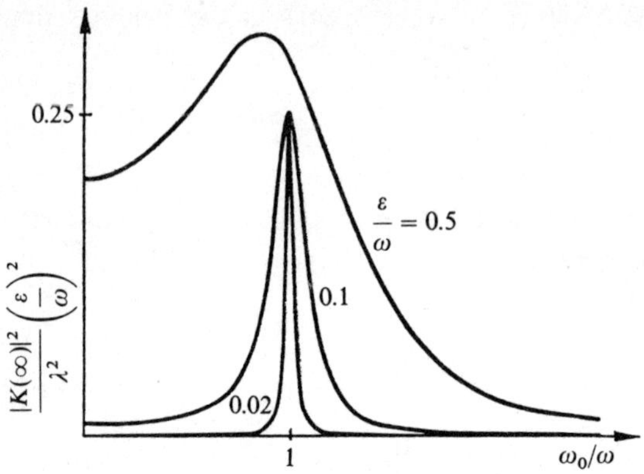

Fig. 11.5. Graph of $|K(\infty)|^2$

when the perturbation is turned on adiabatically, the physical system is found at each time in an eigenvector of $H(t)$ with the eigenvalue obtained from the initial one by continuity.

This result, as well as those mentioned for sudden perturbations, are general and will be discussed more extensively below.

As a second example take the function $f(t)$ to be

$$f(t) = \lambda\sqrt{2M\hbar\omega^3}\, e^{-\varepsilon|t|} \cos\omega_0 t \, , \tag{11.51}$$

where λ is a real dimensionless constant fixing the strength of the interaction, and ε is a positive parameter measuring the damping of this periodic perturbation for large $|t|$. In this case

$$K(t) = \lambda\omega \int_0^t ds\, e^{-\varepsilon s} e^{i\omega s} \cos\omega_0 s \, , \tag{11.52}$$

whence we immediately obtain

$$|K(\infty)|^2 = \frac{\lambda^2\omega^2}{4}\left\{\frac{1}{(\omega+\omega_0)^2+\varepsilon^2} + \frac{1}{(\omega-\omega_0)^2+\varepsilon^2} + \frac{2[(\omega^2-\omega_0^2)+\varepsilon^2]}{[(\omega-\omega_0)^2+\varepsilon^2][(\omega+\omega_0)^2+\varepsilon^2]}\right\} . \tag{11.53}$$

This quantity has a Breit-Wigner behavior as a function of ω_0 in a neighborhood of ω (see Fig. 11.5). The width of the peak is proportional to ε and its height is proportional to ε^{-2}.

Previously we have shown that the transition probability $P_{0\to m}$ has a maximum for $|K|^2$ close to m; thus, if the damping is negligible, in the neighborhood of $\omega_0 = \omega$ the transition $P_{0\to m}(\infty)$ attain two maxima as functions of ω_0, and they move closer to the natural frequency of the oscillator as m increases. (See Fig. 11.6).

The probability that any transition $0 \to m$ with $m \neq 0$ takes place is given by

$$P(\infty) = \sum_{m=1}^{\infty} P_{0\to m}(\infty) = 1 - P_{0\to 0}(\infty) = 1 - e^{-|K(\infty)|^2} \tag{11.54}$$

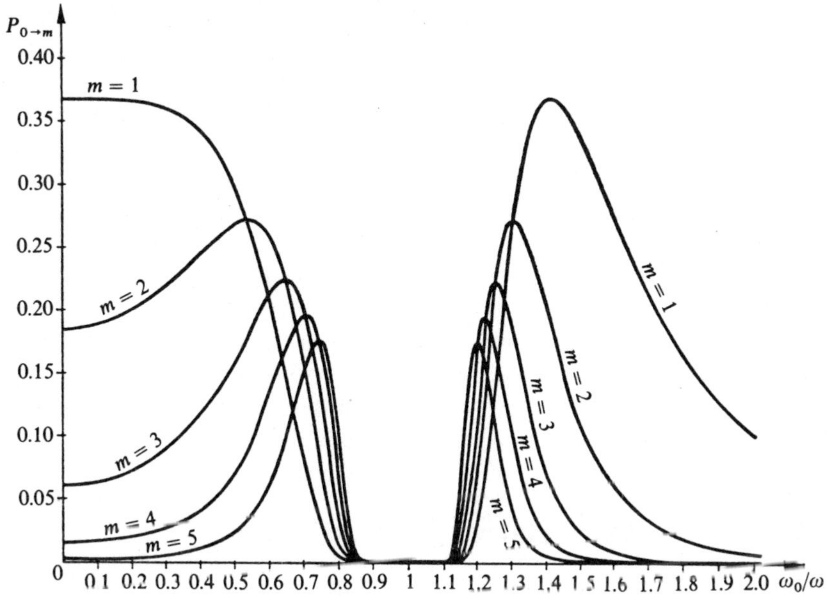

Fig. 11.6. Transition probabilities $P_{0\to m}(\infty)$ as functions of ω_0 for $\lambda = 1$, $\varepsilon \downarrow 0$

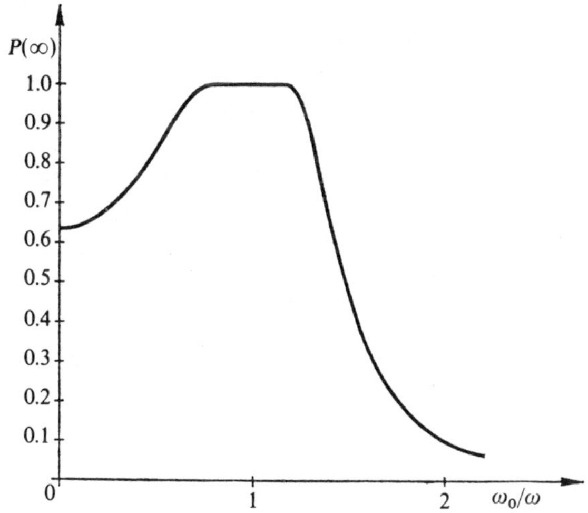

Fig. 11.7. Transition probability from the ground state to any other state as a function of ω_0/ω for $\lambda = 1$ and $\varepsilon \downarrow 0$

(see Fig. 11.7). The maximum of this quantity at $\omega_0 = \omega$ is the narrower, the smaller the intensity λ of the interaction.

Summarizing, we have verified that for negligible damping, when the frequency of the perturbation is changed through a neighborhood of ω, practically all the oscillator modes are excited with the maximum probability. This resonant behavior is also reflected in the average value of H_0 in $|\Psi(t)\rangle$, when $t \to \infty$.

Indeed, (11.50) and (4.82) show that this expectation value is $(|K(\infty)|^2+1/2)\hbar\omega$, with a dispersion of $|K(\infty)|\hbar\omega$. Therefore, as $\varepsilon \downarrow 0$ and ω_0 approaches ω, the energy stored in the state grows, as well as its dispersion, while the relative uncertainty tends to 0. This is the same result one would obtain classically, as follows from (11.45).

Before finishing with these examples we want to consider the case where the interaction (11.51) is very weak ($|\lambda| \ll 1$). If we performed a perturbative computation keeping only linear terms in λ in the transition amplitudes, we would find that the only nonvanishing transition probabilities would be those in (11.40). But we can see that due to the shape of $|K(\infty)|^2$ given in (11.53) such calculation would be completely wrong if ω_0 is such that $|\omega - \omega_0| \lesssim |\lambda|\omega$ and ε is very small. When ω_0 is in the neighborhood of ω the perturbative computation implies that the probabilities $P_{n\to n\pm 1}$ increase in this region, but the details of their behavior cannot be derived using perturbation theory.

11.4 The Interaction Picture

We have seen in Sect. 2.14 that the time evolution of a physical system can be described in a variety of equivalent ways depending on what part of the evolution is attributed to the dynamical variables and what part corresponds to the state vectors. There, we considered in detail two extreme cases: the Schrödinger picture, where the whole time evolution is carried out by the state vectors, and the Heisenberg picture, where the time evolution is contained in the time dependence of the operators associated to the dynamical variables. Now we want to integrate Eq. (11.1) using the solution of (11.3). Since we are assuming that $H_1(t)$ is a small perturbation, it is rather natural to expect that the problem will simplify if we extract the time dependence of $|\Psi(t)\rangle$ due to the presence of H_0 in the Hamiltonian. This new picture is known as the *interaction* or *Dirac picture.* Operators and states in this picture will be labeled by an index I. Taking into account that if $H_1(t) \equiv 0$ then

$$|\Psi(t)\rangle = \mathrm{e}^{-\mathrm{i}(t-t_0)H_0/\hbar}|\Psi(t_0)\rangle\ , \tag{11.55}$$

the relation between the Schrödinger and the interaction picture will be

$$\boxed{\begin{aligned} |\Psi_I(t)\rangle &= \mathrm{e}^{\mathrm{i}(t-t_1)H_0/\hbar}|\Psi(t)\rangle \\ A_I(t) &= \mathrm{e}^{\mathrm{i}(t-t_1)H_0/\hbar}A(t)\mathrm{e}^{-\mathrm{i}(t-t_1)H_0/\hbar} \end{aligned}}\ , \tag{11.56}$$

for state vectors and operators respectively. Notice that t_1 is an arbitrary time where both representations coincide and without loss of generality we take $t_1 = 0$. From (11.56) we obtain

$$\boxed{\mathrm{i}\hbar\frac{d|\Psi_I(t)\rangle}{dt} = H_{1I}(t)|\Psi_I(t)\rangle}\ , \tag{11.57}$$

where $H_{1I}(t)$ is the operator $H_1(t)$ in the interaction picture.

As in Sect. 2.9, the differential equation (11.57), together with the boundary condition at $t = t_0$, is equivalent to the integral equation

$$\boxed{|\Psi_I(t)\rangle = |\Psi_I(t_0)\rangle - \frac{\mathrm{i}}{\hbar}\int_{t_0}^{t} dt' H_{1I}(t')|\Psi_I(t')\rangle}\ , \tag{11.58}$$

having the formal solution

$$|\Psi_I(t)\rangle = P\,\exp\left[-\frac{\mathrm{i}}{\hbar}\int_{t_0}^{t} dt' H_{1I}(t')\right]|\Psi_I(t_0)\rangle\ , \quad t > t_0\ , \tag{11.59}$$

where P is Dyson's time ordering operator. Taking (11.56) into account, we find

$$\begin{aligned}&|\Psi(t)\rangle = U(t,t_0)|\Psi(t_0)\rangle\ ,\\ &U(t,t_0) = \mathrm{e}^{-\mathrm{i}H_0 t/\hbar} P\,\exp\left[-\frac{\mathrm{i}}{\hbar}\int_{t_0}^{t} dt' H_{1I}(t')\right]\mathrm{e}^{\mathrm{i}H_0 t_0/\hbar}\ , \quad t > t_0\ ,\end{aligned} \tag{11.60}$$

giving a compact representation of the expansion (valid for all t)

$$\begin{aligned}U(t,t_0) &= U_0(t,t_0) + \sum_{n=1}^{\infty} U_n(t,t_0)\ ,\\ U_n(t,t_0) &= \frac{1}{(\mathrm{i}\hbar)^n}\int_{t_0}^{t} dt_n \int_{t_0}^{t_n} dt_{n-1}\dots\int_{t_0}^{t_2} dt_1 U_0(t,t_n)\\ &\quad\times H_1(t_n)U_0(t_n,t_{n-1})H_1(t_{n-1})U_0(t_{n-1},t_{n-2})\\ &\quad\dots U_0(t_2,t_1)H_1(t_1)U_0(t_1,t_0)\ ,\end{aligned} \tag{11.61}$$

where $U_0(t_j, t_k)$ is the evolution operator corresponding to H_0 and which we call the free time evolution.

Equation (11.61) is thus a power series expansion in the perturbation $H_1(t)$, and in practice, the evaluation of its terms becomes more and more complicated as the value of n increases and therefore the method is only useful in general if only the first few terms are important.

With regard to the convergence of the series (11.61), we already pointed out in Sect. 2.9 that if $H_1(t)$ is bounded and continuous as a function of t in the strong topology, the series is strongly convergent. It would be desirable to extend the class of operators $H_1(t)$ for which the terms in this series are well defined and give a convergent expansion. Unfortunately, the perturbations which are regular in the sense of Kato, for stationary perturbations, do not have to be regular for time evolution [KA 51], even though one should expect that this series would have an asymptotic meaning for small values of the coupling constant, whenever all the operations performed are well defined in each order of perturbation theory. For some general conditions on $H(t)$ ensuring the existence of the unitary propagator $U(t, t_0)$ see for example [RS 75].

11.5 Transition Probability

Let us assume that H_0 has an orthonormal basis of eigenvectors $\{(|\phi_n\rangle\}$ with eigenvalues $E_n^{(0)}$ of arbitrary multiplicities. If at the initial time $t = t_0$ the system is in the state $|\phi_i\rangle$, one is often interested in knowing the probability to find it in the state $|\phi_f\rangle$, $f \neq i$, at a later time t. We will limit ourselves to the case when H_0 has a discrete spectrum, even though all the arguments can be straightforwardly generalized to operators with a continuous spectrum as well.

According to first principles, the transition amplitude is given by

$$A_{i\to f}(t, t_0) = \langle\phi_f|U(t, t_0)|\phi_i\rangle\ . \tag{11.62}$$

Substituting (11.61) in (11.62) and taking into account that $U_0(t, t_0)$ only changes the phase of $|\phi_i\rangle$, so that it still remains orthogonal to $|\phi_f\rangle$, we obtain

$$A_{i\to f}(t, t_0) = \sum_{n=1}^{\infty} A_{i\to f}^{(n)}(t, t_0) \equiv \sum_{n=1}^{\infty}\langle\phi_f|U_n(t, t_0)|\phi_i\rangle\ . \tag{11.63}$$

Using equation (11.61) for $U_n(t, t_0)$, the orthonormality of the complete set $\{|\phi_k\rangle\}$ and that $U_0(t_i, t_j)$ is diagonal in this basis, the general term in the expansion (11.63) takes the form

$$\begin{aligned} A_{i\to f}^{(n)}(t, t_0) = {} & \frac{1}{(\mathrm{i}\hbar)^n}\int_{t_0}^{t} dt_n \int_{t_0}^{t_n} dt_{n-1}\dots\int_{t_0}^{t_2} dt_1 \sum_{k_{n-1}}\sum_{k_{n-2}}\dots\sum_{k_1} \\ & \times \big[\langle\phi_f|U_0(t, t_n)|\phi_f\rangle\langle\phi_f|H_1(t_n)|\phi_{k_{n-1}}\rangle\langle\phi_{k_{n-1}}|U_0(t_n, t_{n-1})|\phi_{k_{n-1}}\rangle \\ & \times\langle\phi_{k_{n-1}}|H_1(t_{n-1})|\phi_{k_{n-2}}\rangle\langle\phi_{k_{n-2}}|U_0(t_{n-1}, t_{n-2})|\phi_{k_{n-2}}\rangle \\ & \dots\langle\phi_{k_2}|U_0(t_2, t_1)|\phi_{k_1}\rangle\langle\phi_{k_1}|H_1(t_1)|\phi_i\rangle\langle\phi_i|U_0(t_1, t_0)|\phi_i\rangle\big]\ . \end{aligned} \tag{11.64}$$

If we introduce the notation

$$\langle\phi_l|H_1(t)|\phi_r\rangle \equiv V_{lr}(t) \tag{11.65}$$

and use the explicit form of U_0 then

$$\begin{aligned} A_{i\to f}^{(n)}(t, t_0) = {} & \frac{1}{(\mathrm{i}\hbar)^n}\int_{t_0}^{t} dt_n \int_{t_0}^{t_n} dt_{n-1}\dots\int_{t_0}^{t_2} dt_1 \sum_{k_{n-1}}\sum_{k_{n-2}}\dots\sum_{k_1} \\ & \times \Big[\exp[-\mathrm{i}(t - t_n)E_f^{(0)}/\hbar]V_{fk_{n-1}}(t_n) \\ & \times \exp[-\mathrm{i}(t_n - t_{n-1})E_{k_{n-1}}^{(0)}/\hbar]V_{k_{n-1}k_{n-2}}(t_{n-1}) \\ & \times \exp[-\mathrm{i}(t_{n-1} - t_{n-2})E_{k_{n-2}}^{(0)}/\hbar]\dots \\ & \times[\exp[-\mathrm{i}(t_2 - t_1)E_{k_1}^{(0)}/\hbar]V_{k_1 i}(t_1)\exp[-\mathrm{i}(t_1 - t_0)E_i^{(0)}/\hbar]\Big]\ , \end{aligned} \tag{11.66}$$

which is the desired equation. Reading (11.66) from right to left, it has the following interpretation: At time t_0 the system begins in a state $|\phi_i\rangle$ propagating freely until $t_1 \geq t_0$; at this time the interaction induces a transition to the state

$|\phi_{k_1}\rangle$; then the system evolves freely again until $t_2 \geq t_1$, when the interaction induces another transition to the state $|\phi_{k_2}\rangle$, and so on. We have to sum over all possible intermediate states as well as integrate over all possible times where interactions take place, with the constraint $t \geq t_n \geq t_{n-1} \geq \ldots \geq t_1 \geq t_0$. It is useful to represent the terms in (11.63) diagramatically. In Fig. 11.8 we show the lowest order diagrams.

If only the first term in the expansion (11.53) is important, then

$$A_{i\to f}(t,t_0) \simeq A^{(\mathrm{B})}_{i\to f}(t,t_0) \equiv \frac{1}{\mathrm{i}\hbar}\int_{t_0}^{t} dt_1 \times \exp[-\mathrm{i}(t-t_1)E_f^{(0)}/\hbar]V_{fi}(t_1)\exp[-\mathrm{i}(t_1-t_0)E_i^{(0)}/\hbar] , \tag{11.67}$$

and in this *Born approximation* the transition probability becomes

$$\boxed{P^{(\mathrm{B})}_{i\to f}(t,t_0) = \frac{1}{\hbar^2}\left|\int_{t_0}^{t} dt_1\, \mathrm{e}^{\mathrm{i}\omega_{fi}t_1} V_{fi}(t_1)\right|^2} , \tag{11.68}$$

where $\omega_{fi} \equiv (E_f^{(0)} - E_i^{(0)})/\hbar$.

It should be noticed that in this approximation

$$P^{(\mathrm{B})}_{i\to f}(t,t_0) = P^{(\mathrm{B})}_{f\to i}(t,t_0) , \tag{11.69}$$

which is not true in general when higher order terms are included. One should not mistakenly identify (11.69) with the microreversibility principle (8.170), an exact consequence of the theory if the dynamics is time reversal invariant. For finite times, the microreversibility relation is written as

$$P_{i\to f}(t,t_0) = P_{U_T f\to U_T i}(-t_0,-t) , \tag{11.70}$$

which for time independent $H(t)$ becomes

$$P_{i\to f}(t,t_0) = P_{U_T f\to U_T i}(t,t_0) . \tag{11.71}$$

For the problem treated exactly in Sect. 11.2, (11.68) leads immediately to

$$P^{(\mathrm{B})}_{+\to -}(t,0) = \frac{\omega_1^2}{(\omega-\omega_0)^2}\sin^2\frac{(\omega-\omega_0)t}{2} , \tag{11.72}$$

to be compared with (11.18). Both expressions yield essentially the same results if $|t[(\omega-\omega_0)^2+\omega_1^2]^{1/2}| \ll 1$. However we remark that if $\omega_1 > |\omega-\omega_0|$ there will be values of t for which the transition probability (11.72) will be larger than 1, so that the approximation considered would violate unitarity. Equation (11.72) would therefore not be a good approximation to the transition probability and we would have to take into account higher order terms in order to cancel the loss of unitarity.

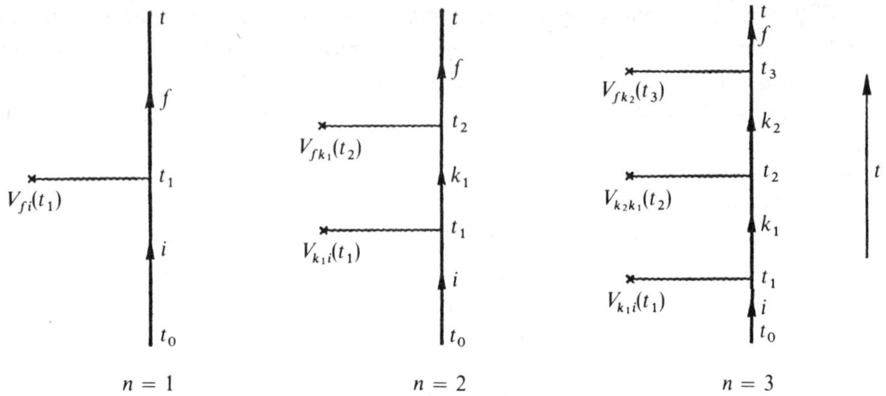

Fig. 11.8. Lowest order diagrams contributing to $A_{i\to f}(t, t_0)$. Time flows upward in these graphs

11.6 Constant Perturbations

We now consider a system with unperturbed Hamiltonian H_0, in a stationary state $|\phi_i\rangle$ and at $t = 0$ we introduce a time-independent perturbation V, namely

$$H_1(t) = \begin{cases} 0, & t < 0 \\ V, & t \geq 0 . \end{cases} \tag{11.73}$$

The probability to find the system in the state $|\phi_f\rangle$ at time $t > 0$ is given to first order of perturbation theory by (11.68); in this case the integration is easily carried out yielding

$$P^{(\mathrm{B})}_{i\to f}(t, t_0) = \frac{4|\langle\phi_f|V|\phi_i\rangle|^2}{\hbar^2\omega_{fi}^2} \sin^2\frac{\omega_{fi}t}{2} . \tag{11.74}$$

This transition probability is plotted as a function of ω_{fi} in Fig. 11.9. If as discussed here $|\phi_i\rangle$ and $|\phi_f\rangle$ correspond to discrete levels, and $E_f^{(0)} \neq E_i^{(0)}$ the transition probability is a periodic function of time with period $2\pi/|\omega_{fi}|$, varying between 0 and $4|V_{fi}|^2/(E_f^{(0)} - E_i^{(0)})^2$. If on the other hand $E_f^{(0)} = E_i^{(0)}$, even though $|\phi_f\rangle \perp |\phi_i\rangle$, the transition probability in the Born approximation rises as t^2. Obviously this approximation does not make any sense when the transition probability is bigger than one: then the higher order effects would necessarily be important, and through various cancellations, they would ensure that the probability never becomes larger than 1, as required by the unitarity of the time evolution operator.

It is also clear that for small times, the perturbation considered causes transitions to all those states $|\phi_f\rangle$ such that $V_{fi} \neq 0$ and that these transition probabilities, which are very small, all grow as t^2. As t increases, it is evident from Fig. 11.9 that the important transitions are those to states $|\phi_f\rangle$ for which

$$|E_f^{(0)} - E_i^{(0)}| \lesssim \frac{2\pi\hbar}{t} . \tag{11.75}$$

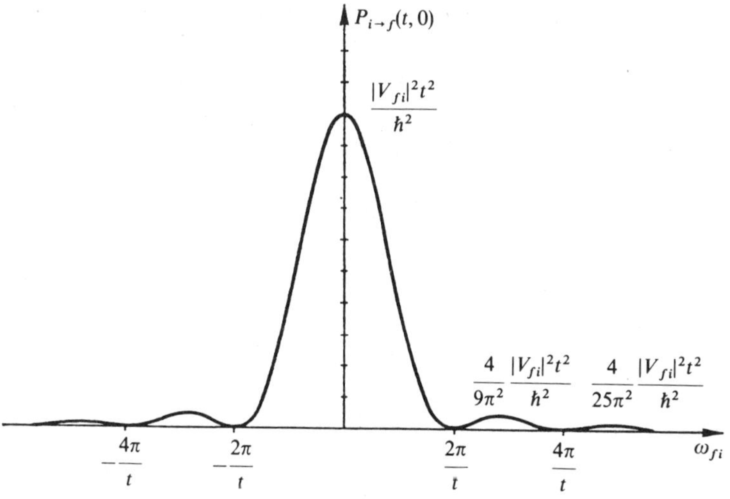

Fig. 11.9. Transition probability to first order of perturbation theory as a function of ω_{fi}. The values for different maxima are indicated

In other words transitions induced by constant perturbations conserve the unperturbed energy up to the terms of order $2\pi\hbar/t$.

It is particularly interesting to analyze what happens when the final state $|\phi_f\rangle$ is not a normalizable state in the point spectrum of H_0 but it belongs to the continuous spectrum with normalization

$$\langle\phi_{f'}|\phi_f\rangle = \frac{1}{N(f)}\delta(f'-f)\,. \tag{11.76}$$

The interest in these situations lies on the fact that they describe decay phenomena: $|\phi_i\rangle$ can for instance represent the state of a particle or more generally, of an atomic or nuclear system bound under the Hamiltonian H_0, and due to the effect of the interaction V it can decay into a set of fragments of continuous energy as seen by H_0.

In this case we do not measure the transition probability from an initial state $|\phi_i\rangle$ to a final state $|\phi_f\rangle$, but the transition probability to a set of final states characterized by the fact that the index f belongs to a subset Δf of its range of values. Using (11.76) the projector onto this set of states will be

$$P_{\Delta f} = \int_{\Delta f} df\, |\phi_f\rangle N(f)\langle\phi_f|\,. \tag{11.77}$$

If instead of labelling the states by the index f we use the enegy E and the remaining necessary quantum numbers are designated by α, the projector becomes

$$P_{\Delta f} = \int_{\Delta\alpha} d\alpha \int_{\Delta E} dE\, |\phi_{E,\alpha}\rangle \varrho_\alpha(E)\langle\phi_{E,\alpha}|\,, \tag{11.78}$$

where $\varrho_\alpha(E)\,dE\,d\alpha = N(f)\,df$ and the integrations (or generalized sums) are over the sets $\Delta\alpha$ and $\Delta E = (E - \Delta E/2, E + \Delta E/2)$ corresponding to the interval Δf considered. Therefore the quantity of interest in this case is

$$P^{(B)}_{i\to\Delta f}(t) = \int_{\Delta\alpha} d\alpha \int_{E-\Delta E/2}^{E+\Delta E/2} dE' \langle\phi_i|V|\phi_{E',\alpha}\rangle \times \varrho_\alpha(E')\langle\phi_{E',\alpha}|V|\phi_i\rangle \frac{\sin^2(\overline{E}'t/2\hbar)}{(\overline{E}'/2)^2}\,, \tag{11.79}$$

with $\overline{E}' \equiv E' - E_i^{(0)}$. If the interval of final energies is sufficiently small, we can think of $\varrho_\alpha(E')$ and $\langle\phi_{E'\alpha}|V|\phi_i\rangle$ as constants, and therefore

$$P^{(B)}_{i\to\Delta f}(t) = \int_{\Delta\alpha} d\alpha |\langle\phi_{E,\alpha}|V|\phi_i\rangle|^2 \varrho_\alpha(E) \int_{E-\Delta E/2}^{E+\Delta E/2} \frac{\sin^2(\overline{E}'t/2\hbar)}{(\overline{E}'/2)^2} dE'. \tag{11.80}$$

We will assume t to be large enough so that the function to be integrated (whose shape is similar to the function plotted in Fig. 11.9) has many oscillations in the interval ΔE, namely:

$$\Delta E \gg \frac{2\pi\hbar}{t}\,. \tag{11.81}$$

We now evaluate the integral over E'. A simple change of variables yields

$$I \equiv \int_{E-\Delta E/2}^{E+\Delta E/2} dE' \frac{\sin^2(\overline{E}'t/2\hbar)}{(\overline{E}'/2)^2} = \frac{2t}{\hbar} \int_{t(E-E_i^{(0)}-\Delta E/2)/2\hbar}^{t(E-E_i^{(0)}+\Delta E/2)/2\hbar} du \frac{\sin^2 u}{u^2}. \tag{11.82}$$

We can distiguish two cases:

1) If the central peak of the integrand is not contained in the integration region (energy non-conservation), then

$$I \simeq \frac{8\hbar}{t(E - E_i^{(0)})^2} \int_{t(E-E_i^{(0)}-\Delta E/2)/2\hbar}^{t(E-E_i^{(0)}+\Delta E/2)/2\hbar} du\, \sin^2 u\,. \tag{11.83}$$

The average value of $\sin^2 u$ in the interval $[0, 2\pi]$ is $1/2$, and the integral runs over an interval of length $t\Delta E/2\hbar \gg 2\pi$. Thus

$$I \simeq \frac{2\Delta E}{(E - E_i^{(0)})^2} \quad \text{and} \tag{11.84}$$

$$P^{(B)}_{i\to\Delta f}(t) \simeq \frac{2\Delta E}{(E - E_i^{(0)})^2} \int_{\Delta\alpha} d\alpha\, |\langle\phi_{E,\alpha}|V|\phi_i\rangle|^2 \varrho_\alpha(E)\,, \quad E \neq E_i^{(0)}. \tag{11.85}$$

2) On the other hand, if the central peak is contained in the integration region (energy conservation), we can extend the integral in (11.82) to $(-\infty, +\infty)$ for all practical purposes, and we obtain $I = (2\pi t/\hbar)(1 + \mathrm{O}(t^{-1}))$, so that

$$P^{(\mathrm{B})}_{i\to\Delta f}(t) \simeq \frac{2\pi t}{\hbar}\int_{\Delta\alpha} d\alpha\, |\langle\phi_{E_i^{(0)},\alpha}|V|\phi_i\rangle|^2 \varrho_\alpha(E_i^{(0)})\,, \quad E = E_i^{(0)}\,. \tag{11.86}$$

Hence for large t and in the Born approximation, the transitions conserving energy are dominant (assuming that the associated matrix elements do not all vanish as a consequence of some selection rule). Thus, we can restrict ourselves to this type of transitions.

The linear growth with t of the transition probability in the Born approximation (11.86) suggests that we define the transition probability $w^{(\mathrm{B})}_{i\to\alpha}$ per unit time and per unit interval of the variable α as

$$w^{(\mathrm{B})}_{i\to\alpha} = \frac{d^2 P^{(\mathrm{B})}_{i\to\alpha}}{d\alpha\, dt}\,, \tag{11.87}$$

where $P^{(\mathrm{B})}_{i\to\alpha}$ denotes the transition probability $i\to\alpha$ and any final energy. Notice that the derivative d/dt automatically selects those final states with energy $E_i^{(0)}$.

It is clear that we cannot expect a similar procedure to work for the exact transition probability $P_{i\to f}$ when $|\phi_i\rangle$ is normalized, for this would violate the unitarity limit.

A consequence of (11.86) and (11.87) is

$$\boxed{w^{(\mathrm{B})}_{i\to\alpha} = \frac{2\pi}{\hbar}|\langle\phi_{E_f^{(0)},\alpha}|V|\phi_i\rangle|^2 \varrho_\alpha(E_f^{(0)})}\,, \quad E_f^{(0)} = E_i^{(0)}\,, \tag{11.88}$$

known as *Fermi's golden rule* (compare with (8.72) and its Born approximation). The quantity $\varrho_\alpha(E)$ is called the *density of final states*. One should not forget that this formula is valid only if the time t is sufficiently large so that (11.81) is satisfied, and sufficiently small so that (11.86) is much smaller than unity, and the first order of perturbation theory can be trusted. Summarizing, Fermi's golden rule is applicable and useful only when the relations

$$t\Delta E \gg 2\pi\hbar\,, \quad w^{(\mathrm{B})}_{i\to\alpha} t \ll 1 \tag{11.89}$$

are simultaneously satisfied.

From the preceeding analysis we conclude that the depletion probability of the initial state $|\phi_i\rangle$ per unit time is constant whenever the inequalities (11.89) are satisfied. In other words, starting with an ensemble of $N(0)$ quantum system all prepared in the state $|\phi_i\rangle$, the population would have a constant rate of decay under the conditions (11.89):

$$N(t) = N(0)\exp(-\Gamma_i t/\hbar)\,, \tag{11.90}$$

where Γ_i satisfies

$$\Gamma_i/\hbar \simeq \int d\alpha\, w^{(\mathrm{B})}_{i\to\alpha}\,. \tag{11.91}$$

The quantity Γ_i is called the *width* of the state $|\phi_i\rangle$ and its *(mean) lifetime* is simply $\tau_i \equiv \hbar/\Gamma_i$.

The exponential law (11.90) has been justified here in the Born approximation. It can be shown [SI 72a] that it should necessarily fail for very small and very large times. We will not discuss here the definition and properties of unstable states, whose study was initiated by *Wigner* and *Weisskopf* [WW 30] and explained in detail in [GW 64]. But the reader interested in a critical and rigorous account of these issues is referred to [SI 73], where among other things it is shown that the formula (11.91) is correct for a large class of interactions and there is a thorough discussion of the confusion often found in the literature concerning the corrections to the Born approximation in computing lifetimes.

At the end of this chapter we present an explicitly soluble model which illuminates some of the problems just enumerated.

11.7 Turning On Perturbations Adiabatically

In the previous section we assumed the interaction to have a time dependence given by (11.73). One might think that the result obtained would be influenced by the sharp introduction of the interaction at $t = 0$. However we will show that the same physical results are obtained for large interactions times when the perturbation is turned on adiabatically.

More concretely, suppose $H_1(t)$ has the form

$$H_1(t) = \mathrm{e}^{\varepsilon t} V \, , \tag{11.92}$$

where ε is a positive constant which will become 0 at the end of the computation. To first order in perturbation theory and assuming that $|\phi_i\rangle$ has been prepared in the far past when the perturbation was negligible, the transition probability is

$$P^{(\mathrm{B})}_{i\to f}(t) = \frac{1}{\hbar^2}|\langle\phi_f|V|\phi_i\rangle|^2 \left| \int_{-\infty}^{t} dt_1 \; \mathrm{e}^{\mathrm{i}\omega_{fi}t_1}\mathrm{e}^{\varepsilon t_1} \right|^2 \tag{11.93}$$

and consequently

$$P^{(\mathrm{B})}_{i\to f}(t) = \frac{1}{\hbar^2}|\langle\phi_f|V|\phi_i\rangle|^2 \frac{\exp(2\varepsilon t)}{\omega_{fi}^2 + \varepsilon^2} \, . \tag{11.94}$$

This probability is represented as a function of ω_{fi} in Fig. 11.10. Notice the qualitative similarity between the curve in Fig. 11.10 and the one in Fig. 11.9. The role played there by the quantity $2\pi/t$ is analogous to the role played here by the parameter ε. This is not unreasonable, since in the previous cases the interaction V was acting during the time interval $[0, t]$ and now the perturbation is practically V if $|t| \lesssim 1/\varepsilon$.

If the final state is part of the continuum, the expression equivalent to (11.80) is now

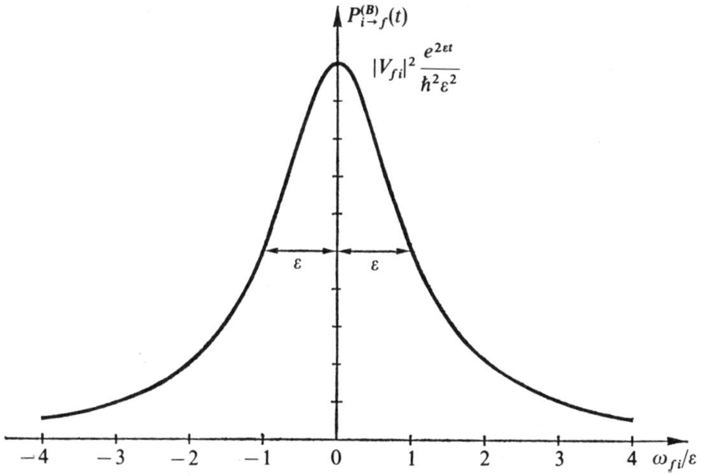

Fig. 11.10. The transition probability to first order of perturbation theory as a function of ω_{fi}

$$P^{(B)}_{i\to f}(t) = \int_{\Delta\alpha} d\alpha |\langle\phi_{E,\alpha}|V|\phi_i\rangle|^2 \varrho_\alpha(E) \int_{E-\Delta E/2}^{E+\Delta E/2} dE' \frac{\exp(2\varepsilon t)}{(E'-E_i^{(0)})^2 + (\hbar\varepsilon)^2} \, . \tag{11.95}$$

If as in the previous section we focus our attention to the case when the maximum of the integrand is in the integration interval (energy conservation) and if furthermore

$$\Delta E \gg \hbar\varepsilon \tag{11.96}$$

[similar to (11.81)], we can extend the integration over E' to the interval $(-\infty, +\infty)$ and obtain

$$P^{(B)}_{i\to\Delta\alpha}(t) = \frac{\pi}{\hbar\varepsilon} e^{2\varepsilon t} \int_{\Delta\alpha} d\alpha \, |\langle\phi_{E_i^{(0)},\alpha}|V|\phi_i\rangle|^2 \varrho_\alpha(E_i^{(0)}) \, . \tag{11.97}$$

Hence, the transition probability per unit time defined in (11.87) is once again determined by (11.88) after $\varepsilon \downarrow 0$. We thus see that Fermi's golden rule is not very sensitive to the particular form of introducing the interaction.

Obviously, there are cases when it is not enough to keep only the first order in the expansion (11.63). In particular, it could happen that $V_{fi} = 0$ and therefore there is no transition at first order but it may be induced via second or high order terms. To second order the transition probability is then

$$P^{(2)}_{i\to f}(t) = \frac{1}{\hbar^4} \left| \sum_k V_{fk} V_{ki} \int_{-\infty}^{t} dt_2 \exp[(\mathrm{i}\omega_{fk} + \varepsilon)t_2] \int_{-\infty}^{t_2} dt_1 \exp[(\mathrm{i}\omega_{ki} + \varepsilon)t_1] \right|^2 \tag{11.98}$$

and therefore

$$P^{(2)}_{i\to f}(t) = \frac{1}{\hbar^2}\left|\sum_k \frac{V_{fk}V_{ki}}{E_i^{(0)} - E_k^{(0)} + \mathrm{i}\hbar\varepsilon}\right|^2 \frac{\exp(4\varepsilon t)}{\omega_{fi}^2 + 4\varepsilon^2}\,, \tag{11.99}$$

leading to

$$w^{(2)}_{i\to\alpha} = \frac{2\pi}{\hbar}\left|\sum_k \frac{\langle\phi_f|V|\phi_k\rangle\langle\phi_k|V|\phi_i\rangle}{E_i^{(0)} - E_k^{(0)} + \mathrm{i}0}\right|^2 \varrho_\alpha(E_f^{(0)})\,, \quad E_f^{(0)} = E_i^{(0)} \tag{11.100}$$

using arguments analogous to those we followed for the Born approximation.

The last formula can be interpreted by saying that the transition from the state $|\phi_i\rangle$ to $|\phi_f\rangle$ takes place through the intermediate state $|\phi_k\rangle$. The sum runs over all the intermediate states, and in general $E_k^{(0)} \neq E_i^{(0)}$. One should notice that when $E_k^{(0)}$ belongs to the continuous spectrum and $E_k^{(0)} = E_i^{(0)}$ the term i0 is fundamental for the integral over energies in (11.100) to be well defined. The amplitude whose square modulus appears in (11.100) is usually called the second order matrix element since it plays the same role as the first order term $\langle\phi_f|V|\phi_i\rangle$ in Fermi's golden rule.

The equation (11.100) can also be obtained introducing the perturbation as in the previous section, and then taking the limit $t \to \infty$ in the sense of distributions.

Taking all the amplitudes to all orders of perturbation theory leads by the same procedure to the *platinum rule*

$$w_{i\to\alpha} = \frac{2\pi}{\hbar}\left|V_{fi} + \sum_k \frac{V_{fk}V_{ki}}{E_{ik}^{(0)} + \mathrm{i}0} + \sum_{k_1k_2} \frac{V_{fk_2}V_{k_2k_1}V_{k_1i}}{(E_{ik_2}^{(0)} + \mathrm{i}0)(E_{ik_1}^{(0)} + \mathrm{i}0)} + \dots\right|^2$$
$$\times\, \varrho_\alpha(E_f^{(0)})\,, \quad E_f^{(0)} = E_i^{(0)}\,, \tag{11.101}$$

where $E_{kl}^{(0)} = E_k^{(0)} - E_l^{(0)}$.

11.8 Periodic Perturbations

In Sects. 11.6,7 we have considered constant perturbations introduced in two different ways. We now wish to study to first order in perturbation theory the effect of a harmonic interaction of frequency ω which we introduce adiabatically for simplicity, even though the final result will be the same if we introduced it suddenly at $t = 0$. Suppose then that

$$H_1(t) = \left[V\mathrm{e}^{-\mathrm{i}\omega t} + V^\dagger \mathrm{e}^{\mathrm{i}\omega t}\right]\mathrm{e}^{\varepsilon t}\,, \tag{11.102}$$

where V is a time-independent operator and $\varepsilon > 0$. Obviously in the limit $\omega \to 0$ we should recover the previous results. To first order of pertubation theory the transition probability becomes

$$P^{(B)}_{i\to f}(t) = \frac{1}{\hbar^2}\left|\int_{-\infty}^{t} dt_1 \left[V_{fi}\exp\left[i(\omega_{fi}-\omega-i\varepsilon)t_1\right]\right.\right.$$
$$\left.\left. + V^\dagger_{fi}\exp\left[i(\omega_{fi}+\omega-i\varepsilon)t_1\right]\right]\right|^2 , \tag{11.103}$$

and therefore

$$P^{(B)}_{i\to f}(t) = \frac{\exp(2\varepsilon t)}{\hbar^2}\left\{\frac{|V_{fi}|^2}{(\omega_{fi}-\omega)^2+\varepsilon^2} + \frac{|V^\dagger_{fi}|^2}{(\omega_{fi}+\omega)^2+\varepsilon^2}\right.$$
$$\left. + \text{interference terms}\right\} . \tag{11.104}$$

In Fig. 11.11 we represent this probability as a function ω, assuming that $V_{fi} = V^\dagger_{fi}$ for simplicity. From the figure we immediately see that if $\varepsilon \ll |\omega_{fi}|$ the transition probability is only important for those final states satisfying $\omega_{fi} = \pm\omega$, namely

$$E^{(0)}_f = E^{(0)}_i \pm \hbar\omega + O(\hbar\varepsilon) \tag{11.105}$$

and the same occurs in the general case (11.104).

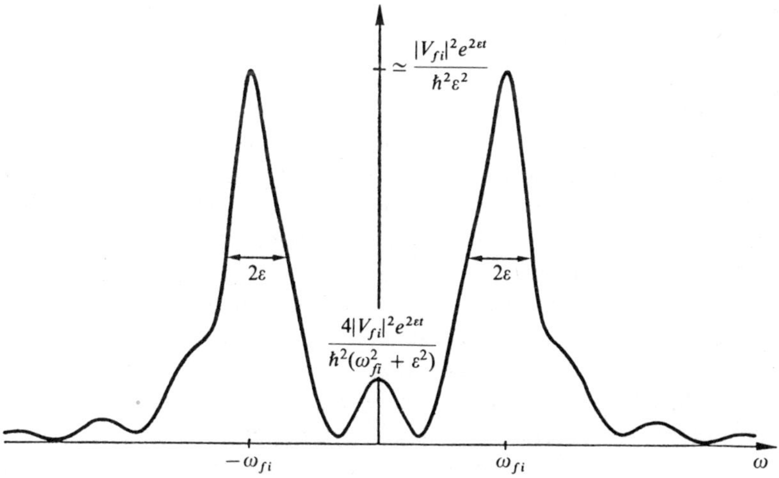

Fig. 11.11. Representation of $P_{i\to f}(t)$ as a function of the external frequency ω

The conservation of energy suggested by (11.105) for the transition from $|\phi_i\rangle$ to $|\phi_f\rangle$, both with well defined energies, has been obtained in the Born approximation. The examples considered in Sects. 11.2, 3 indicated that such conservation does not hold necessarily for the exact solution to the problem, in which we may have a nontrivial dependence on ω different from a δ-function which our first order computations cannot reproduce. Such resonant behavior depends generally on the strength of the interaction.

When the term in (11.105) with the + sign dominates we say that the system absorbes an energy $\hbar\omega$ from the external source, while in the opposite case we say that the system emits this amount of energy.

If the transition is to a state in the continuum, arguments entirely similar to those used previously lead to a transition probability

$$P^{(B)}_{i\to\Delta\alpha}(t) = \int_{\Delta\alpha} d\alpha\, |\langle\phi_{E_f^{(0)},\alpha}|V|\phi_i\rangle|^2 \varrho_\alpha(E_f^{(0)}) \int_{-\infty}^{+\infty} dE' \times \frac{\exp(2\varepsilon t)}{(E'-E_i^{(0)}-\hbar\omega)^2+\hbar^2\varepsilon^2} \tag{11.106}$$

when $E_f^{(0)} = E_i^{(0)} + \hbar\omega$; we have neglected the interference and extended the integration to $(-\infty,+\infty)$ assuming $|\omega_{fi}| \gg \varepsilon$, $\Delta E \gg \hbar\varepsilon$. Manipulations by now familiar yield

$$w^{(B)}_{i\to\alpha} = \frac{2\pi}{\hbar}|\langle\phi_{E_f^{(0)},\alpha}|V|\phi_i\rangle|^2 \varrho_\alpha(E_f^{(0)}) \,, \quad E_f^{(0)} = E_i^{(0)} + \hbar\omega \,, \tag{11.107}$$

and similar expressions hold when the system emits energy. We only have to replace $V \to V^\dagger$ and $+ \to -$.

11.9 Sudden Perturbations

In the following we derive some important results for time-dependent perturbations in two extreme situations: a sudden and an adiabatic introduction of the interaction. We noticed before that either way of introducing the interaction does not alter the transition probabilities per unit time, at large times, between stationary states of H_0. However their effects are quite different if one is interested in the time evolution of the initial wave function through the transient period, and in particular if we want to calculate the transition probability from an initial state which is an eigenvalue of H_0 to an eigenstate of H_1, the Hamiltonian reached at the end of the transient period. Suppose that at the initial time t_0 the unperturbed Hamiltonian of the system is H_0 and we vary it continuously until it takes the form H_1 at time t_1. It is useful to parametrize time by a variable τ, so that $t = t_0 + T\tau$, with $T = t_1 - t_0$. We denote the Hamiltonian $H(t)$ at time t by $\tilde{H}(\tau)$ and, obviously,

$$\tilde{H}(0) = H_0 \,, \quad \tilde{H}(1) = H_1 \,. \tag{11.108}$$

Writing $U(t,t_0) \equiv U_T(\tau)$, the evolution operator satisfies

$$U_T(\tau) = 1 - \frac{\mathrm{i}}{\hbar}T\int_0^\tau d\tau' \tilde{H}(\tau')U_T(\tau') \,. \tag{11.109}$$

The problem of the time evolution of the system after this transient regime consists of determining $U_T(1)$. The general results we will obtain are related to the behavior of $U_T(1)$ in the limiting cases $T \to 0$ (*sudden* perturbation) and $T \to \infty$ (*adiabatic* perturbation).

If $T \to 0$, (11.109) suggests that

$$\boxed{\lim_{T\to 0} U_T(1) = I} \tag{11.110}$$

(this is obviously true if $\tilde{H}(\tau)$ is bounded in $[0, 1]$). In other words, the change in the Hamiltonian is so fast that the system has no time to adjust, and the wave function is the same at $\tau = 0$ and at $\tau = 1$. If T is sufficiently small, we may expect $U_T(1) \approx I$. This is known as the *sudden approximation*. If the initial normalized state is $|\phi_0\rangle$, in this approximation $U(t_1, t_0)|\phi_0\rangle \approx |\phi_0\rangle$ and an estimate of the error is obtained by computing the probability p_0 for a transition to any other state to take place,

$$p_0 = \langle \phi_0 | U_T^\dagger(1) Q_0 U_T(1) | \phi_0 \rangle \, , \tag{11.111}$$

where $Q_0 = I - |\phi_0\rangle\langle\phi_0|$ is the projection operator onto the subspace orthogonal to $|\phi_0\rangle$. The sudden approximation will be acceptable as long as $p_0 \ll 1$. A procedure to estimate p_0 consists of calculating the first nontrivial contribution to the expansion of $U_T(1)$ in powers of T:

$$U_T(1) = I - \frac{\mathrm{i}}{\hbar} T \int_0^1 d\tau_1 \tilde{H}(\tau_1) + \left(\frac{\mathrm{i}}{\hbar}\right)^2 T^2 \int_0^1 d\tau_1 \int_0^{\tau_1} d\tau_2 \tilde{H}(\tau_1)\tilde{H}(\tau_2) + \dots \, . \tag{11.112}$$

If $\overline{H}$ is the average value of the Hamiltonian in the interval considered,

$$\overline{H} = \int_0^1 d\tau \, \tilde{H}(\tau) = \frac{1}{T} \int_{t_0}^{t_1} dt \, H(t) \, , \tag{11.113}$$

replacing (11.112) in (11.111) leads to

$$p_0 = \frac{T^2}{\hbar^2} \left[\langle \phi_0 | \overline{H}^2 | \phi_0 \rangle - \langle \phi_0 | \overline{H} | \phi_0 \rangle^2 \right] \mathrm{O}(T^3) \, . \tag{11.114}$$

Since the quantity in brackets is nothing but the square of the uncertainty of the operator $\overline{H}$ in the state $|\phi_0\rangle$, we have

$$p_0 = \frac{T^2 (\Delta_{\phi_0} \overline{H})^2}{\hbar^2} + \mathrm{O}(T^3) \, , \tag{11.115}$$

and the sudden approximation will hold if

$$T \ll \frac{\hbar}{\Delta_{\phi_0} \overline{H}} \, . \tag{11.116}$$

When $H(t)$ is of the form $H(t) = H_0 + H_I(t)$, where only $H_I(t)$ depends on time and $|\phi_0\rangle$ is an eigenstate of H_0, then $\Delta_{\phi_0} \overline{H} = \Delta_{\phi_0} \overline{H}_I$, with $\overline{H}_I$ defined as $\overline{H}$, and (11.116) becomes

$$T \ll \frac{\hbar}{\Delta_{\phi_0} \overline{H}_I} . \tag{11.117}$$

Thus, for example, for a harmonic oscillator driven by a perturbation of the type (11.47) we have $\overline{H}_I = -[2M\hbar\omega^3]^{1/2}(x/2)$. If initially the system is in the eigenstate $|\psi_n\rangle$ one easily verifies using (4.49) that $\Delta_{\psi_n} \overline{H}_I = (\hbar\omega/2)\sqrt{2n+1}$ and, therefore, the sudden approximation is applicable only if $\omega T \ll 2/\sqrt{2n+1}$ i.e., if T is much smaller than the time ($\sim \hbar/\Delta_{\psi_n} \overline{H}$) characteristic of the state $|\psi_n\rangle$ under the mean dynamics.

11.10 The Adiabatic Theorem

In the same way than in the limit $T \to 0$ we found the result (11.110) which provided the basis for the sudden approximation, in the opposite case $T \to \infty$ we have the *adiabatic theorem*, which underlies the *adiabatic approximation*. The adiabatic theorem, which we will prove in this section, has played an extraordinary important role in the development of the quantum theory [JA 66]. The proof we present follows [KA 50], [FR 55], as well as [ME 59]. An important improvement can be found in [AS 87].

Suppose that $\tilde{H}(\tau)$, $0 \le \tau \le 1$, has a purely discrete spectrum, with eigenvalues $E_1(\tau), E_2(\tau), \ldots, E_n(\tau), \ldots$. The projection operators onto the corresponding subspaces are respectively $P_1(\tau), P_2(\tau), \ldots, P_n(\tau), \ldots$. We assume that the eigenvalues are piecewise differentiable in the parameter τ, and that there is no level crossing throughout the transition, in other words:

$$E_i(\tau) \neq E_j(\tau) , \quad i \neq j , \quad \tau \in [0,1] . \tag{11.118}$$

We further require $dP_j(\tau)/d\tau$ and $d^2P_j(\tau)/d\tau^2$ to exist and to be bounded and piecewise continuous in the interval $0 \le \tau \le 1$. Under these conditions it is possible to prove the adiabatic theorem:

$$U_T(\tau)P_i(0) - P_i(\tau)U_T(\tau) = \mathrm{O}\left(\frac{1}{T}\right) , \quad T \to \infty , \quad i = 1, 2, \ldots . \tag{11.119}$$

Notice that if initially the system is in the state $|\phi_i\rangle$ so that $H(0)|\phi_i\rangle = E_i(0)|\phi_i\rangle$, then (11.119) implies

$$\lim_{T\to\infty} U_T(\tau)|\phi_i\rangle = P_i(\tau) \lim_{T\to\infty} U_T(\tau)|\phi_i\rangle , \quad i = 1, 2, \ldots \tag{11.120}$$

and in the limit $T \to \infty$ the state $U_T(\tau)|\phi_i\rangle$ belongs to the subspace generated by the eigenvectors with eigenvalue $E_i(\tau)$, for all values of $\tau \in [0,1]$.

Even though this theorem makes sense only for discrete eigenvalues, it is not necessary to require that the whole spectrum of $\tilde{H}(\tau)$ be discrete. If the spectrum has a continuous part, the theorem (11.119) can still be proved [KA 50] if for its discrete part the relations (11.118) as well as the differentiability conditions on the projection operators $P_i(\tau)$ hold. We can also do without the restriction

(11.118) although then the conclusion (11.119) weakens to be O(1) instead of $O(1/T)$.

The proof of the adiabatic theorem is trivial if the subspaces associated to each of the eigenvalues remain constant in time:

$$P_i(\tau) = P_i(0) \equiv P_i \,, \quad i = 1, 2, \dots \,. \tag{11.121}$$

In this case

$$\tilde{H}(\tau) = \sum_i E_i(\tau) P_i \,, \tag{11.122}$$

and since each of the projection operators P_j comutes with $\tilde{H}(\tau)$ we obtain

$$U_T(\tau) P_j = P_j U_T(\tau) \,, \tag{11.123}$$

valid for all T, so that (11.119) follows trivially. It is important to notice that in this case the evolution operator is simply

$$U_T(\tau) = \exp\left[-\frac{\mathrm{i}T}{\hbar}\int_0^\tau d\tau_1 \tilde{H}(\tau_1)\right] \tag{11.124}$$

or, equivalently,

$$U_T(\tau) = \sum_i \exp[-\mathrm{i}T\alpha_i(\tau)/\hbar] P_i \,, \quad \text{where} \tag{11.125}$$

$$\alpha_i(\tau) \equiv \int_0^\tau d\tau_1 E_i(\tau_1) \,. \tag{11.126}$$

In the general case the proof is considerably more involved and it consists of three steps: first we change to the evolution picture where the Hamiltonian has time-independent spectral projections, i.e., we go over to a time-dependent reference frame following the axes which diagonalize $\tilde{H}(\tau)$. In this picture, the time evolution generator contains an immediately integrable contribution, which we eliminate by going over to a second picture and finally one shows that the remaining evolution operator differs from the identity by terms $O(1/T)$.

Let us write the evolution operator U_T as a product of three unitary transformations

$$U_T(\tau) = V(\tau) S_T(\tau) W_T(\tau) \,. \tag{11.127}$$

As the generator $K(\tau)$ of $V(\tau)$ in the sense:

$$\mathrm{i}\hbar \frac{dV(\tau)}{d\tau} = K(\tau) V(\tau) \,, \quad V(0) = I \,, \tag{11.128}$$

we choose the self-adjoint operator defined by

$$K(\tau) = \mathrm{i}\hbar \sum_j \frac{dP_j(\tau)}{d\tau} P_j(\tau) \,. \tag{11.129}$$

To prove that $K(\tau)$ is self-adjoint it is enough to take into account that $P_j(\tau) = P_j^2(\tau)$ implies

$$\frac{dP_j(\tau)}{d\tau} = P_j(\tau)\frac{dP_j(\tau)}{d\tau} + \frac{dP_j(\tau)}{d\tau}P_j(\tau)\,, \tag{11.130}$$

and since $\sum_j P_j(\tau) = I$ we have $\sum_j dP_j(\tau)/d\tau = 0$.

The choice (11.129) for $K(\tau)$ guarantees that in the V picture the projection operators $P_{jV}(\tau) \equiv V^\dagger(\tau)P_j(\tau)V(\tau)$ diagonalizing $\tilde{H}_V(\tau)$, are constant:

$$P_{jV}(\tau) = P_j(0)\,. \tag{11.131}$$

Indeed, taking into account Eq. (11.128) and its adjoint

$$\mathrm{i}\hbar\frac{dP_{jV}(\tau)}{d\tau} = V^\dagger(\tau)\left[\mathrm{i}\hbar\frac{dP_j(\tau)}{d\tau} + P_j(\tau)K(\tau) - K(\tau)P_j(\tau)\right]V(\tau)\,, \tag{11.132}$$

and using (11.129) together with the self-adjointness of $K(\tau)$:

$$\frac{dP_{jV}(\tau)}{d\tau} = V^\dagger(\tau) \times\left[\frac{dP_j(\tau)}{d\tau} - \sum_k P_j(\tau)P_k(\tau)\frac{dP_k(\tau)}{d\tau} - \sum_k \frac{dP_k(\tau)}{d\tau}P_k(\tau)P_j(\tau)\right]V(\tau)\,. \tag{11.133}$$

Now, the orthogonality of the projection operators for different eigenvalues and (11.130) imply $dP_{jV}(\tau)/d\tau = 0$, and since $P_{jV}(0) = P_j(0)$, the equality (11.131) is proved. An immediate corollary is

$$\tilde{H}_V(\tau) = \sum_j E_j(\tau)P_j(0)\,. \tag{11.134}$$

On the other hand, the evolution operator in the new picture $U_T^{(V)}(\tau) \equiv V^\dagger(\tau)U_T(\tau)$ satisfies

$$\mathrm{i}\hbar\frac{dU_T^{(V)}(\tau)}{d\tau} = [T\tilde{H}_V(\tau) - K_V(\tau)]U_T^{(V)}(\tau)\,, \quad U_T^{(V)}(0) = I\,. \tag{11.135}$$

Since $\tilde{H}_V(\tau)$ and $K_V(\tau)$ are independent of T, it is to be expected that in the $T \to \infty$ limit the first term of the right hand side in (11.135) dominates. According to the program outlined at the beginning, we define $S_T(\tau)$ via

$$\mathrm{i}\hbar\frac{dS_T(\tau)}{d\tau} = T\tilde{H}_V(\tau)S_T(\tau)\,, \quad S_T(0) = I\,. \tag{11.136}$$

From (11.134) and following the steps which led to (11.125) we have

$$S_T(\tau) = \sum_j \exp[-\mathrm{i}T\alpha_j(\tau)/\hbar]P_j(0) \tag{11.137}$$

with the definition (11.126).

Finally we change to a last picture with operator $S_T^\dagger(\tau)V^\dagger(\tau)U_T(\tau) \equiv W_T(\tau)$, and (−) generator $K_T^{(W)}(\tau) \equiv S_T^\dagger(\tau)K_V(\tau)S_T(\tau) = S_T^\dagger(\tau)V^\dagger(\tau)K(\tau)V(\tau)S_T(\tau)$:

$$\mathrm{i}\hbar\frac{dW_T(\tau)}{d\tau} = -K_T^{(W)}(\tau)W_T(\tau)\,, \quad W_T(0) = I\,, \tag{11.138}$$

equivalent to the integral equation

$$W_T(\tau) = I + \frac{\mathrm{i}}{\hbar}\int_0^\tau d\tau_1 K_T^{(W)}(\tau_1)W_T(\tau_1)\,. \tag{11.139}$$

Now we prove that in the limit $T \to \infty$

$$W_T(\tau) = I + \mathrm{O}(1/T)\,. \tag{11.140}$$

We begin by considering the operator $F_T(\tau)$

$$F_T(\tau) = \int_0^\tau d\tau_1 K_T^{(W)}(\tau_1)\,. \tag{11.141}$$

Any bounded operator (and in particular $F_T(\tau)$), can be written as

$$F_T(\tau) = \sum_{i,j} P_i(0)F_T(\tau)P_j(0) \equiv \sum_{i,j} F_{T,ij}(\tau)\,. \tag{11.142}$$

Using (11.137) we obtain

$$\begin{aligned} F_{T,ij}(\tau) &= \int_0^\tau d\tau_1 \exp\{\mathrm{i}T[\alpha_i(\tau_1) - \alpha_j(\tau_1)]/\hbar\}P_i(0)K_V(\tau_1)P_j(0) \\ &\equiv \int_0^\tau d\tau_1 \exp\{\mathrm{i}T[\alpha_i(\tau_1) - \alpha_j(\tau_1)]/\hbar\}K_{V,ij}(\tau_1) \end{aligned} \tag{11.143}$$

and since (11.131) implies that $P_j(0)V^\dagger(\tau) = V^\dagger(\tau)P_j(\tau)$ then:

$$\begin{aligned} F_{T,ij}(\tau) = &\int_0^\tau d\tau_1 \exp\{\mathrm{i}T[\alpha_i(\tau_1) - \alpha_j(\tau_1)]/\hbar\} \\ &\times V^\dagger(\tau_1)P_i(\tau_1)K(\tau_1)P_j(\tau_1)V(\tau_1)\,. \end{aligned} \tag{11.144}$$

From (11.129) and (11.130) it is easy to derive $P_i(\tau)K(\tau)P_i(\tau) = 0$ and thereby $F_{T,ii}(\tau) = 0$. Let $i \neq j$; since $K_{V,ij}(\tau)$ is a continuous function of τ, our assumptions imply that $\alpha_i(\tau) - \alpha_j(\tau)$ is a continuous nonvanishing monotonic function of τ; after integrating by parts in (11.143) we obtain

$$\begin{aligned} F_{T,ij}(\tau) = \frac{\hbar}{\mathrm{i}T}\Bigg[&\exp\{\mathrm{i}T[\alpha_i(\tau_1) - \alpha_j(\tau_1)]/\hbar\}\frac{K_{V,ij}(\tau_1)}{E_i(\tau_1) - E_j(\tau_1)}\Bigg|_0^\tau \\ &- \int_0^\tau d\tau_1 \exp\{\mathrm{i}T[\alpha_i(\tau_1) - \alpha_j)\tau_1)]/\hbar\}\frac{d}{d\tau_1}\frac{K_{V,ij}(\tau_1)}{E_i(\tau_1) - E_j(\tau_1)}\Bigg]\,, \end{aligned} \tag{11.145}$$

whence $F_{T,ij}(\tau)$, $i \neq j$, converges asymptotically towards 0 as $1/T$. Summarizing, as $T \to \infty$ we have:

$$F_T(\tau) = \mathrm{O}(1/T) \, . \tag{11.146}$$

Integration by parts turns (11.139) into:

$$W_T(\tau) = I + \frac{\mathrm{i}}{\hbar} F_T(\tau) W_T(\tau) - \frac{\mathrm{i}}{\hbar} \int_0^\tau d\tau_1 \, F_T(\tau_1) \frac{dW_T(\tau_1)}{d\tau_1} \, , \tag{11.147}$$

and using (11.138)

$$W_T(\tau) = I + \frac{\mathrm{i}}{\hbar} F_T(\tau) W_T(\tau) + \frac{1}{\hbar^2} \int_0^\tau d\tau_1 \, F_T(\tau_1) K_T^{(W)}(\tau_1) W_T(\tau_1) \, . \tag{11.148}$$

Since the last two terms in this equation contain the factor $F_T(\tau)$, (11.140) is proven. From $U_T(\tau) = V(\tau) S_T(\tau) W_T(\tau)$ we obtain for $T \to \infty$

$$U_T(\tau) = V(\tau) S_T(\tau) \left[I + \mathrm{O}\left(\frac{1}{T} \right) \right] \, . \tag{11.149}$$

Finally (11.137) implies $S_T(\tau) P_j(0) = P_j(0) S_T(\tau)$ and hence $V(\tau) S_T(\tau) P_j(0) = V(\tau) P_j(0) S_T(\tau) = P_j(\tau) V(\tau) S_T(\tau)$. This concludes the proof of the adiabatic theorem (11.119).

11.11 The Adiabatic Approximation

When T is sufficiently large, we can replace $U(t_1, t_0)$ to a first approximation by

$$\boxed{U(t_1, t_0) = U_T(1) \simeq V(1) S_T(1)} \, , \tag{11.150}$$

and this is the *adiabatic approximation*. If the initial normalized state is $|\phi_0\rangle$, in this approximation $U(t_1, t_0)|\phi_0\rangle \approx V(1) S_T(1) |\phi_0\rangle$. As before we can estimate the error by computing the probability p_0 of finding the system at time t_1 in a state different from $V(1) S_T(1) |\phi_0\rangle$:

$$p_0 = \langle \phi_0 | U^\dagger(t_1, t_0) V(1) S_T(1) Q_0 S_T^\dagger(1) V^\dagger(1) U(t_1, t_0) | \phi_0 \rangle \, , \tag{11.151}$$

and as before $Q_0 = I - |\phi_0\rangle\langle\phi_0|$. This quantity may be rewritten as

$$p_0 = \langle \phi_0 | W_T^\dagger(1) Q_0 W_T(1) | \phi_0 \rangle \, . \tag{11.152}$$

Solving (11.139) iteratively, and keeping only the first order term, we find

$$p_0 \simeq \frac{1}{\hbar^2} \langle \phi_0 | F_T(1) Q_0 F_T(1) | \phi_0 \rangle = \frac{1}{\hbar^2} (\Delta_{\phi_0} F_T)^2 \, , \tag{11.153}$$

where $\Delta_{\phi_0} F_T$ is the uncertainty of the self-adjoint operator $F_T(1)$ in the state $|\phi_0\rangle$. Hence, the adiabatic approximation is applicable when

$$\Delta_{\phi_0} F_T \ll \hbar \, . \tag{11.154}$$

This condition is not easy to verify in practice, because in general we do not have enough explicit information about the operator $F_T(1)$.

A different way of writing (11.154), more amenable to interpretation, is the following: assume that the initial normalized state $|\phi_1\rangle$ is an eigenstate of H_0, with eigenvalue $E_1(0)$. Equation (11.153) may be recast as

$$p_1 \simeq \frac{1}{\hbar^2} \sum_{j \neq 1} \|F_{T,j1}(1)\phi_1\|^2 . \tag{11.155}$$

Using (11.144) we obtain

$$F_{T,j1}(1)|\phi_1\rangle = \mathrm{i}\hbar \int_0^1 d\tau \exp\{\mathrm{i}T[\alpha_j(\tau_1) - \alpha_1(\tau)]/\hbar\} V^\dagger(\tau) P_j(\tau) \times \frac{dP_1(\tau)}{d\tau} P_1(\tau) V(\tau) |\phi_1\rangle , \tag{11.156}$$

and since $P_j(\tau)V(\tau) = V(\tau)P_j(0)$:

$$F_{T,j1}(1)|\phi_1\rangle = \mathrm{i}\hbar \int_0^1 d\tau \exp\{\mathrm{i}T[\alpha_j(\tau) - \alpha_1(\tau)]/\hbar] P_j(0) V^\dagger(\tau) \times \frac{dP_1(\tau)}{d\tau} V(\tau)|\phi_1\rangle . \tag{11.157}$$

Introducing an orthonormal basis $\{\phi_{j\alpha}\rangle\}$ diagonalizing $H(0)$, the vectors $|\phi_{j\alpha}(\tau)\rangle \equiv V(\tau)|\phi_{j\alpha}\rangle$ diagonalize $H(\tau)$ and we can write

$$\langle\phi_{j\alpha}|F_{T,j1}(1)|\phi_1\rangle = \mathrm{i}\hbar \int_0^1 d\tau \, \exp\{\mathrm{i}T[\alpha_j(\tau) - \alpha_1(\tau)]/\hbar\} \times \langle\phi_{j\alpha}(\tau)|\frac{dP_1(\tau)}{d\tau}|\phi_1(\tau)\rangle . \tag{11.158}$$

Since $\langle\phi_{j\alpha}(\tau)|(dP_1(\tau)/d\tau)|\phi_1(\tau)\rangle = \langle\phi_{j\alpha}(\tau)|(d/d\tau)\phi_1(\tau)\rangle$, $j \neq 1$, we finally obtain

$$p_1 \simeq \sum_{\alpha, j \neq 1} \left| \int_0^1 d\tau \, \exp\{\mathrm{i}T[\alpha_j(\tau) - \alpha_1(\tau)]/\hbar\} \langle\phi_{j\alpha}(\tau)|\frac{d\phi_1(\tau)}{d\tau}\rangle \right|^2 \tag{11.159}$$

and the adiabatic approximation for $|\phi_1\rangle$ will hold only if $p_1 \ll 1$. We can interpret this condition by saying that the axes which diagonalize $\tilde{H}(\tau)$ rotate with frequencies negligible with respect to the transition frequencies characteristic of the system.

As an example, consider the case of a forced harmonic oscillator driven by a perturbation of the form (11.47). Then

$$\tilde{H}(\tau) = \hbar\omega \left[a^\dagger a + \frac{1}{2} - \tau a^\dagger - \tau a\right] , \quad 0 \leq \tau \leq 1 . \tag{11.160}$$

Defining new creation and annihilation operators

$$b^\dagger(\tau) = a^\dagger - \tau \ , \quad b(\tau) = a - \tau \ , \tag{11.161}$$

which are obtained by a unitary transformation of $a, a^\dagger$, equivalent to a translation, we can write

$$\tilde{H}(\tau) = \hbar\omega \left[b^\dagger(\tau) b(\tau) + \frac{1}{2} - \tau^2 \right] . \tag{11.162}$$

Choosing phases conveniently so that $V(\tau)|n;0\rangle = |n;\tau\rangle$, the eigenvectors and eigenvalues are

$$|n;\tau\rangle = \frac{1}{\sqrt{n!}} [b^\dagger(\tau)]^n |\underline{\tau}\rangle \ , \quad E_n(\tau) = \hbar\omega \left[n + \frac{1}{2} - \tau^2 \right] , \tag{11.163}$$

where $|\underline{\tau}\rangle$ represents the coherent state

$$|\underline{\tau}\rangle = \mathrm{e}^{-\tau^2/2} \mathrm{e}^{\tau a^\dagger} |0\rangle \ . \tag{11.164}$$

If the initial state is $|\phi_1\rangle \equiv |n\rangle$, we easily find $\alpha_m(\tau) - \alpha_n(\tau) = \hbar\omega(m-n)\tau$, and a simple computation shows

$$\langle m;\tau| \frac{d}{d\tau} |n;\tau\rangle = \sqrt{n+1}\,\delta_{m,n+1} - \sqrt{n}\,\delta_{m,n-1} \ . \tag{11.165}$$

Hence

$$p_n \simeq n \left| \int_0^1 d\tau \, \mathrm{e}^{-\mathrm{i}\omega T \tau} \right|^2 + (n+1) \left| \int_0^1 d\tau \, \mathrm{e}^{\mathrm{i}\omega T \tau} \right|^2 \tag{11.166}$$

and therefore

$$p_n \simeq \frac{2(2n+1)}{\omega^2 T^2} (1 - \cos\omega T) \ . \tag{11.167}$$

In conclusion the adiabatic approximation holds if $\omega T \gg \sqrt{2n+1}$, as was found in Sect. 11.3 with considerably less detail.

A renewal of theoretical and experimental interest in the adiabatic processes has taken place in the last few years. According to the adiabatic approximation, if the final Hamiltonian H_1 of an adiabatic transient coincides with the initial one ($H_1 = H_0$) the system returns up to a *phase* factor to the initial vector state (provided that it is a non-degenerate eigenstate of H_0). In a seminal paper Berry [BE 84] has pointed out that this phase has not only a *dynamical* component due to $S_T(1)$ but it also has a further contribution, known nowadays as *geometric, Berry, topological* or *holonomic phase.* This extra piece arises from $V(1)$ and has the remarkable property of depending only on the path followed by $H(t)$ in the *parameter space* (for example, the space of the components of $B(t)$ in (11.4)) and not on how $H(t)$ traverses this path (compatible with adiabaticity). Berry's phase has a nice geometric interpretation as the curvature of a connection in the space of unit rays [SI 83], and its surprisingly universal emergence transcends the adiabatic [AA 87], cyclic and unitary [SB 88] processes. Its phenomenological manifestations through interferences have been observed, for instance, with neutrons in helicoidal magnetic fields [BD 87] and with polarized photons in twisted optical fibers [TC 86].

11.12 The Decay Law for Unstable Quantum Systems

Given the scarcity of general rigorous results in the treatment of time-dependent perturbations [SI 71], it very interesting to construct models incorporating real characteristics of physical systems which at the same time admit explicit solutions. With one of these models, we discuss below the the validity of the exponential decay law for unstable quantum states.

At time $t = 0$ we prepare a physical system in an unstable and normalized state $|B(0)\rangle \equiv |B\rangle$, which evolves under the action of a time-independent Hamiltonian H until some time $t > 0$. At this moment we perform a measurement of the system to find out whether it has decayed; namely, whether $|B(t)\rangle \perp |B(0)\rangle$. The probability $P(t)$ to find the system in the original state is

$$P(t) = |A(t)|^2 = |\langle B|e^{-iHt/\hbar}|B\rangle|^2 , \tag{11.168}$$

with $A(t) \equiv \langle B(0)|B(t)\rangle$. It is often assumed that $P(t)$ is of the form $P(t) = e^{-t/\tau}$, where τ is the lifetime of the system in the state $|B\rangle$. We first show that the exponential law for $P(t)$ contradicts the general results of quantum mechanics. Next we explicitly compute $P(t)$ for the decay of an unstable state in a soluble model.

For $t \downarrow 0$ and assuming $|B\rangle \in D(H)$ we have

$$A(t) = 1 - \frac{it}{\hbar}\langle B|H|B\rangle + O(t^2) . \tag{11.169}$$

Since H is self-adjoint, $\langle B|H|B\rangle$ is real and hence

$$P(t) = 1 + O(t^2) , \quad t \downarrow 0 \tag{11.170}$$

and therefore, the exponential decay law should fail for sufficiently small times.

On the other hand it can be shown [SI 72a] that if $P(t) = O(e^{-t/\tau})$, $t \to \infty$, then the spectral measure $(-\infty, \lambda] \to \|E_\lambda|B\rangle\|^2$, where $\{E_\lambda\}$ is the resolution of the identity for H, has as support the whole real line and it is absolutely continuous. Thus if the Hamiltonian H is bounded below or if $|B\rangle$ has a non-vanishing projection onto some bound state of H, the exponential decay law will not hold for sufficiently large times. [The idea of the proof is simple [SI 78]: if $|A(t)| = O(e^{-t/\tau})$, $\tau > 0$, for $t \to \infty$, since $A(t) = A^*(-t)$ we will also have $|A(t)| = O(\exp[-|t|/\tau])$ for $|t| \to \infty$. However

$$A(t) = \int_{\sigma(H)} e^{-i\lambda t/\hbar} d\|E_\lambda|B\rangle\|^2 . \tag{11.171}$$

Furthermore the behavior of $|A(t)|$ at large times implies that the Fourier transform $\hat{A}(\lambda)$ of $A(t)$ is analytic in the strip $|\mathrm{Im}\lambda| < \hbar/\tau$. The uniqueness of this transform together with (11.171) shows that $(d/d\lambda)\|E_\lambda|B\rangle\|^2$ must be analytic in the same strip. Since $\sigma(H)$ is assumed to be bounded on the left, this analyticity would require $A(t) \equiv 0$, in contradiction with $A(0) = 1$].

It is precisely the possibility of regenerating the state $|B\rangle$ in the evolution process which explains the deviations from the pure exponential law. Indeed, given t, t' we have

$$A(t+t') = A(t')A(t) + \sum_{|a\rangle \perp |B\rangle} \langle B|\mathrm{e}^{-\mathrm{i}t'H/\hbar}|a\rangle\langle a|\mathrm{e}^{-\mathrm{i}tH/\hbar}|B\rangle \, . \tag{11.172}$$

If the later regeneration of $|B\rangle$ were not possible, i.e. if $\langle B|\exp(-\mathrm{i}t'H/\hbar)|a\rangle = 0$, we would have $A(t+t') = A(t')A(t)$, and, the exponential law would hold.

A typical example of an unstable state is the $2s^2$ state of the helium atom, where each electron is in the $2s$ state of the hydrogen-like ion He^+. When the electrostatic repulsion between the electrons and the nucleus recoil are neglected this atomic state is bound and stable (in the absence of interactions with the electromagnetic field) and it lies on the absolutely continuous part of the energy spectrum. When we include Coulomb repulsion between the electrons the state can decay into a state in the continuum, giving rise to a self-ionization process (see Sect. 14.1). The rigorous study of this problem is extraordinarily complicated and it has not been carried out up to date. We are going to introduce a very simplified model which incorporates some of the features of the physical situation mentioned and which can be solved explicitly.

Let $\mathcal{H}$ be the Hilbert space defined by pairs $\{\alpha, f(x)\}$ where α is an arbitrary complex number and f is a square integrable function in the interval $[0,\infty)$. The square norm is defined by

$$\|\{\alpha, f\}\|^2 \equiv |\alpha|^2 + \int_0^\infty dx\,|f(x)|^2 \, . \tag{11.173}$$

Define a Hamiltonian operator H_0 as

$$H_0 : \{\alpha, f(x)\} \to \{\alpha, xf(x)\} \, . \tag{11.174}$$

Clearly, H_0 is self-adjoint with continuous spectrum $[0,1) \cup (1,\infty)$ and with a single non-degenerate eigenvalue equal to 1. Let $|B\rangle$ be the state $\{1,0\}$ and let $|a\rangle$ be the (improper) state $\{0, \delta(x-a)\}$. As interaction V we take

$$V : \{\alpha, f(x)\} \to \{\langle v|f\rangle, \alpha v(x)\} \tag{11.175}$$

where $v \in L^2(0,\infty)$ is sufficiently smooth (Hölder continuity is enough [MA 68]) so that the spectrum of the total Hamiltonian $H \equiv H_0 + V$ presents no pathologies.

The evolution operator $U(t) = \exp(-\mathrm{i}Ht)$ (in $\hbar = 1$ units), can be represented in $D(H^2)$ and for $t \geq 0$ as

$$U(t) = \frac{\mathrm{i}}{2\pi}\int_{\Gamma_\varepsilon} \mathrm{e}^{-\mathrm{i}tz}\frac{1}{z-H}(c-H)^2\frac{dz}{(c-z)^2}\,, \tag{11.176}$$

where Γ_ε, with $\varepsilon > 0$, is the infinite contour formed by two lines parallel to the real axes: $z = \tau + \mathrm{i}\varepsilon$, $\tau \in \mathbb{R}$ traversed to the right, and $z = \tau - \mathrm{i}\varepsilon$, $\tau \in \mathbb{R}$ traversed towards the left. In (11.176) c is any complex number satisfying $|\mathrm{Im}\, c| > \varepsilon$. The previous equation can be formally derived using Cauchy's theorem (for a rigorous proof see [DS 64]).

From (8.94) we have

$$\frac{1}{z-H} = \frac{1}{z-H_0} + \frac{1}{z-H_0}V\frac{1}{z-H_0} + \frac{1}{z-H_0}V\frac{1}{z-H_0}V\frac{1}{z-H_0} + \dots\,. \tag{11.177}$$

Taking into account (11.175) and (11.177) we obtain

$$\begin{aligned} \langle B|\frac{1}{z-H}|B\rangle &= \left[z-1+\int_0^\infty dy\frac{|v(y)|^2}{y-z}\right]^{-1}\,, \\ \langle x|\frac{1}{z-H}|B\rangle &= \frac{v(x)}{z-x}\left[z-1+\int_0^\infty dy\frac{|v(y)|^2}{y-z}\right]^{-1}\,, \end{aligned} \tag{11.178}$$

and, therefore,

$$\begin{aligned} A(t) &= \langle B|\mathrm{e}^{-\mathrm{i}tH}|B\rangle \\ &= \frac{\mathrm{i}}{2\pi}\int_{\Gamma_\varepsilon} dz\,\mathrm{e}^{-\mathrm{i}tz}\left[\frac{2c-z-1}{(c-z)^2} + \left[z-1+\int_0^\infty dy\frac{|v(y)|^2}{y-z}\right]^{-1}\right]\,. \end{aligned} \tag{11.179}$$

On the other hand the unitarity of $U(t)$ implies that $P(t) = |A(t)|^2$ should satisfy

$$P(t) = 1 - \int_0^\infty dx\,|\langle x|\mathrm{e}^{-\mathrm{i}Ht}|B\rangle|^2\,. \tag{11.180}$$

It follows from (11.178) in the Born approximation:

$$\langle x|\frac{1}{z-H}|B\rangle \simeq \langle x|\frac{1}{z-H}|B\rangle_{\mathrm{Born}} = \frac{v(x)}{(z-x)(z-1)} \tag{11.181}$$

and using now (11.176):

$$\langle x|U(t)|B\rangle \simeq \langle x|U(t)|B\rangle_{\mathrm{Born}} = \frac{v(x)}{x-1}[\mathrm{e}^{-\mathrm{i}tx} - \mathrm{e}^{-\mathrm{i}t}]\,. \tag{11.182}$$

Therefore

$$|\langle x|U(t)|B\rangle_{\mathrm{Born}}|^2 = \frac{\sin^2[(x-1)t/2]}{[(x-1)/2]^2}|v(x)|^2 \underset{t\to\infty}{\sim} 2\pi|v(1)|^2 t\delta(1-x) \tag{11.183}$$

and hence

$$P_{\mathrm{Born}}(t) \underset{t\to\infty}{\sim} 1 - 2\pi|v(1)|^2 t \equiv 1 - \Gamma_{\mathrm{Born}} t\,. \tag{11.184}$$

The lifetime of $|B\rangle$ in this approximation is defined as $1/\Gamma_{\rm Born}$, and we approximate $P(t)$ as $P_{\rm Born}(t) \simeq \exp(-\Gamma_{\rm Born} t)$, which is usually assumed to be valid for weak interactions.

Since the interaction V in this model is separable and of finite rank, and since we assume $v(x)$ to be smooth, H does not contain any continuous singular part in its spectrum and $\sigma_{\rm a.c.}(H) = [0, \infty)$. With regard to the point spectrum, we consider the eigenvalue equation

$$H\{\alpha, f\} = \lambda\{\alpha, f\} \, . \tag{11.185}$$

Its solution is $\{1, v(x)/(\lambda - x)\}$ with λ satisfying

$$\lambda = 1 - \int_0^\infty dy \, \frac{|v(y)|^2}{y - \lambda} \, . \tag{11.186}$$

The number of solutions to (11.186) depends on the properties of v. Assuming for simplicity that $v(x) \neq 0$, $\forall x \in (0, \infty)$ and that $v(0) = 0$, then there is a single solution λ_0 of (11.186) giving rise to a normalizable eigenstate, with negative energy λ_0, if and only if

$$\int_0^\infty dy \frac{|v(y)|^2}{y} > 1 \, . \tag{11.187}$$

If this integral is smaller than 1 there are no eigenstates. Finally, when it equals 1, there is an eigenstate with zero energy if $v(x)$ tends to 0 with x faster than $x^{1/2}$.

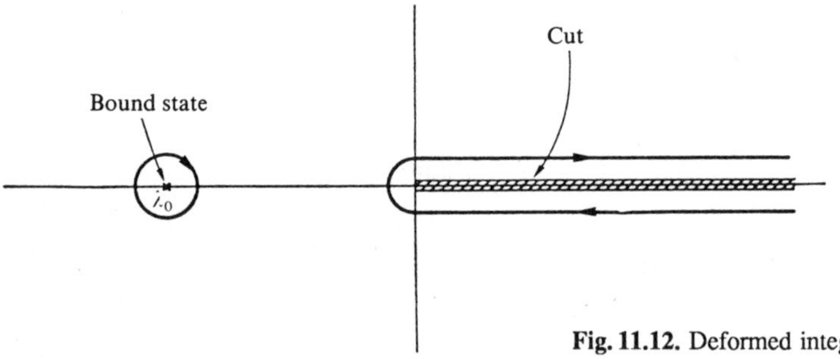

Fig. 11.12. Deformed integration contour Γ'_ε

To calculate $A(t)$ it is convenient to deform the integration contour in (11.179) to Γ'_ε shown in Fig. 11.12. The contour around λ_0 appears only under the assumption (11.187). Decomposing the amplitude $A(t)$ as a sum of the contributions from the cut and the possible bound state, letting $\varepsilon \downarrow 0$ and taking into account

$$\frac{1}{x \pm {\rm i}0} = {\rm PV}\frac{1}{x} \mp {\rm i}\pi\delta(x) \, , \tag{11.188}$$

we obtain

$$A(t) = A_{\rm B}(t) + A_{\rm c}(t) \; ,$$
$$A_{\rm B}(t) = A_{\rm B}(0){\rm e}^{-{\rm i}\lambda_0 t} \; , \quad A_{\rm c}(t) = \int_0^\infty dy\, F(y){\rm e}^{-{\rm i}yt} \; ,$$
$$A_{\rm B}(0) = \left\{ 1 + \int_0^\infty dy \frac{|v(y)|^2}{(y-\lambda_0)^2} \right\}^{-1} \; , \tag{11.189}$$
$$F(y) = |v(y)|^2 \left\{ \left[y - 1 + {\rm PV} \int_0^\infty d\tau \frac{|v(\tau)|^2}{\tau - y} \right]^2 + \pi^2 |v(y)|^4 \right\}^{-1} \; .$$

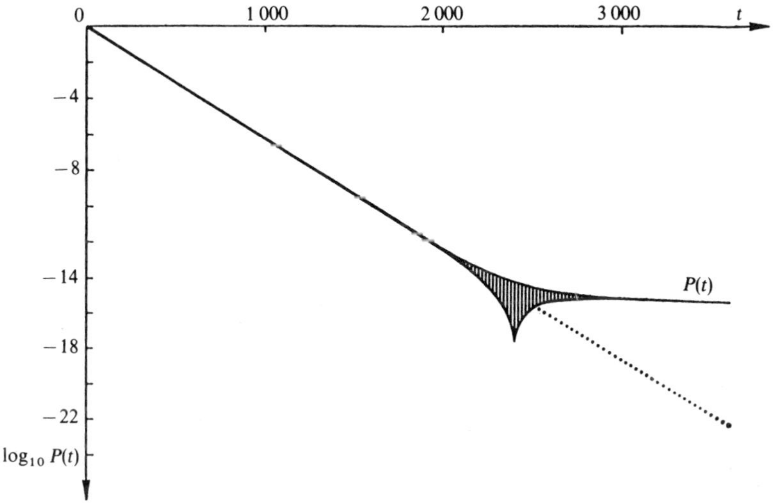

Fig. 11.13. Representation of $P(t)$ for $g^2 = 0.01$. The shadowed region contains many oscillations of $P(t)$. The dotted line represents $P_{\rm Born}(t)$

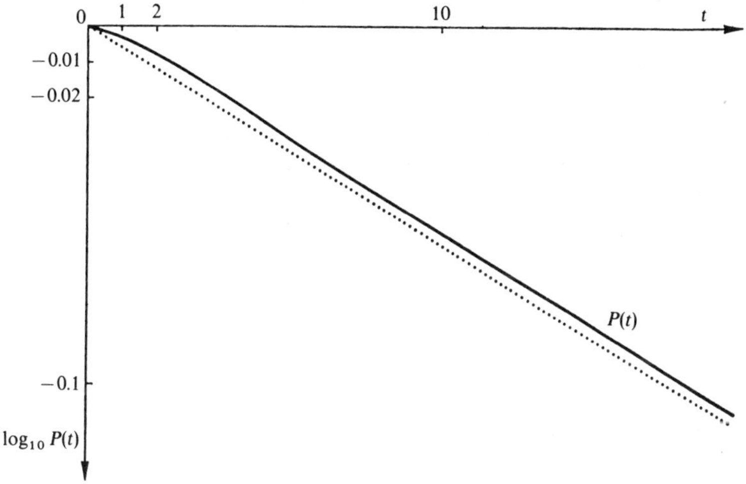

Fig. 11.14. Magnified detail of Fig. 11.13 for small t

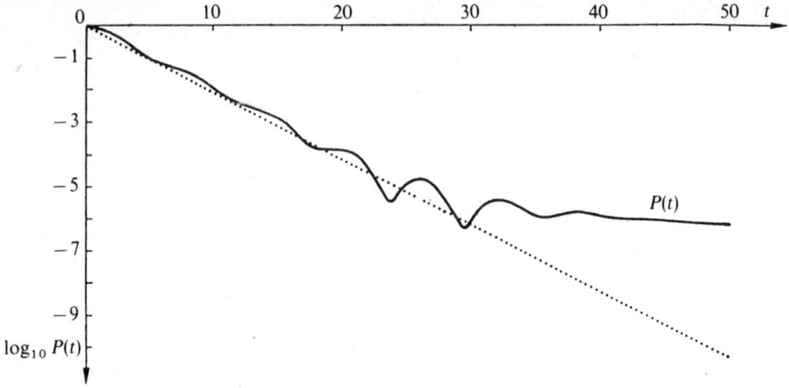

Fig. 11.15. Representation of $P(t)$ for $g^2 = 0.3$. The dotted line denotes an approximate probability generalizing $P_{\text{Born}}(t)$ when g^2 is not very small [GP 77]

Notice that in general $A_c(t) \to 0$ when $t \to \infty$ (Riemann-Lebesgue lemma) and, therefore, if (11.187) is satisfied, $P(t) \to |A_B(0)|^2 \neq 0$ for large times.

The computation of $A_c(t)$ requires a double integration which in general is rather difficult to carry out. However [GP 77] it can be expressed through known functions if we take

$$|v(x)|^2 = \frac{\sqrt{2}g^2}{\pi} \frac{\sqrt{x}}{1+x^2} \tag{11.190}$$

where g plays the role of the coupling constant. For such an interaction λ_0 is the solution to

$$\lambda - 1 + \frac{g^2}{1+\lambda^2}\left[1 - \lambda - \sqrt{-2\lambda}\right] = 0 \tag{11.191}$$

in the complex plane cut along $[0, \infty)$. The condition (11.187) is now $g^2 > 1$. In Fig. 11.13, 14, 15 and 11.16 we represent $P(t)$ for different values of the coupling constant. If $g^2 \ll 1$ we see that the exponential decay law is approximately valid until times very large with respect to $1/\Gamma_{\text{Born}}$.

The effect of repeated measurements of the system with the corresponding successive collapses of the states has been considered by various authors [FO 76] in trying to explain an exponential decay law for all $t > 0$. The effect that (11.170) for small times might have in the observability of the proton lifetime has also been discussed in some detail [CM 82].

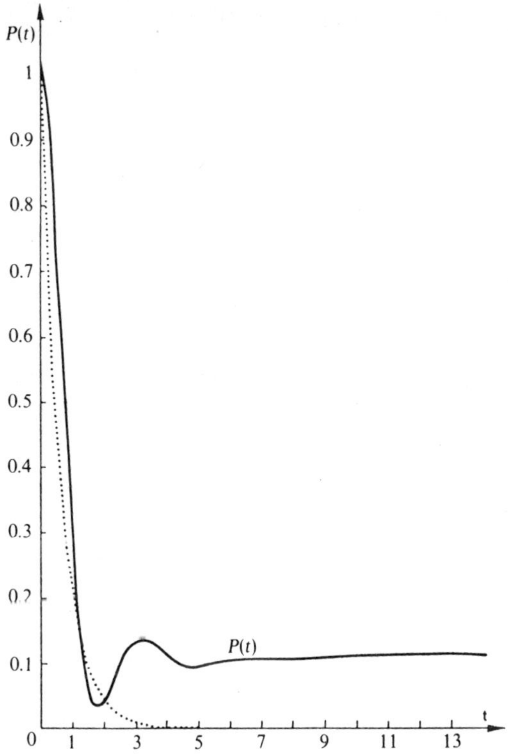

Fig. 11.16. Same for $g^2 = 2$

12. Particles in an Electromagnetic Field

12.1 Introduction

It is well known [JA 75] that in classical electrodynamics, an electromagnetic field is described by the electric and magnetic fields $\boldsymbol{E}(\boldsymbol{r};t)$, $\boldsymbol{B}(\boldsymbol{r};t)$. These fields satisfy Maxwell's equations, which in unrationalized Gaussian units become

$$\begin{aligned} &\nabla \times \boldsymbol{E}(\boldsymbol{r};t) + \frac{1}{c}\frac{\partial \boldsymbol{B}(\boldsymbol{r};t)}{\partial t} = 0\,, \quad \nabla \cdot \boldsymbol{B}(\boldsymbol{r};t) = 0\,, \\ &\nabla \cdot \boldsymbol{E}(\boldsymbol{r};t) = 4\pi\varrho_{\mathrm{e}}(\boldsymbol{r};t)\,, \\ &\nabla \times \boldsymbol{B}(\boldsymbol{r};t) - \frac{1}{c}\frac{\partial \boldsymbol{E}(\boldsymbol{r};t)}{\partial t} = \frac{4\pi}{c}\boldsymbol{J}_{\mathrm{em}}(\boldsymbol{r};t)\,, \end{aligned} \tag{12.1}$$

where $\varrho_{\mathrm{e}}(\boldsymbol{r};t)$ and $\boldsymbol{J}_{\mathrm{em}}(\boldsymbol{r};t)$ are respectively the charge and current densities, related by the continuity equation

$$\frac{\partial \varrho_{\mathrm{e}}(\boldsymbol{r};t)}{\partial t} + \nabla \cdot \boldsymbol{J}_{\mathrm{em}}(\boldsymbol{r};t) = 0\,. \tag{12.2}$$

The first pair of Maxwell equations imply that the electromagnetic fields can be (locally) expressed in terms of the scalar $\phi(\boldsymbol{r};t)$ and vector $\boldsymbol{A}(\boldsymbol{r};t)$ potentials in the form

$$\boldsymbol{E}(\boldsymbol{r};t) = -\frac{1}{c}\frac{\partial \boldsymbol{A}(\boldsymbol{r};t)}{\partial t} - \nabla\phi(\boldsymbol{r};t)\,, \quad \boldsymbol{B}(\boldsymbol{r};t) = \nabla \times \boldsymbol{A}(\boldsymbol{r};t)\,. \tag{12.3}$$

For given fields $\boldsymbol{E}$ and $\boldsymbol{B}$, the potentials ϕ and $\boldsymbol{A}$ are not uniquely determined. If we consider the new potentials

$$\phi'(\boldsymbol{r};t) = \phi(\boldsymbol{r};t) - \frac{1}{c}\frac{\partial \chi(\boldsymbol{r};t)}{\partial t}\,, \quad \boldsymbol{A}'(\boldsymbol{r};t) = \boldsymbol{A}(\boldsymbol{r};t) + \nabla\chi(\boldsymbol{r};t)\,, \tag{12.4}$$

where $\chi(\boldsymbol{r};t)$ is an arbitrary function, we obtain the same fields, as is easily seen using (12.3). The transformations (12.4) are known as *gauge* transformations (of the second kind). Conversely, it is possible to show that all potentials yielding the same electromagnetic fields are related by a gauge transformation. It is always possible to choose a gauge transformation in such a way that the potentials ϕ and $\boldsymbol{A}$ satisfy the Lorentz gauge condition

$$\nabla \cdot \boldsymbol{A}(\boldsymbol{r};t) + \frac{1}{c}\frac{\partial \phi(\boldsymbol{r};t)}{\partial t} = 0\,. \tag{12.5}$$

This condition does not completely determine the potentials, because we can still perform gauge transformations (12.4), as long as the gauge function χ satisfies the equation

$$\Box\chi(\boldsymbol{r};t) \equiv \frac{1}{c^2}\frac{\partial^2\chi(\boldsymbol{r};t)}{\partial t^2} - \Delta\chi(\boldsymbol{r};t) = 0 \, . \tag{12.6}$$

In particular, if the fields $\boldsymbol{E}, \boldsymbol{B}$ are constant and uniform, i.e., the fields are independent of $\boldsymbol{r}$ and t, a possible choice of potentials satisfying the Lorentz gauge condition is

$$\phi(\boldsymbol{r}) = -\boldsymbol{r}\cdot\boldsymbol{E} \, , \quad \boldsymbol{A}(\boldsymbol{r}) = \tfrac{1}{2}\boldsymbol{B}\times\boldsymbol{r} \, . \tag{12.7}$$

We also know that the motion of a non-relativistic particle of mass M and charge q in the presence of electric and magnetic fields $\boldsymbol{E}, \boldsymbol{B}$ is described by the Lorentz equation

$$\frac{d\boldsymbol{p}}{dt} = q\left[\boldsymbol{E}(\boldsymbol{r};t) + \frac{1}{c}\boldsymbol{v}\times\boldsymbol{B}(\boldsymbol{r};t)\right] \, , \tag{12.8}$$

with $\boldsymbol{p} = M\boldsymbol{v}$. In this equation only the electric and magnetic fields appear, thus the trajectories are completely independent of the gauge chosen, namely, of the representation of the potentials ϕ and $\boldsymbol{A}$. Similar results are obtained when one considers an arbitrary system of charges and currents.

Since it is the starting point of the quantum formulation, we need to construct the Hamiltonian describing the motion of a particle in the presence of electromagnetic fields. It is well-known that the Lagrangian for a particle coupled to E.M. fields is

$$L(\boldsymbol{r},\boldsymbol{v};t) = \frac{1}{2}M\boldsymbol{v}^2 - q\left[\phi(\boldsymbol{r};t) - \frac{1}{c}\boldsymbol{v}\cdot\boldsymbol{A}(\boldsymbol{r};t)\right] \, . \tag{12.9}$$

The momentum $\boldsymbol{p}$ conjugate to $\boldsymbol{r}$ is defined by $p_i = \partial L/\partial \dot{x}_i$, and it is given by

$$\boldsymbol{p} = M\boldsymbol{v} + \frac{q}{c}\boldsymbol{A}(\boldsymbol{r};t) \, . \tag{12.10}$$

Hence the momentum contains a term depending explicitly on the vector potential $\boldsymbol{A}$ together with the standard mechanical momentum which we label throughout this chapter by $\boldsymbol{\pi} = M\boldsymbol{v}$ to distinguish it from the dynamical momentum $\boldsymbol{p}$. From (12.9) and (12.10) we obtain the Hamiltonian for the problem to be

$$H(\boldsymbol{r},\boldsymbol{p};t) = \frac{1}{2M}\left[\boldsymbol{p} - \frac{q}{c}\boldsymbol{A}(\boldsymbol{r};t)\right]^2 + q\phi(\boldsymbol{r};t) \, , \tag{12.11}$$

and from (12.11) one can derive the Hamilton equations of motion. For these the invariance of the trajectories under changes of gauge is less evident. However it is enough to recall that Hamilton's equations are completely equivalent to the Lorentz equation, to realize that the former are also gauge invariant. Hence

$$\boldsymbol{r}'(t) = \boldsymbol{r}(t) \, , \quad \boldsymbol{\pi}'(t) = \boldsymbol{\pi}(t) \, , \tag{12.12}$$

where the unprimed quantities refer to those obtained using the potentials ϕ and $\boldsymbol{A}$ and those primed refer to quantities computed with ϕ' and $\boldsymbol{A}'$ related to the previous ones by the gauge transformation (12.4). These results together with the definition (12.10) immediately imply that the basic quantities in the Hamiltonian formalism $\boldsymbol{r}(t)$, $\boldsymbol{p}(t)$ change under gauge transformations according to

$$\begin{aligned} \boldsymbol{r}'(t) &= \boldsymbol{r}(t) , \\ \boldsymbol{p}'(t) &= \boldsymbol{p}(t) + \frac{q}{c}\nabla\chi(\boldsymbol{r}(t); t) . \end{aligned} \tag{12.13}$$

Thus, in the Hamiltonian formulation the value of the basic dynamical variables at each instant of time depends on the gauge chosen. This is not surprising since the Hamilton equations derived from (12.11) contain $\phi(\boldsymbol{r}; t)$ and $\boldsymbol{A}(\boldsymbol{r}; t)$ explicitly.

Given the phase space associated to a gauge ϕ, $\boldsymbol{A}$ for a given electromagnetic field, a quantity $F(\boldsymbol{r}, \boldsymbol{p}, \phi, \boldsymbol{A}; t)$ defined on the particle trajectories $\boldsymbol{r}(t)$, $\boldsymbol{p}(t)$ in this phase space is gauge independent, and therefore a candidate for a physical observable if

$$F(\boldsymbol{r}, \boldsymbol{p}, \phi, \boldsymbol{A}; t) \to F(\boldsymbol{r}', \boldsymbol{p}', \phi', \boldsymbol{A}'; t) = F(\boldsymbol{r}, \boldsymbol{p}, \phi, \boldsymbol{A}; t) . \tag{12.14}$$

In particular all quantities of the form $F(\boldsymbol{r}, \boldsymbol{p} - (q/c)\boldsymbol{A}(\boldsymbol{r}; t); t) = F(\boldsymbol{r}, \boldsymbol{\pi}; t)$ will be gauge independent. The Hamiltonian function is gauge dependent since as a consequence of (12.11) and (12.13) it changes under (12.4) by

$$H(\boldsymbol{r}, \boldsymbol{p}; t) \to H(\boldsymbol{r}, \boldsymbol{p}; t) - \frac{q}{c}\frac{\partial\chi(\boldsymbol{r}; t)}{\partial t} . \tag{12.15}$$

12.2 The Schrödinger Equation

We now discuss the quantization of a spinless particle in the presence of an electromagnetic field describable in terms of globally defined scalar and vector potentials. According to postulate VI, the commutation relations of the position and momentum operators $\boldsymbol{x}, \boldsymbol{p}$ are

$$\begin{aligned} [x_i, p_j] &= \mathrm{i}\hbar\delta_{ij} , \\ [x_i, x_j] &= [p_i, p_j] = 0 , \end{aligned} \tag{12.16}$$

independently of the gauge chosen. Therefore in the position representation and in any gauge, the operator $\boldsymbol{x}$ is simply multiplication by $\boldsymbol{x}$ [see (12.13)], and the operator $\boldsymbol{p}$ can be chosen as $-\mathrm{i}\hbar\nabla$ with a convenient although gauge dependent choice of the phases of the basis $\{|\boldsymbol{x}\rangle\}$. Taking this into account and using the classical Hamiltonian given in (12.1), the Schrödinger equation becomes

$$\boxed{\mathrm{i}\hbar\frac{\partial\Psi(\boldsymbol{r};t)}{\partial t} = \left\{\frac{1}{2M}\left[-\mathrm{i}\hbar\nabla - \frac{q}{c}\boldsymbol{A}(\boldsymbol{r};t)\right]^2 + q\phi(\boldsymbol{r};t)\right\}\Psi(\boldsymbol{r};t)} \tag{12.17}$$

or equivalently

$$\begin{aligned}\mathrm{i}\hbar\frac{\partial\Psi(\boldsymbol{r};t)}{\partial t} =& -\frac{\hbar^2}{2M}\Delta\Psi(\boldsymbol{r};t) + \frac{\mathrm{i}\hbar q}{Mc}\boldsymbol{A}(\boldsymbol{r};t)\cdot\nabla\Psi(\boldsymbol{r};t)\\ &+ \frac{\mathrm{i}\hbar q}{2Mc}[\nabla\cdot\boldsymbol{A}(\boldsymbol{r};t)]\Psi(\boldsymbol{r};t) + \frac{q^2}{2Mc^2}\boldsymbol{A}^2(\boldsymbol{r};t)\Psi(\boldsymbol{r};t)\\ &+ q\phi(\boldsymbol{r};t)\Psi(\boldsymbol{r};t)\,. \end{aligned} \tag{12.18}$$

As in Chap. 3 we can derive then the probability density $\varrho(\boldsymbol{r};t)$, and the probability current density $\boldsymbol{J}(\boldsymbol{r};t)$. They are given by

$$\begin{aligned}\varrho(\boldsymbol{r};t) &= |\Psi(\boldsymbol{r};t)|^2\,,\\ \boldsymbol{J}(\boldsymbol{r};t) &= \frac{\hbar}{2\mathrm{i}M}\Big\{\Psi^*(\boldsymbol{r};t)(\nabla\Psi(\boldsymbol{r};t)) - (\nabla\Psi^*(\boldsymbol{r};t))\Psi(\boldsymbol{r};t)\\ &\quad -\frac{2\mathrm{i}q}{\hbar c}\boldsymbol{A}(\boldsymbol{r};t)|\Psi(\boldsymbol{r};t)|^2\Big\}\,.\end{aligned} \tag{12.19}$$

To solve (12.17) in a concrete situation, it is necessary to choose for the description of the electromagnetic field some potentials $\phi(\boldsymbol{r};t)$ and $\boldsymbol{A}(\boldsymbol{r};t)$; in other words, we have to fix the gauge. And we have to make sure that the physical results obtained are gauge invariant. This question is not as evident as in classical mechanics because now we do not have anything equivalent to the Lorentz equation of motion. Let $\Psi(\boldsymbol{r};t)$ be the wave function for the state $|\Psi(t)\rangle$ in the gauge where the potentials are $\phi(\boldsymbol{r};t)$ and $\boldsymbol{A}(\boldsymbol{r};t)$, and let $\Psi'(\boldsymbol{r};t)$ be the wave function of the same state $|\Psi(t)\rangle$ in a new gauge $\phi'(\boldsymbol{r};t)$ and $\boldsymbol{A}'(\boldsymbol{r};t)$, related to the former by (12.4). The gauge invariance of the theory requires in particular that for any state $|\Psi(t)\rangle$ the conditions

$$\begin{aligned}&\langle\Psi'(.\,;t)|\boldsymbol{r}|\Psi'(.\,;t)\rangle = \langle\Psi(.\,;t)|\boldsymbol{r}|\Psi(.\,;t)\rangle\,,\\ &\langle\Psi'(.\,;t)| -\mathrm{i}\hbar\nabla - \frac{q}{c}\boldsymbol{A}'(.\,;t)|\Psi'(.\,;t)\rangle\\ &\quad = \langle\Psi(.\,;t)| -\mathrm{i}\hbar\nabla - \frac{q}{c}\boldsymbol{A}(.\,;t)|\Psi(.\,;t)\rangle\,,\end{aligned} \tag{12.20}$$

should be satisfied, and hence the unitary operator T_χ (taking into account that the set of gauges associated to the same electromagnetic field is connected) which implements the symmetry

$$\Psi(.\,;t) \to \Psi'(.\,;t) = T_\chi\Psi(.\,;t) \tag{12.21}$$

must satisfy the relations

$$\begin{aligned}&T_\chi^\dagger\boldsymbol{r}T_\chi = \boldsymbol{r}\,,\\ &T_\chi^\dagger\left[-\mathrm{i}\hbar\nabla - \frac{q}{c}\boldsymbol{A}'(\boldsymbol{r};t)\right]T_\chi = -\mathrm{i}\hbar\nabla - \frac{q}{c}\boldsymbol{A}(\boldsymbol{r};t)\,.\end{aligned} \tag{12.22}$$

The first of these implies

$$\boldsymbol{r}T_\chi = T_\chi \boldsymbol{r} , \tag{12.23}$$

so that the operator we are looking for should have the form

$$T_\chi = \exp\left[\mathrm{i}F_\chi(\boldsymbol{r};t)\right] \tag{12.24}$$

The second condition in (12.22) can be written using (12.4) as

$$T_\chi^\dagger(-\mathrm{i}\hbar\nabla)T_\chi = -\mathrm{i}\hbar\nabla + \frac{q}{c}\nabla\chi(\boldsymbol{r};t) , \tag{12.25}$$

which with (12.24) yields

$$\nabla F_\chi(\boldsymbol{r};t) = \frac{q}{\hbar c}\nabla\chi(\boldsymbol{r};t) \tag{12.26}$$

and hence

$$F_\chi(\boldsymbol{r};t) = \frac{q}{\hbar c}\chi(\boldsymbol{r};t) + \varphi(t) . \tag{12.27}$$

Here $\varphi(t)$ is an arbitrary function of time. Choosing conveniently the global phase of T_χ we may assume $\varphi(t) \equiv 0$, so that the unitary operator T_χ becomes

$$T_\chi = \exp\left[\frac{\mathrm{i}q}{\hbar c}\chi(\boldsymbol{r};t)\right] , \tag{12.28}$$

and the transformation law of the basis vectors $\{|\boldsymbol{r}\rangle\}$ and of the wave functions will be

$$\begin{aligned} |\boldsymbol{r}\rangle \rightarrow |\boldsymbol{r}\rangle' &= \exp\left[-\frac{\mathrm{i}q}{\hbar c}\chi(\boldsymbol{r};t)\right] |\boldsymbol{r}\rangle , \\ \Psi(\boldsymbol{r};t) &= \exp\left[\frac{\mathrm{i}q}{\hbar c}\chi(\boldsymbol{r};t)\right] \Psi(\boldsymbol{r};t) , \end{aligned} \tag{12.29}$$

with $\Psi'(\boldsymbol{r};t) = {}'\langle\boldsymbol{r}|\Psi(t)\rangle$, $\Psi(\boldsymbol{r};t) = \langle\boldsymbol{r}|\Psi(t)\rangle$; namely, under a gauge transformation the wave function changes by a space-time dependent local phase and therefore in general it cannot be removed.

Let $F(\boldsymbol{r},\boldsymbol{p},\phi,\boldsymbol{A};t)$ be a classical physical observable which therefore satisfies (12.14). The average values of the operator representing this observable should be gauge invariant and thus they must satisfy

$$\langle\Psi(t)|F(\boldsymbol{r}',\boldsymbol{p}',\phi',\boldsymbol{A}';t)|\Psi(t)\rangle = \langle\Psi(t)|F(\boldsymbol{r},\boldsymbol{p},\phi,\boldsymbol{A};t)|\Psi(t)\rangle . \tag{12.30}$$

Since $|\Psi(t)\rangle$ is arbitrary, we obtain

$$T_\chi F(\boldsymbol{r},-\mathrm{i}\hbar\nabla,\phi,\boldsymbol{A};t)T_\chi^\dagger = F\left(\boldsymbol{r},-\mathrm{i}\hbar\nabla,\phi - \frac{1}{c}\frac{\partial\chi}{\partial t},\boldsymbol{A}+\nabla\chi;t\right) \tag{12.31}$$

as the gauge invariance condition for an observable F. In particular

$$\begin{aligned}
&T_\chi \boldsymbol{r} T_\chi^\dagger = \boldsymbol{r}\,, \quad T_\chi(-\mathrm{i}\hbar\nabla)T_\chi^\dagger = -\mathrm{i}\hbar\nabla - \frac{q}{c}\nabla\chi\,, \quad T_\chi \boldsymbol{\pi} T_\chi^\dagger = \boldsymbol{\pi}\,,\\
&T_\chi \boldsymbol{L} T_\chi^\dagger = T_\chi(\boldsymbol{r}\times(-\mathrm{i}\hbar\nabla))T_\chi^\dagger = \boldsymbol{L} - \frac{q}{c}\boldsymbol{r}\times\nabla\chi\,,\\
&T_\chi \boldsymbol{\Lambda} T_\chi^\dagger = T_\chi(\boldsymbol{r}\times\boldsymbol{\pi})T_\chi^\dagger = \boldsymbol{\Lambda}\,,\\
&T_\chi H T_\chi^\dagger = T_\chi\left(\frac{\boldsymbol{\pi}^2}{2M} + q\phi\right)T_\chi^\dagger = \frac{\boldsymbol{\pi}^2}{2M} + q\left(\phi - \frac{1}{c}\frac{\partial\chi}{\partial t}\right) + \frac{q}{c}\frac{\partial\chi}{\partial t}\\
&\qquad = \left(\frac{\boldsymbol{\pi}^2}{2M} + q\phi'\right) + \frac{q}{c}\frac{\partial\chi}{\partial t}
\end{aligned} \tag{12.32}$$

so that $\boldsymbol{r}, \boldsymbol{\pi}, \boldsymbol{\Lambda}$, are gauge invariant observables, while $\boldsymbol{p}, \boldsymbol{L}$ and H are not. Consequently any statement concerning the eigenvalues and eigenfunctions of operators such as $\boldsymbol{p}, \boldsymbol{L}$ and H will be in general gauge dependent and do not have an intrinsic meaning. In practice however, we generally choose the simplest possible gauges, so that the validity of particular statements extend to more restricted classes of gauges. For example, in time-independent electromagnetic fields one often chooses $\phi, \boldsymbol{A}$ to be time independent, so that for this particularly restricted class of electromagnetic potentials, the Hamiltonian is gauge invariant. Otherwise the spectrum would have a trivial shift of levels if $\chi(\boldsymbol{r};t) = \text{const}\times t$, while when χ is an arbitrary function of $\boldsymbol{r}, t$, H would become time dependent and its spectrum might change drastically. This implies that the interpretation of $H(t)$ as the energy of the system only makes sense when it does not depend explicitly on time, and this is something which can always be achieved for static electromagnetic fields with appropriate gauges. Thus, concerning postulate V (Chap. 2) we should remark that even though there it was stated that $H(t)$ is a physical observable, it might nonetheless have no intrinsic meaning.

Finally it is easy to verify that the Schrödinger equation is gauge invariant since the change in $H(t)$ is compensated by the local change of phase in the wave function. Therefore in the new gauge (12.17) becomes

$$\mathrm{i}\hbar\frac{\partial\Psi'(\boldsymbol{r};t)}{\partial t} = \left\{\frac{1}{2M}\left[-\mathrm{i}\hbar\nabla - \frac{q}{c}\boldsymbol{A}'(\boldsymbol{r};t))\right]^2 + q\phi'(\boldsymbol{r};t)\right\}\Psi'(\boldsymbol{r};t)\,. \tag{12.33}$$

It is also straightforward to show that the current and probability densities defined by (12.19) are independent of the gauge chosen.

The structure of the Schrödinger equation (12.17) for a spinless particle in the presence of an E.M. field follows from the requirement that it should be not only invariant under global changes of phase (or *gauge transformations of the first kind*) $\Psi \to \Psi' \equiv \exp(\mathrm{i}q\chi/\hbar c)\Psi$, χ constant, but also under space-time dependent changes of phase. Under this local change of phase, the free Schrödinger equation becomes

$$\mathrm{i}\hbar\frac{\partial\Psi'(\boldsymbol{r};t)}{\partial t} = \left\{\frac{1}{2M}\left[-\mathrm{i}\hbar\nabla - \frac{q}{c}(\nabla\chi)\right]^2 - \frac{q}{c}\frac{\partial\chi}{\partial t}\right\}\Psi'(\boldsymbol{r};t)\,. \tag{12.34}$$

This immediately suggests that in order to keep the formal invariance of the Schrödinger equation under local phase changes, also called *gauge transformations of the second kind*, it suffices to introduce into the equation the potentials

ϕ and $\boldsymbol{A}$, coupled as indicated in (12.17), and transforming according to (12.4). Local phase changes therefore induce local gauge transformations.

Mathematical conditions which guarantee the self-adjointness of the Hamiltonian (12.17) have been extensively studied (see for example [LS 81], [SI 82]). It is possible to show that for $\boldsymbol{A}, \phi$ real,

$$H(t) \equiv H_{\boldsymbol{A},\phi} \equiv \frac{1}{2M}\left[-\mathrm{i}\hbar\nabla - \frac{q}{c}\boldsymbol{A}(\boldsymbol{r};t)\right]^2 + q\phi(\boldsymbol{r};t) \tag{12.35}$$

is essentially self-adjoint in $C_0^\infty(\mathbb{R}^3)$ as long as

$$A_j(.\,;t) \in L^4_{\mathrm{loc}}(\mathbb{R}^3)\,, \quad \nabla\cdot\boldsymbol{A}(.\,;t) \in L^2_{\mathrm{loc}}(\mathbb{R}^3)\,, \tag{12.36}$$

and $\phi = \phi_1 + \phi_2$ measurable, with

$$\begin{aligned} &\phi_{1,+}(.\,;t) \in L^2_{\mathrm{loc}}(\mathbb{R}^3)\,, \quad \phi_{1,-}(.\,;t) \in L^2(\mathbb{R}^3) + L^\infty(\mathbb{R}^3)\,,\\ &|\phi_2(.\,;t)| \le C(t)(1 + |\boldsymbol{x}|)\,, \end{aligned} \tag{12.37}$$

where $\phi_\pm \equiv (\phi \pm |\phi|)/2$.

It is clear that these conditions are satisfied in the special case (12.7). The presence of the magnetic potential in (12.35) produces on the spectrum an enhancement of diamagnetic origin. More concretely

$$\inf\sigma(H_{\boldsymbol{A},\phi}) \ge \inf\sigma(H_{\boldsymbol{0},\phi})\,, \tag{12.38}$$

as follows from Kato's inequality

$$-\Delta|\psi| \le \mathrm{Re}\left[(\mathrm{sgn}\,\psi)(-\mathrm{i}\nabla - \frac{q}{\hbar c}\boldsymbol{A})^2\psi\right]\,, \quad \forall\psi \in C_0^\infty(\mathbb{R}^3)\,. \tag{12.39}$$

For static potentials (ϕ and $\boldsymbol{A}$ independent of t) and under the mathematical conditions pointed out before, it is shown [LE 83] that if $\boldsymbol{A}' = \boldsymbol{A} + \nabla\chi$, with χ time independent, $H_{\boldsymbol{A},\phi}$ and $H_{\boldsymbol{A}',\phi}$ are unitarily equivalent: $T_\chi H_{\boldsymbol{A},\phi} T_\chi^\dagger = H_{\boldsymbol{A}',\phi}$. Moreover, it is always possible to choose $\boldsymbol{A}'$ to satisfy $\nabla\cdot\boldsymbol{A}' = 0$ (Coulomb gauge) without violating the mathematical conditions quoted. Finally, gauge invariance allows us to derive interesting spectral conditions. Qualitatively they imply that the introduction of static magnetic fields decreasing sufficiently fast at large distances does not modify the essential spectrum of $H_{\boldsymbol{0},\phi}$ nor the possible existence of an infinite number of points in the discrete spectrum.

12.3 Uncertainty Relations

Since according to (12.10) the velocity operator is given by

$$\boldsymbol{v} = \frac{1}{M}\boldsymbol{p} - \frac{q}{Mc}\boldsymbol{A}(\boldsymbol{r};t)\,, \tag{12.40}$$

using the commutation relations (12.16) we immediately obtain

$$
\begin{aligned}
[x_i, v_i] &= \frac{i\hbar}{M}\delta_{ij} , \quad [x_i, x_j] = 0 , \\
[v_i, v_j] &= \frac{i\hbar q}{M^2 c}\varepsilon_{ijk} B_k .
\end{aligned}
\tag{12.41}
$$

The last relation implies that in the presence of an external magnetic field, the velocity operators in two different directions will not commute in general. If $\hat{\boldsymbol{n}}$ and $\hat{\boldsymbol{n}}'$ are unit vectors, then

$$
[\hat{\boldsymbol{n}} \cdot \boldsymbol{v}, \hat{\boldsymbol{n}}' \cdot \boldsymbol{v}] = \frac{i\hbar q}{M^2 c}(\hat{\boldsymbol{n}} \times \hat{\boldsymbol{n}}') \cdot \boldsymbol{B} ,
\tag{12.42}
$$

which immediately implies the uncertainty relation

$$
\Delta(\hat{\boldsymbol{n}} \cdot \boldsymbol{v})\Delta(\hat{\boldsymbol{n}}' \cdot \boldsymbol{v}) \geq \left| \frac{\hbar q}{2M^2 c}(\hat{\boldsymbol{n}} \times \hat{\boldsymbol{n}}') \cdot \boldsymbol{B} \right| .
\tag{12.43}
$$

To estimate the right-hand side, notice that

$$
\left| \frac{\hbar q}{2M^2 c} \right| = 1.0180627(45)10^7\,\mathrm{cm}^2\mathrm{s}^{-2}\mathrm{G}^{-1} \left| \frac{q}{e} \right| \left(\frac{m_e}{M} \right)^2 ,
\tag{12.44}
$$

where e is the electron charge and m_e its mass.

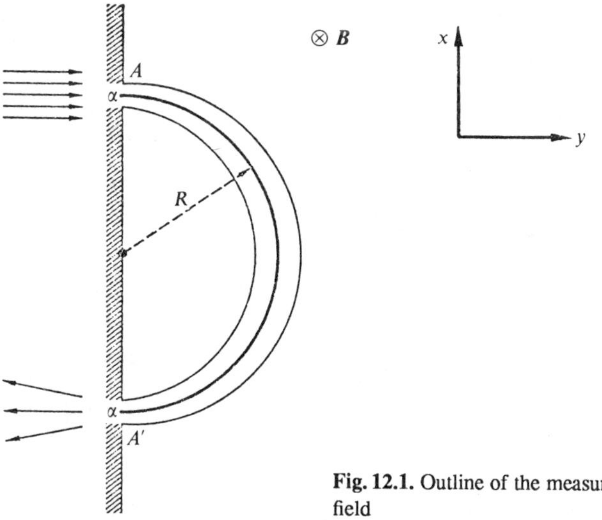

Fig. 12.1. Outline of the measurement of v_x, v_y in a magnetic field

Equation (12.43) can be given an operational justification. If we send particles of charge q towards a slit A as depicted in Fig. 12.1, the uncertainties when they exit through the slit A' are $\Delta v_x \sim \hbar/Ma$, $\Delta v_y \sim (qB/Mc)\Delta R \sim (qB/Mc)a$, so that $\Delta v_x \Delta v_y \sim (qB/M^2c)\hbar$. The first uncertainty is produced by the diffraction through the slit A and the second is due to the fact that since we do not know through which point of the slit A' the particle exits, there is an uncertainty in the curvature radius inside the magnetic field.

In 1931 Dirac [DI 31] emphasized the especially important role played by a "monopolar" magnetic field $\boldsymbol{B}(\boldsymbol{r}) = g\hat{\boldsymbol{r}}/r^2$ produced by a point-like magnetic charge of strength g located at the origin of coordinates. For this field there exists no global choice of the vector potential in $\mathbb{R}^3 - \{0\}$; otherwise, if C_1 and C_2 are two infinitesimally small circles centered upon $(0, 0, R)$ and $(0, 0, -R)$, with radius ε and parallel to the $z = 0$ plane, the circulation along them with appropriate orientations of the hypothetical globally defined $\boldsymbol{A}$ would satisfy as a consequence of Stokes theorem:

$$\int_{C_1} d\boldsymbol{x} \cdot \boldsymbol{A} - \int_{C_2} d\boldsymbol{x} \cdot \boldsymbol{A} = \int_{\Gamma_{1,2}} d\boldsymbol{S} \cdot (\nabla \times \boldsymbol{A}) = \int_{\Gamma_{1,2}} d\boldsymbol{S} \cdot g\frac{\hat{\boldsymbol{r}}}{r^2} , \qquad (12.45)$$

where $\Gamma_{1,2}$ is the region of the sphere of radius R left after we remove two infinitesimal polar caps with boundaries C_1, C_2. In the limit $\varepsilon \to 0$, the left-hand side would converge to zero, whereas the right-hand side approaches $4\pi g$.

We can nevertheless construct $\boldsymbol{A}_+$ defined in $\mathbb{R}^3 - \{x = y = 0, z \leq 0\}$ given by the 1-form

$$\boldsymbol{A}_+(\boldsymbol{x}) \cdot d\boldsymbol{x} \equiv g(1 - \cos\theta)\, d\phi . \qquad (12.46)$$

The sets of points eliminated $\{x = y = 0, z \leq 0\}$ is known as the Dirac *string*. This choice of potential must be completed with

$$\boldsymbol{A}_-(\boldsymbol{x}) \cdot d\boldsymbol{x} \equiv -g(1 + \cos\theta)\, d\phi , \qquad (12.47)$$

where the string is now directed along the semiaxis $z \geq 0$, and in this way we cover all of $\mathbb{R}^3 - \{0\}$. In the overlapping region $(|x| + |y| > 0)$, the potentials $\boldsymbol{A}_\pm$ are locally equivalent up to a gauge tranformation

$$\boldsymbol{A}_+(\boldsymbol{x}) \cdot d\boldsymbol{x} - \boldsymbol{A}_-(\boldsymbol{x}) \cdot d\boldsymbol{x} = d(2g\phi) . \qquad (12.48)$$

Notice that the angle ϕ is a multivalued function. Nevertheless, the gauge transformation T_χ for a particle of electric charge q moving in such monopole field will be single valued, making the Dirac string unobservable as long as

$$\frac{qg}{\hbar c} = \frac{n}{2} , \quad n \text{ an integer} . \qquad (12.49)$$

This equation is known as the Dirac *quantization condition.*

According to this relation, if there were a single monopole in the Universe, all electric charges should be integer multiples of an elementary unit. This would provide an explanation to the surprising experimental finding that

$$\big||q_{\text{electron}}| - |q_{\text{proton}}|\big| \lesssim 10^{-21} |q_{\text{electron}}| . \qquad (12.50)$$

An alternative and independent explanation for the quantization of electric charge is found in the theories of grand unification. In these models the fact that the electromagnetic gauge invariance is unbroken implies quite surprisingly the existence of super-heavy magnetic monopoles (the 't Hooft-Polyakov monopole) with a more complex structure than the point-like Dirac monopole. The apparent inevitability of monopoles to explain the quantization of charge is in sharp contrast with the difficulty of observing them, in spite of numerous experimental attempts to detect their existence. Interested readers can find further information about these issues in [AC 72], [CO 83] and [PR 84].

We now return to the canonical commutation relations studied earlier in this section. For a particle of charge q in the presence of a monopole magnetic field $\boldsymbol{B}$, it is straightforward to prove that the operator

$$\boldsymbol{j} \equiv \boldsymbol{r} \times \boldsymbol{\pi} - \frac{qg}{c}\hat{\boldsymbol{r}} \tag{12.51}$$

satisfies the angular momentum commutation relations as well as

$$[j_j, x_k] = i\hbar\varepsilon_{jkl}x_l \,, \quad [j_j, \pi_k] = i\hbar\varepsilon_{jkl}\pi_l \,. \tag{12.52}$$

This is not surprising; $\boldsymbol{j}$ is the total angular momentum of a system composed of a charged particle of charge q moving in the magnetic field produced by a Dirac magnetic monopole. The term $\boldsymbol{r} \times \boldsymbol{\pi}$ is the orbital part $\boldsymbol{\Lambda}$, and the rest represents the contribution $\boldsymbol{j}_{\text{em}}$ to the angular momentum due to the static electric and magnetic fields produced by the charge and the magnetic pole.

From (12.52) and (12.49) we derive

$$\hat{\boldsymbol{r}} \cdot \boldsymbol{j}_{\text{em}} = \hat{\boldsymbol{r}} \cdot \boldsymbol{j} = -\frac{qg}{c} = -\frac{n}{2}\hbar \,. \tag{12.53}$$

This is a surprising result when n is odd (for the electromagnetic field is classical and the particles were assumed to be spinless) and suggests that in the presence of monopoles the distinction between bosons and fermions is quite subtle [CO 83].

12.4 The Aharonov-Bohm Effect

We now discuss a very peculiar effect which appears in the study of a particle in the presence of an electromagnetic field. This effect was predicted by *Aharonov* and *Bohm* [AB 59] and experimentally observed by *Chambers* [CH 60]. We begin by studying the propagator $G(\boldsymbol{x}, t; \boldsymbol{x}', t')$ of a particle in a magnetic field described by the potentials $\phi(\boldsymbol{x}; t) \equiv 0$ and $\boldsymbol{A}(\boldsymbol{x})$. This propagator is given by (3.173) with the action associated to the Lagrangian (12.9):

$$G(\boldsymbol{x}, t; \boldsymbol{x}', t') = \lim_{\varepsilon \downarrow 0} \frac{1}{A} \int \frac{d^3x_1}{A} \cdots \frac{d^3x_N}{A} \exp\left\{ \frac{i\varepsilon}{\hbar} \sum_{j=1}^{N-1} \right.$$
$$\left. \times \left[\frac{M}{2}\left(\frac{\boldsymbol{x}_j - \boldsymbol{x}_{j-1}}{\varepsilon}\right)^2 + \frac{q}{c}\left(\frac{\boldsymbol{x}_j - \boldsymbol{x}_{j-1}}{\varepsilon}\right) \cdot \boldsymbol{A}\left(\frac{\boldsymbol{x}_j + \boldsymbol{x}_{j-1}}{2}\right)\right]\right\} \,, \tag{12.54}$$

where $A \equiv (2\pi i\hbar\varepsilon/M)^{3/2}$. As explained there, this corresponds to summing the quantities

$$\exp\left[\frac{i}{\hbar}S_0(\boldsymbol{x},t;\boldsymbol{x}',t')\right]\exp\left[\frac{iq}{\hbar c}\int_{\boldsymbol{x}'}^{\boldsymbol{x}} d\boldsymbol{l}\cdot\boldsymbol{A}(\boldsymbol{y})\right] \tag{12.55}$$

over all paths from $(\boldsymbol{x}',t')$ to $(\boldsymbol{x},t)$, where $S_0(\boldsymbol{x},t;\boldsymbol{x}',t')$ is the classical action for such a path in the absence of an electromagnetic field and where the line integral is taken along the path. The relative phase between two paths c_1 and c_2 from $(\boldsymbol{x}',t')$ to $(\boldsymbol{x},t)$ varies due to the electromagnetic interactions by the amount

$$\frac{q}{\hbar c}\int_{c_1} d\boldsymbol{l}\cdot\boldsymbol{A}(\boldsymbol{y}) - \frac{q}{\hbar c}\int_{c_2} d\boldsymbol{l}\cdot\boldsymbol{A}(\boldsymbol{y}) = \frac{q}{\hbar c}\Phi\,, \tag{12.56}$$

where Φ is the magnetic flux through any surface bounded by the paths c_1 and c_2 with the appropriate orientation. If the flux were an integer multiple of $\Phi_0 \equiv 2\pi\hbar c/q$, i.e., $\Phi = n\Phi_0$, then the magnetic field would not produce any change in the relative phase between both paths.

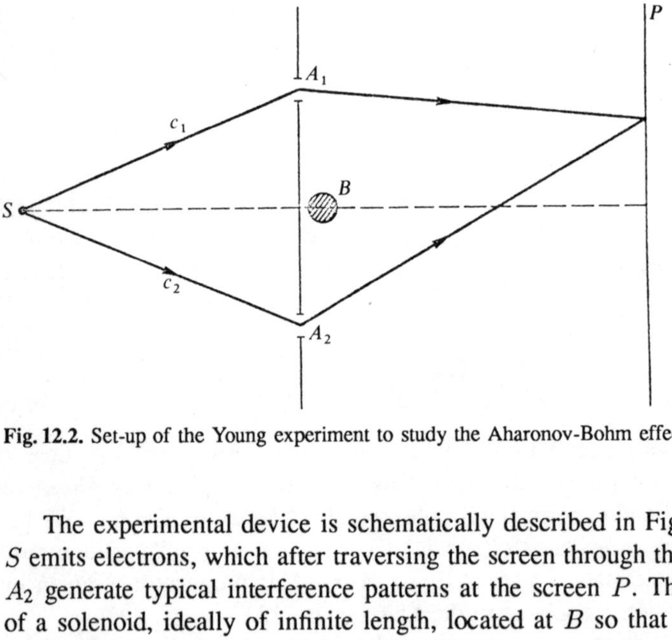

Fig. 12.2. Set-up of the Young experiment to study the Aharonov-Bohm effect

The experimental device is schematically described in Fig. 12.2. The source S emits electrons, which after traversing the screen through the openings A_1 and A_2 generate typical interference patterns at the screen P. The novelty consists of a solenoid, ideally of infinite length, located at B so that the magnetic field is completely confined to its interior, and it is perpendicular to the plane of the drawing. We also assume the cross section of the solenoid to be sufficiently small so that the diffracted electrons have a negligible probability to penetrate its interior. When we increase the magnetic field of the solenoid, the relative phase of the particles passing through A_1 with respect to those passing through A_2 will change according to (12.56), and thus the interference pattern produced on the screen will change, even though the electrons never passed through the region where $B \neq 0$. This notable effect, predicted by Aharonov and Bohm, shows that for the local effects on the wave functions, it is the electromagnetic potentials and not the fields which are the fundamental quantities in quantum mechanics. Notice also that the interference pattern depends on $A(x)$ only through Φ and it is independent of the gauge choice.

Even though the field B vanishes outside of the solenoid, there is no gauge where $A = 0$ in that region. Locally the gauge can vanish, but not globally because this region is not simply connected (in the idealized situation where the solenoid is infinitely long). A classical particle outside the solenoid would move freely; this is not the case in quantum mechanics for if we simply rotate it around the solenoid, the particle would "feel" the field inside it. In terms of wave functions, if $\Psi_0(x;t)$ represents a free wave packet, the function

$$\Psi(x;t) \equiv \exp\left[\frac{iq}{\hbar c}\int_{x_0}^{x} dl \cdot A(y)\right] \Psi_0(x;t) \tag{12.57}$$

(where the integration is over any path outside the solenoid and joining the fiducial point x_0 with x) satisfies the Schrödinger equation in the presence of the potential A locally. However, the wave function $\Psi(x;t)$ is not globally single valued unless the flux Φ is an integer multiple of the elementary flux Φ_0. This elementary flux coincides with the one produced by an elementary Dirac monopole with respect to a particle of charge q, i.e., with magnetic charge $g = \hbar c/2q$. The ineffectiveness of the solenoid in this case is equivalent to the unobservability of the Dirac string.

A detailed discussion of the theoretical aspects of this effect as well as of the various experiments exhibiting it can be found in [OP 85], [GM 86].

12.5 Spin-1/2 Particles in an E.M. Field

From the arguments of previous sections we can write down the Schrödinger equation for a particle of mass M and charge q in the presence of an external electromagnetic field and under the action of a mechanical potential $V(r)$:

$$i\hbar\frac{\partial\Psi(r;t)}{\partial t} = \left\{\frac{1}{2M}\left[-i\hbar\nabla - \frac{q}{c}A(r;t)\right]^2 + q\phi(r;t) + V(r)\right\}\Psi(r;t) \tag{12.58}$$

The naive extension of (12.58) for spin-1/2 particles, taking $\Psi(r;t)$ to be a *spinor* [a two-component wave function like (3.36)] leads to wrong results when we try to study, for example, the effect of magnetic fields on atomic electrons. This is because the electrons are particles with spin 1/2 with a magnetic moment whose orientation changes in the presence of magnetic fields. This dynamical aspect is not incorporated in Eq. (12.58). If we compare this equation with the one for vanishing electromagnetic fields $A(r;t) = \phi(r;t) = 0$, we notice that (12.58) can be obtained from the latter via the substitutions

$$\boxed{i\hbar\frac{\partial}{\partial t}\to i\hbar\frac{\partial}{\partial t}-q\phi(\boldsymbol{r};t)\,,\quad -i\hbar\nabla\to -i\hbar\nabla-\frac{q}{c}\boldsymbol{A}(\boldsymbol{r};t)}\ . \tag{12.59}$$

This prescription to introduce electromagnetic interactions is known as *minimal coupling*.

If, as we have just discussed, we implement the substitution (12.59) in the equation

$$i\hbar\frac{\partial\Psi(\boldsymbol{r};t)}{\partial t}=\left[\frac{1}{2M}(-i\hbar\nabla)^2+V(\boldsymbol{r})\right]\Psi(\boldsymbol{r};t)\,, \tag{12.60}$$

we obviously obtain (12.58). However if the particle has spin 1/2 the components Ψ_+ and Ψ_- of the spinor $\Psi(\boldsymbol{r};t)$ are not coupled in (12.58). But (B.12) implies that (12.60) is equivalent to

$$i\hbar\frac{\partial\Psi(\boldsymbol{r};t)}{\partial t}=\left\{\frac{1}{2M}[(-i\hbar\nabla)\cdot\boldsymbol{\sigma}]^2+V(\boldsymbol{r})\right\}\Psi(\boldsymbol{r};t)\,, \tag{12.61}$$

and if we introduce now the electromagnetic field through minimal coupling

$$i\hbar\frac{\partial\Psi(\boldsymbol{r};t)}{\partial t}=\left\{\frac{1}{2M}\left[\boldsymbol{\sigma}\cdot\left(-i\hbar\nabla-\frac{q}{c}\boldsymbol{A}(\boldsymbol{r};t)\right)\right]^2+q\phi(\boldsymbol{r};t)+V(\boldsymbol{r})\right\}\Psi(\boldsymbol{r};t)\,, \tag{12.62}$$

which does not coincide with (12.58). Indeed, using (B.12) we find

$$\begin{aligned}
&\left[\boldsymbol{\sigma}\cdot\left(-i\hbar\nabla-\frac{q}{c}\boldsymbol{A}\right)\right]^2\\
&=\left(-i\hbar\nabla-\frac{q}{c}\boldsymbol{A}\right)^2+i\varepsilon_{ijk}\sigma_k\left(-i\hbar\nabla_i-\frac{q}{c}A_i\right)\left(-i\hbar\nabla_j-\frac{q}{c}A_j\right)\\
&=\left(-i\hbar\nabla-\frac{q}{c}\boldsymbol{A}\right)^2-\frac{\hbar q}{c}\varepsilon_{ijk}\sigma_k(\nabla_iA_j+A_i\nabla_j)\,.
\end{aligned} \tag{12.63}$$

Due to the antisymmetry of ε_{ijk}, and using the notation $(\nabla_iA_j)\equiv\partial A_j/\partial x_i$, we obtain

$$\left[\boldsymbol{\sigma}\cdot\left(-i\hbar\nabla-\frac{q}{c}\boldsymbol{A}\right)\right]^2=\left(-i\hbar\nabla-\frac{q}{c}\boldsymbol{A}\right)^2-\frac{\hbar q}{c}\varepsilon_{ijk}\sigma_k(\nabla_iA_j)\,. \tag{12.64}$$

Since $\varepsilon_{ijk}\sigma_k(\nabla_iA_j)=\boldsymbol{\sigma}\cdot\boldsymbol{B}$, we finally have

$$\boxed{\begin{aligned}i\hbar\frac{\partial\Psi(\boldsymbol{r};t)}{\partial t}=\Big\{&\frac{1}{2M}\left[-i\hbar\nabla-\frac{q}{c}\boldsymbol{A}(\boldsymbol{r};t)\right]^2-\frac{\hbar q}{2Mc}\boldsymbol{\sigma}\cdot\boldsymbol{B}(\boldsymbol{r};t)\\&+q\phi(\boldsymbol{r};t)+V(\boldsymbol{r})\Big\}\Psi(\boldsymbol{r};t)\,,\end{aligned}} \tag{12.65}$$

namely:

$$i\hbar\frac{\partial\Psi(\boldsymbol{r};t)}{\partial t}=\left\{-\frac{\hbar^2}{2M}\Delta+\frac{i\hbar q}{Mc}\boldsymbol{A}(\boldsymbol{r};t)\cdot\nabla+\frac{i\hbar q}{2Mc}(\nabla\cdot\boldsymbol{A}(\boldsymbol{r};t))+\frac{q^2}{2Mc}\right.$$
$$\left.\times\,\boldsymbol{A}^2(\boldsymbol{r};t)-\frac{\hbar q}{2Mc}\boldsymbol{\sigma}\cdot\boldsymbol{B}(\boldsymbol{r};t)+q\phi(\boldsymbol{r};t)+V(\boldsymbol{r})\right\}\Psi(\boldsymbol{r};t)\,. \qquad (12.66)$$

This equation differs from (12.58) in the term proportional to $\boldsymbol{\sigma}\cdot\boldsymbol{B}$. This is the so-called *Pauli equation* and the new term couples the two components of the wave function.

Pauli introduced this coupling by phenomenological reasoning, but it in fact follows from the relativistic treatment of the problem using the Dirac equation. For this reason it was believed that the spin was an essentially relativistic phenomenon. However it was later shown [GS 61] that a formulation with only Galilean invariance also gives rise to the spin of the electron with the same value for the intrinsic magnetic moment as the Dirac equation.

To better understand the meaning of the term $\boldsymbol{\sigma}\cdot\boldsymbol{B}$, we study the Hamiltonian of a particle in an electrostatic potential and with a constant uniform magnetic field: $\phi=\phi(\boldsymbol{r})$, $\boldsymbol{A}=\boldsymbol{B}\times\boldsymbol{r}/2$. Taking into account that $\nabla\cdot\boldsymbol{A}=0$ and that $\boldsymbol{A}\cdot\nabla=(\mathrm{i}/2\hbar)\boldsymbol{B}\cdot\boldsymbol{L}$, where $\boldsymbol{L}$ is the orbital angular momentum operator, we have

$$H=-\frac{\hbar^2}{2M}\Delta-\frac{q}{2Mc}\boldsymbol{B}\cdot(\boldsymbol{L}+2\boldsymbol{S})+\frac{q^2}{2Mc^2}\boldsymbol{A}^2(\boldsymbol{r})+q\phi(\boldsymbol{r})\,. \qquad (12.67)$$

It is well known classically [JA 75] that a particle of charge q with kinematical orbital angular momentum $\boldsymbol{\Lambda}$ has associated with it a magnetic dipole moment $\boldsymbol{\mu}=(q/2Mc)\boldsymbol{\Lambda}$ which can be written as

$$\boldsymbol{\mu}=\frac{q}{2Mc}\boldsymbol{L}-\frac{q^2}{4Mc^2}[r^2\boldsymbol{B}-(\boldsymbol{r}\cdot\boldsymbol{B})\boldsymbol{r}]\,, \qquad (12.68)$$

that is, $\boldsymbol{\mu}$ is the sum of a permanent magnetic moment $(q/2Mc)\boldsymbol{L}$ and an induced magnetic moment vanishing with $\boldsymbol{B}$. The interaction energy of $\boldsymbol{\mu}$ with $\boldsymbol{B}$ is

$$-\frac{q}{2Mc}\boldsymbol{L}\cdot\boldsymbol{B}+\frac{1}{2}\frac{q^2}{4Mc^2}[r^2B^2-(\boldsymbol{r}\cdot\boldsymbol{B})^2]$$
$$=-\frac{q}{2Mc}\boldsymbol{L}\cdot\boldsymbol{B}+\frac{q^2}{2Mc^2}\left[\frac{1}{2}\boldsymbol{B}\times\boldsymbol{r}\right]^2\,, \qquad (12.69)$$

which coincides with the second and third terms in (12.67) (except for spin).

Hence $2(q/2Mc)\boldsymbol{S}$ plays the role of an intrinsic permanent magnetic moment due to the spin, twice as large as the moment that one would obtain if $\boldsymbol{S}$ were an orbital angular momentum.

The quadratic term in (12.67), called the diamagnetic interaction, is usually negligible with respect to the paramagnetic interactions [second term in the r.h.s. of (12.67)]; the relative magnitude is $\xi\sim|qr^2B/4c\hbar|$ and for particles carrying an electron charge it is $\xi\sim3.8\times10^6\,(r/\mathrm{cm})^2B/\mathrm{G}$. It is therefore very small, in general, if r has atomic dimensions. Nonetheless the diamagnetic term can be very important when the motion is not confined (for example in scattering situations) or when some selection rule forces the vanishing to first order of the paramagnetic contribution.

The Pauli equation (12.65) gives a good description of the electron in an electromagnetic field, but when one tries to apply it to other spin-1/2 particles which undergo strong interactions, such as the proton or the neutron, it is necessary to modify it. In general for spin-1/2 particles the correct equation is

$$\boxed{\begin{aligned} \mathrm{i}\hbar\frac{\partial \Psi(\boldsymbol{r};t)}{\partial t} &= \Big\{-\frac{\hbar^2}{2M}\Delta + \mathrm{i}\frac{\hbar q}{Mc}\boldsymbol{A}(\boldsymbol{r};t)\cdot\boldsymbol{\nabla} + \frac{\mathrm{i}\hbar q}{2Mc}[\boldsymbol{\nabla}\cdot\boldsymbol{A}(\boldsymbol{r};t)] \\ &+ \frac{q^2}{2Mc^2}\boldsymbol{A}^2(\boldsymbol{r};t) - \beta\frac{1}{\hbar}\boldsymbol{S}\cdot\boldsymbol{B}(\boldsymbol{r};t) + q\phi(\boldsymbol{r};t) + V(\boldsymbol{r})\Big\}\Psi(\boldsymbol{r};t)\end{aligned}} \tag{12.70}$$

for an intrinsic magnetic moment $\boldsymbol{\mu} = \beta\boldsymbol{S}/\hbar$. For electrons, protons and neutrons one writes

$$\beta_{\mathrm{e}} = -g_{\mathrm{e}}\frac{\hbar|e|}{2m_{\mathrm{e}}c}\,,\quad \beta_{\mathrm{p}} = g_{\mathrm{p}}\frac{\hbar|e|}{2m_{\mathrm{p}}c}\,,\quad \beta_{\mathrm{n}} = g_{\mathrm{n}}\frac{\hbar|e|}{2m_{\mathrm{p}}c}\,. \tag{12.71}$$

The quantities

$$\begin{aligned}\mu_{\mathrm{B}} &\equiv \frac{\hbar|e|}{2m_{\mathrm{e}}c} = 0.57883785(95)\times 10^{-14}\,\mathrm{MeV\,G^{-1}}\,,\\ \mu_{\mathrm{N}} &\equiv \frac{\hbar|e|}{2m_{\mathrm{p}}c} = 3.1524515(53)\times 10^{-18}\,\mathrm{MeV\,G^{-1}}\,,\end{aligned} \tag{12.72}$$

with dimensions of magnetic moments, are called the *Bohr* and *nuclear magnetons*, respectively. The dimensionless numbers g are called *gyromagnetic ratios* or *factors* and their experimental values are

$$\begin{aligned}\frac{g_{\mathrm{e}}}{2} &= 1.001159652209(31)\,,\quad \frac{g_{\mathrm{p}}}{2} = 2.7928444(11)\,,\\ \frac{g_{\mathrm{n}}}{2} &= -1.91304308(54)\,.\end{aligned} \tag{12.73}$$

Theoretically it is possible to compute g_{e} using the electroweak theory with hadronic corrections. The result is

$$\left(\frac{g_{\mathrm{e}}}{2}\right)_{\mathrm{theor.}} = 1.001159652460(136)\,. \tag{12.74}$$

Notice that in the Pauli equation $(g_{\mathrm{e}}/2) = 1$ and thus it is quite suitable to describe electrons in an electromagnetic field. On the other hand due to our ignorance of the strong interactions, it is not yet possible to compute g_{p} and g_{n}, although the quark model predicts $g_{\mathrm{p}}/g_{\mathrm{n}} = -3/2$, to be compared with the experimental value $g_{\mathrm{p}}/g_{\mathrm{n}} \approx -1.46$.

Finally we should warn the reader that when one says that μ is the intrinsic magnetic moment of the particle it is meant that this is the expectation value of the third component of the magnetic moment operator $\boldsymbol{\mu}$ in the spin state $|ss\rangle$. In the cases discussed in this section $\boldsymbol{\mu} = \beta\boldsymbol{S}/\hbar$ and the magnetic moment is precisely $\beta/2$. It equals $-g_{\mathrm{e}}/2$ for the electron in units of Bohr magnetons, and $g_{\mathrm{p}}/2\,(g_{\mathrm{n}}/2)$ for the proton (neutron) in nuclear magnetons.

12.6 A Particle in a Constant Uniform Magnetic Field

If the spin-1/2 particle is subject only to the action of a constant uniform magnetic field B along the Oz direction, we can choose as electromagnetic potentials

$$\phi = 0 \,, \quad A_1 = -By \,, \quad A_2 = A_3 = 0 \,, \tag{12.75}$$

so that (12.70) becomes

$$\mathrm{i}\hbar \frac{\partial \Psi(\boldsymbol{r};t)}{\partial t} = \left\{ -\frac{\hbar^2}{2M}\Delta - \frac{\mathrm{i}\hbar q}{Mc} By \frac{\partial}{\partial x} + \frac{q^2}{2Mc^2} B^2 y^2 - \frac{1}{\hbar}\beta B S_z \right\} \Psi(\boldsymbol{r};t) \,. \tag{12.76}$$

The stationary state problem is

$$\left\{ -\frac{\hbar^2}{2M}\Delta - \frac{\mathrm{i}\hbar q}{Mc} By \frac{\partial}{\partial x} + \frac{q^2}{2Mc^2} B^2 y^2 - \frac{1}{\hbar}\beta B S_z \right\} \psi(\boldsymbol{r}) = E\psi(\boldsymbol{r}) \,. \tag{12.77}$$

Taking into account that p_x, p_z and S_z commute with the Hamiltonian, it is convenient to search for solutions to (12.77) which are also eigenstates of these constants of motion with eigenvalues $p_x, p_z, m\hbar$, respectively:

$$\psi(\boldsymbol{r}) = \frac{1}{2\pi\hbar} \exp\left[\frac{\mathrm{i}}{\hbar}(xp_x + zp_z) \right] \phi(y)\chi^m_{1/2} \,, \tag{12.78}$$

where $\chi^m_{1/2}$ is the normalized spin wave function obeying $S_z\chi^m_{1/2} = m\hbar\chi^m_{1/2}$. The function $\phi(y)$ must satisfy

$$\left[-\frac{\hbar^2}{2M}\frac{d^2}{dy^2} + \frac{1}{2M}(p_x^2 + p_z^2) + \frac{qB}{Mc} yp_x + \frac{q^2B^2}{2Mc^2}y^2 - m\beta B \right] \phi(y) = E\phi(y) \,, \tag{12.79}$$

and therefore

$$\frac{d^2\phi(y)}{dy^2} + \frac{2M}{\hbar^2}\left[E + \beta m B - \frac{p_z^2}{2M} - \frac{q^2B^2}{2Mc^2}(y - y_0)^2 \right] \phi(y) = 0 \,, \tag{12.80}$$

with $y_0 \equiv -cp_x/qB$.

Comparing this equation with that of the one-dimensional harmonic oscillator, we easily derive the form of the eigenvalues to be

$$E_{n,m,p_z,p_x} = \left(n + \frac{1}{2} \right) \frac{|q|\hbar B}{Mc} + \frac{1}{2M}p_z^2 - \beta m B \,, \quad n = 0,1,2,\ldots \,. \tag{12.81}$$

Notice the (infinite) degeneracy with respect to p_x. The levels (12.81) are called the Landau levels. According to the results of Sect. 10.2, these levels are identical to those we would obtain using $\boldsymbol{A} = \boldsymbol{B} \times \boldsymbol{r}/2$ instead of (12.75). The normalized spinorial eigenfunctions are

$$\psi_{n,m,p_z,p_x} = \frac{1}{2\pi\hbar}\left[\frac{\sqrt{|q|B/\hbar c}}{\sqrt{\pi}2^n n!}\right]^{1/2} \exp\left[\frac{\mathrm{i}}{\hbar}(xp_x + zp_z)\right]$$
$$\times \exp\left[-\frac{|q|B}{2\hbar c}(y-y_0)^2\right] H_n\left[\sqrt{\frac{|q|B}{\hbar c}}(y-y_0)\right] \chi^m_{1/2} \,. \quad (12.82)$$

We now compare this result with the classical one. When we solve the classical equation

$$M\frac{d^2\boldsymbol{r}}{dt^2} = \frac{q}{c}\boldsymbol{v}\times\boldsymbol{B} \quad (12.83)$$

we obtain

$$\begin{aligned} x &= x_0 + \frac{Mcv_\mathrm{T}}{|q|B}\sin\left(\frac{qB}{Mc}t+\alpha\right) , \\ y &= y_0 + \frac{Mcv_\mathrm{T}}{|q|B}\cos\left(\frac{qB}{Mc}t+\alpha\right) , \qquad z = z_0 + v_z t , \end{aligned} \quad (12.84)$$

where v_T is the modulus of the projection of $\boldsymbol{v}$ onto the plane perpendicular to $\boldsymbol{B}$.

Classically the particle moves uniformly along the Oz axis, and this has an immediate quantum analog, since along this axis the motion is not quantized. It is described by a plane wave contributing the amount $p_z^2/2M = Mv_z^2/2$ to the energy.

The classical trajectory projected into the plane perpendicular to $\boldsymbol{B}$ is a circle centered upon (x_0, y_0) with radius $Mcv_\mathrm{T}/|q|B$ and the particle moves with angular velocity $|q|B/Mc$. Notice that x_0 and y_0 can be written as

$$x_0 = x + \frac{Mc}{qB}v_y = x + \frac{cp_y}{qB} , \quad y_0 = y - \frac{Mc}{qB}v_x = -\frac{cp_x}{qB} . \quad (12.85)$$

In quantum theory, the operators $x + cp_y/qB$ and $-cp_x/qB$ do not commute, and therefore if we fix the value of the second, the first operator (the abscissa x_0) is totally undetermined. Since we have fixed the value of p_x, the motion along Ox is a plane wave. Finally, the periodicity of the classical motion in the Oxy plane is reflected in a harmonic motion along Oy with the cyclotron frequency, as (12.80) shows.

The previous analysis persists virtually unchanged if we add to the magnetic field a constant electrostatic field $\boldsymbol{E}$ orthogonal to $\boldsymbol{B}$. This simple problem is very relevant in the study of the quantum Hall effect [KA 84].

Neglecting spin, which is irrelevant for the following considerations, the spectrum of energies (12.81) is essential, running over the interval $[E_*, \infty)$, with $E_* \equiv |q|\hbar B/2Mc$. Restricting the Hamiltonian to the Oxy plane, the spectrum is pure point with infinitely degenerate isolated levels. In this sense we can say that the magnetic field $B \neq 0$ is confining in directions orthogonal to it, and we expect that the introduction of a scalar potential $\phi(\boldsymbol{r})$ restricting the motion along the Oz axis suffices to make (a part of) the spectrum discrete. Indeed [HE 81], if $\phi \in L^2(\mathbb{R}^3) + L^\infty_\varepsilon(\mathbb{R}^3)$, then $\sigma_{\text{ess}}(H_{A,\phi}) = \sigma_{\text{ess}}(H_{A,0}) = [E_*, \infty)$. If furthermore, $\phi(\boldsymbol{r}) \leq 0$ and it does not vanish identically for $|\boldsymbol{r}| \geq R$, then $\sigma_{\text{disc}}(H_{A,\phi}) \neq \emptyset$. When such ϕ is cylindrically symmetric about $\boldsymbol{B}$, there exist infinitely many levels in the discrete part of the energy spectrum.

When these arguments are applied to simply negative ions ($Z + 1$ electrons) the surprising conclusion is reached that for any value of the constant magnetic field $B \neq 0$ acting on the ion, there are always an infinite number of bound states.

12.7 The Fine Structure of Hydrogen-Like Atoms

Next we describe in a simple but non-rigorous way various corrections to the spectrum of hydrogen-like atoms. The reduced Hamiltonian of an electron in the electrostatic potential generated by the nucleus is

$$H_0 = -\frac{\hbar^2}{2\mu}\Delta - \frac{Ze^2}{r} , \qquad (12.86)$$

where μ is the reduced mass of the electron. The most important corrections to the previous Hamiltonian are generated by relativistic effects and the electron spin. The relativistic kinetic energy for a free electron is

$$K = [m_e^2 c^4 + \boldsymbol{p}^2 c^2]^{1/2} - m_e c^2 . \qquad (12.87)$$

Since for light hydrogen-like atoms $v/c \simeq Z\alpha \ll 1$, it suffices to consider only the first few terms in the expansion in powers of $|\boldsymbol{p}|/m_e$,

$$K = \frac{\boldsymbol{p}^2}{2\mu} - \frac{1}{2}\frac{1}{\mu c^2}\left(\frac{\boldsymbol{p}^2}{2\mu}\right) + \cdots , \qquad (12.88)$$

where we have replaced m_e by μ, according to the discussion given in Sect. 6.6. The first term in this expansion corresponds to the first term in H_0. The second term in (12.88) represents a correction to the kinetic energy in the Hamiltonian:

$$H_K = -\frac{1}{2}\frac{1}{\mu c^2}\left(-\frac{\hbar^2}{2\mu}\Delta\right)^2 . \qquad (12.89)$$

On the other hand the relativistic quantum-mechanical study of one-electron atoms shows that the interaction of the electron with the nucleus is non-local and this effect can be described by adding to H_0 the so-called *Darwin term*

$$H_D = \frac{\hbar^2}{8\mu^2 c^2}(\Delta V)(\boldsymbol{r}) , \qquad (12.90)$$

and since $V(r) = -Ze^2/r$, and $\Delta(1/r) = -4\pi\delta(\boldsymbol{r})$, it becomes

$$H_{\mathrm{D}} = \frac{\pi Z e^2 \hbar^2}{2\mu^2 c^2}\,\delta(\boldsymbol{r})\,. \tag{12.91}$$

Finally, since the electron, under the action of the electrostatic potential $\phi(r)$, is subject to an electric field $\boldsymbol{E} = -\boldsymbol{\nabla}\phi = -(d\phi(r)/dr)\hat{\boldsymbol{r}}$, in the reference frame where the electron is at rest it will also observe a magnetic field $\boldsymbol{B} = \boldsymbol{E} \times \boldsymbol{v}/c$, where $\boldsymbol{v}$ is the velocity of the electron. Taking into account the expression for $\boldsymbol{E}$,

$$\boldsymbol{B} = -\frac{1}{m_{\mathrm{e}}c}\frac{d\phi(r)}{dr}\frac{1}{r}\,\boldsymbol{L}\,, \tag{12.92}$$

where $\boldsymbol{L}$ is the electron's orbital angular momentum. The intrinsic magnetic moment of the electron $(e/m_{\mathrm{e}}c)\boldsymbol{S}$ in the presence of this $\boldsymbol{B}$ produces the coupling energy $-\boldsymbol{\mu}\cdot\boldsymbol{B}$, which one should add to the previous terms:

$$-\boldsymbol{\mu}\cdot\boldsymbol{B} = \frac{1}{m_{\mathrm{e}}^2c^2}\frac{1}{r}\left(\frac{dV(r)}{dr}\right)(\boldsymbol{L}\cdot\boldsymbol{S})\,, \quad V = e\phi\,. \tag{12.93}$$

This computation would be correct if the electron rest frame were an inertial frame; since it is actually an accelerated frame *Thomas* [TH 26] and *Frenkel* [FR 26] showed that this difference accounts for an extra factor of 1/2 [JA 75], so that the change in the Hamiltonian will actually be

$$H_{LS} = \frac{1}{2}\frac{1}{(\mu c)^2}\frac{1}{r}\frac{dV(r)}{dr}(\boldsymbol{L}\cdot\boldsymbol{S})\,, \tag{12.94}$$

where $V(r) = -Ze^2/r$ and the substitution $m_{\mathrm{e}} \to \mu$ has been made again. The energy H_{LS} is known as the *spin-orbit coupling*.

Summarizing; the Hamiltonian including all these corrections becomes

$$H = H_0 + H_1 = H_0 + H_{\mathrm{K}} + H_{LS} + H_{\mathrm{D}}\,,$$

$$\boxed{\begin{aligned} H_0 &= -\frac{\hbar^2}{2\mu}\Delta - \frac{Ze^2}{r}\,, \quad H_{\mathrm{K}} = -\frac{1}{2\mu c^2}\left(-\frac{\hbar^2}{2\mu}\Delta\right)^2\,, \\ H_{LS} &= \frac{Ze^2}{2\mu^2c^2}\frac{1}{r^3}(\boldsymbol{L}\cdot\boldsymbol{S})\,, \quad H_{\mathrm{D}} = \frac{\pi Ze^2\hbar^2}{2\mu^2c^2}\,\delta(\boldsymbol{r})\,. \end{aligned}} \tag{12.95}$$

We now compute the effect of the perturbation H_1 on the bound state spectrum of H_0. As an eigenstate basis of H_0 we can take $\{|nlm_lm_s\rangle\}$, where n is the principal quantum number, l is the orbital angular momentum and m_l, m_s are the components along the Oz-axis of the orbital and spin angular momenta, respectively (in units of $\hbar$). Recall that the eigenvalues of H_0 are degenerate with respect to the possible values of l, m_l, and m_s. Obviously, a basis equivalent to the previous one is generated by the states $|nljm\rangle$, where j is the total angular momentum and m its third component. Both bases are related by

$$|nljm\rangle = \sum_{\sigma} C(l1/2j; m-\sigma, \sigma, m)|nl\,m-\sigma\,\sigma\rangle \ . \tag{12.96}$$

The advantage of using the basis associated to $\boldsymbol{L}^2, \boldsymbol{S}^2, \boldsymbol{J}^2, J_z$ is that these operators commute with H_1 and its matrix representation will be diagonal; this is not the case if one uses the basis associated to $\boldsymbol{L}^2, L_z, \boldsymbol{S}^2, S_z$, since H_{LS} does not commute with L_z, S_z. Taking this into account and studying the effect of H_1 in the discrete spectrum of H_0 to first order in perturbation theory, it is enough to compute $\langle nljm|H_1|nljm\rangle$. Since $\boldsymbol{J} = \boldsymbol{L} + \boldsymbol{S}$, we have

$$\boldsymbol{S}\cdot\boldsymbol{L} = \tfrac{1}{2}[\boldsymbol{J}^2 - \boldsymbol{L}^2 - \boldsymbol{S}^2] \tag{12.97}$$

and therefore, if $l > 0$,

$$\langle njlm|H_{LS}|nljm\rangle = \frac{Ze^2\hbar^2}{4\mu^2c^2}\left[j(j+1) - l(l+1) - \frac{3}{4}\right]\langle nljm|r^{-3}|nljm\rangle \ . \tag{12.98}$$

In the computation of the matrix element of r^{-3} the spin plays no role, and using (6.99) we find

$$\langle nljm|H_{LS}|nljm\rangle = \frac{Ze^2\hbar^2}{4\mu^2c^2}\frac{1}{l(l+1/2)(l+1)n^3a^3}\left[j(j+1) - l(l+1) - \frac{3}{4}\right] , \tag{12.99}$$

where $a \equiv \hbar/[\mu(Z\alpha c)]$. Since the unperturbed energies are $E_{nl} = -\mu(Z\alpha c)^2/2n^2$, the previous result can also be written as

$$\langle nljm|H_{LS}|nljm\rangle$$
$$= -E_{nl}\frac{(Z\alpha)^2}{n(2l+1)}\begin{cases}\dfrac{1}{l+1} , & j = l + \dfrac{1}{2} \\ -\dfrac{1}{l} , & j = l - \dfrac{1}{2}\end{cases} , \quad l > 0 \ . \tag{12.100}$$

For states $l = 0$ the average value of H_{LS} obviously vanishes due to the factor $\boldsymbol{S}\cdot\boldsymbol{L}$.

We now include the contribution of H_{K}. Since

$$\left(\frac{p^2}{2\mu} - \frac{Ze^2}{r}\right)|nljm\rangle = E_{nl}|nljm\rangle \ , \tag{12.101}$$

we have

$$\langle nljm|H_{\mathrm{K}}|nljm\rangle = -\frac{1}{2\mu c^2}\langle nljm|\left(E_{nl} + \frac{Ze^2}{r}\right)^2|nljm\rangle \tag{12.102}$$

and using (6.99)

$$\langle nljm|H_{\mathrm{K}}|nljm\rangle = -\frac{1}{2\mu c^2}\left[E_{nl}^2 + \frac{2Ze^2E_{nl}}{an^2} + \frac{2Z^2e^4}{(2l+1)n^3a^2}\right] \ . \tag{12.103}$$

Hence

$$\langle nljm|H_{\mathrm{K}}|nljm\rangle = E_{nl}\frac{(Z\alpha)^2}{n^2}\left(\frac{n}{l+1/2}-\frac{3}{4}\right) . \qquad (12.104)$$

Finally the Darwin term gives a contribution

$$\langle nljm|H_{\mathrm{D}}|nljm\rangle = \frac{\pi Z e^2\hbar^2}{2\mu^2c^2}|\psi_{nlm}(0)|^2 , \qquad (12.105)$$

which is non-vanishing only when $l = 0$:

$$\langle nljm|H_{\mathrm{D}}|nljm\rangle = -E_{nl}\frac{(Z\alpha)^2}{n} , \quad l = 0 . \qquad (12.106)$$

Collecting all the corrections obtained we finally have for all values of l and to first order of perturbation theory the fine structure of the energy levels of one-electron atoms:

$$E_{nlj} = E_{nl}\left\{1+\frac{(Z\alpha)^2}{n^2}\left[\frac{n}{j+1/2}-\frac{3}{4}\right]\right\} , \qquad (12.107)$$

a result already mentioned for the hydrogen atom in Chaps. 1 and 6.

12.8 One-Electron Atoms in a Magnetic Field

Next we study a hydrogen-like atom in the presence of a constant uniform magnetic field, whose effect is described by adding to the Hamiltonian (12.95) a new term

$$H_{\mathrm{M}} = -\frac{e}{2m_{\mathrm{e}}c}\boldsymbol{B}\cdot(\boldsymbol{L}+2\boldsymbol{S}) , \qquad (12.108)$$

where $g_{\mathrm{e}}/2 = 1$ has been assumed and the diamagnetic term $(e^2/8m_{\mathrm{e}}c^2)(\boldsymbol{r}\times\boldsymbol{B})^2$ has been neglected because $r \sim 10^{-8}$ cm and the magnetic fields produced in laboratories satisfy $B \lesssim 10^6$ G.

We now evaluate the change in the bound state spectrum of H_0 produced by the perturbation $H_1 + H_{\mathrm{M}}$, with H_1 given in (12.95), to first order in perturbation theory. Choosing the coordinate frame in such a way that $\boldsymbol{B}$ is directed along the positive z-axis:

$$H_{\mathrm{M}} = -\frac{eB}{2m_{\mathrm{e}}c}(L_z+2S_z) = -\frac{eB}{2m_{\mathrm{e}}c}(J_z+S_z) . \qquad (12.109)$$

In the basis $\{|nljm\rangle\}$ previously introduced, the changes in the energies are given by the eigenvalues of a matrix of order $2n^2$, for each fixed n, with matrix elements $\langle nl'j'm'|H_1 + H_{\mathrm{M}}|nljm\rangle$. Obviously H_1 as well as the term in H_{M} proportional to J_z are diagonal in this representation, while the term S_z of H_{M}, although diagonal in l and m, is not diagonal in j so that the only non-vanishing matrix elements are of the form $\langle nlj'm|H_1 + H_{\mathrm{M}}|nljm\rangle$. This, together with (12.109) yields

$$\langle nlj'm|H_1 + H_{\mathrm{M}}|nljm\rangle = \delta_{jj'}\left\{E_{nl}\frac{(Z\alpha)^2}{n^2}\left(\frac{n}{j+1/2}-\frac{3}{4}\right)+\mu_{\mathrm{B}}Bm\right\}$$
$$+\mu_{\mathrm{B}}B\langle nlj'm|\frac{1}{\hbar}S_z|nljm\rangle\ , \qquad (12.110)$$

and using (12.96) we immediately obtain

$$\langle nlj'm|\frac{1}{\hbar}S_z|nljm\rangle = \sum_\sigma \sigma C(l1/2j';m-\sigma,\sigma)C(l1/2j;m-\sigma,\sigma). \qquad (12.111)$$

Hence, if $l > 0$ and $l_\pm \equiv l \pm 1/2$,

$$\begin{aligned}\langle nll_\pm m|\frac{1}{\hbar}S_z|nll_\pm m\rangle &= \pm\frac{m}{2l_+}\ ,\\ \langle nll_\pm m|\frac{1}{\hbar}S_z|nll_\mp m\rangle &= -\frac{1}{2l_+}\sqrt{l_+^2-m^2}\ ,\end{aligned} \qquad (12.112)$$

while if $l = 0$ and therefore $j = 1/2$, we have

$$\langle n\,0\,1/2\,m|\frac{1}{\hbar}S_z|n\,0\,1/2\,m\rangle = m\ . \qquad (12.113)$$

In conclusion, if $l \neq 0$, then

$$\begin{aligned}&\langle nll_\pm m|H_1 + H_{\mathrm{M}}|nll_\pm m\rangle\\ &\quad = E_{nl}\frac{(Z\alpha)^2}{n^2}\left(\frac{n}{l_\pm+1/2}-\frac{3}{4}\right)+\mu_{\mathrm{B}}Bm\frac{2l_\pm+1}{2l_+}\ ,\\ &\langle nll_\pm m|H_1 + H_{\mathrm{M}}|nll_\mp m\rangle = -\frac{\mu_{\mathrm{B}}B}{2l_+}\sqrt{l_+^2-m^2}\ ,\end{aligned} \qquad (12.114)$$

and if $l = 0$

$$\langle n\,0\,1/2m|H_1 + H_{\mathrm{M}}|n\,0\,1/2m\rangle = E_{nl}\frac{(Z\alpha)^2}{n^2}\left(n-\frac{3}{4}\right)+2\mu_{\mathrm{B}}Bm\ . \qquad (12.115)$$

On the other hand, since $H_1 + H_{\mathrm{M}}$ is diagonal with respect to n, l, m, it is enough to diagonalize each of the submatrices (at most of order 2) corresponding to fixed values of these quantum numbers. The submatrices associated to $n, l, \pm(l+1/2)$ are one-dimensional and give levels with energies

$$E_{nl,\pm(l+1/2)} = E_{nl} + E_{nl}\frac{(Z\alpha)^2}{n^2}\left(\frac{n}{l+1}-\frac{3}{4}\right) \pm \mu_{\mathrm{B}}B(l+1)\ . \qquad (12.116)$$

In contrast the submatrices for n, l, m with $m \neq \pm(l+1/2)$ are two by two matrices giving rise to the secular equation

$$\begin{vmatrix} E_{nl}\frac{(Z\alpha)^2}{n^2}\left(\frac{n}{l+1}-\frac{3}{4}\right)+2\mu_B Bm\frac{l+1}{2l+1}-\varepsilon & -\frac{\mu_B B}{2l+1}\sqrt{l_+^2-m^2} \\ -\frac{\mu_B B}{2l+1}\sqrt{l_+^2-m^2} & E_{nl}\frac{(Z\alpha)^2}{n^2}\left(\frac{n}{l}-\frac{3}{4}\right)+2\mu_B Bm\frac{l}{2l+1}-\varepsilon \end{vmatrix} = 0\,. \tag{12.117}$$

Let Δ_{nl} be the energy separation between the fine structure levels $E_{nl,l+1/2}$, $E_{nl,l-1/2}$, i.e.,

$$\Delta_{nl} = |E_{nl}|\frac{(Z\alpha)^2}{n}\frac{1}{l(l+1)}\,. \tag{12.118}$$

Defining

$$\varepsilon' = \varepsilon - E_{nl}\frac{(Z\alpha)^2}{n^2}\left(\frac{n}{l}-\frac{3}{4}\right)\,, \tag{12.119}$$

(12.117) becomes

$$\begin{vmatrix} \Delta_{nl}+2m\frac{l+1}{2l+1}\mu_B B-\varepsilon' & -\frac{\mu_B B}{2l+1}\sqrt{l_+^2-m^2} \\ -\frac{\mu_B B}{2l+1}\sqrt{l_+^2-m^2} & 2m\frac{l}{2l+1}\mu_B B-\varepsilon' \end{vmatrix} = 0 \tag{12.120}$$

with roots

$$\varepsilon'^{\pm} = \frac{1}{2}\Delta_{nl} + m\mu_B B \pm \frac{1}{2}\left[\Delta_{nl}^2 + \frac{4m}{2l+1}\Delta_{nl}\mu_B B + \mu_B^2 B^2\right]^{1/2}\,. \tag{12.121}$$

Therefore, the perturbed energy levels are

$$\begin{aligned} E^{\pm}_{nlm} &= E_{nl} + \left[\frac{1}{2} - (l+1)\left(1-\frac{3l}{4n}\right)\right]\Delta_{nl} + m\mu_B B \\ &\quad \pm\frac{1}{2}\left[\Delta_{nl}^2 + \frac{4m}{2l+1}\Delta_{nl}\mu_B B + \mu_B^2 B^2\right]^{1/2} \quad \text{if} \quad m \neq \pm(l+1/2) \\ E_{nl,\pm(l+1/2)} &= E_{nl} - l\left[1 - \frac{3(l+1)}{4n}\right]\Delta_{nl} \pm \mu_B B(l+1)\,. \end{aligned} \tag{12.122}$$

In particular for the $n = 2$, $l = 1$ levels, we have

$$\begin{aligned} \frac{E^{\pm}_{21m} - E_{21}}{\Delta_{21}} &= -\frac{3}{4} + m\left(\frac{\mu_B B}{\Delta_{21}}\right) \pm \frac{1}{2}\left[1 + \frac{4}{3}m\left(\frac{\mu_B B}{\Delta_{21}}\right) + \left(\frac{\mu_B B}{\Delta_{21}}\right)^2\right]^{1/2}, \\ &\text{if} \quad m = \pm 1/2\,, \quad \text{and} \\ \frac{E_{21,\pm 3/2} - E_{21}}{\Delta_{21}} &= -\frac{1}{4} \pm 2\left(\frac{\mu_B B}{\Delta_{21}}\right)\,. \end{aligned} \tag{12.123}$$

The terms on the left in this equation are represented in Fig. 12.3 as functions of $\mu_B B/\Delta_{21}$. Notice that for $Z = 1$ the quantity $\mu_B B/\Delta_{21}$ is equal to

$$\frac{\mu_B B}{\Delta_{21}} = 1.278 \times 10^{-4}\, B/\mathrm{G}\,. \tag{12.124}$$

There are two interesting limiting cases. First we suppose that the magnetic field is sufficiently weak so that $|\langle H_M\rangle| \ll |\langle H_1\rangle|$. The eigenstates of H_0+H_1 are $|nljm\rangle$ with energies E_{nlj} given approximately by (12.107). When we introduce the small perturbation H_M the fine structure sublevels never mix and each of them splits into $(2j+1)$ levels with energies

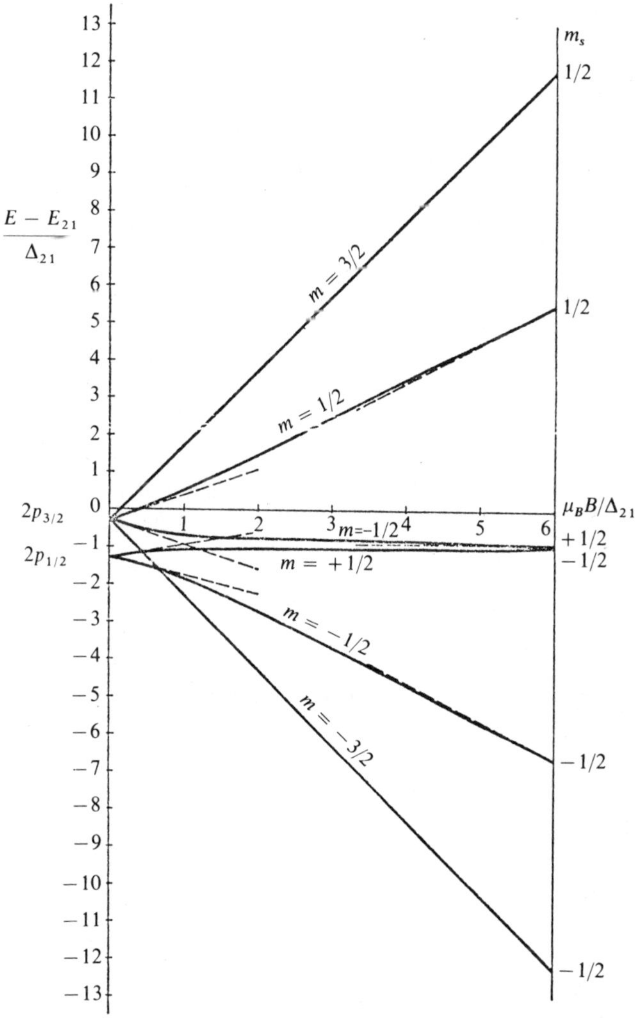

Fig. 12.3. Splitting of the $2p_{3/2}$ and $2p_{1/2}$ levels of hydrogen in the presence of a magnetic field. The Zeeman and Paschen-Back limits are represented by the broken lines

$$E_{nl(l\pm1/2)m} = E_{nl} + E_{nl}\,\frac{(Z\alpha)^2}{n^2}\left[\frac{n}{l_\pm + 1/2} - \frac{3}{4}\right] + m\,\frac{2l_\pm + 1}{2l_+}\,\mu_\text{B}B \tag{12.125}$$

or equivalently

$$E_{nl(l\pm1/2)m} = E_{nl} - \left[\frac{l(l+1)}{l_\pm + 1/2} - \frac{3l(l+1)}{4n}\right]\Delta_{nl} + m\,\frac{2l_\pm + 1}{2l_+}\,\mu_\text{B}B\,. \tag{12.126}$$

Notice that the new levels are equidistant in energy with a level splitting independent not only of the principal quantum number n but also of the nuclear charge Z. This separation is proportional to the external magnetic field. This effect is known in the literature as the *Zeeman effect.*

Expanding the function $E^{\pm}_{nlm}$ given in (12.122) in powers of $\mu_\text{B}B/\Delta_{nl}$ and keeping only linear terms, we immediately obtain that $E^{\pm}_{nlm} \to E_{nl(l\pm1/2)m}$ if $m \neq \pm(l+1/2)$, while $E_{nl(l+1/2),\pm(l+1/2)} = E_{nl,\pm(l+1/2)}$ for all B. In Fig. 12.3 and for small values of $\mu_\text{B}B/\Delta_{nl}$, we also represent the splitting of levels given by (12.126).

The other extreme case corresponds to magnetic fields so intense that $|\langle H_1\rangle| \ll |\langle H_\text{M}\rangle| \ll |\langle H_0\rangle|$. Neglecting H_1 the eigenstates of H_0 with energy E_{nl} can be chosen in the form $|nlm_lm_s\rangle$, and in this basis the operator H_M is diagonal; thus the perturbed energy levels are

$$E_{nmm_s} = E_{nl} + (m + m_s)\mu_\text{B}B\,, \tag{12.127}$$

where $m = m_l + m_s$ and $2l+3$ levels appear for each n and l. This effect is known as the *Paschen-Back effect.* Taking the limit $\Delta_{nl} \to 0$ in (12.122) we easily obtain

$$\begin{aligned} &E^{\pm}_{nlm} \to E_{nm,\pm1/2}\,, \quad m \neq \pm\left(l + \tfrac{1}{2}\right)\,, \\ &E_{nl,\pm(l+1/2)} \to E_{n,\pm(l+1/2),\pm1/2}\,. \end{aligned} \tag{12.128}$$

The levels (12.127) also appear in Fig. 12.3 in the region where $\mu_\text{B}B/\Delta_{21}$ is large. Notice that the Paschen-Back effect is a limiting effect since in practice one cannot reach stable magnetic fields larger than 10^5 G, and for these $\mu_\text{B}B/\Delta_{nl} \lesssim 10$, when $n = 2$, $l = 1$. Even for such intense fields it is justified to use the first order of perturbation theory for low values of n.

We finish by briefly discussing how the emission spectrum is affected by a magnetic field. We concentrate our attention to the levels $1s_{1/2}$, $2p_{1/2}(\simeq 2s_{1/2})$ and $2p_{3/2}$, and the associated transitions $2p_{3/2} \to 1s_{1/2}$, $2p_{1/2} \to 1s_{1/2}$ which in the absence of an external magnetic field would give rise to two spectral lines as indicated in Fig. 12.4. After introducing the field, the levels $1s_{1/2}$, $2p_{1/2}$, $2s_{1/2}$, and $2p_{3/2}$ split respectively in 2, 2, 2 and 4 levels and one would expect that $2\times4+2\times2+2\times2 = 16$ lines would appear instead of the previous two, but this does not occur due to selection rules. First, one has to take into account that the only important transitions are the electric dipole transitions (see Chap. 15) with the selection rules $\Delta m_s = 0$, $\Delta m = \Delta m_l = 0, \pm 1$. The lines with $\Delta m = 0$ are called π-lines and those with $|\Delta m| = 1$ σ-lines, corresponding to different types of polarization of the emitted light. On the other hand, for the value of l we have the selection rule $|\Delta l| = 1$. The combination of these rules implies that only ten lines appear, giving rise to the so-called *anomalous Zeeman effect*. These lines, however, do not all behave in exactly the same way, since when $\mu_B B/\Delta_{21}$ grows, m_s becomes a better quantum number and the limiting selection rule $\Delta m_s = 0$ suppresses four of them; those indicated by dotted lines in Fig. 12.4. Furthermore, the energy of the transitions $E_{21\,3/2} \to E_{10\,1/2}$ and $E^-_{21\,1/2} \to E_{10\,-1/2}$, $E^+_{21\,1/2} \to E_{10\,1/2}$ and $E^-_{21\,-1/2} \to E_{10\,-1/2}$, and $E^+_{21\,-1/2} \to E_{10\,1/2}$ and $E_{21\,-3/2} \to E_{10\,-1/2}$ tend to coincide in the limit of large magnetic fields. Therefore, in this limit, we only observe a triplet (*normal Zeeman effect*) consisting of the central π-line and with respect to it, two equidistant σ-lines. The energy difference between the σ and π lines is $\mu_B B$, corresponding to the Larmor frequency $\omega_L \equiv \mu_B B/\hbar$.

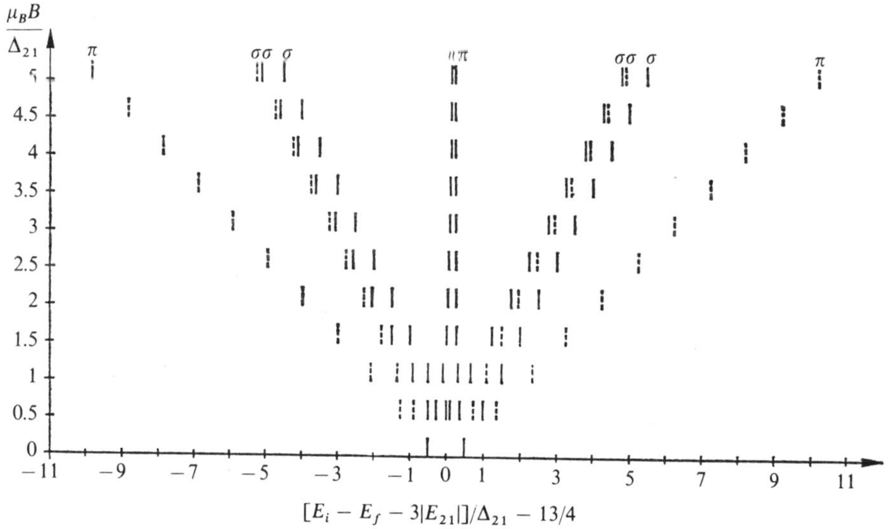

Fig. 12.4. Emission lines of hydrogen between the levels $1s_{1/2}, 2s_{1/2}, 2p_{1/2}$ and $2p_{3/2}$ for different intensities of the external magnetic field

Even though the structure of the anomalous Zeeman effect depends on the spectral line considered, the normal Zeeman effect is identical in all cases, the energy of the different levels is given by (12.127) and the electric dipole transitions among them, subject to $\Delta m_s = 0$, $\Delta m_l = 0, \pm 1$, always give rise to a central π-line, and with respect to it, two equidistant σ-lines.

The relatively small intensity ($\sim 10^5$ G) of stable magnetic fields achieved in laboratories was the justification to neglect the diamagnetic interaction in our analysis. The situation changes drastically in neutron stars, with surface fields of the order of 10^{12} G. It is therefore interesting to study the effect of very intense magnetic fields in hydrogen and helium atoms. We only consider the hydrogen atom. In atomic units the Hamiltonian is

$$H_B = -\frac{1}{2}\Delta - \frac{1}{r} + \frac{1}{2c}\boldsymbol{B}\cdot\boldsymbol{L} + \frac{1}{8}\mathcal{B}^2 r_\perp^2 \, , \tag{12.129}$$

where $\mathcal{B} = B/c$, $r_\perp = x^2 + y^2$ if we take the magnetic field along the z-direction, and we choose $\boldsymbol{A} = \boldsymbol{B} \times \boldsymbol{r}/2$. As stated in Sect. 12.6, the essential spectrum of H_B does not change when we remove the Coulomb potential. Taking into account (12.81), we obtain $\sigma_{\text{ess}}(H_B) = [\mathcal{B}/2, \infty)$. It is known that the ground state of (12.129) is in the $L_z = 0$ sector, and the ground energy $E_0(B)$ is non-degenerate [AH 78]. The Rayleigh-Schrödinger series for $E_0(B)$ with $H_B - H_{B=0}$ as perturbation is

$$E_0(B) = \frac{1}{2}\left[-1 + \frac{1}{2}\mathcal{B}^2 - \frac{53}{96}\mathcal{B}^4 + \frac{5581}{2302}\mathcal{B}^6 - \frac{21\,577\,397}{1\,105\,920}\mathcal{B}^8 + \cdots\right] \tag{12.130}$$

and it is very likely divergent: numerical estimates suggest that if $2E_0(B) = \sum_{n=0}^{\infty} a_{2n}\mathcal{B}^{2n}$, then

$$a_{2n} = (-1)^{n+1}\left(\frac{4}{\pi}\right)^{5/2} \pi^{-2n}\Gamma(2n + 3/2)[1 + \mathrm{O}(n^{-1})] \, . \tag{12.131}$$

In any case, it is known that the series (12.130) is Borel summable in a neighborhood of $B = 0$ to the correct energy value of the level. Finally the ionization energy behaves asymptotically as

$$\frac{\mathcal{B}}{2} - E_0(B) \sim \frac{1}{2}[\ln\mathcal{B} - 2\ln\ln\mathcal{B} - \gamma - \ln 2]^2 \tag{12.132}$$

when $B \to \infty$ (γ = Euler's constant).

There are many numerical computations for this problem [HO 85]. For a treatment with Padé approximants of (12.130) incorporating the dominant behavior in (12.132) and approximately valid for all values of B, see [GP 76].

13. Systems of Identical Particles

13.1 Introduction

In the framework of classical mechanics there is no need to treat systems containing identical particles in any special way, for even though all their characteristic properties are the same, it is always possible to identify each one of them. It suffices in principle to measure the position and momentum of each particle at a given instant of time with infinite precision. With these values as initial conditions and using the equations of motion, each particle can be identified at any other time.

The situation changes radically in quantum mechanics, where we know it is not possible to define trajectories as in the classical case. If we consider a system of identical particles at some instant of time, the wave function representing its state is given by means of products of individual wave packets which overlap in general; thus if at a later instant of time a particle is localized, it is impossible to determine which one it was among those in the initial configuration. It is this overlap of wave packets, almost always inevitable under time evolution, that gives rise to the fundamental difference between the classical and the quantum treatment of distinguishability of identical particles.

This chapter deals with systems of identical particles, and introduces a formalism suited to their study.

13.2 Symmetrization of Wave Functions

Saying that a system is composed of N identical particles is equivalent to saying that there is no observable capable of distinguishing between them. Therefore, any observable, and in particular the Hamiltonian of the system must contain a completely symmetrical dependence on the coordinates (positions and spins) of these particles. Equivalently, if A is *any* observable of the system, and P is any permutation in S_N (permutation group on N elements), we will say that the particles are *identical*, if and only if

$$\boxed{[A,P]=0} \quad \forall P \in S_N \, . \tag{13.1}$$

In (13.1) we use the same symbol for the permutation and the operator implementing it in the Hilbert space $\mathcal{H}_N = \mathcal{H}_1 \otimes \mathcal{H}_1 \otimes \ldots \otimes \mathcal{H}_1$ associated to the N-particle system. Next we define this operator.

Let ξ_i be the coordinates of the ith particle (for instance, its position and the third component of its spin); if $\psi(\xi_1, \xi_2, \ldots \xi_N)$ represents the wave function of the system, the action of a permutation P is defined by

$$\boxed{(P\psi)(\xi_1, \ldots, \xi_N) = \psi(\xi_{P(1)}, \ldots, \xi_{P(N)})} \, . \tag{13.2}$$

With the simplified notation $|f_1, f_2, \ldots, f_N\rangle \equiv |f_1\rangle \otimes \ldots \otimes |f_N\rangle \equiv |f_1\rangle \ldots |f_N\rangle$:

$$\boxed{P|f_1, \ldots, f_N\rangle = |f_{P^{-1}(1)}, \ldots, f_{P^{-1}(N)}\rangle} \, , \tag{13.3}$$

since as P is unitary

$$\begin{aligned} \langle \xi_1, \ldots, \xi_N | P | \psi \rangle &= \langle \psi | P^{-1} | \xi_1, \ldots, \xi_N \rangle^* \\ &= \langle \psi | \xi_{P(1)}, \ldots, \xi_{P(N)} \rangle^* = \langle \xi_{P(1)}, \ldots \xi_{P(N)} | \psi \rangle \, . \end{aligned} \tag{13.4}$$

It is easy to convince oneself that (13.3) defines an action of S_N:

$$(P_2 P_1)|f_1, \ldots, f_N\rangle = P_2(P_1|f_1, \ldots, f_N\rangle) \, , \tag{13.5}$$

for if $|g_i\rangle \equiv |f_{P_1^{-1}(i)}\rangle$, then $|g_{P_2^{-1}(i)}\rangle = |f_{(P_2P_1)^{-1}(i)}\rangle$. If $\psi(\xi_1, \ldots, \xi_N)$ is a solution of the time-independent Schrödinger equation

$$(H\psi)(\xi_1, \xi_2, \ldots, \xi_N) = E\psi(\xi_1, \xi_2, \ldots, \xi_N) \, , \tag{13.6}$$

the wave function $(P\psi)(\xi_1, \xi_2, \ldots, \xi_N)$, where P is a permutation in S_N, is also a solution with the same energy. This is an easy consequence of (13.1) and (13.6). Since there are $N!$ permutations, we can obtain in this way $N!$ wave functions. Some of them will be linear combinations of the others, but in general, we will find several linearly independent solutions with the same energy. Hence, almost all the energy levels of a Hamiltonian of identical particles will be highly degenerate. This degeneracy is known as *exchange degeneracy*.

If $\psi(\xi_1, \xi_2, \ldots, \xi_N)$ is the wave function of a system of N identical particles, and A is any observable, it follows immediately from (13.1) that

$$\langle \psi | A | \psi \rangle = \langle P\psi | A | P\psi \rangle \, . \tag{13.7}$$

On the other hand if at time t_0 the wave function is $\Psi(\xi_1, \xi_2, \ldots, \xi_N; t_0)$, and $U(t, t_0)$ is the time evolution operator, satisfying

$$[U(t, t_0), P] = 0 \tag{13.8}$$

for all permutations $P \in S_N$ (because it is generated by the energy operator), we have

$$(PU(t, t_0)\Psi)(\xi_1, \xi_2, \ldots, \xi_N; t_0) = (U(t, t_0)P\Psi)(\xi_1, \xi_2, \ldots, \xi_N; t_0) \tag{13.9}$$

and therefore the state obtained at time t from the time evolution of the state $(P\Psi)(\xi_1, \xi_2, \ldots, \xi_N; t_0)$ coincides with $(P\Psi)(\xi_1, \xi_2, \ldots, \xi_N; t)$. From this we conclude that dynamical states represented by vectors differing only by a permutation of the coordinates of the identical particles, cannot be distinguished by any observation at any time.

To further clarify the previous arguments, consider a system formed by three identical particles, and the six normalized wave functions $\psi(ijk)$, where $i \neq j \neq k$ takes the values 1, 2, 3. To simplify the notation we use i instead of ξ_i. Out of these functions we can construct

$$\psi_S(123) = \frac{1}{\sqrt{6}}[\psi(123)+\psi(132)+\psi(231)+\psi(213)+\psi(312)+\psi(321)]\ , \tag{13.10}$$

$$\psi_A(123) = \frac{1}{\sqrt{6}}[\psi(123)-\psi(132)+\psi(231)-\psi(213)+\psi(312)-\psi(321)]\ , \tag{13.11}$$

$$\begin{aligned} \psi_{M_1}(123) &= \frac{1}{2\sqrt{3}}[2\psi(123) - \psi(132) + 2\psi(213) - \psi(231) \\ &\quad - \psi(312) - \psi(321)]\ , \\ \psi_{M_2}(123) &= \tfrac{1}{2}[\psi(132) - \psi(231) + \psi(312) - \psi(321)]\ , \end{aligned} \tag{13.12}$$

$$\begin{aligned} \psi'_{M_1}(123) &= \frac{1}{2\sqrt{3}}[2\psi(123) + \psi(132) - 2\psi(213) - \psi(231) \\ &\quad - \psi(312) + \psi(321)]\ , \\ \psi'_{M_2}(123) &= \tfrac{1}{2}[\psi(132) + \psi(231) - \psi(312) - \psi(321)]\ , \end{aligned} \tag{13.13}$$

generating the same space as the six starting wave functions. This does not mean that the dimension of this space is six; if $\psi(123)$ was totally symmetric, only $\psi_S(123)$ would be different from 0. The numerical factors preceeding the parentheses in the previous formulae are simply normalization constants when $\psi(ijk)$ is orthogonal to $\psi(123)$ for $(i,j,k) \neq (1,2,3)$. It is straightforward to verify that $(P\psi_S)(123) = \psi_S(123)$ and $(P\psi_A)(123) = \pm\psi_A(123)$, according to whether the permutation is even or odd, and therefore ψ_S, ψ_A are totally symmetric or antisymmetric functions, respectively. Furthermore under any $P \in S_3$ the space generated by the two functions (13.12) is stable and the same holds for the subspace spanned by the functions (13.13), although for an arbitrary permutation P, $P\psi_{M_1} \neq \lambda\psi_{M_1}$ in general. Therefore, if ψ_{M_1} were physically realizable, the same physical state would be represented by essentially different wave functions ψ_{M_1}, $P\psi_{M_1}$.

This implies that the Hilbert space $\mathcal{H}_3$ of a system of three identical particles can be decomposed as the direct sum of four subspaces: $\mathcal{H}_3 = \mathcal{H}_S \oplus \mathcal{H}_A \oplus \mathcal{H}_M \oplus \mathcal{H}_{M'}$, whose functions transform under the group S_3, as do those in (13.10, 11, 12) and (13.13), respectively. The previous decomposition of H_3 as an orthogonal direct sum of $\mathcal{H}_S$, $\mathcal{H}_A$, and $\mathcal{H}_{\mathcal{S}} \equiv \mathcal{H}_M \oplus \mathcal{H}_{M'}$, corresponds to the classification of the wave functions of three identical particles in different symmetry types: total symmetry in $\mathcal{H}_S$, total antisymmetry in $\mathcal{H}_A$, and mixed symmetry corresponding to the Young diagram (2,1) in $\mathcal{H}_{\mathcal{S}} = \mathcal{H}_M \oplus \mathcal{H}_{M'}$. Moreover $\mathcal{H}_M$, $\mathcal{H}_{M'}$, are stable under S_3, and they give rise to two equivalent representations of S_3. The decomposition of $\mathcal{H}_{\mathcal{S}}$ in the subspaces $\mathcal{H}_M$ and $\mathcal{H}_{M'}$ is not unique. From the commutativity of the observables A with any permutation, one derives [GM 62] that the action of the observable A does not change the symmetry type. In other words, the matrix elements of A between states of different symmetry vanish, and therefore between $\mathcal{H}_{\mathrm{S}}$, $\mathcal{H}_{\mathrm{A}}$, $\mathcal{H}_{\mathcal{S}}$, we have a superselection rule. Furthermore, if A is an observable it is a direct consequence of $[A, P] = 0$ that the eigenvalues of A in the subspace $\mathcal{H}_{\mathcal{S}}$ are at least doubly degenerate (exchange degeneracy).

If $\mathcal{H}(a)$ denotes the subspace of solutions of $A|\psi\rangle = a|\psi\rangle$, then $\mathcal{H}(a) = \mathcal{H}_{\mathrm{S}}(a) \oplus \mathcal{H}_{\mathrm{A}}(a) \oplus \mathcal{H}_{\mathcal{S}}(a)$. In principle, and barring any accidental degeneracies, only one of the summands will be non-trivial. In other words, the eigenvalue equations $A|\psi\rangle = a|\psi\rangle$ will be solvable at most for a single symmetry type, for fixed a.

We started discussing the case $N = 3$, because it is the first case which contains special difficulties. For $N = 2$, every function $\psi(12)$ is a sum of a symmetric plus an antisymmetric function, so that $\mathcal{H}_2 = \mathcal{H}_{\mathrm{S}} \oplus \mathcal{H}_{\mathrm{A}}$ and neither mixed symmetries nor exchange degeneracies appear.

In the case of a system of N identical particles, and using standard techniques [HA 62] in the study of the permutation group it can be shown [GM 62] that $\mathcal{H}_N$ is a direct sum of subspaces $\mathcal{H}_N = \mathcal{H}_{\mathrm{S}} \oplus \mathcal{H}_{\mathrm{A}} \oplus \mathcal{H}_{\mathcal{S}_1} \oplus \mathcal{H}_{\mathcal{S}_2} \oplus \ldots$, all of them stable under the action of S_N and the observables. $\mathcal{H}_{\mathrm{S}}$ and $\mathcal{H}_{\mathrm{A}}$ are the spaces of totally symmetric and antisymmetric functions, respectively, and functions in these subspaces generate one-dimensional representations of S_N . In the remaining subspaces, $\mathcal{H}_{\mathcal{S}_1}, \mathcal{H}_{\mathcal{S}_2}, \ldots$, the basis functions have a more complex symmetry and they originate higher-dimensional irreducible representations of S_N. Each subspace $\mathcal{H}_{\mathcal{S}_i}$ contains several equivalent irreducible representations of S_N. As before between these subspaces there is a superselection rule and furthermore, it is impossible to know whether the physical system is found in $\psi \in \mathcal{H}_{\mathcal{S}_i}$ or in $P\psi \in \mathcal{H}_{\mathcal{S}_i}$. Apart from the wave functions in $\mathcal{H}_{\mathrm{S}}$ and $\mathcal{H}_{\mathrm{A}}$ all others show exchange degeneracy.

A very important physical consequence of this fact is that if we admit that a physical system of N identical particles possesses some complete set of commuting observables, its pure state cannot belong to any of the subspaces of mixed symmetry. They should necessarily be either totally symmetric or totally antisymmetric. The reason is the following: if $\{A_1, \ldots, A_r\}$ is a complete set of compatible observables, and $a_1, \ldots, a_r$ are the eigenvalues of $A_1, \ldots, A_r$, respectively, then the state $|a_1, \ldots, a_r\rangle$, is essentially unique. But if $|a_1, \ldots, a_r\rangle \in \mathcal{H}_{\mathcal{S}_i}$ we can find some permutation P so that $P|a_1, \ldots, a_r\rangle \neq e^{i\alpha}|a_1, \ldots, a_r\rangle$, for all α, and at the same time $P|a_1, \ldots, a_r\rangle$ is an eigenstate of $A_1, \ldots, A_r$ with eigenvalues $a_1, \ldots, a_r$, leading to a contradiction.

Apart from these fundamental theoretical reasons, there is abundant experimental evidence (at least for the more familiar elementary particles) which support the exclusion of the so-called *parastatistics* based on the mixed symmetries. We are inevitably led to the

Symmetrization Principle: The pure states of a system of identical particles must be totally symmetric or antisymmetric under the exchange of any two of them.

On the other hand, in the relativistic quantum theory of local fields it is possible to establish [SW 64b] starting with very general considerations and in particular with the previous principle the

Spin-Statistics Theorem: The pure states of a system of identical particles are totally symmetric (antisymmetric) if their spin is an integer (half-odd integer).

This theorem complements the symmetrization principle and will be adopted in the following.

The origin of the name of this theorem is the following: we will later show that systems of identical particles with totally symmetric wave functions obey *Bose-Einstein statistics* (these particles are called *bosons*) while a system of particles with totally antisymmetric wave functions obey the *Fermi-Dirac statistics* (such particles are called *fermions*).

The consideration of non-trivial topologies for the configuration space of a system of identical particles, opens the possibility of *fractional statistics*, interpolating between the two previous ones. These circumstances appear for example for "particles" composed of a particle of charge q restricted to move in two dimensions, and an orthogonal tube of magnetic flux Φ. In this case the exchange of two of these "particles" produces a phase factor $\exp(i\theta)$, with $\theta = q\Phi/\hbar c$. Only for $\theta = 0, \pi$ do we recover the ordinary statistics [WI 82], [WU 84].

The experimental data confirm that electrons (atomic and molecular structures), nucleons (nuclear structures), ^{3}He (low temperature behavior), all of them with spin $1/2$, are fermions. Similarly for the photons (radiation theory) and ^{4}He (low temperature properties), which are respectively particles with spin 1 and 0, their experimental behavior confirms their bosonic character. The experimental evidence of the spin-statistics connection for other types of particles is, in general, much weaker or totally nonexistent [GR 64].

We now have to face a puzzle: in the universe there is a large number of electrons, for example, and it would seem necessary, according to the previous arguments, to always take all of them into account and to use for their description totally antisymmetric wave functions. Actually it is not necessary to carry out such a radical antisymmetrization of the wave function. Indeed suppose that there were only two electrons sufficiently far apart so that we can neglect the overlap between their wave functions and their mutual interactions. If $\psi_1(\xi_1)$ and $\psi_2(\xi_2)$ are the normalized wave functions of each one of them, the approximately normalized and antisymmetrized total wave function is

$$\psi(\xi_1\xi_2) = \frac{1}{\sqrt{2}}[\psi_1(\xi_1)\psi_2(\xi_2) - \psi_1(\xi_2)\psi_2(\xi_1)] \, . \tag{13.14}$$

We may now ask for example what is the density (average number per unit volume) of electrons at a point ξ; i.e., with a position and a given value of the third component of spin. This density is given by

$$\varrho(\xi) = \int d\xi_2 \, |\psi(\xi, \xi_2)|^2 + \int d\xi_1 \, |\psi(\xi_1, \xi)|^2 \, , \tag{13.15}$$

where the integrals are ordinary ones for the space variables and discrete sums over the spin indices. Taking into account (13.14) we obtain

$$\varrho(\xi) = |\psi_1(\xi)|^2 + |\psi_2(\xi)|^2 - 2\mathrm{Re}[\psi_1^*(\xi)\psi_2(\xi) \int d\xi_1 \, \psi_2^*(\xi_1)\psi_1(\xi_1)] \, . \tag{13.16}$$

Since $\psi_1(\xi)$ and $\psi_2(\xi)$ are localized, respectively, in practically disjoint regions R_1 and R_2, the last term in (13.16) is negligible, and if $\xi \in R_1$ we have $\varrho(\xi) \simeq |\psi_1(\xi)|^2$, a result that we would have obtained had we completely ignored the second electron. The argument we have just presented concerning only two particles can be extended without difficulty to more general situations and to other physical observables with the same conclusion: the symmetrization or antisymmetrization rule should only be applied to those particles that are relevant in the problem under consideration. Notice that in the classical limit, the wave functions of the particles are very localized in space and therefore no symmetrization effects appear for classical systems.

We have already discussed in Sect. 5.2 the concept of an elementary particle as an object whose structure is unknown or even if it is known, it is not relevant in the problem of interest. Thus, for example, in problems where only very small energies are relevant, molecules and atoms can be taken as particles; if the relevant energies are in the order of electron-volts, the atomic structure is important and one should consider the atoms as composed of electrons and nuclei, which could in turn be treated as particles. If the energies increased up to the MeV range, the nuclear structure becomes important. With the energies attained today, there is no evidence of infrastructure for leptons or quarks (constituents of the nucleon). These particles are called elementary with the understanding that such terminology can be as unfortunate as when applied to atoms. One might think that the spin-statistics connection applies only to elementary particles, namely, to particles with no known structure. This would be very unsatisfactory. Actually the principle is applicable to any type of particles, elementary or otherwise.

We illustrate this statement by considering two hydrogen atoms. Let ξ_i, η_i be the electron and proton coordinates, respectively, with $i = 1, 2$. The correctly antisymmetrized wave function for the system of the two electrons and the two protons is

$$\psi_A(\xi_1\eta_1; \xi_2\eta_2) = \frac{1}{\sqrt{2}}[\psi_1(\xi_1\eta_1; \xi_2\eta_2) - \psi_2(\xi_1\eta_1; \xi_2\eta_2)] \, , \tag{13.17}$$

with

$$\begin{aligned} \psi_1(\xi_1\eta_1;\xi_2\eta_2) &= N[\psi(\xi_1\eta_1;\xi_2\eta_2)+\psi(\xi_2\eta_2;\xi_1\eta_1)]\,, \\ \psi_2(\xi_1\eta_1;\xi_2\eta_2) &= N[\psi(\xi_1\eta_2;\xi_2\eta_1)+\psi(\xi_2\eta_1;\xi_1\eta_2)]\,, \end{aligned} \tag{13.18}$$

where N is a normalization constant and ψ is an arbitrary function. When both atoms are sufficiently far apart (the only case where the internal structure could be neglected) the overlap between ψ_1 and ψ_2 is negligible, and for any local observable B (an operator which transforms a wave function concentrated in a region of space into another one concentrated practically in the same region), we have

$$\langle\psi_A|B|\psi_A\rangle \simeq \langle\psi_1|B|\psi_1\rangle \simeq \langle\psi_2|B|\psi_2\rangle\,. \tag{13.19}$$

In this situation, and for such observables (in practice all of them), we can therefore describe the wave function ψ_A of the system, to a very good approximation, by means of either ψ_1 or ψ_2. We thus see that for hydrogen atoms, which have integer spin, the wave function can be taken as the product of the individual wave functions for each atom symmetrized under the exchange $(\xi_1,\eta_1) \leftrightarrow (\xi_2,\eta_2)$ when they are sufficiently far apart. The extension to the general case is very similar: for identical composite systems whose structure is of no interest the correct rule is to take the totally symmetric wave function if each system contains an even number of fermions and any number of bosons (and therefore its spin is an integer). On the other hand we should use totally antisymmetric wave functions if such system contains an odd number of fermions and any number of bosons (i.e., if its total spin is half-odd integral). This is in perfect agreement with the principle of spin-statistics connection.

An important consequence of our previous argument is the *exclusion principle*, valid only for fermions. For a system of fermions, the wave function is totally antisymmetric, and therefore

$$(P_{ij}\psi)(\xi_1,\ldots,\xi_i,\ldots,\xi_j,\ldots,\xi_N) = -\psi(\xi_1,\ldots,\xi_j,\ldots,\xi_i,\ldots,\xi_N)\,, \tag{13.20}$$

where P_{ij} is the permutation exchanging ξ_i and ξ_j. From (13.20) we see that the wave function vanishes whenever $\xi_i = \xi_j$. This means that the probability of finding two electrons with the same value of their coordinates vanishes (exclusion principle).

Two identical fermions cannot be found in the same position in space if they have the same spin projection. Similarly they cannot have the same momentum if they have the same helicity. In general two identical fermions cannot occupy the same state for it is impossible to construct a totally antisymmetric wave function for such a system which does not vanish identically.

The symmetrization principle and the connection between spin and statistics refers to identical and physically indistinguishable particles. In practice, however, it is often convenient to generalize the mathematical formalism to treat on the same footing particles which could be considered indistinguishable only after some interaction is neglected. For example, in Sect. 7.15 we discussed isospin, considering protons and neutrons as different aspects of the nucleon. Given a system of protons and neutrons, it is only necessary to antisymmetrize the wave function with respect to all protons and independently over all neutrons. But without adding any extra physical hypotheses, that is, without generalizing the previous principles, it is possible to *formally* extend the exclusion principle to a system of nucleons with the simple purpose of unifying different formalisms, but without any physical transcendence. To a given wave function of a system of protons and neutrons, appropriately antisymmetrized over the protons and over the neutrons, we can always associate a physical equivalent and totally antisymmetric wave function of nucleons: it is enough to add an extra variable for each particle (the third component of isospin), replacing the symbols p, n, and after this is done, we antisymmetrize the resulting wave function in all its variables, including the new one. More concretely, let $\mathcal{H}_{Z,N}$ be the subspace of $\mathcal{H}_{Z+N}$ generated by those states where the first Z variables refer to protons and the last N to neutrons and which are separately antisymmetrized in both sets of variables. Let us call $\mathcal{H}_{Z+N,\mathrm{A}}$ the totally antisymmetric part of $\mathcal{H}_{Z+N}$, obtained from $\mathcal{H}_{Z+N}$ by applying the antisymmetrizer

$$A = \frac{1}{(Z+N)!} \sum_{P \in S_{Z+N}} (-1)^{\pi(P)} P \,, \tag{13.21}$$

where $\pi(P)$ is the parity of the permutation P. Notice that $A = A^\dagger = A^2$. It is easy to show that the mapping

$$|\psi\rangle \in \mathcal{H}_{Z,N} \to |\tilde{\psi}\rangle \equiv \binom{Z+N}{Z}^{1/2} A|\psi\rangle \in \mathcal{H}_{Z+N,\mathrm{A}} \tag{13.22}$$

is a one-to-one isometry from $\mathcal{H}_{Z,N}$ onto $\mathcal{H}_{Z+N,\mathrm{A}}$. This is the correspondence mentioned, which associates a totally antisymmetrized nucleon state to the initial state of protons and neutrons. With regard to observables, given an observable X in $\mathcal{H}_{Z,N}$ which respects the indistinguishability of protons and neutrons, the observable associated in the language of nucleons should leave invariant the physical predictions; such an observable is obtained by means of the correspondence (13.22):

$$X \to \tilde{X} \equiv \binom{Z+N}{Z} AKXKA \,, \tag{13.23}$$

where K is the orthogonal projection from $\mathcal{H}_{Z+N}$ onto $\mathcal{H}_{Z,N}$. Notice that $[\tilde{X}, P] = 0$, $\forall P \in S_{Z+N}$. Indeed:

$$\begin{aligned} \langle\tilde{\psi}|\tilde{X}|\tilde{\psi}\rangle &= \binom{Z+N}{Z}^2 \langle A\psi|AKXKA|A\psi\rangle \\ &= \binom{Z+N}{Z}^2 \langle A\psi|KXK|A\psi\rangle \,. \end{aligned} \tag{13.24}$$

When we expand the matrix element $\langle A\psi|KXK|A\psi\rangle$ we will get a factor $1/(Z+N)!^2$ from the operators A [see (13.21)]; on the other hand each $(Z+N)!\,A|\psi\rangle$ contains $\binom{Z+N}{Z}$ different vectors with a structure similar to that of $|\psi\rangle$ and with a coefficient $Z!N!$. All these vectors are annihilated by K, with the exception of the one where all the proton indices precede the neutron indices. Then,

$$\langle\tilde{\psi}|\tilde{X}|\tilde{\psi}\rangle = \binom{Z+N}{Z}^2 \frac{1}{(Z+N)!^2}(Z!N!)^2\langle\psi|X|\psi\rangle = \langle\psi|X|\psi\rangle\ , \qquad (13.25)$$

which was to be shown.

13.3 Non-Interacting Identical Particles

Consider a system of N identical particles without mutual interaction and whose Hamiltonian H_0 has the form

$$H_0 = \sum_{i=1}^{N} H_0(i) = \sum_{i=1}^{N}\left[\frac{1}{2M}p_i^2 + V(i)\right]\ . \qquad (13.26)$$

Assuming for simplicity the spectrum of $H_0(i)$ to be discrete, let $\psi_{a_1}(i)$, $\psi_{a_2}(i), \psi_{a_3}(i), \dots$ be the normalized single-particle eigenfunctions (or briefly the *orbitals*), corresponding to eigenstates of $H_0(i)$ with eigenvalues $E_{a_1}, E_{a_2}, E_{a_3}, \dots$, respectively:

$$H_0(1)\psi_{a_i}(1) = E_{a_i}\psi_{a_i}(1)\ . \qquad (13.27)$$

The wave function

$$\Psi_{a_1a_2\dots a_N}(1,2,\dots N) \equiv \psi_{a_1}(1)\psi_{a_2}(2)\psi_{a_N}(N) \qquad (13.28)$$

is an eigenfunction of H_0 given in (13.26) with eigenvalue

$$E_{a_1a_2\dots a_N} = E_{a_1} + E_{a_2} + \dots + E_{a_N}\ . \qquad (13.29)$$

It is often said that in this state the orbitals $a_1, a_2, \dots, a_N$ (not necessarily all different) are occupied. In accord with the spin-statistics theorem, the only physically admissible wave functions are those totally symmetric in the case of bosons and totally antisymmetric in the case of fermions. For fermions, the wave functions will be linear combinations of functions of the form

$$\boxed{\begin{aligned}&\Psi^{\mathrm{A}}_{a_1a_2\dots a_N}(1,2,\dots N)\\ &\quad = \frac{1}{\sqrt{N!}}\sum_{P\in S_N}(-1)^{\pi(P)}\psi_{a_1}(P1)\psi_{a_2}(P2)\dots\psi_{a_N}(PN)\end{aligned}}\ . \qquad (13.30)$$

Obviously, in this case all the occupied orbitals in (13.30) should be different, for otherwise the wave function would vanish identically. To make sure that the normalization factor has been chosen correctly it is enough to use the properties of the antisymmetrizer A, and arguments similar to those use at the end of the previous section. A more detailed proof is:

$$\begin{aligned}&\int d\tau\,(\Psi^{\mathrm{A}}_{a_1a_2\ldots a_N})^*(1,2,\ldots,N)\Psi^{\mathrm{A}}_{a_1a_2\ldots a_N}(1,2,\ldots,N)\\ &\quad=\frac{1}{N!}\sum_P\sum_{P'}(-1)^{\pi(P)+\pi(P')}\\ &\quad\times\int d\tau\,\psi^*_{a_1}(P1)\ldots\psi^*_{a_N}(PN)\psi_{a_1}(P'1)\ldots\psi_{a_N}(P'N)\,,\end{aligned}\tag{13.31}$$

where $\int d\tau$ includes the integration over all space coordinates and a sum over all spin coordinates or other quantum numbers required to completely characterize a single-particle state. The integrals on the right-hand side all vanish except when $P' = P$, and then the integral is equal to 1. Therefore the right-hand side is $N!^{-1}\sum_P 1 = 1$.

The normalized antisymmetric wave function (13.30) for N fermions can be written as an $N\times N$ determinant known as the *Slater determinant*:

$$\Psi^{\mathrm{A}}_{a_1a_2\ldots a_N}(1,2,\ldots,N)=\frac{1}{\sqrt{N!}}\begin{vmatrix}\psi_{a_1}(1) & \psi_{a_1}(2) & \ldots & \psi_{a_1}(N)\\ \psi_{a_2}(1) & \psi_{a_2}(2) & \ldots & \psi_{a_2}(N)\\ \ldots & \ldots & \ldots & \ldots\\ \psi_{a_N}(1) & \psi_{a_N}(2) & \ldots & \psi_{a_N}(N)\end{vmatrix}\,.\tag{(13.32)}$$

The exchange of two coordinates is equivalent to the exchange of two columns, which changes the sign of the determinant, making explicit the required antisymmetry. On the other hand the exchange of two orbitals correspond to the exchange of two rows and it also leads to the change of sign of the wave function. Thus (13.30) can also be written as

$$\Psi^{\mathrm{A}}_{a_1a_2\ldots a_N}(1,2,\ldots,N)=\frac{1}{\sqrt{N!}}\sum_P(-1)^{\pi(P)}\psi_{a_{P1}}(1)\psi_{a_{P2}}(2)\ldots\psi_{a_{PN}}(N)\,,\tag{13.33}$$

where instead of permuting the coordinates we have permuted the orbitals.

In the case of bosons, the totally symmetric and normalized basis wave functions will be

$$\boxed{\begin{aligned}\Psi^{\mathrm{S}}_{a_1a_2\ldots a_N}(1,2,\ldots,N)&=\left[\frac{1}{N!N_1!N_2!\ldots}\right]^{1/2}\\ &\times\sum_P\psi_{a_1}(P1)\psi_{a_2}(P2)\ldots\psi_{a_N}(PN)\end{aligned}}\tag{13.34}$$

and now two or more orbitals can be the same. We denote by $N_1, N_2, N_3, \ldots,$ the number of particles in the orbitals $a_1, a_2, a_3, \ldots,$ respectively, with N_1+N_2+

$N_3 + \ldots = N$. The correctness of the normalization factor can be proved along the lines of the argument for fermions.

In the sum (13.34) not all terms are necessarily different; each one appears repeated $N_1!N_2!\ldots$ times. If we consider the sum to be restricted only to different terms (13.34) can be written as

$$\Psi^{S}_{a_1\ldots a_N}(1,2,\ldots,N) = \sqrt{\frac{N_1!N_2!\ldots}{N!}} \sum_P{}' \psi_{a_1}(P1)\ldots\psi_{a_N}(PN) , \qquad (13.35)$$

where $\sum'$ denotes that restriction on the summation symbol.

It is obvious that for both bosons and fermions it is only necessary to give the occupation numbers $(N_1, N_2, N_3, \ldots)$, where $N_1 + N_2 + N_3 + \ldots = N$, to characterize the state. For fermions each occupation number is either 0 or 1, while for bosons they can take any integer value ≥ 0. The energy of the state $(N_1, N_2, N_3, \ldots)$ is therefore

$$E = N_1E_1 + N_2E_2 + N_3E_3 + \ldots , \qquad (13.36)$$

with $E_i \equiv E_{a_i}$. We assume in the following that the orbitals are labelled in increasing order of energies $E_1 \leq E_2 \leq E_3 \ldots$. It is clear that the energy spectrum for bosons and fermions are completely different due to the different values of the occupation numbers in both cases. Thus, for example, in the ground state for a system of N bosons, the only occupied orbitals are those with energy E_1, and therefore the energy of the state is $E = NE_1$. If, on the other hand, we consider the ground state for a system of fermions, the exclusion principle requires that the first N states are occupied and all others are empty. The energy of the state is $E = E_1 + E_2 + E_3 + \ldots + E_N$ (always $\geq NE_1$).

A system of N identical particles, with $N \gg 1$, where all the mutual interactions have been neglected, is known as a perfect gas. In situations of thermodynamic equilibrium, all properties of these gases can be derived from the thermodynamic potential [LL 80]:

$$\begin{aligned} \Omega &= \sum_i \Omega_i , \\ \Omega_i &= -k_B T \ln \sum_{N_i} \{\exp[(\mu - E_i)/k_B T]\}^{N_i} . \end{aligned} \qquad (13.37)$$

T is the absolute temperature, μ the chemical potential per particle, and Ω_i is the thermodynamic potential corresponding to the a_i orbital. The sum $\sum_{N_i}$ runs over all possible occupation numbers of the orbital a_i.

In the often considered case where the potential V [in (13.26)] is only due to the impenetrability of the cavity confining the gas, the discretization of the energy levels is guaranteed by the finite size of the cavity.

We consider the two possible cases separately:

1) *Fermions*. The values of N_i are 0 and 1, and, therefore,

$$\Omega_i = -k_B T \ln\{1 + \exp[(\mu - E_i)/k_B T]\} . \tag{13.38}$$

The average number of particles in the orbital a_i is given by

$$\boxed{\bar{N}_i = -\frac{\partial \Omega_i}{\partial \mu} = \frac{1}{\exp[-(\mu - E_i)/k_B T] + 1}} , \tag{13.39}$$

which is precisely the distribution function for a perfect gas obeying the Fermi-Dirac statistics. The chemical potential is determined as a function of T and N through the equation

$$N = \sum_i \{\exp[(E_i - \mu)/k_B T] + 1\}^{-1} . \tag{13.40}$$

2) *Bosons*. The possible values of N_i are all non-negative integers, and therefore,

$$\Omega_i = -k_B T \ln \sum_{N_i=0}^{\infty} \{\exp[(\mu - E_i)/k_B T]\}^{N_i} . \tag{13.41}$$

The series giving Ω converges only if $\exp[(\mu - E_i)/k_B T] < 1$ for all energies E_i. When we deal with a perfect gas of free bosons (except for the constraint of a finite volume), $E_i \simeq 0$ and therefore the corresponding chemical potential cannot be positive. From (13.41) we obtain

$$\Omega_i = -k_B T \ln\{1 - \exp[(\mu - E_i)/k_B T]\} , \tag{13.42}$$

and the average occupation numbers are

$$\boxed{\bar{N}_i = \frac{1}{\exp[(E_i - \mu)/k_B T] - 1}} . \tag{13.43}$$

This is the distribution function for a perfect gas obeying Bose-Einstein statistics. The chemical potential is determined as a function of N and T through the equation

$$N = \sum_i \exp\{[(E_i - \mu)/k_B T] - 1\}^{-1} . \tag{13.44}$$

This justifies the name of fermions and bosons given to particles of half-odd integral and integral spins, respectively.

13.4 Fermi Gas

In many branches of physics (atomic and nuclear physics, solid state, stellar evolution, etc.,) the perfect gases of electrons or nucleons play a very important role. Since both particles have spin $1/2$, their properties are identical from a statistical point of view, and we will study only a gas of electrons. Furthermore we will only consider the case of free non-relativistic particles. The energy of an electron under these assumptions, is given by $E = p^2/2M$, with $M \equiv m_e$.

The average number $N(p)\Delta p$ of particles with momenta between p and $p+\Delta p$, $\Delta p \ll p$, will be the product of the number of possible states with their momentum in that interval, $n(p)\Delta p$, times the average occupation number of one of them. Assuming the gas to be confined to a cavity of volume V, bounded by a surface of area S, completely impenetrable to the particles, it can be shown rigorously [ST 68] that in the limit $\Delta p/\hbar \gg S/V$ (and therefore, $p/\hbar \gg S/V$),

$$n(p) = 2\frac{V}{(2\pi\hbar)^3}4\pi p^2 , \tag{13.45}$$

where the factor "2" comes from the two possible spin orientations. The condition $p/\hbar \gg S/V$ is equivalent to requiring that the wavelength associated to the electron is much smaller than the dimension of the cavity. Since (13.45) implies that $n(p)$ is independent of the specific shape of the cavity, a cubic cavity of side $L = V^{1/3}$ is often used to derive (13.45) with the wave functions vanishing at the walls as boundary conditions: the space part of the wave function, satisfying the free Schrödinger equation and vanishing at the walls, can be written in the form

$$\psi_k(\boldsymbol{r}) = A \sin(k_x x)\sin(k_y y)\sin(k_z z) , \tag{13.46}$$

with $k_x L = n_x\pi$, $k_y L = n_y\pi$, $k_z L = n_z\pi$, choosing conveniently the coordinate axes. The wave vector $\boldsymbol{k}$ is quantized, and it occupies the nodes of a cubic lattice whose unit cell has volume $(\pi/L)^3$. We only have to consider the cases $k_x, k_y, k_z > 0$, since all others produce the same states. The number of states with momenta between p and $p+\Delta p$ will be equal to twice the number of lattice points enclosed in the positive octant of a spherical shell between the radii p and $p+\Delta p$. When $(\Delta p)L \gg \hbar$, this number is very approximately equal to the ratio between the volume of the shell and that of the unit cell. In this way we obtain (13.45). An alternative and equally standard procedure consists of assuming the gas to be enclosed in a cubic cavity, but with periodic boundary conditions at the walls. The results obtained are identical. This is not very surprising if we take into account that the boundary conditions should only affect those phenomena occurring close to the surface, but not the bulk properties. An easy mnemonic to remember when deriving (13.45) is to take into account that it can also be obtained assuming that in the phase space of a particle of spin s (in this case $s = 1/2$), there are $2s + 1$ states in the elementary volume $(2\pi\hbar)^3$.

Combining (13.39) and (13.45):

$$N(p) = \frac{2V}{(2\pi\hbar)^3} \frac{4\pi p^2}{\exp[(p^2/2M - \mu)/k_B T] + 1} \tag{13.47}$$

and using (13.40) the function $\mu = \mu(N, T)$ is determined by

$$\frac{N}{V} = \frac{1}{\pi^2\hbar^3} \int_0^\infty dp \frac{p^2}{\exp[(p^2/2M - \mu)/k_B T] + 1} . \tag{13.48}$$

Notice that (13.48) can be formally obtained from (13.40) via the substitution

$$\frac{1}{V}\sum_i \ldots \to \frac{2s+1}{(2\pi\hbar)^3} \int d^3p \ldots . \tag{13.49}$$

Defining the number of electrons per unit volume $n_e \equiv N/V$, the quantity $\alpha \equiv -\mu/k_B T$ and the integration variable $u = p^2/2Mk_B T$, the previous equation becomes

$$n_e = \frac{(2Mk_B T)^{3/2}}{2\pi^2\hbar^3} \int_0^\infty du \frac{u^{1/2}}{e^{\alpha+u} + 1} . \tag{13.50}$$

In Fig. 13.1 we represent $f(\varepsilon) \equiv [1 + e^\alpha \exp(\varepsilon/k_B T)]^{-1}$ as a function of ε for an electron density $n_e = 6.022 \times 10^{23}\,\mathrm{cm}^{-3}$ at different temperatures.

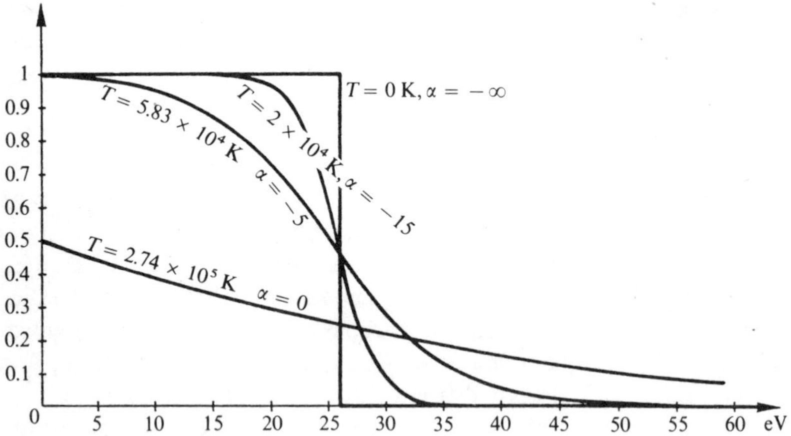

Fig. 13.1. Representation of $f(\varepsilon) \equiv [1+e^\alpha \exp(\varepsilon/k_B T)]^{-1}$ as a function of ε at different temperatures for the value of n_e given in the text

Converting the sum in (13.37) into an integral as we did in the previous csae, and using (13.38), the thermodynamic potential becomes

$$\Omega = -\frac{2V}{(2\pi\hbar)^3} k_B T \int_0^\infty dp\, 4\pi p^2 \ln\{1 + \exp[(\mu - p^2/2M)/k_B T]\} \tag{13.51}$$

and after integration by parts we obtain

$$\Omega = -\frac{Vk_B T(2Mk_B T)^{3/2}}{3\pi^2\hbar^3}\int_0^\infty du\frac{u^{3/2}}{e^{\alpha+u}+1} . \tag{13.52}$$

Following a similar procedure, the total energy of the gas is

$$E = \frac{Vk_B T(2Mk_B T)^{3/2}}{2\pi^2\hbar^3}\int_0^\infty du\frac{u^{3/2}}{e^{\alpha+u}+1} . \tag{13.53}$$

Taking into account that $\Omega = -PV$, where P is the pressure, Eqs. (13.52) and (13.53) imply

$$PV = \frac{2}{3}E , \tag{13.54}$$

valid for a free electron gas in the non-relativistic approximation. If the conditions are such that $\exp[(\mu - p^2/2M)/k_B T] \ll 1$, then the Fermi statistics coincide with Boltzmann statistics, $E = 3Nk_B T/2$ and (13.54) becomes the Clapeyron equation.

It is interesting to find out what happens in the limit $T \to 0$. In this case (13.40) requires that $\mu > 0$, and from (13.47) we obtain

$$N(p) = \begin{cases} \dfrac{2V4\pi p^2}{(2\pi\hbar)^3} , & p < p_F \\ 0 , & p > p_F \end{cases} \tag{13.55}$$

where $p_F = (2M\mu)^{1/2}$ is called the *Fermi momentum*. In other words all the states with momentum $p < p_F$ are occupied forming the *Fermi sea* and those with $p > p_F$ are empty. In this case we have a totally degenerate Fermi gas. Using (13.48) we can determine p_F as a function of n_e, and taking into account that now the region of integration is the interval $(0, p_F)$ we immediately obtain

$$p_F = (3\pi^2\hbar^3 n_e)^{1/3} . \tag{13.56}$$

Similarly, we can obtain for the energy:

$$E = \frac{V}{10\pi^2\hbar^3 M}p_F^5 = \frac{3}{10}(3\pi^2)^{2/3}\frac{\hbar^2}{M}n_e^{3/2}N \tag{13.57}$$

and using (13.54) the pressure becomes

$$P = \frac{1}{15\pi^2\hbar^3 M}p_F^5 = \frac{1}{5}(3\pi^2)^{2/3}\frac{\hbar^2}{M}n_e^{5/3} . \tag{13.58}$$

These relations are valid when $T \to 0$, but they can be approximately applied when T is such that $\mu \gg k_B T$; i.e., when

$$k_B T \ll \frac{(3\pi^2)^{2/3}}{2}\frac{\hbar^2}{M}n_e^{2/3} . \tag{13.59}$$

Since p_F is the maximum value of the electron momentum when $T \to 0$ and it can also be used to estimate the momentum for other cases, the non-relativistic approximation is applicable only if $p_F \ll Mc$; namely, when

$$n_e \ll \frac{M^3c^3}{3\pi^2\hbar^3} . \tag{13.60}$$

As an application, we study what happens in the interior of stars whose principal constituent is hydrogen. Due to the extraordinarily high temperatures in such inners, all atoms are ionized, and therefore, the number of electrons per cm^3 is of the order of $n_e = \varrho N_A$, where ϱ is the matter density expressed in $\mathrm{g\,cm^{-3}}$ and N_A is Avogadro's number. Taking into account (13.60), the non-relativistic approximation is valid if $\varrho \ll 10^6\,\mathrm{g\,cm^{-3}}$. In the sun, the interior density is $\varrho \simeq 10^2\,\mathrm{g\,cm^{-3}}$ and therefore the non-relativistic approximation is perfectly applicable, but it ceases to work for white dwarfs ($\varrho \simeq 10^6$–$10^7\,\mathrm{g\,cm^{-3}}$), and in other stars in the last stages of their evolution. On the other hand, the inequality (13.59) is equivalent to $\varrho \gg 6 \times 10^{-9}(T/\mathrm{K})^{3/2}\,\mathrm{g\,cm^{-3}}$. In the sun's interior $T \approx 10^7\,\mathrm{K}$ and the inequality is definitely not satisfied. In this particular case the electron gas can be treated using Boltzmann statistics. In white dwarfs $T \simeq 10^6\,\mathrm{K}$ and the inequality is amply satisfied. In the interior of these stars, the electron gas is degenerate, and its pressure provides the basic mechanism stabilizing them against gravitational collapse.

In cases far from either complete degeneracy or classical behavior, the value of μ as a function of N and T has to be derived by numerical integration or using other approximation methods [CH 39].

Another quantity of interest is the specific heat of an electron gas at low temperatures. To compute it, we start with (13.52), taking into account that $\alpha < 0$ and $|\alpha|$ is very large. Using the change of variables $u - |\alpha| = z$, we obtain,

$$\Omega = -\frac{Vk_BT(2Mk_BT)^{3/2}}{3\pi^2\hbar^3}\int_{-|\alpha|}^{\infty} dz \frac{(|\alpha| + z)^{3/2}}{e^z + 1} . \tag{13.61}$$

Separating the integration interval into $(-|\alpha|, 0)$ and $(0, \infty)$, changing $z \to -z$ in the first integral and taking into account $(e^{-z} + 1)^{-1} = 1 - (e^z + 1)^{-1}$, we immediately obtain

$$\Omega = -\frac{Vk_BT(2Mk_BT)^{3/2}}{3\pi^2\hbar^3}\left\{\int_0^{|\alpha|} dz\,(|\alpha| - z)^{3/2} - \int_0^{|\alpha|} dz \frac{(|\alpha| - z)^{3/2}}{e^z + 1} + \int_0^{\infty} dz \frac{(|\alpha| + z)^{3/2}}{e^z + 1}\right\} . \tag{13.62}$$

Since $|\alpha|$ is very large we can extend the second integral to infinity and hence

$$\Omega \simeq -\frac{Vk_BT(2Mk_BT)^{3/2}}{3\pi^2\hbar^3}\left\{\frac{2}{5}|\alpha|^{5/2} + \int_0^{\infty} dz \frac{(|\alpha| + z)^{3/2} - (|\alpha| - z)^{3/2}}{e^z + 1}\right\} . \tag{13.63}$$

From $|\alpha| = \mu/k_BT$, to obtain the first few terms in the expansion of Ω in powers of T, it is enough to expand the numerator of the integral in powers of z and keep only the first non-vanishing term; therefore

$$\Omega = -\frac{2^{5/2}M^{3/2}V}{15\pi^2\hbar^3}\mu^{5/2} - \frac{2^{3/2}M^{3/2}V}{\pi^2\hbar^3}\mu^{1/2}(k_B T)^2 \int_0^\infty dz \frac{z}{e^z+1} + \dots \quad (13.64)$$

The integral in (13.64) is equal to $\pi^2/12$ and the entropy becomes

$$S = -\left(\frac{\partial\Omega}{\partial T}\right)_\mu = \frac{2^{1/2}M^{3/2}V}{3\hbar^3}\mu^{1/2}k_B^2 T + \dots \,. \quad (13.65)$$

Since to 0th order in temperature, $\mu = 3^{2/3}\pi^{4/3}\hbar^2 n_e^{2/3}/2M$, we obtain for the specific heat at constant volume:

$$C_V = \left(\frac{\pi}{3}\right)^{2/3} \frac{Mk_B^2}{\hbar^2} NT \frac{1}{n_e^{2/3}} + \dots \,. \quad (13.66)$$

Therefore, the specific heat C_V for a perfect electron gas at low temperatures depends linearly on the temperature. Notice that the first term in (13.66) appears both in C_V and C_P. This is because $S \propto T$ and $C_P - C_V \propto T^3$ at low temperatures due to Nernst's theorem.

For metals at very low temperatures, the specific heat due to ions obeys a T^3 law (the Debye law) and the specific heat due to free electrons in the metal varies as T according to (13.66); consequently, for all metals at very low temperatures $C_V \propto T$, in agreement with experimental data [KI 66].

13.5 Bose Gas

As an example of a Bose gas we consider electromagnetic radiation in thermodynamic equilibrium: blackbody radiation. We will see how the application of the previous ideas yields Planck's spectral distribution, the historical starting point for quantum theory, as discussed in Chap. 1.

Since the interaction between photons is almost negligible and these particles have spin 1, the radiation of a blackbody can be considered as a perfect Bose gas. Due to the essentially non-interacting character of photons, thermal equilibrium can only be achieved in the presence of matter. The mechanism maintaining equilibrium is absorption and emission of photons by matter. For this reason, the total number of photons present, N, is a variable quantity which should be determined by requiring that the free energy, for given values of V and T, should be a minimum: $\partial F/\partial N = 0$. Since the chemical potential is given by $\mu = (\partial F/\partial N)_{V,T}$, we obtain for a photon gas $\mu = 0$, and therefore, according to (13.43), the number of photons with energy $E_i = \hbar\omega_i$ is

$$\boxed{\bar{N}_i = \{\exp[\hbar\omega_i/k_B T] - 1\}^{-1}} \,. \quad (13.67)$$

From this we immediately derive the number of photons with angular frequency between ω and $\omega + \Delta\omega$ to be $N(\omega)\Delta\omega$, where

$$N(\omega) = \frac{V}{\pi^2 c^3} \frac{\omega^2}{\exp(\hbar\omega/k_B T) - 1} . \tag{13.68}$$

and we have taken into account that the momentum is $p = \hbar\omega/c$ and that the photon has only two possible states of polarization. The energy of radiation with frequencey ν, per unit frequency and unit volume, is thus the well-known *Planck formula* (1.6)

$$u(\nu, T) = \frac{8\pi\nu^2}{c^3} k_B T \frac{h\nu/k_B T}{\exp(h\nu/k_B T) - 1} . \tag{13.69}$$

Similarly, the energy of radiation with wavelength λ, per unit wavelength and unit volume, is obtained from the previous formula simply by recalling that $\lambda = c/\nu$:

$$\tilde{u}(\lambda, T) = \frac{16\pi^2 c\hbar}{\lambda^5} \frac{1}{\exp(2\pi c\hbar/k_B T\lambda) - 1} . \tag{13.70}$$

From (13.69) we easily derive:

$$\begin{aligned} \hbar\omega \ll k_B T \,, \quad u(\nu, T) &\simeq \frac{8\pi\nu^2}{c^3} k_B T \,, \\ \hbar\omega \gg k_B T \,, \quad u(\nu, T) &\simeq \frac{8\pi\nu^2}{c^3} h\nu\, \mathrm{e}^{-h\nu/k_B T} \,, \end{aligned} \tag{13.71}$$

which are respectively the *Rayleigh-Jeans* and *Wien laws* (1.4) and (1.2).

The function $\tilde{u}(\lambda, T)$ has a maximum for $\lambda_{\max}$ determined by the equation $\partial\tilde{u}(\lambda, T)/\partial\lambda = 0$, namely

$$\frac{2\pi c\hbar}{k_B T\lambda_{\max}} = 5(1 - \exp[-2\pi c\hbar/k_B T\lambda_{\max}]) . \tag{13.72}$$

Therefore $2\pi c\hbar/k_B T\lambda_{\max} = 4.965114232\ldots$, and hence

$$\boxed{\lambda_{\max} T = 0.2897790(98)\ \mathrm{cm\ K}} \,, \tag{13.73}$$

which is the famous *Wien's displacement law*.

Since $\mu = 0$, the free energy coincides with the thermodynamic potential. Combining (13.37) and (13.42) we obtain

$$\Omega = F = V \frac{k_B T}{\pi^2 c^3} \int_0^\infty d\omega\, \omega^2 \ln\{1 - \exp[-\hbar\omega/k_B T]\} . \tag{13.74}$$

Defining $x = \hbar\omega/k_B T$ and integrating by parts:

$$\Omega = -V \frac{k_B^4 T^4}{3\pi^2 c^3 \hbar^3} \int_0^\infty dx\, x^3 \frac{1}{\mathrm{e}^x - 1} . \tag{13.75}$$

The last integral equals $\pi^4/15$, and therefore,

$$\Omega = F = -\frac{4\sigma}{3c}VT^4 , \tag{13.76}$$

where σ is the Stefan-Boltzmann constant

$$\sigma = \frac{\pi^2 k_B^4}{60\hbar^3 c^2} = 5.67032(81) \times 10^{-5}\,\mathrm{erg\,cm^{-2}\,s^{-1}\,K^{-4}} . \tag{13.77}$$

The entropy is

$$S = -\frac{\partial F}{\partial T} = \frac{16\sigma}{3c}VT^3 , \tag{13.78}$$

and since the total energy of the radiation satisfies $E = F + TS$, we obtain

$$E = \frac{4\sigma}{c}VT^4 . \tag{13.79}$$

This, together with $\Omega = -PV$, finally yields

$$PV = \tfrac{1}{3}E . \tag{13.80}$$

We now study what happens to a material body in thermal equilibrium with the surrounding radiation. The body absorbs and emits photons continuously and eventually an isotropic and stationary distribution is obtained. The energy flux of the blackbody radiation with frequencies between ν and $\nu + d\nu$ through a solid angle $d\Omega$ per unit volume is given by

$$\frac{c}{4\pi}u(\nu, T)\, d\Omega\, d\nu \tag{13.81}$$

and the incident radiation energy per unit time on the unit surface of the body, forming an angle θ with the normal, will be the previous quantity multiplied by $\cos\theta$. The absorptivity $A(\nu, \theta)$ of the body is defined in such a way that the electromagnetic energy absorbed per unit time, surface, frequency and solid angle is

$$\frac{c}{4\pi}u(\nu, T)\cos\theta A(\nu, \theta) . \tag{13.82}$$

Similarly, let $J(\nu, \theta)$ be the emissivity or electromagnetic energy emitted per unit of time, surface, frequency, and solid angle. If we assume that the reflection takes place without varying either the angle θ or the frequency ν and, furthermore, that the radiation which is not reflected is completely absorbed, then the absorbed radiation (13.82) should be compensated by the emitted radiation, and, therefore,

$$\frac{J(\nu, \theta)}{A(\nu, \theta)} = \frac{2h}{c^2}\frac{\nu^3}{\exp(h\nu/k_B T) - 1}\cos\theta . \tag{13.83}$$

This is known as Kirchhoff's law: the ratio between the emissivity and the absorptivity is a universal function. For a blackbody, by definition, $A = 1$, and the total energy emitted per unit time and unit surface is

$$J = 2\pi \int_0^\infty d\nu \int_0^{\pi/2} d\theta \sin\theta \cos\theta \frac{2h}{c^2} \frac{\nu^3}{\exp(h\nu/k_B T) - 1} \tag{13.84}$$

and hence

$$\boxed{J = \sigma T^4} \;. \tag{13.85}$$

In other words, the total energy emitted per unit time and unit surface by a blackbody is proportional to the fourth power of the absolute temperature.

If the body scatters light, it is possible to give a more restricted formulation of Kirchhoff's law and still prove the validity of (13.85) [LL 80].

13.6 Creation and Annihilation Operators

We proved previously that the state of a system with N identical particles is completely specified by the occupation numbers $(N_1, N_2, \ldots, N_i \ldots)$ corresponding to the normalized single-particle states $|1\rangle$, $|2\rangle$, ..., $|i\rangle$, ..., respectively. From now on we will represent these states by the kets $|(N_1, N_2, \ldots, N_i, \ldots)\rangle$. In the study of this system it is very useful to introduce the formalism of creation and annihilation operators, which we present first for bosons and then for fermions. To present this formalism, we need to introduce the *Fock space* $\mathcal{H}_\Phi \equiv \bigoplus_0^\infty \mathcal{H}_{(N)}$, where $\mathcal{H}_{(N)}$ ($N \geq 1$) is the Hilbert space of physically realizable pure states of a system of N identical particles, and $\mathcal{H}_{(0)}$ ($= \mathbb{C}$) is the one-dimensional space spanned by a vector $|0\rangle$ of unit norm, representing the state known as the *vacuum*, with all occupation numbers equal to zero: $|0\rangle = |(0, 0, \ldots 0, \ldots)\rangle$. We recall that $\mathcal{H}_{(N)} = \mathcal{H}_{N,\mathrm{S}}$ for bosons, and $\mathcal{H}_{(N)} = \mathcal{H}_{N,\mathrm{A}}$ for fermions.

1) Bosons. Let a_i be an operator in $\mathcal{H}_\Phi$ such that

$$\boxed{a_i|(N_1, \ldots, N_i, \ldots)\rangle = \sqrt{N_i}\,|(N_1, \ldots, N_i - 1, \ldots)\rangle} \;. \tag{13.86}$$

This operator is called an *annihilation operator*, because it annihilates a particle in the state $|i\rangle$. From (13.86) we immediately derive that $a_i|0\rangle = 0$, and that the adjoint operator $a_i^\dagger$ satisfies

$$\boxed{a_i^\dagger|(N_1, \ldots, N_i, \ldots)\rangle = \sqrt{N_i + 1}\,|(N_1, \ldots, N_i + 1, \ldots)\rangle} \tag{13.87}$$

and for this reason it is called a *creation operator*.

The operator $\mathcal{N}_i = a_i^\dagger a_i$ is called the particle *number operator* for the state $|i\rangle$ because

$$\boxed{\mathcal{N}_i|(N_1, \ldots, N_i, \ldots)\rangle = N_i|(N_1, \ldots, N_i, \ldots)\rangle} \tag{13.88}$$

as a simple consequence of (13.86) and (13.87). From these definitions we obtain the commutation relations

$$[a_i, a_j^\dagger] = \delta_{ij} \,, \quad [a_i, a_j] = [a_i^\dagger, a_j^\dagger] = 0 \tag{13.89}$$

as well as

$$\boxed{|(N_1, \ldots, N_i, \ldots)\rangle = [N_1! \ldots N_i! \ldots]^{-1/2} (a_1^\dagger)^{N_1} \ldots (a_i^\dagger)^{N_i} \ldots |0\rangle} \,. \tag{13.90}$$

These states are normalized as well as symmetrized. The symbol $|(N_1 \ldots)\rangle$ indicates precisely this symmetrization to distinguish it from

$$|N_1 N_2 \ldots\rangle \equiv \underbrace{|1\rangle|1\rangle \ldots |1\rangle}_{N_1} \; \underbrace{|2\rangle \ldots |2\rangle \ldots |2\rangle}_{N_2} \ldots \,.$$

The basis $|\psi_1\rangle = |1\rangle$, $|\psi_2\rangle = |2\rangle, \ldots$, of normalized single-particle states used previously is arbitrary, in principle. If we use a different basis $|g_1\rangle$, $|g_2\rangle, \ldots$, the new creation and annihilation operators a_{g_i}, $a_{g_i}^\dagger$ can be expressed in terms of a_{ψ_j}, $a_{\psi_j}^\dagger$, using the appropriate change of bases with the result

$$a_{g_i} = \sum_{j=1}^{\infty} \langle g_i | \psi_j \rangle a_{\psi_j} \,, \quad a_{g_i}^\dagger = \sum_{j=1}^{\infty} \langle \psi_j | g_i \rangle a_{\psi_j}^\dagger \,. \tag{13.91}$$

These equations make manifest the antilinear (linear) dependence of a ($a^\dagger$) with the state it annihilates (creates). This suggests the definition [compatible with (13.91)]

$$\boxed{a_f \equiv \sum_{j=1}^{\infty} \langle f | \psi_j \rangle a_{\psi_j} \,, \quad a_f^\dagger \equiv \sum_{j=1}^{\infty} \langle \psi_j | f \rangle a_{\psi_j}^\dagger \,,} \tag{13.92}$$

for any vector $|f\rangle$, not necessarily normalized, of the single-particle Hilbert space. From the commutation relations (13.89) and (13.92), we obtain

$$\boxed{[a_j, a_g] = 0 \,, \quad [a_f^\dagger, a_g^\dagger] = 0 \,, \quad [a_f, a_g^\dagger] = \langle f | g \rangle} \,. \tag{13.93}$$

In particular, in the basis $|\xi\rangle$, the non-trivial commutation relation is

$$[a_\xi, a_{\xi'}^\dagger] = \delta(\xi - \xi') \,. \tag{13.94}$$

A symmetrized state of N particles will take the form

$$|\Psi_N\rangle = \frac{1}{\sqrt{N!}} \int d\xi_1 \ldots d\xi_N \, \Psi_N(\xi_1, \ldots, \xi_N) a_{\xi_N}^\dagger \ldots a_{\xi_1}^\dagger |0\rangle \tag{13.95}$$

where the normalization of the totally symmetric wave function implies the normalization of the state. From (13.94) we obtain the action of the annihilation operator a_ξ on $|\Psi_N\rangle$:

$$(a_\xi \Psi_N)(\xi_1,\ldots,\xi_{N-1}) = \sqrt{N}\Psi_N(\xi_1,\ldots,\xi_{N-1},\xi) \,. \tag{13.96}$$

Consider now an interacting system of N identical bosons whose Hamiltonain can be written in the form

$$\begin{aligned} H^{(N)} &= H_0^{(N)} + H_1^{(N)} \\ H_0^{(N)} &= \sum_{i=1}^{N} F_{1(i)} \,, \quad H_1^{(N)} = \sum_{j>i=1}^{N} F_{2(ij)} \,. \end{aligned} \tag{13.97}$$

Here F_1 is a single-particle observable, and F_2 is a two-particle observable. $F_{1(i)}$ indicates that F_1 acts on the space of the particle i, and $F_{2(ij)}$ that F_2 connects the particles i and j. We now show that in this formalism the Hamiltonian $H^{(N)}$ is the restriction, to the totally symmetric subspace of the N-particle Hilbert space, of the Hamiltonian

$$\boxed{H = H_0 + H_1 = \sum_{ij} \langle i|F_1|j\rangle a_i^\dagger a_j + \frac{1}{4}\sum_{ijkl} \langle (ij)|F_2|(kl)\rangle a_j^\dagger a_i^\dagger a_l a_k} \,, \tag{13.98}$$

where the sums runs over all possible states in the basis, and

$$|(kl)\rangle \equiv a_k^\dagger a_l^\dagger |0\rangle \,, \quad \langle (ij)| \equiv \langle 0|a_i a_j \,. \tag{13.99}$$

Notice that

$$\langle (ij)|F_2|(kl)\rangle = \langle ji|F_2|kl\rangle + \langle ij|F_2|kl\rangle \,. \tag{13.100}$$

If F_2 is furhtermore a local observable in the sense that

$$\langle \xi_1', \xi_2'|F_2|\xi_2, \xi_2\rangle = F_2(\xi_1, \xi_2)\delta(\xi_1' - \xi_1)\delta(\xi_2' - \xi_2) \,, \tag{13.101}$$

then, since $F_2(\xi_1,\xi_2) = F_2(\xi_2,\xi_1)$, we have

$$\begin{aligned} \langle (ij)|F_2|(kl)\rangle = &\int d\xi_1 d\xi_2\, \psi_j^*(\xi_1)\psi_i^*(\xi_2)F_2(\xi_1,\xi_2)\psi_k(\xi_1)\psi_l(\xi_2) \\ &+ \int d\xi_1 d\xi_2\, \psi_j^*(\xi_2)\psi_i^*(\xi_1)F_2(\xi_1,\xi_2)\psi_k(\xi_1)\psi_l(\xi_2) \,. \end{aligned} \tag{13.102}$$

To prove the equivalence alluded to between (13.97) and (13.98), consider the term H_0. Let $|\Psi_n\rangle, |\Psi_N'\rangle$ be two totally symmetrized N-particle states. Then, from (13.98):

$$\langle \Psi_N'|H_0|\Psi_N\rangle = \sum_{ij} \langle i|F_1|j\rangle \langle a_i \Psi_N'|a_j \Psi_N\rangle \,. \tag{13.103}$$

On the other hand, from (13.92) we have

$$a_j = \int d\xi\, \psi_j^*(\xi) a_\xi \,, \tag{13.104}$$

where a_ξ is the annihilator operator for a particle in the state $|\xi\rangle$. Consequently, using (13.96) there results

$$\langle\Psi'_N|H_0|\Psi_N\rangle = N\sum_{ij}\int d\xi\, d\xi' d\xi_1\dots d\xi_{N-1}\,\psi_j^*(\xi)\langle i|F_1|j\rangle \\ \times\psi_i(\xi')\Psi'^*_N(\xi_1,\dots,\xi_{N-1},\xi')\Psi_N(\xi_1,\dots,\xi_{N-1},\xi)\,. \qquad (13.105)$$

Since

$$\sum_{ij}\psi_j^*(\xi)\langle i|F_1|j\rangle\psi_i(\xi') = \langle\xi'|F_1|\xi\rangle \qquad (13.106)$$

and the functions $\psi_{N'}, \psi'_N$ are symmetric in their arguments, we finally obtain

$$\langle\Psi'_N|H_0|\Psi_N\rangle = \sum_{i=1}^{N}\int d\xi_1\dots d\xi_i\dots d\xi_N d\xi'_i\,\Psi'^*_N(\xi_1,\dots,\xi'_i,\dots,\xi_N)\langle\xi'_i|F_1|\xi_i \\ \times\Psi_N(\xi_1,\dots,\xi_i,\dots,\xi_n) = \langle\Psi'_N|\sum_{i=1}^{N}F_{1(i)}|\Psi_N\rangle = \langle\Psi'_N|H_0^{(N)}|\Psi_N\rangle. \qquad (13.107)$$

An entirely similar argument can be used to prove that $H_1^{(N)}$ is the restriction of H_1 to the symmetric N-particle space. It is also clear that the relation between (13.97) and (13.98) is true regardless of the nature of the observables F_1, F_2.

Frequently, one takes the single-particle wave functions $\psi_i(\xi)$ as eigenfunctions of the single-particle Hamiltonian F_1, and if E_i is the corresponding eigenvalue, (13.98) becomes

$$H = \sum_i E_i a_i^\dagger a_i + \frac{1}{4}\sum_{ijkl}\langle(ij)|F_2|(kl)\rangle a_j^\dagger a_i^\dagger a_l a_k\,. \qquad (13.108)$$

If ther is no mutual interaction between the particles, i.e., if $F_2 \equiv 0$, then a basis of eigenstates of H is generated by vectors of the form $|(N_1, N_2, \dots,)\rangle$, with eigenvalues $E = \sum_i E_i N_i$.

The annihilation a_ξ and creation operators $^\dagger_\xi$ introduced above are usually written as $\Psi(\xi)$ and $\Psi^\dagger(\xi)$ and they are called *field operators*, coming from the *"second quantization"* of the wave function. This does not mean that a new physical quantization is necessary, but it provides a very useful mathematical artifice to extend to systems with an indefinite number of identical particles formal expressions obtained for systems with one or a few particles. From (13.92), (13.93), (13.94) we easily obtain

$$\boxed{\begin{aligned} &\Psi(\xi) = \sum_i \psi_i(\xi)a_i\,, && \Psi^\dagger(\xi) = \sum_i \psi_i^*(\xi)a_i^\dagger\,, \\ &a_i = \int d\xi\,\psi_i^*(\xi)\Psi(\xi)\,, && a_i^\dagger = \int d\xi\,\psi_i(\xi)\Psi^\dagger(\xi)\,, \\ &[\Psi(\xi),\Psi(\xi')] = [\Psi^\dagger(\xi),\Psi^\dagger(\xi')] = 0\,, && [\Psi(\xi),\Psi^\dagger(\xi')] = \delta(\xi-\xi')\,. \end{aligned}} \qquad (13.109)$$

If the particles are subject to an external potential $V_1(\xi)$, and they interact via two-body potentials $V_2(\xi_1, \xi_2)$, the Hamiltonian for a system with an undetermined number of particles is derived from (13.98) to be:

$$H = \int d\xi\, \Psi^\dagger(\xi) \left[-\frac{\hbar^2 \Delta}{2M} + V_1(\xi) \right] \Psi(\xi) + \frac{1}{2} \int d\xi\, d\xi'\, \Psi^\dagger(\xi')\Psi^\dagger(\xi) V_2(\xi, \xi')\Psi(\xi)\Psi(\xi') \,. \tag{13.110}$$

The first term is analogous to the expectation value of a single-particle operator; and a similar interpretation can be applied to the second. The factor "1/2" is due to the indistinguishability of the two configurations ξ, ξ' and ξ', ξ; and since we do not restrict the integration, both are automatically taken into account.

2) Fermions. The annihilation operator is defined as an operator satisfying

$$\boxed{a_i|(N_1, \dots, N_i \dots)\rangle = (-1)^{\nu_i} N_i |N_1, \dots, 1 - N_i, \dots)\rangle} \,, \tag{13.111}$$

where $\nu_i = \sum_{k=1}^{i-1} N_k$; notice that (13.111) is consistent for fermions, since for fermions N_i can only be either 0 or 1. Thus the creation operator is

$$\boxed{a_i^\dagger|(N_1, \dots, N_i \dots)\rangle = (-1)^{\nu_i}(1 - N_i)|(N_1, \dots, 1 - N_i, \dots)\rangle} \,. \tag{13.112}$$

As before, the vacuum $|0\rangle$ is annihilated by all the a_i's, and the number operator is $\mathcal{N}_i = a_i^\dagger a_i$. From (13.111) and (13.112) we obtain the anticommutation relations

$$\boxed{\begin{aligned} \{a_i, a_j^\dagger\} &= \delta_{ij} \,, \\ \{a_i, a_j\} &= \{a_i^\dagger, a_j^\dagger\} = 0 \,, \end{aligned}} \tag{13.113}$$

where $\{A, B\} \equiv BA \equiv [A, B]_+$. One of the implications of (13.113) is $(a_i^\dagger)^2 = 0$, reflecting the fact that it is not possible to create two fermions in the same state. From the definitions (13.111) and (13.112) we derive

$$\boxed{|(N_1, \dots, N_i \dots)\rangle = (a_1^\dagger)^{N_1} \dots (a_i^\dagger)^{N_i} \dots |0\rangle} \,. \tag{13.114}$$

These states are normalized, and in this case the parentheses in the left-hand side indicate antisymmetrization. Starting with (13.91), all the equations introduced in the study of a system of bosons continue to hold for fermions with the exception of (13.93, 94) and (13.109), where the commutators must be replaced by anticommutators; and (13.100, 102) where the "+" sign must be replaced by a "−" sign.

3) General Remarks. Using the definitions (13.86) and (13.111), we have given an irreducible realization (the Fock representation) of the (anti)commutation relations (13.89) and (13.113). These realizations are characterized by the existence of a state vector, the vacuum $|0\rangle$, annihilated by all the annihilation operators. The application to this state of creation operators generates the Hilbert space of the representation [see (13.90) and (13.114)]. This space is just the Fock space $\mathcal{H}_\Phi$.

It is natural to ask whether any irreducible representation of the rules (13.89) or (13.113) is unitarily equivalent to the Fock representation. To answer this question in the bosonic case, it is necessary to give a more rigorous definition of the operators $a_i, a_i^\dagger$, because they are unbounded, as follows from the rules (13.89). As with the discussion of the canonical commutation relations in Sect. 2.12, we can express the commutation relations in Weyl's finite form. But the difference now is that the presence of an infinite number of degrees of freedom leads to the non-existence of a unique irreducible representation. Only the requirement of a vacuum restores uniqueness. The same happens for the anticommutation relations, even though the operators $a_i, a_i^\dagger$ are now bounded.

We can give cxplicit representations that are different from the Fock representation. Starting first with the bosonic case, define

$$\begin{aligned} a_i' &= a_i - \lambda_i\,, \quad \lambda_i \in \mathbb{C}\,,\\ a_i'^\dagger &= a_i^\dagger - \lambda_i^*\,. \end{aligned} \tag{13.115}$$

These new operators obviously satisfy the commutation relations (13.89) (even in Weyl's finite form). Suppose there is an associated normalized vacuum $|0'\rangle \in \mathcal{H}_\Phi$:

$$a_i'|0'\rangle = 0\,, \quad \text{i.e.,} \quad a_i|0'\rangle = \lambda_i|0'\rangle\,. \tag{13.116}$$

In the formalism of coherent states this $|0'\rangle$ should be

$$|0'\rangle = \prod_1^\infty \exp(-|\lambda_i|^2/2)\exp(\lambda_i a_i^\dagger)|0\rangle\,. \tag{13.117}$$

However, as long as $\sum_1^\infty |\lambda_i|^2 = \infty$, the vector $|0'\rangle$ would be orthogonal to all vectors in the Fock space, and it would therefore vanish as a consequence. This is in contradiction with the condition $\langle 0'|0'\rangle = 1$. Indeed, it suffices to convince oneself that for N large enough:

$$\begin{aligned} &\langle 0|a_{i_1}\dots a_{i_M}\prod_1^N \exp(-|\lambda_i|^2/2)\exp(\lambda_i a_i^\dagger)|0\rangle\\ &= \langle 0|\left[\prod_1^N \exp(-|\lambda_i|^2/2)\exp(\lambda_i a_i^\dagger)\right](a_{i_1}+\lambda_{i_1})\dots(a_{i_M}+\lambda_{i_M})|0\rangle\\ &= \left[\prod_1^N \exp(-|\lambda_i|^2/2)\right]\lambda_{i_1}\dots\lambda_{i_M} \to 0\,, \quad \text{if } N\to\infty\,. \end{aligned} \tag{13.118}$$

In the fermionic case, it is possible to realize a finite number N of pairs $a_i, a_i^\dagger$ in a Hilbert space of finite dimension 2^N. Thus for example, for $N = 1$ we can take

$$a = \begin{pmatrix} 0 & 1 \\ 0 & 0 \end{pmatrix} , \quad a^\dagger = \begin{pmatrix} 0 & 0 \\ 1 & 0 \end{pmatrix} . \tag{13.119}$$

It is clear that the family of operators

$$\begin{aligned} a_i' &= a_i^\dagger , \\ a_i'^\dagger &= a_i , \quad 1 \leq i < \infty , \end{aligned} \tag{13.120}$$

also satisfies (13.113), but they are not unitarily equivalent to the Fock representation, for the state $|(1,1,1,\ldots)\rangle$ (which would be annihilated by all a_i', $i = 1,2,\ldots$) does not belong to $\mathcal{H}_\Phi$. (Note that $|(1,1,\ldots)\rangle$ is orthogonal to all subspaces $\mathcal{H}_{(N)}$.)

In general, it can be shown [WS 55] that the commutation and anticommutation relations for an infinite number of degrees of freedom have an infinite number of inequivalent irreducible representations, and it is the requirement of a vacuum that forces uniqueness.

13.7 Correlation Functions

We now study other features of a perfect gas of particles, bosons or fermions. The discussion is simplified by assuming the system to be enclosed in a cubic cavity of volume V. As mentioned in Sect. 13.4, the properties of the gas far from the surface of the box are independent of the boundary conditions. We choose periodic boundary conditions for the quantization of single-particle states:

$$\psi_{\boldsymbol{p}\sigma}(\boldsymbol{r}\sigma') = \frac{1}{\sqrt{V}} e^{i\boldsymbol{p}\cdot\boldsymbol{r}/\hbar} \delta_{\sigma\sigma'} , \tag{13.121}$$

where σ, σ' stand for the third components of spin, and $\boldsymbol{p} = (2\pi\hbar/L)\boldsymbol{n}$, with $\boldsymbol{n}$ a vector of integers.

We want to compute the relative probability density $G^\varrho(\boldsymbol{r}\sigma, \boldsymbol{r}'\sigma')$ that in the gas with N identical particles described by the density matrix ϱ a particle is found at $(\boldsymbol{r}, \sigma)$ and another one at $(\boldsymbol{r}', \sigma')$. The field operator is now

$$\Psi(\boldsymbol{r}, \sigma) = \sum_{\boldsymbol{p}} \frac{1}{\sqrt{V}} e^{i\boldsymbol{p}\cdot\boldsymbol{r}/\hbar} a_{\boldsymbol{p}\sigma} . \tag{13.122}$$

The commutator (or anticommutator depending on the case) of this operator with its adjoint is given by

$$[\Psi(vvv, \sigma), \Psi^\dagger(\boldsymbol{r}', \sigma')]_\mp = \delta_{\sigma\sigma'} \frac{1}{V} \sum_{\boldsymbol{p}} e^{i\boldsymbol{p}\cdot(\boldsymbol{r}-\boldsymbol{r}')/\hbar} . \tag{13.123}$$

Since [SC 57]

$$\frac{1}{V}\sum_{p} e^{\mathrm{i}p\cdot(r-r')/\hbar} = \sum_{n} \delta(r - r' + nL) , \tag{13.124}$$

and as we are only interested in what happens inside the given box and not in its "repetitions", only the term $\delta(r - r')$ in (13.124) is relevant, and we obtain for the field operators the standard commutation relations.

The quantity of interest is

$$G^{\varrho}(r\sigma, r'\sigma') \equiv \mathrm{Tr}\{\Psi^{\dagger}(r,\sigma)\Psi^{\dagger}(r',\sigma')\Psi(r',\sigma')\Psi(r,\sigma)\varrho\} \tag{13.125}$$

and it is known as the *two-particle correlation function.*

Indeed, a simple computation using (13.96) shows that for $\varrho = |\Psi_0\rangle\langle\Psi_0|$:

$$G^{(0)}(r\sigma, r'\sigma') = N(N-1)\int d\xi_1 \ldots d\xi_{N-2}\, |\Psi_0(\xi_1, \ldots, \xi_{N-2}, r'\sigma, r\sigma)|^2 \tag{13.126}$$

with $\xi \equiv r\sigma$. The number $N(N-1)$ is the number of ordered pairs in the gas. In the general case

$$\varrho = \sum_{n} \lambda_n |\Psi_n\rangle\langle\Psi_n| , \quad \lambda_n \geq 0 , \quad \sum_{n} \lambda_n = 1 , \tag{13.127}$$

and

$$G^{\varrho} = \sum_{n} \lambda_n G^{(n)} . \tag{13.128}$$

We will presently see that in some cases the right-hand side in (13.125) is only a functin of $r - r'$, the basic reason being the translational invariance of ϱ. We will then write $G^{\varrho}(r\sigma, r'\sigma') \equiv G^{\varrho}_{\sigma\sigma'}(r - r')$.

Taking (13.122) into account,

$$\begin{aligned} G^{\varrho}(r\sigma, r'\sigma') = \frac{1}{V^2}\sum_{p_1}\sum_{p_2}\sum_{p_3}\sum_{p_4} \exp\left[-\frac{\mathrm{i}}{\hbar}(p_1 - p_4)\cdot r - \frac{\mathrm{i}}{\hbar}(p_2 - p_3)\cdot r'\right] \\ \times \mathrm{Tr}\left\{a^{\dagger}_{p_1\sigma} a^{\dagger}_{p_2\sigma'} a_{p_3\sigma'} a_{p_4\sigma}\varrho\right\} . \end{aligned} \tag{13.129}$$

a) *Fermions.* For simplicity we limit the computation to a gas of spin-1/2 particles at temperature $T = 0$ in their ground state $\varrho = |\Psi_0\rangle\langle\Psi_0|$. It is obvious now that the matrix element of (13.129) vanishes unless the particles annihilated have momentum smaller than the Fermi momentum, and every particle that is annihilated is also created; thus

$$\begin{aligned} \langle\Psi_0| a^{\dagger}_{p_1\sigma} a^{\dagger}_{p_2\sigma'} a_{p_3\sigma'} a_{p_4\sigma} |\Psi_0\rangle = \theta(p_{\mathrm{F}} - |p_3|)\theta(p_{\mathrm{F}} - |p_4|) \\ \times \left[\delta_{p_1p_4}\delta_{p_2p_3} - \delta_{p_1p_3}\delta_{p_2p_4}\delta_{\sigma\sigma'}\right] . \end{aligned} \tag{13.130}$$

Replacing this quantity in (13.129) gives

$$G^{(0)}_{\sigma\sigma'}(\boldsymbol{r}-\boldsymbol{r}') = \frac{1}{V^2}\left[\left(\sum_{|\boldsymbol{p}|\le p_{\mathrm{F}}}\right)^2 - \delta_{\sigma\sigma'}\left|\sum_{|\boldsymbol{p}|\le p_{\mathrm{F}}} \mathrm{e}^{-\mathrm{i}\boldsymbol{p}\cdot(\boldsymbol{r}-\boldsymbol{r}')/\hbar}\right|^2\right] . \tag{13.131}$$

Taking the limit $V \to \infty$,

$$G^{(0)}_{\sigma\sigma'}(\boldsymbol{r}-\boldsymbol{r}') = \frac{1}{(2\pi\hbar)^6}\left[\frac{16\pi^2 p_{\mathrm{F}}^6}{9} - \delta_{\sigma\sigma'}\left|\int d^3p\, \mathrm{e}^{-\mathrm{i}\boldsymbol{p}\cdot(\boldsymbol{r}-\boldsymbol{r}')/\hbar}\theta(p_{\mathrm{F}} - |\boldsymbol{p}|)\right|^2\right] . \tag{13.132}$$

The last integral can be easily carried out in spherical coordinates. Defining $x \equiv p_{\mathrm{F}}|\boldsymbol{r}-\boldsymbol{r}'|/\hbar$ we obtain

$$G^{(0)}_{\sigma\sigma'}(\boldsymbol{r}-\boldsymbol{r}') = \frac{16\pi^2 p_{\mathrm{F}}^6}{9(2\pi\hbar)16}\left[1 - \delta_{\sigma\sigma'} 9\left(\frac{\sin x - x\cos x}{x^3}\right)^2\right] , \tag{13.133}$$

and using (13.56)

$$G^{(0)}_{\sigma\sigma'}(\boldsymbol{r}-\boldsymbol{r}') = \frac{n^2}{4}\left[1 - \delta_{\sigma\sigma'} 9\left(\frac{\sin x - x\cos x}{x^3}\right)^2\right] , \tag{13.134}$$

where n is the number of particles per unit volume. If $\sigma \neq \sigma'$, the correlation function is a constant, and there are no correlation effects. On the other hand, if $\sigma = \sigma'$, these effects appear. In Fig. 13.2 we represent this correlation function. From the figure we see that the correlation function is important for $|\boldsymbol{r}-\boldsymbol{r}'| \lesssim 3\hbar/p_{\mathrm{F}}$, and that the exclusion principle can produce large correlations for particles with identical spin quantum numbers. It is as though at short distances identical fermions in the same state of spin would repel each other, and this effective repulsion is a consequence of the antisymmetrization of the wave function. It is easy to verify using (13.134) that for $x \ge 5$ the correlation function multiplied by $4/n^2$ is always of the form $1-\varepsilon$ with $\varepsilon \le 8 \times 10^{-3}$, and ε decreasing as x increases. Similar results survive for fermions with spin higher than 1/2.

To realize the importance of this effect, we analyze qualitatively some facts related to stellar evolution [WE 70]. Consider a homogeneous spherical star of radius R composed of N totally ionized hydrogen atoms. ITs gravitational energy is $U_{\mathrm{G}} = -3GM^2/5R$, where G is Newton's constant, and M is the mass of the star. If we increase by an amount dR the radius of the star, its volume changes by $dV = 4\pi R^2 dR$ and the increment in its gravitational energy is $dU_{\mathrm{G}} = (3GM^2/5R^2)dR$. The total energy of the star, U, changes by $dU = -P\,dV + dU_{\mathrm{G}}$, where P is the pressure of the gas. If the star were in stable equilibrium, its energy would be a minimum, so that $dU = 0$ and therefore

$$3PV = \frac{3}{5}\frac{GM^2}{R} . \tag{13.135}$$

For non-relativistic particles the kinetic energy E and the pressure are related by equation (13.54), $E = 3PV/2$, and (13.135) implies that

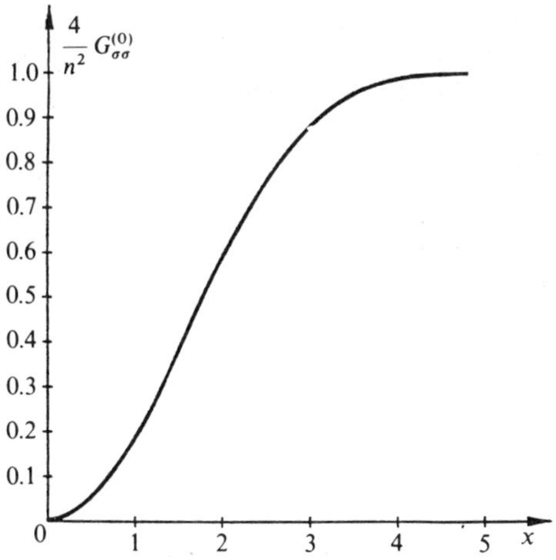

Fig. 13.2. Two-particle correlation function for spin-1/2 particles

$$E = \frac{3}{10}\frac{GM^2}{R} \,. \tag{13.136}$$

Initially the star has a large radius, the average distance between the electron (or protons) d is very large, and the characteristic correlation effects of the Fermi-Dirac distribution will be negligible. Then $E = (3/2)k_\mathrm{B}T(N+N) = 3k_\mathrm{B}TN$ as an immediate consequence of (13.50) and (13.52) in the $\alpha \to \infty$ limit. As $V \sim Nd^3$, $M \simeq N(m_\mathrm{p} + m_\mathrm{e}) \simeq Nm_\mathrm{p}$, the equality (13.136) leads to

$$k_\mathrm{B}T \sim \left(\frac{N}{N_*}\right)^{2/3}\frac{\hbar c}{d} \,, \tag{13.137}$$

where $N_* = \alpha_\mathrm{G}^{-3/2} = 2.20 \times 10^{57}$ and we have neglected numerical factors of order unity. This simply means that the kinetic energy per particle is related to the gravitational energy per particle through the virial theorem. Since according to this theorem $U = U_\mathrm{G}/2$, the gas cloud will tend to contract when it emits energy in the form of radiation; d will decrease, and its temperature will rise.

At some point in this collapse process d will become so small, that the effects of degenerate electrons will become important. This happens when $d \simeq \hbar/p_\mathrm{F}$. In this situation the kinetic energy of an electron is of the order of $\hbar^2/m_\mathrm{e}d^2$, and (13.137) must be replaced by a relation of the form

$$k_\mathrm{B}T + \frac{\hbar^2}{m_\mathrm{e}d^2} \sim \left(\frac{N}{N_*}\right)^{2/3}\frac{\hbar c}{d} \tag{13.138}$$

(the contribution of protons to this effect is still negligible). From (13.138) we immeaditealy derive that the temperature will increase until it reaches $k_\mathrm{B}T_\mathrm{max} \sim (N/N_*)^{4/3}m_\mathrm{e}c^2$. From this moment on the star will contract and cool until $T \simeq 0\,\mathrm{K}$ for $d = d_\mathrm{min} \sim (N_*/N)^{2/3}\hbar/m_\mathrm{e}c$. During the cooling from T_max to $T \simeq 0\,\mathrm{K}$, the star radiates a large amount of energy. Stars in this phase are known as *white dwarfs*.

It is obvious that our mass of gas deserves the name "star" only if $k_B T_{max}$ is big enough to produce nuclear reactions (hydrogen fusion, and after this fuel is exhausted, fusion of other elements generated by nucleosynthesis): this occurs if $k_B T_{max} \gtrsim m_e c^2/50$, and from this $N \gtrsim N_*/10 \simeq 10^{56}$. If $N < 10^{56}$, the history of the mass of gas can be described in terms of the opposing effects due to the gas pressure on the one hand and gravitation on the other. If the planet Jupiter ($N \simeq 10^{54}$) were somewhat more massive, it would actually be a star, and we would be living in a double star system Sun-Jupiter.

This picture of star evolution is very simplistic. It can be shown that if the mass of the star is of the order of one and a half times the mass of the Sun, the degeneration pressure due to the electron gas would not be enough to stop the gravitational collapse, and the star would continue to contract until it is stabilized by the degeneracy pressure of the nucleon gas, thus reaching densities of the order of nuclear matter (*neutron stars*). If the mass of the star is of the order of three or four times the mass of the Sun, not even the degeneracy pressure of the neutron gas can stop the gravitational collapse, and it is expected that the star will eventually form a *black hole*.

b) *Bosons*. We assume spin-zero bosons and that the gas is described by a pure state $|\Psi_0\rangle$ with well-defined occupation numbers N_p. For the matrix element (13.129) not to vanish, either of the following conditions should be satisfied:

$$\begin{aligned} &\text{1st} \quad \boldsymbol{p}_1 = \boldsymbol{p}_2 = \boldsymbol{p}_3 = \boldsymbol{p}_4 \;, \\ &\text{2nd} \quad \boldsymbol{p}_3 \neq \boldsymbol{p}_4 \;, \quad \boldsymbol{p}_1 = \boldsymbol{p}_3 \;, \quad \boldsymbol{p}_2 = \boldsymbol{p}_4 \;, \\ &\text{3rd} \quad \boldsymbol{p}_3 \neq \boldsymbol{p}_4 \;, \quad \boldsymbol{p}_1 = \boldsymbol{p}_4 \;, \quad \boldsymbol{p}_2 = \boldsymbol{p}_3 \;, \end{aligned}$$

and therefore, in terms of occupation numbers, we can write

$$\begin{aligned} \langle\Psi_0|a^\dagger_{p_1} a^\dagger_{p_2} a_{p_3} a_{p_4}|\Psi_0\rangle &= \delta_{p_2 p_4}\delta_{p_1 p_4}\delta_{p_2 p_4} N_{p_4}(N_{p_4}-1) \\ &+ (1-\delta_{p_3 p_4})(\delta_{p_1 p_4}\delta_{p_2 p_3} + \delta_{p_1 p_3}\delta_{p_2 p_4}) N_{p_3} N_{p_4} \;. \end{aligned} \tag{13.139}$$

Substituting this into (13.129) gives

$$\begin{aligned} G^{(0)}(\boldsymbol{r}-\boldsymbol{r}') = n^2 &+ \frac{1}{V^2}\left|\sum_{\boldsymbol{p}} \mathrm{e}^{-\mathrm{i}\boldsymbol{p}\cdot(\boldsymbol{r}-\boldsymbol{r}')/\hbar} N_{\boldsymbol{p}}\right|^2 \\ &- \frac{1}{V^2}\sum_{\boldsymbol{p}} N_{\boldsymbol{p}}(N_{\boldsymbol{p}}+1) \;, \end{aligned} \tag{13.140}$$

where n is the number of bosons per unit volume. If $N_p = N\delta_{pp_0}$, then

$$G^{(0)}(\boldsymbol{r}-\boldsymbol{r}') = \frac{N(N-1)}{V^2} \;, \tag{13.141}$$

as one would expect, since the probability density for finding a particle is N/V, and for finding a second one at the same time is $(N-1)/V$. Notice now that the correlation is negligible if $N \gg 1$. On the other hand, if for fixed V N is largeand the particles are distributed among many states so that $N_p/N \gg 1$, then the last term in (13.140) is negligible compared with the first two, and we obtain

$$G^{(0)}(\boldsymbol{r}-\boldsymbol{r}') \simeq n^2 + \frac{1}{V^2}\left|\sum_{\boldsymbol{p}} \mathrm{e}^{-\mathrm{i}\boldsymbol{p}\cdot(\boldsymbol{r}-\boldsymbol{r}')/\hbar} N_{\boldsymbol{p}}\right|^2 , \tag{13.142}$$

showing that the particles are subject to an attractive exchange interaction whose explicit form depends on the value of $N_{\boldsymbol{p}}$. If, for instance, we assume

$$N_{\boldsymbol{p}} = \frac{(2\pi\hbar)^3\alpha^{3/2}}{\pi^{3/2}2^{3/2}} n \exp\left[-\alpha(\boldsymbol{p}-\boldsymbol{p}_0)^2/2\right] \tag{13.143}$$

(normalized to represent n particles per unit volume), then we obtain for (13.142)

$$G^{(0)}(\boldsymbol{r}-\boldsymbol{r}') \simeq n^2[1+\exp(-(\boldsymbol{r}-\boldsymbol{r}')^2/\hbar^2\alpha)] \tag{13.144}$$

and there is a sharp correlation effect at short distances.

As an illustration of the importance of this effect in the case of photons, consider a pure photonic state of the form $|\Psi_0\rangle = |(N_{\boldsymbol{k}}, N_{\boldsymbol{k}'})\rangle$, where $N_{\boldsymbol{k}}, N_{\boldsymbol{k}'}$ are the number of photons present with two given momenta $\hbar\boldsymbol{k}$, $\hbar\boldsymbol{k}'$. We assume that these $N_{\boldsymbol{k}} + N_{\boldsymbol{k}'}$ photons have the same helicity, left unspecified. If we place two detectors D_1, D_2 at two points $\boldsymbol{r}, \boldsymbol{r}'$ the relative probability density P_{12} of detecting simultaneously one of these photons at D_1 and the other at D_2 is given by

$$P_{12}(\boldsymbol{r}-\boldsymbol{r}) = \frac{\langle\Psi_0|\Psi^\dagger(\boldsymbol{r})\Psi^\dagger(\boldsymbol{r}')\Psi(\boldsymbol{r}')\Psi(\boldsymbol{r})|\Psi_0\rangle}{\langle\Psi_0|\Psi_0\rangle} . \tag{13.145}$$

An elementary computations shows:

$$\begin{aligned} P_{12}(\boldsymbol{r}-\boldsymbol{r}') &= N_{\boldsymbol{k}}(N_{\boldsymbol{k}}-1) + N_{\boldsymbol{k}'}(N_{\boldsymbol{k}'}-1) \\ &\quad + 2N_{\boldsymbol{k}}N_{\boldsymbol{k}'}\{1+\cos[(\boldsymbol{k}-\boldsymbol{k}')\cdot(\boldsymbol{r}-\boldsymbol{r}')]\} . \end{aligned} \tag{13.146}$$

Similarly, the relative probability $P_1(\boldsymbol{r})$ of detecting a photon at D_1 is

$$P_1(\boldsymbol{r}) = \frac{\langle\Psi_0|\Psi^\dagger(\boldsymbol{r})\Psi(\boldsymbol{r})|\Psi_0\rangle}{\langle\Psi_0|\Psi_0\rangle} = N_{\boldsymbol{k}} + N_{\boldsymbol{k}'} , \tag{13.147}$$

and the same value is obtained for $P_2(\boldsymbol{r}')$. Assuming $N_{\boldsymbol{k}}, N_{\boldsymbol{k}'} \gg 1$, we obtain

$$P_{12}(\boldsymbol{r}-\boldsymbol{r}') = P_1(\boldsymbol{r})P_2(\boldsymbol{r}') + 2\sqrt{P_1(\boldsymbol{r})P_2(\boldsymbol{r}')}\cos[(\boldsymbol{k}-\boldsymbol{k}')\cdot(\boldsymbol{r}-\boldsymbol{r}')] . \tag{13.148}$$

The second term of the right-hand side in (13.148) reflects the bosonic character of the photons; at short distances it becomes positive, showing the tendency of bosons to stay together. This correlation for photons given in (13.148) is known as the *Handbury-Brown* and *Twiss effect* [HT 56], and it has been observed experimentally.

13.8 Superfluidity

As an application of the theory developed previously, we analyze the phenomenon of superfluidity discovered by *Kapitza* [KA 38], which becomes manifest as the absence of viscosity in ^{4}He at temperatures close to absolute zero. In 1941 *Landau* [LA 41] gave an explanation of superfluidity in terms of quantum liquids and, in 1957, *Bogolyubov* [BO 57] provided the microscopic theory of this phenomenon that we briefly explain in this section.

Atoms of ^{4}He have zero spin and are therefore bosons. Ignoring their internal structure, if $\boldsymbol{r}_i$ is the position of the ith atom, the Hamiltonian for a system with N atoms is

$$H = -\frac{\hbar^2}{2M}\sum_{i=1}^{N}\Delta_i + \sum_{i>j=1}^{N} W(|\boldsymbol{r}_i - \boldsymbol{r}_j|) , \tag{13.149}$$

where W is a weak interaction between the atoms that will not be specified. Using the single-particle wave functions of a free particle in a box of volume $V = L^3$ with periodic boundary conditions, then in the formalism of creation and annihilation operators the Hamiltonian is given by:

$$H = \sum_{\boldsymbol{p}} \varepsilon(\boldsymbol{p}) a_{\boldsymbol{p}}^\dagger a_{\boldsymbol{p}} + \frac{1}{4}\sum_{\boldsymbol{p}_1\boldsymbol{p}_2}\sum_{\boldsymbol{p}_1'\boldsymbol{p}_2'} \langle(\boldsymbol{p}_1\boldsymbol{p}_2)|W|(\boldsymbol{p}_2'\boldsymbol{p}_1')\rangle a_{\boldsymbol{p}_2}^\dagger a_{\boldsymbol{p}_1}^\dagger a_{\boldsymbol{p}_1'} a_{\boldsymbol{p}_2'} , \tag{13.150}$$

where $\varepsilon(\boldsymbol{p}) = \boldsymbol{p}^2/M$ and

$$\langle(\boldsymbol{p}_1\boldsymbol{p}_2)|W|(\boldsymbol{p}_2'\boldsymbol{p}_1')\rangle = \frac{1}{V^2}\int_V d^3r_1 d^3r_2 \,\{\exp[\mathrm{i}(\boldsymbol{p}_1' - \boldsymbol{p}_1)\cdot \boldsymbol{r}_1/\hbar]$$
$$\times \exp[\mathrm{i}(\boldsymbol{p}_2' - \boldsymbol{p}_2)\cdot \boldsymbol{r}_2/\hbar] + (\boldsymbol{p}_1 \leftrightarrow \boldsymbol{p}_2)\}\, W(|\boldsymbol{r}_1 - \boldsymbol{r}_2|) . \tag{13.151}$$

Introducing center of mass and relative coordinates $\boldsymbol{R} = (\boldsymbol{r}_1 + \boldsymbol{r}_2)/2$ and $\boldsymbol{r} = \boldsymbol{r}_1 - \boldsymbol{r}_2$, we easily obtain

$$\langle(\boldsymbol{p}_1\boldsymbol{p}_2)|W|\boldsymbol{p}_2'\boldsymbol{p}_1')\rangle = \frac{1}{V}[\hat{W}(|\boldsymbol{p}_1' - \boldsymbol{p}_1|)\delta_{\boldsymbol{p}_1+\boldsymbol{p}_2,\boldsymbol{p}_1'+\boldsymbol{p}_2'} + (\boldsymbol{p}_1 \leftrightarrow \boldsymbol{p}_2)] \tag{13.152}$$

with

$$\hat{W}(|\boldsymbol{p}_1' - \boldsymbol{p}_1|) = \frac{4\pi\hbar}{|\boldsymbol{p}_1' - \boldsymbol{p}_1|}\int_0^\infty dr\, rW(r)\sin\left[\frac{1}{\hbar}|\boldsymbol{p}_1' - \boldsymbol{p}_1|r\right]$$

as long as we assume the range of W to be much smaller than L. Thus

$$H = \sum_{\boldsymbol{p}} \varepsilon(\boldsymbol{p}) a_{\boldsymbol{p}}^\dagger a_{\boldsymbol{p}} + \frac{1}{2V}\sum_{\substack{\boldsymbol{p}_1\boldsymbol{p}_2\boldsymbol{p}_1'\boldsymbol{p}_2' \\ \boldsymbol{p}_1+\boldsymbol{p}_2=\boldsymbol{p}_1'+\boldsymbol{p}_2'}} \hat{W}(|\boldsymbol{p}_1' - \boldsymbol{p}_1|) a_{\boldsymbol{p}_2}^\dagger a_{\boldsymbol{p}_1}^\dagger a_{\boldsymbol{p}_1'} a_{\boldsymbol{p}_2'} , \tag{13.153}$$

where the conservation of momentum is a simple consequence of the discrete translational invariance of the problem.

If $W \equiv 0$, the ground state of the system at $T = 0$ would be characterized by the occupation numbers $N_p = N\delta_{0,p}$, in other words, all atoms would condense into the state of momentum $p = 0$. Since W is weak, in the ground state $N_0 (\equiv N_{p=0})$ must be close to N. We will assume from now on that we are dealing with a system with $N \to \infty$, $V \to \infty$, but with constant density N/V; N_0 close to N means that $(N - N_0)/\sqrt{N} \to 0$. In this limit, when we apply the operators a_0 and $a_0^\dagger$ to the ground state and the first excited states, we obtain basically the original states multiplied by $\sqrt{N_0}$, as far as their macroscopic properties are concerned, and therefore it is legitimate to treat a_0 and $a_0^\dagger$ as complex numbers, such that $a_0 = a_0^\dagger = \sqrt{N_0}$. Separating those terms in (13.153) containing a_0 and $a_0^\dagger$ from the rest, and using the previous arguments, we obtain

$$\begin{aligned} H = {} & \sum_p{}' \frac{p^2}{2M} a_p^\dagger a_p + \frac{N_0^2}{2V}\hat{W}(0) + \frac{N_0}{2V}\hat{W}(0)\sum_p{}' 2a_p^\dagger a_p \\ & + \frac{N_0}{2V}\sum_p{}' \hat{W}(p)\left[a_p^\dagger a_{-p}^\dagger + a_p a_{-p} + 2a_p^\dagger a_p\right] + H_1 \,, \end{aligned} \tag{13.154}$$

where the prime indicates that the term $p = 0$ is omitted, and H_1 contains at most an operator a or $a^\dagger$ corresponding to $p = 0$. It is then obvious that the order of magnitude of H_1 relative to $H - H_1$ is at most $(N - N_0)/\sqrt{N_0}$, and it can be neglected in the limit considered. On the other hand,

$$\begin{aligned} & \frac{N_0^2}{2V}\hat{W}(0) + \frac{N_0}{2V}\hat{W}(0)\sum_p{}' 2a_p^\dagger a_p \\ & = \frac{N_0}{2V}\hat{W}(0)[N_0 + 2N - 2N_0] \simeq \frac{N^2\hat{W}(0)}{2V} \end{aligned} \tag{13.155}$$

and the original Hamiltonian has been reduced in this approximation to

$$\begin{aligned} H = {} & \frac{N^2\hat{W}(0)}{2V} + \sum_p{}' \frac{p^2}{2M} a_p^\dagger a_p \\ & + \frac{N}{2V}\sum_p{}' \hat{W}(p)\left[a_p^\dagger a_{-p}^\dagger + a_p a_{-p} + 2a_p^\dagger a_p\right] \,, \end{aligned} \tag{13.156}$$

In principle it would seem that the problem has been simplified enough to make the methods of ordinary perturbation theory applicable. This unfortunately is not so. It is possible to show [BR 59] that the methods developed in Chap. 10 are not applicable in general to a system with a large number of particles, due to the enourmous number of possible intermediate states which would make the perturbative series non-convergent, even if the interactions are very weak. A useful technique in some cases – in particular in the problem at hand – is to introduce new operators via the definition

$$a_p = u(p)A_p + v(p)A_{-p}^\dagger \,, \tag{13.157}$$

where $u(p)$ and $v(p)$ are real functions of $p = |\boldsymbol{p}|$ satisfying $u^2(p) - v^2(p) = 1$, to be determined later on. Then

$$A_{\boldsymbol{p}} = u(p)a_{\boldsymbol{p}} - v(p)a^\dagger_{-\boldsymbol{p}} \tag{13.158}$$

and we immediately verify that $A_{\boldsymbol{p}}$ and $A^\dagger_{\boldsymbol{p}}$ satisfy the commutation relations for bosonic annihilation and creation operators. The states annihilated and created by them are known as *quasiparticle* states. Writing (13.156) in terms of the new operators, we have

$$\begin{aligned} H = {} & \frac{N^2\hat{W}(0)}{2V} + \sum_{\boldsymbol{p}}{}' \left[\left(\frac{\boldsymbol{p}^2}{2M} + \frac{N\hat{W}(p)}{V}\right) v^2(p) + \frac{N\hat{W}(p)}{V} v(p)u(p)\right] \\ & + \sum_{\boldsymbol{p}}{}' \left\{\frac{\boldsymbol{p}^2}{2M}[u^2(p) + v^2(p)] + \frac{N\hat{W}(p)}{V}[u(p) + v(p)]^2\right\} A^\dagger_{\boldsymbol{p}} A_{\boldsymbol{p}} \\ & + \sum_{\boldsymbol{p}}{}' \left\{\frac{\boldsymbol{p}^2}{2M} u(p)v(p) + \frac{N\hat{W}(p)}{V}[u(p) + v(p)]^2\right\} \left(A^\dagger_{\boldsymbol{p}} A^\dagger_{-\boldsymbol{p}} + A_{\boldsymbol{p}} A_{-\boldsymbol{p}}\right) . \end{aligned} \tag{13.159}$$

We choose the functions $u(p)$ and $v(p)$ so that

$$\frac{\boldsymbol{p}^2}{2M} u(p)v(p) + \frac{N\hat{W}(p)}{2V}[u(p) + v(p)]^2 = 0 . \tag{13.160}$$

For this to be satisfied, it is enough to take

$$u(p) = \frac{1}{\sqrt{1 - D^2(p)}} , \quad v(p) = \frac{D(p)}{\sqrt{1 - D^2(p)}} , \tag{13.161}$$

with

$$\begin{aligned} D(p) &= \frac{V}{N\hat{W}(p)} \left[E(p) - \frac{\boldsymbol{p}^2}{2M} - \frac{N\hat{W}(p)}{V}\right] , \\ E(p) &= \left[\frac{p^4}{4M^2} + \frac{Np^2}{MV}\hat{W}(p)\right]^{1/2} , \end{aligned} \tag{13.162}$$

and assuming that $p^2/4M + N\hat{W}(p)/V \geq 0$. With this choice we obtain

$$\begin{aligned} H &= H_0 + \sum_{\boldsymbol{p}} E(p) A^\dagger_{\boldsymbol{p}} A_{\boldsymbol{p}} , \\ H_0 &= \frac{N^2\hat{W}(0)}{2V} + \frac{1}{2}\sum_{\boldsymbol{p}}{}' \left\{E(p) - \left[\frac{\boldsymbol{p}^2}{2M} + \frac{N\hat{W}(p)}{V}\right]\right\} . \end{aligned} \tag{13.163}$$

Since H_0 is a constant quantity, we can straightforwardly write down the eigenvectors and eigenvalues of H:

$$|(\dots n(\boldsymbol{p})\dots)\rangle = \left[\prod_{\boldsymbol{p}} n(\boldsymbol{p})!\right]^{-1/2} \prod_{\boldsymbol{p}} (A_{\boldsymbol{p}}^{\dagger})^{n(\boldsymbol{p})}|0\rangle\ ,$$
$$E(\dots n(\boldsymbol{p})\dots) = H_0 + \sum_{\boldsymbol{p}} E(\boldsymbol{p})n(\boldsymbol{p})\ , \tag{13.164}$$

where $|0\rangle$ is the quasiparticle vacuum, and $n(\boldsymbol{p})$ is the number of quasiparticles created by $A_{\boldsymbol{p}}^{\dagger}$ in the state considered. (The transformation from the old operators $a, a^{\dagger}$ to the new quasiparticle operators (13.158) may be implemented unitarily as long as the interaction is sufficiently regular, and decays fast for large values of the momentum. For details concerning the Bogolyubov canonical transformations used in this case and in the BCS theory described in the next section, see for example [LA 74].

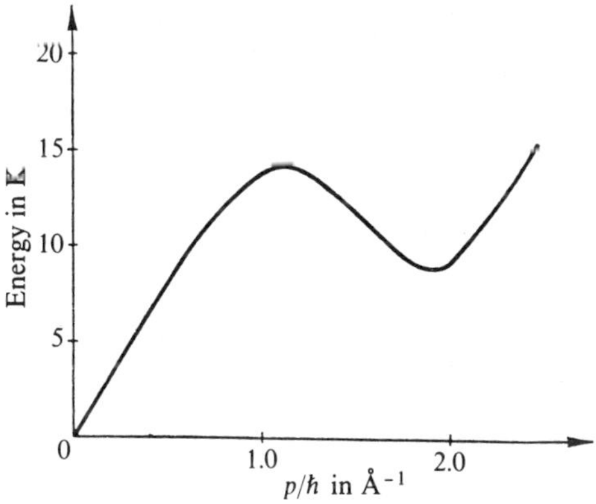

Fig. 13.3. Energy of the quasiparticles as a function of momentum

It is interesting to point out that the quasiparticles have a well-defined momentum, for the momentum operator is given by

$$\boldsymbol{P} = \sum_{\boldsymbol{p}} \boldsymbol{p} a_{\boldsymbol{p}}^{\dagger} a_{\boldsymbol{p}} = \sum_{\boldsymbol{p}} \boldsymbol{p} A_{\boldsymbol{p}}^{\dagger} A_{\boldsymbol{p}}\ , \tag{13.165}$$

as follows from (13.157). Hence we can say that the quasiparticle created by $A_{\boldsymbol{p}}^{\dagger}$ has momentum $\boldsymbol{p}$ and the equation (13.162) for $E(\boldsymbol{p})$ gives its energy as a function of the momentum. The ground state of the system is the quasiparticle vacuum, and it describes the state of the liquid at the absolute zero of temperature. For small momenta,

$$E(p) \sim (\hat{W}(0)N/MV)^{1/2} p\ , \quad p \to 0\ . \tag{13.166}$$

From this we obtain the speed of sound at absolute zero $v_0 = [\hat{W}(0)N/MV]^{1/2}$. On the other hand, for large momenta, $\hat{W}(p) \to 0$, and we have

$$E(p) = \frac{p^2}{2M} + \frac{N}{V}\hat{W}(p) + \dots\ , \tag{13.167}$$

namely, the quasiparticle energies coincide with those of particles. The detailed structure of $E(p)$ can be computed theoretically [FE 69], [FC 56] and measured experimentally by analyzing the inelastic scattering of slow neutrons by the ^{4}He atoms of the liquid. Its behavior is given in Fig. 13.3. One should notice that the spectrum satisfies

$$\min \frac{E(p)}{p} \equiv v_{cr} > 0 . \tag{13.168}$$

Geometrically $E(p)/p$ is the slope of the straight line joining the origin with the point of abscissa $p \neq 0$ in the curve $E(p)$ in the plane (p, E). The minimum obviously occurs when this line is tangential to the curve, and from Fig. 13.3 we easily see that $v_{cr} \neq 0$.

We now want to show that it is precisely this fact which explains the superfluidity of ^{4}He. Consider liquid ^{4}He flowing at aboluste zero along a capillary with constant velocity $\boldsymbol{V}$. If the liquid has viscosity, part of its kinetic energy will be converted into internal energy. To analyze the generation of this increase in internal energy, it is convenient to choose a coordinate system moving with the liquid. In this system, the helium is initially at rest, and the capillary is moving with velocity $-\boldsymbol{V}$. When there is viscosity, the helium must begin to move, and for this to occur, it is necessary that first some elementary excitations appear altering the internal state of the liquid. Assume that an excitation with momentum $\boldsymbol{p}$ and energy $E(p)$ appears. Then the energy of the liquid in the reference frame where it is initially at rest is $E(p)$, and it has momentum $\boldsymbol{p}$. Going back to the frame where the capillary is at rest, the energy E' and momentum $\boldsymbol{p}'$ of the liquid in these coordinates are

$$E' = E(p) + \boldsymbol{p} \cdot \boldsymbol{V} + \frac{\bar{M}\boldsymbol{V}^2}{2} , \quad \boldsymbol{p}' = \boldsymbol{p} + \bar{M}\boldsymbol{V} , \tag{13.169}$$

where $\bar{M}$ is the mass of the liquid. The term $\bar{M}\boldsymbol{V}^2/2$ is the original kinetic energy of the moving liquid and therefore the energy of the excitation is $E(p) + \boldsymbol{p} \cdot \boldsymbol{V}$. For the liquid's velocity to diminish as a result of viscosity, it is necessary that this excitation energy be negative, i.e. $E(p) + \boldsymbol{p} \cdot \boldsymbol{V} < 0$. For a given value of $\boldsymbol{p}$, the minimum of $E(p) + \boldsymbol{p} \cdot \boldsymbol{V}$ takes place when $\boldsymbol{p}$ and $\boldsymbol{V}$ are antiparallel, and therefore the following should hold:

$$E(p) - p|\boldsymbol{V}| < 0 , \tag{13.170}$$

namely $|\boldsymbol{V}| > E(p)/p$; hence, as long as $|\boldsymbol{V}| < v_{cr}$, we will have superfluidity. Since the liquid cannot have excitations, there is no viscosity, and it flows without friction. It can be shown that for liquid helium, superfluidity appears at temperatures $T < 2.18\,\mathrm{K}$ [LL 80], [BO 67]. In these references one can find a more thorough discussion of the topics presented in this section.

13.9 Superconductivity

As a final example, we consider superconductivity. This phenomenon was discovered by *Kamerling-Onnes* [KA 11] in 1911. He observed that when mercury was cooled below a critical temperature $T_c = 4.15\,\mathrm{K}$, there was a sharp drop in its resistivity, and that for all $T < T_c$ is value remained compatible with zero. Today we know a large number of superconducting substances, and for all of them, there is a critical temperature below which the resistivity is negligible. One of the most amazing phenomena related to superconducting substances is the possibility of generating persistent currents, also discovered by *Kammerline-Onnes*. This effect can be described as follows: consider a ring of superconducting material at $T > T_c$ in the presence of a magnetic field $\boldsymbol{B}$ orthogonal to the plane of the ring. Next we cool the conductor until it reaches a temperature $T < T_c$, and then we turn off the magnetic field. This creates a current in the ring, whose flux equals the previously existing one. Since the resistivity is null, this current will persist, i.e., its intensity will not decrease with time.

For years the search for an appropriate explanation of superconductivity attracted many physicists. As high points in the development of the theory, we should mention the works of *Gorter* and *Casimir* [GC 34], *London* and *London* [LL 35], and *Pippard* [PI 53]. All of them gave macroscopic theories of this phenomenon. The first one to realize that superconductivity in a metal is related to the interactions between electrons and *phonons* (the quanta associated to the vibrations of the crystal lattice) was *Fröhlich* [FR 50]. This discovery kindled the attention for a microscopic explanation of superconductivity, which culminated in the first satisfactory theory, due to *Bardeen, Cooper* and *Schrieffer* [BC 57] and known today as the BCS theory. We will only sketch the important features of this theory, following the method developed by *Bogolyubov* [BO 58]. The starting point is the following Hamiltonian for a gas of electrons in a volume V of a conductor:

$$\begin{aligned} H &= H_0 + H_1 \,, \\ H_0 &= \sum_{p,\sigma} \varepsilon(p) a^\dagger(\boldsymbol{p}\sigma) a(\boldsymbol{p}\sigma) \,, \\ H_1 &= -\frac{1}{2V} \sum_{pp'\sigma} V_{p,p'} a^\dagger(\boldsymbol{p}\sigma) a^\dagger(-\boldsymbol{p}-\sigma) a(-\boldsymbol{p}'-\sigma) a(\boldsymbol{p}'\sigma) \,, \end{aligned} \tag{13.171}$$

where $a^\dagger(\boldsymbol{p}\sigma)$ is the operator creating an electron of momentum $\boldsymbol{p}$ and third component of spin σ. Furthermore, $\varepsilon(p) = p^2/2M - \mu$, where $M \equiv m_e$ and μ is the chemical potential determined by the condition $\langle \sum_{p\sigma} a^\dagger(\boldsymbol{p}\sigma) a(\boldsymbol{p}\sigma) \rangle = N$, N being the number of particles in the system. On the other hand, $V_{p,p'}$ is the potential between the electrons due to the exchange of phonons and it satisfies $V_{p,p'} = V_{p',p}$.

The justification of this Hamiltonian is discussed in [BC 57]. To find the energy spectrum of this Hamiltonian it is convenient to introduce new operators $A(\boldsymbol{p})$ and $B(\boldsymbol{p})$ defined by

$$
\begin{aligned}
a(\boldsymbol{p}1/2) &= u(\boldsymbol{p})A(\boldsymbol{p}) + v(\boldsymbol{p})B^\dagger(\boldsymbol{p}) \ , \\
a(-\boldsymbol{p}-1/2) &= u(\boldsymbol{p})B(\boldsymbol{p}) - v(\boldsymbol{p})A^\dagger(\boldsymbol{p}) \ ,
\end{aligned}
\tag{13.172}
$$

where we assume that $u(\boldsymbol{p})$ and $v(\boldsymbol{p})$ are real functions satisfying $u(\boldsymbol{p}) = u(-\boldsymbol{p})$, $v(\boldsymbol{p}) = v(-\boldsymbol{p})$ and $u^2(\boldsymbol{p}) + v^2(\boldsymbol{p}) = 1$. Under these conditions the transformation (13.172) is canonical; i.e. $A(\boldsymbol{p})$ and $B(\boldsymbol{p})$, given by

$$
\begin{aligned}
A(\boldsymbol{p}) &= u(\boldsymbol{p})a(\boldsymbol{p}1/2) - v(\boldsymbol{p})a^\dagger(-\boldsymbol{p}-1/2) \ , \\
B(\boldsymbol{p}) &= u(\boldsymbol{p})a(-\boldsymbol{p}-1/2) + v(\boldsymbol{p})a^\dagger(\boldsymbol{p}1/2) \ ,
\end{aligned}
\tag{13.173}
$$

satisfy the anticommutation relations

$$
\{A(\boldsymbol{p}), A^\dagger(\boldsymbol{p}')\} = \delta_{\boldsymbol{p}\boldsymbol{p}'} \ , \quad \{B(\boldsymbol{p}), B^\dagger(\boldsymbol{p}')\} = \delta_{\boldsymbol{p}\boldsymbol{p}'}
\tag{13.174}
$$

and all other anticommutators vanish. Next we substitute (13.172) in (13.171) and use the anticommutation relations to move all the annihilation operators to the right. A simple computation yields

$$
\begin{aligned}
H &= E_0 + H_0' + H_1' + H_2' \ , \\
E_0 &= 2\sum_{\boldsymbol{p}} \varepsilon(p)v^2(\boldsymbol{p}) - \frac{1}{V}\sum_{\boldsymbol{p}\boldsymbol{p}'} V_{\boldsymbol{p},\boldsymbol{p}'}u(\boldsymbol{p})v(\boldsymbol{p})u(\boldsymbol{p}')v(\boldsymbol{p}') \ , \\
H_0' &= \sum_{\boldsymbol{p}} \left\{ \varepsilon(p)[u^2(\boldsymbol{p}) - v^2(\boldsymbol{p})] + \frac{2u(\boldsymbol{p})v(\boldsymbol{p})}{V}\sum_{\boldsymbol{p}'} V_{\boldsymbol{p},\boldsymbol{p}'}u(\boldsymbol{p}')v(\boldsymbol{p}') \right\} \\
&\quad \times [A^\dagger(\boldsymbol{p})A(\boldsymbol{p}) + B^\dagger(\boldsymbol{p})B(\boldsymbol{p})] \ , \\
H_1' &= \sum_{\boldsymbol{p}} \left\{ 2\varepsilon(p)u(\boldsymbol{p})v(\boldsymbol{p}) - \frac{u^2(\boldsymbol{p}) - v^2(\boldsymbol{p})}{V}\sum_{\boldsymbol{p}'} V_{\boldsymbol{p},\boldsymbol{p}'}u(\boldsymbol{p}')v(\boldsymbol{p}') \right\} \\
&\quad \times [A^\dagger(\boldsymbol{p})B^\dagger(\boldsymbol{p}) + B(\boldsymbol{p})A(\boldsymbol{p})] \ ,
\end{aligned}
\tag{13.175}
$$

and H_2' contains all those terms with three or four creation and annihilation operators; in the computation of the lowest energy states its contribution is negligible, since it requires the excitation of two quasiparticles [BO 58]. Next we fix $u(\boldsymbol{p})$ and $v(\boldsymbol{p})$ so that $H_1' \equiv 0$. This is achieved if

$$
2\varepsilon(p)u(\boldsymbol{p})v(\boldsymbol{p}) = \frac{u^2(\boldsymbol{p}) - v^2(\boldsymbol{p})}{V}\sum_{\boldsymbol{p}'} V_{\boldsymbol{p},\boldsymbol{p}'}u(\boldsymbol{p}')v(\boldsymbol{p}') \ .
\tag{13.176}
$$

Taking

$$
\begin{aligned}
u(\boldsymbol{p}) &= \frac{1}{\sqrt{2}}\left[1 + \frac{\varepsilon(p)}{\sqrt{\Delta^2(\boldsymbol{p}) + \varepsilon^2(p)}}\right]^{1/2} \ , \\
v(\boldsymbol{p}) &= \frac{1}{\sqrt{2}}\left[1 - \frac{\varepsilon(p)}{\sqrt{\Delta^2(\boldsymbol{p}) + \varepsilon^2(p)}}\right]^{1/2}
\end{aligned}
\tag{13.177}
$$

the condition (13.176) implies that $\Delta(\boldsymbol{p})$ is determined by the integral equation

$$\Delta(\boldsymbol{p}) = \frac{1}{2V} \sum_{\boldsymbol{p}'} \frac{V_{\boldsymbol{p},\boldsymbol{p}'} \Delta(\boldsymbol{p}')}{\sqrt{\Delta^2(\boldsymbol{p}') + \varepsilon^2(p')}} . \tag{13.178}$$

The Hamiltonian of the problem is reduced to

$$\begin{aligned} H &= E_0 + H_0' , \\ E_0 &= \sum_{\boldsymbol{p}} \frac{1}{\sqrt{\Delta^2(\boldsymbol{p}) + \varepsilon^2(p)}} \Big\{ \varepsilon(p) \left[\sqrt{\Delta^2(\boldsymbol{p}) + \varepsilon^2(p)} - \varepsilon(p) \right] \\ &\quad - \tfrac{1}{2} \Delta^2(\boldsymbol{p}) \Big\} , \\ H_0' &= \sum_{\boldsymbol{p}} \sqrt{\Delta^2(\boldsymbol{p}) + \varepsilon(p)} [A^\dagger(\boldsymbol{p}) A(\boldsymbol{p}) + B^\dagger(\boldsymbol{p}) B(\boldsymbol{p})] , \end{aligned} \tag{13.179}$$

where $\Delta(\boldsymbol{p})$ is the solution of (13.178).

Since we are only trying to give an idea of the method, suppose $T = 0$, and write $\mu \equiv p_0^2/2M$. Notice that all energies are referred to the chemical potential as origin, due to our choice of $\varepsilon(p)$. We will also adopt the simplifying hypothesis

$$V_{\boldsymbol{p},\boldsymbol{p}'} = \begin{cases} C & \text{if } |\varepsilon(p)| \, , \, |\varepsilon(p')| \le \hbar\omega_c \\ 0 & \text{otherwise} \end{cases} , \tag{13.180}$$

where $\hbar\omega_c \ll p_0^2/2M$; i.e., the interaction only takes place between electrons with momenta in a thin shell around the momentum p_0. It is consistent with (13.178) and (13.179) to take

$$\Delta(\boldsymbol{p}) = \begin{cases} \Delta & \text{if } |\varepsilon(p)| \le \hbar\omega_c \\ 0 & \text{in all other cases} \end{cases} , \tag{13.181}$$

and in the limit as V goes to infinity, the constant Δ is determined by the equation

$$\Delta = \Delta \frac{C}{2(2\pi\hbar)^3} \int d^3p \frac{1}{\sqrt{\Delta^2 + \varepsilon^2(p)}} , \tag{13.182}$$

where the integration runs over all those values of $\boldsymbol{p}$ satisfying $|\varepsilon(p)| \le \hbar\omega_c$; we thus obtain

$$\Delta = \Delta \frac{C}{2\pi^2\hbar^3} \int_{\sqrt{p_0^2 - 2M\hbar\omega_c}}^{\sqrt{p_0^2 + 2M\hbar\omega_c}} dp\, p^2 \left[\Delta^2 + \frac{1}{4M^2}(p^2 - p_0^2)^2 \right]^{-1/2} . \tag{13.183}$$

Introducing the change of variables $p = p_0 + x$ and taking into account that $p_0^2 \gg 2M\hbar\omega_c$:

$$\Delta \simeq \Delta \frac{Cp_0^2}{4\pi^2\hbar^3} \int_{-M\hbar\omega_c/p_0}^{M\hbar\omega_c/p_0} dx \left[\Delta^2 + \frac{p_0^2}{M^2} x^2 \right]^{-1/2} , \tag{13.184}$$

and therefore

$$\Delta \simeq \Delta \frac{CMp_0}{4\pi^2\hbar^3} \ln \frac{\sqrt{\Delta^2 + \hbar^2\omega_c^2} + \hbar\omega_c}{\sqrt{\Delta^2 + \hbar^2\omega_c^2} - \hbar\omega_c} . \tag{13.185}$$

Once we find the solution to the previous equation, the energy spectrum becomes

$$E(N_A(\boldsymbol{p}), N_B(\boldsymbol{p})) = E_0 + \sum_{p} \sqrt{\varepsilon^2(p) + \Delta^2(\boldsymbol{p})}[N_A(\boldsymbol{p}) + N_B(\boldsymbol{p})] , \tag{13.186}$$

where E_0 is given by (13.179), and $N_A(\boldsymbol{p})$ and $N_B(\boldsymbol{p})$ are the quasiparticle numbers of each type.

We now analyze the possible solutions of (13.185). We always have the normal solution $\Delta \equiv 0$. The ground state corresponds to the absence of quasi-particles, and its energy is $E_0 = 0$. Moreover $p_0 = p_{\mathrm{F}}$, and $u(\boldsymbol{p})$ and $v(\boldsymbol{p})$ can be chosen as

$$\begin{aligned} u(\boldsymbol{p}) &= 1 , \quad v(\boldsymbol{p}) = 0 , \quad \text{if} \quad \varepsilon(p) > 0 , \\ u(\boldsymbol{p}) &= 1 , \quad v(\boldsymbol{p}) = 1 , \quad \text{if} \quad \varepsilon(p) < 0 , \end{aligned} \tag{13.187}$$

where $\varepsilon(p) \gtrless 0$ implies $|\boldsymbol{p}| \gtrless p_{\mathrm{F}}$. If $|\boldsymbol{p}| > p_{\mathrm{F}}$, then $A(\boldsymbol{p}) = a(\boldsymbol{p}1/2)$ and $B(\boldsymbol{p}) = a(-\boldsymbol{p} - 1/2)$ and therefore they annihilate fermions outside the Fermi sea. Otherwise, if $|\boldsymbol{p}| < p_{\mathrm{F}}$ then $A(\boldsymbol{p}) = -a^\dagger(-\boldsymbol{p} - 1/2)$ and $B(\boldsymbol{p}) = a^\dagger(\boldsymbol{p}1/2)$, i.e., they create fermions inside the Fermi sea, or, in other words, they annihilate holes in the sea. In this case we have that the lowest energy state is $E(0, 0) = 0$ with all the states $|\boldsymbol{p}| < p_{\mathrm{F}}$ occupied and all states with $|\boldsymbol{p}| > p_{\mathrm{F}}$ empty. In the excited states one or more electrons from the sea occupy states with $|\boldsymbol{p}| > p_{\mathrm{F}}$. If the system is sufficiently large, starting from 0 the spectrum of excited states is practically continuous.

If C is positive and sufficiently large, or equivalently, if there is a strongly attractive interaction between fermions with opposite spin and opposite momenta close to the surface of the Fermi sea (we say then that the electrons form a *Cooper pair*), there is a second solution to (13.185) given by

$$\Delta = 2\hbar\omega_c \frac{\mathrm{e}^{-D/C}}{1 - \mathrm{e}^{-2D/C}} , \qquad D \equiv \frac{2\pi^2\hbar^3}{Mp_0} \tag{13.188}$$

and known as the superconducting solution. It is worth stressing that this solution is not analytic at $C = 0$, and cannot be obtained by means of perturbative methods. This explains in part the difficulty in arriving at the BCS theory. Now the state of lowest energy is $E(0, 0) = E_0 < 0$, and it is therefore the ground state of the system. As a consequence of keeping fixed the number of electrons, it is possible to show that between the ground state and the first excited state there is a forbidden energy gap 2Δ. Indeed:

$$A^\dagger(\boldsymbol{p})A(\boldsymbol{p}) - B^\dagger(\boldsymbol{p})B(\boldsymbol{p}) = a^\dagger(\boldsymbol{p}1/2)a(\boldsymbol{p}1/2) - a^\dagger(-\boldsymbol{p} - 1/2)a(-\boldsymbol{p} - 1/2) . \tag{13.189}$$

In the ground state $\langle a^\dagger(\boldsymbol{p}1/2)a(\boldsymbol{p}1/2)\rangle = \langle a^\dagger(-\boldsymbol{p} - 1/2)a(-\boldsymbol{p} - 1/2)\rangle$, in other words, the electrons appear in paired states. If we want to break one of these pairs, we automatically obtain two unpaired electrons and $N_A + N_B = 2$, so that the first excited state has an energy 2Δ with respect to the ground state. The existence of this forbidden band gives an explanation to the various physical phenomena related to superconductivity. The reader interested in more details is referred to [LY 62], [RI 65] and [SC 64], as well as to the fundamental original papers collected in [BO 68].

Before concluding this section, we would like to point out that an entirely similar phenomenon of strong pairing between nucleons explains the large energetic distance (in energy) between the ground state and the first excited states in even-even nuclei (nuclei with Z and N even).

Since 1986 the situation has drmatically changed with the discovery of high-T_c superconductivity ($T_c \sim 40$–125 K), for which no satisfactory theory exists at present.

14. Atoms

14.1 Introduction

In previous chapters we described hydrogen-like atoms, and we have also considered the helium atom. In this chapter we will give some ideas about the problems encountered in trying to study arbitrary atoms. Due to the complexity of this problem, we will not present a detailed account, and refer the interested reader to the excellent book of *Condon* and *Shortley* [CS 57] and to the books by *Bethe* and *Jackiw* [BJ 68], *Kuhn* [KU 62] and *Slater* [SL 60], among others. For a rigorous analysis of these systems we recommend [TH 81], [VW 81]. So far we have been using arbitrary units. In order to simplify the form of the equations in this chapter and the next we will adopt *natural units* where $\hbar = c = 1$. In this system, any physical quantity can be expressed in powers of length, mass, or energy.

We consider an atom consisting of a point-like nucleus of charge $Z|e|$ surrounded by N electrons; the approximate Hamiltonian is

$$\boxed{H = -\frac{1}{2M}\Delta_{s} - \frac{1}{2m_{\mathrm{e}}}\sum_{i=1}^{N}\Delta_{s_i} - \sum_{i=1}^{N}\frac{Ze^2}{|s_i - s|} + \sum_{1\leq i<j}^{N}\frac{e^2}{|s_i - s_j|}}\ , \tag{14.1}$$

where m_{e} and M are respectively the masses of the electron and the nucleus, and $\boldsymbol{s}, \boldsymbol{s}_1, \boldsymbol{s}_2, \dots \boldsymbol{s}_N$ are the coordinates of the nucleus and the electrons. Instead of the $3(N+1)$ coordinates introduced, we can use center of mass and relative coordinates defined by

$$\boldsymbol{R} = \frac{M\boldsymbol{s} + \sum m_{\mathrm{e}}\boldsymbol{s}_i}{M + Nm_{\mathrm{e}}}\ , \qquad \boldsymbol{r}_i = \boldsymbol{s}_i - \boldsymbol{s}\ , \tag{14.2}$$

and the Hamiltonian becomes

$$H = -\frac{1}{2(M+Nm_{\mathrm{e}})}\Delta - \frac{1}{2\mu}\sum_{i=1}^{N}\boldsymbol{\Delta}_i - \frac{1}{M}\sum_{1\leq i<j}^{N}\boldsymbol{\nabla}_i\cdot\boldsymbol{\nabla}_j$$
$$-\sum_{i=1}^{N}\frac{Ze^2}{r_i} + \sum_{1\leq i<j}^{N}\frac{e^2}{r_{ij}}\ , \tag{14.3}$$

where $\Delta \equiv \Delta_R$, $\boldsymbol{\nabla}_i \equiv \boldsymbol{\nabla}_{r_i}$, $r_i = |\boldsymbol{r}_i|$, $r_{ij} = |\boldsymbol{r}_i - \boldsymbol{r}_j|$ and $\mu = m_{\mathrm{e}}M/(M+m_{\mathrm{e}})$ is the reduced mass of the electron with respect to the nucleus. Separating as usual the motion of the center of mass, the Hamiltonian describing the relative motion (which we will denote with the same symbol) is

$$H = -\frac{1}{2\mu}\sum_{i=1}^{N}\Delta_i - \sum_{i=1}^{N}\frac{Ze^2}{r_i} + \sum_{1\leq i<j}^{N}\frac{e^2}{r_{ij}} - \frac{1}{M}\sum_{1\leq i<j}^{N}\nabla_i\cdot\nabla_j \,. \tag{14.4}$$

Since $M \gg \mu$, the last term (known as the Hughes-Eckart term) can be treated in perturbation theory, and it is possible to show [CS 57], [KU 62] that its effect in physically interesting situations is very small; thus we will generally neglect it.

It is convenient to take into account that the last term in (14.4), the appearance of the reduced mass in the first, and the terms due to the finite distribution of charge in the nucleus – neglected in (14.4) – produce a shift in the atomic energy levels known as the isotopic shift. To observe it, a mixture of different isotopes of the same chemical element are used; since the mass of the nucleus, the reduced mass of the electron and the nuclear charge distribution are different for each isotope, there will be a splitting of the level, which in principle permits an experimental study of the isotopic shift [KU 62]. This effect is easier to observe in atoms with vanishing nuclear spin; otherwise the splitting of levels produced by the isotopic shift would be superimposed to the hyperfine splitting due to the interaction of the nuclear magnetic moment with the magnetic field produced by the electrons, which is of the same order of magnitude.

From a mathematical point of view, it can be shown [KA 80], [RS 75] that the potential term in the Hamiltonian H is H_0-infinitesimal, where H_0 is the kinetic term of H. Hence H is self-adjoint in the domain of H_0. With regard to the reduced Hamiltonian (14.4), essentially self-adjoint in $C_0^\infty(\mathbb{R}^{3N})$, the celebrated theorem of Hunziker-Van Winter-Zhislin [HU 66] [RS 78] (see Sect. 8.22) guarantees that the essential spectrum of H in (14.4) is $[\Sigma, \infty)$, where the threshold value Σ represents the lowest energy of the system "nucleus + $(N-1)$ electrons". It is also known that H does not have a continuous singular spectrum [SI 72]. With respect to the discrete spectrum of H, it can be shown that it contains infinitely many levels (with Σ as an accumulation point) as long as $N = Z(\neq 0)$ (neutral atoms), or $0 < N < Z$ (positively charged ions). Physically this is not very surprising, for $N-1$ electrons do not completely screen the nuclear charge, and on the last electron acts still at large distances a long range attractive potential. For $N > Z$ this argument does not apply, and *Lieb* [LI 84] has elegantly shown that if we neglect the Hughes-Eckart term, the maximum number N_c of particles of charge $-|e|$ that can be bound to a nucleus of charge $Z|e|$ necessarily satisfies the inequality $N_c < 2Z + 1$, independently of their masses and statistics (fermions and/or bosons). Possibly this is not the best bound for identical fermions; in this case it is rigorously known that $N_c/Z \sim 1$, $Z \to \infty$, and $N_c - Z \simeq 1$ is suspected. In any case, Lieb's inequality is enough to show that the ion H^{--} does not exist.

When in (14.4) we neglect the Hughes-Eckart term and the electrostatic repulsion between electrons, there may appear point-spectrum levels in $(\Sigma, 0)$ superimposed on the absolutely continuous spectrum. For example, in the case of He without the terms mentioned, σ_p is $E_{n,m} = -\mu(Z\alpha)^2(n^{-2} + m^{-2})/2$, $n, m \geq 1$ and $E_{2,2}$ is the lowest level lying on the continuum. However the electrostatic repulsion induces a self-ionization process and some of the eigenstates corresponding to such levels decay (Auger effect) [SI 73].

It is evident that the exact solution to the stationary Schrödinger equation with the Hamiltonian (14.4), even without the last term, is in general impossible, and it will be necessary to turn to approximation methods. For large atoms, it is to be expected that the ith electrom moves approximately under the influence of an average potential $V_i(\boldsymbol{r}_i)$ due to the nucleus and the other electrons. The problem is therefore to find this average potential and the associated eigenfunctions. Once this problem is solved we treat the interactions neglected in the average potential as perturbations. Mathematically, the method proposed consists of writing the Hamiltonian as

$$\begin{aligned} H &= H_0 + H_1 , \\ H_0 &= -\frac{1}{2\mu}\sum_{i=1}^{N} \Delta_i + \sum_{i=1}^{N} V_i(\boldsymbol{r}_i) , \\ H_1 &= -\sum_{i=1}^{N} \frac{Ze^2}{r_i} + \sum_{1\leq i<j}^{N} \frac{e^2}{r_{ij}} - \sum_{i=1}^{N} V_i(\boldsymbol{r}_i) , \end{aligned} \tag{14.5}$$

where the average potentials (generally with spherical symmetry) V_i will be estimated via physical considerations to give the most significative contribution to the interaction, and once the spectral problem for H_0 is solved, we treat H_1 as a perturbation.

We saw in Sect. 12.7 how the relativistic effects (corrections to the kinetic energy, spin-orbit coupling and Darwin term) accounted for the fine structure of the hydrogen-like atoms, as a consequence of the relativistic formulation of the problem. Even though such a formulation is lacking for complex atoms, it is, however, possible to include these corrections approximately [SL 60] and in so doing one obtains a very complicated Hamiltonian. For the sake of simplicity, we will only include the spin-orbit couplings of each electron. For complex atoms these are generally the leading corrections. Therefore, we take for the Hamiltonian the expression

$$H = H_0 + H_1 + H_2 + H_3 , \tag{14.6}$$

where the "free" Hamiltonian is

$$\boxed{H_0 = \sum_{i=1}^{N} \left[-\frac{1}{2\mu}\Delta_i + V_i(r_i) \right]} \tag{14.6}$$

and the remaining terms are

$$\boxed{H_1 = \sum_{i=1}^{N} \left[-\frac{Ze^2}{r_i} - V_i(r_i) \right]} \,, \tag{14.8}$$

$$\boxed{H_2 = \sum_{i=1}^{N} \xi_i(r_i) \boldsymbol{L}_i \cdot \boldsymbol{S}_i \,, \quad \xi(r) \equiv \frac{1}{2m_\mathrm{e}^2} \frac{1}{r} \frac{dV(r)}{dr}} \,, \tag{14.9}$$

$$\boxed{H_3 = \sum_{i>j=1}^{N} \frac{e^2}{r_{ij}}} \,. \tag{14.10}$$

After posing the problem in this simplified form we begin by giving an idea of how to solve it.

14.2 The Thomas-Fermi Method

Using the fact that electrons obey the Fermi-Dirac statistics, *Thomas* [TH 26a] and *Fermi* [FE 27] developed a simple method to compute the approximate form of the average potential $V(r)$ acting on the atomic electrons. As we will see later, this model is also useful as the starting point of more sophisticated calculations.

Consider a system of N electrons confined to a cavity of volume V. As discussed in Sect. 13.4, at zero temperature these electrons are distributed filling the Fermi sphere in momentum space with radius

$$p_\mathrm{F} = (3\pi^2 \varrho)^{1/3} \,, \tag{14.11}$$

where $\varrho = N/V$. Their largest kinetic energy is therefore

$$E_\mathrm{F} = \frac{(3\pi^2)^{2/3}}{2m_\mathrm{e}} \varrho^{2/3} \,. \tag{14.12}$$

Returning to the atom, let us suppose that it has a large number of electrons N so that their kinetic energies will generally be large, and the average potential $V(r)$ due to the other electrons and the nucleus will not vary much over a distance of the order of the associated wavelengths. Thus we can assume that locally the system can be treated as a perfect gas and that local equations like (14.12) still hold. We also assume that the electrons interact enough with each other to establish thermal equilibrium between different regions of the atom, so that the maximal energy is the same in all points:

$$\boxed{\frac{(3\pi^2)^{2/3}}{2m_\mathrm{e}} \varrho(r)^{2/3} + V(r) = C \,, \quad \text{if } \varrho(r) > 0} \,, \tag{14.13}$$

where C is a constant. On the other hand, the potential energy can be determined from the electron density via the Poisson equation

$$\Delta V(r) = 4\pi e^2[Z\delta(\boldsymbol{r}) - \varrho(r)] , \tag{14.14}$$

where $Z|e|$ is the charge of the point nucleus. From (14.13) and (14.14), and for $r \neq 0$ such that $\varrho(r) > 0$, we obtain

$$\frac{1}{r}\frac{d^2}{dr^2}[r(C - V(r))] = \frac{4e^2}{3\pi}(2m_e)^{3/2}[C - V(r)]^{3/2} . \tag{14.15}$$

Implementing the changes of variables

$$V(r) = C - \frac{Ze^2}{r}\Phi(x) , \quad r = Z^{-1/3}bx , \tag{14.16}$$

where

$$b = \frac{1}{2}\left(\frac{3\pi}{4}\right)^{2/3} a_0 = 0.885341\ldots a_0 \tag{14.17}$$

and a_0 is the Bohr radius, Eq. (14.15) becomes

$$\boxed{x^{1/2}\frac{d^2\Phi(x)}{dx^2} = \Phi(x)^{3/2} \quad \text{if } x > 0 \quad \text{and } \Phi(x) > 0} , \tag{14.18}$$

which should be supplemented with appropriate boundary conditions at $x = 0$. Taking into account that as $r \to 0$ the potential should agree with the nuclear potential $-Ze^2/r$, in this limit $r[C - V(r)]$ should approach Ze^2, and hence a first boundary condition is

$$\boxed{\Phi(0) = 1} . \tag{14.19}$$

The solutions to the non-linear Thomas-Fermi equation (14.18) with the boundary condition (14.19) behave asymptotically in a neighborhood of $x = 0$ in the form

$$\Phi(x) = 1 - Ax + \frac{4}{3}x^{3/2} - \frac{2A}{5}x^{5/2} + \frac{1}{3}x^3 + \ldots , \tag{14.20}$$

where A is a constant not specified by this expansion. In particular $\Phi(x)$ has a finite slope at the origin.

If the first zero of $\Phi(x)$ occurs at $x = x_0$ ($\neq 0$ and finite), this point cannot be a minimum, for otherwise $\Phi(x_0) = \Phi'(x_0) = 0$, and the existence and uniqueness theorem for (14.18) would imply that $\Phi(x)$ vanishes identically. Hence the equation considered is no longer valid to the right of x_0, points where $\Phi(x)$ would be negative. For $\Phi(x) < 0$ the electron density must vanish [use (14.13)] and (14.14) implies that for $x > x_0$ the correct equation is

$$\frac{d^2\Phi(x)}{dx^2} = 0 . \tag{14.21}$$

The radius r_N of the sphere containing the N electrons in an atom of nuclear charge $Z|e|$ is given by the smallest solution of r_N of the equation

$$4\pi \int_0^{r_N} dr\, r^2 \varrho(r) = N \,. \tag{14.22}$$

Using (14.16) and (14.18) and defining $x_N = Z^{1/3} r_N / b$ gives

$$\int_0^{x_N} dx\, x\Phi''(x) = \frac{N}{Z} \tag{14.23}$$

and from this

$$\Phi(x_N) - x_N \Phi'(x_N) = \frac{Z-N}{Z} \,. \tag{14.24}$$

If we are dealing with atoms or ions of net charge $(Z-N)|e|$, then $\varrho(r_N) = 0$, $x_0 = x_N$, and from (14.24) we obtain

$$\Phi(x_0) = 0 \,, \quad x_0 \Phi'(X_0) = -\frac{Z-N}{Z} \,. \tag{14.25}$$

Equation (14.18) reveals that in the region of physical interest $\Phi(x)$ is convex, and hence $x_0\Phi'(x_0) \leq 0$; from this we conclude that the Thomas-Fermi model cannot be applied to ions with negative total charge. For the case of neutral atoms the radius, if finite, is determined according to (14.25) by

$$\Phi(x_0) = 0 \,, \quad x_0 \Phi'(x_0) = 0 \tag{14.26}$$

and from the arguments presented before, this implies $x_0 = \infty$. Therefore in the Thomas-Fermi model a neutral atom has infinite radius. A numerical computation [KM 55] shows that for neutral atoms (the only case considered here) the constant appearing in (14.20) is $A = 1.588070972\ldots$ and $\Phi(x)$ at large distances behaves like

$$\Phi(x) = \frac{144}{x^3} + \ldots \,, \quad x \to \infty \,. \tag{14.27}$$

[The function $144/x^3$ is an exact solution of (14.16) even though it is singular at the origin]. In Table 14.1 we give some values of $\Phi(x)$ in this case, and in Fig. 14.1 its graphical representation.

An analytic expression approximating $\Phi(x)$ for neutral atoms, with an error not bigger than 3 per 1000, is the following [LA 55]:

$$\begin{aligned}\Phi(x) = [1 &+ 0.02747x^{1/2} + 1.243x - 0.1486x^{3/2} \\ &+ 0.2302x^2 + 0.007298x^{5/2} + 0.006944x^3]^{-1} \,, \quad \forall x \geq 0 \,.\end{aligned} \tag{14.28}$$

The constant C has to be determined through Eq. (14.13). If R is the atomic radius, $\varrho(R) = 0$ and $C = V(R) = -(Z-N)e^2/R$. Thus for neutral atoms, $C = 0$, and for positive ions $C < 0$; the case $C > 0$ corresponding to negative ions cannot be analyzed by this method, as explained above.

Table 14.1. Numerical values of $\Phi(x)$ for neutral atoms

x	$\Phi(x)$	x	$\Phi(x)$	x	$\Phi(x)$
0	1.000	1.4	0.333	6	0.059 4
0.02	0.972	1.6	0.298	7	0.046 1
0.04	0.947	1.8	0.268	8	0.036 6
0.06	0.924	2.0	0.243	9	0.029 6
0.08	0.902	2.2	0.221	10	0.024 3
0.10	0.882	2.4	0.202	11	0.020 3
0.2	0.793	2.6	0.185	12	0.017 1
0.3	0.721	2.8	0.170	13	0.014 5
0.4	0.660	3.0	0.157	14	0.012 5
0.5	0.607	3.2	0.145	15	0.010 8
0.6	0.561	3.4	0.134	20	0.005 8
0.7	0.521	3.6	0.125	25	0.003 5
0.8	0.485	3.8	0.116	30	0.002 3
0.9	0.453	4.0	0.108	40	0.001 1
1.0	0.424	4.5	0.091 9	50	0.000 63
1.2	0.374	5.0	0.078 8	60	0.000 39

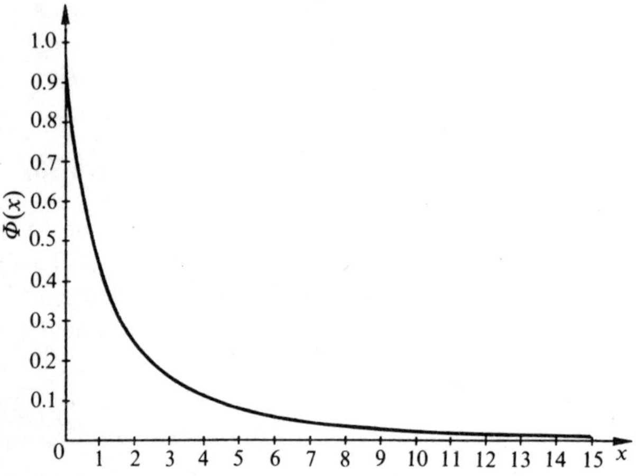

Fig. 14.1. Graph of $\Phi(x)$ for neutral atoms

It is useful to emphasize that the density is given by

$$\varrho(r) = \frac{32}{9\pi^3 a_0^3} Z^2 \left[\frac{\Phi(x)}{x}\right]^{3/2} , \tag{14.29}$$

where the scale length for each atom is proportional to $Z^{-1/3}$. To determine the sphere of radius $r_{Z/2}$, containing half of the electrons, we have to solve (14.23) or (14.24) with $N = Z/2$. We thus find $r_{Z/2} \simeq 1.682 a_0 Z^{-1/3}$. Similarly, one finds for the radius of the sphere containing all but one electron $r_{Z-1} \sim 4 \times 3^{2/3} b \simeq 7.37 a_0$ when $Z \to \infty$; for practical cases $25 \lesssim Z \lesssim 100$, r_{Z-1} varies between 2–3 Å.

An interesting application of the model is the computation of the value of Z for which a bound state of orbital angular momentum l appears for the first time. For this to happen, the effective potential

$$V_l(r) = V(r) + \frac{l(l+1)}{2m_e r^2} \tag{14.30}$$

must be negative in some interval of values of r. Using (14.16, 17) and that $C = 0$ for neutral atoms, the corresponding region in x should satisfy

$$(3\pi/4)^{2/3} Z^{2/3} x\Phi(x) > l(l+1) . \tag{14.31}$$

For the sake of estimates, we may replace this inequality by

$$(3\pi/4)^{2/3} Z^{2/3} \max\,[x\Phi(x)] > (l+1/2)^2 \tag{14.32}$$

where we have replaced $l(l+1)$ by $(l+1/2)^2$, as is common in the semiclassical approximation. From (14.32) we obtain

$$Z > \frac{(2l+1)^3}{6\pi\{\max\,[x\Phi(x)]\}^{3/2}} . \tag{14.33}$$

From the values of Table 14.1 we see that $x\Phi(x)$ has a broad maximum of height 0.487, and therefore $Z > 0.156(2l+1)^3$. It is to be expected that the symbol $>$ can be changed to an equality if we "slightly" increase the coefficient. Hence, with $Z = 0.17(2l+1)^3$ we obtain

$$\begin{array}{lllll} l = 1 & 2 & 3 & 4 \\ Z = 4.6 & 21.3 & 58.3 & 123.9 \end{array} \tag{14.34}$$

Rounding off these numbers, we have $Z = 5, 21, 58, 124$: the first three agree with experimental results, and the last one predicts that a g electron would be bounded first to a nucleus with $Z = 124$ [RM 72].

Dirac [DI 30] improved the Thomas-Fermi model by introducing the exchange interaction, and then (14.18) must be replaced by

$$\frac{d^2\Phi(x)}{dx^2} = x[x^{-1/2}\Phi^{1/2}(x) + \beta]^3 , \tag{14.35}$$

where $\beta = (b/2a_0)^{1/2}\, Z^{-2/3}/\pi$. For neutral atoms this equation has to be integrated with the conditions $\Phi(0) = 1$, $\Phi(x_0)/x_0 = \Phi'(x_0) = \beta^2/16$, where x_0 is the point where the pressure vanishes. *Latter* [LA 55] solved the radial Schrödinger equation for the potential obtained with the Thomas-Fermi-Dirac model. The resulting eigenvalues compare very favorably with experimental data, and the wave functions obtained provide a good starting point for the Hartree-Fock method, to be studied in the next section. A traditionally good reference for the Thomas-Fermi model is [GO 56].

Recently [LS 73], [LS 77a], [LI 81], the rigorous bases for the Thomas-Fermi model have been laid down for atoms and also for molecules and crystals. Considering only the case of atoms, the total energy for a system of N electrons in this model, including the kinetic and electrostatic energies, is

$$E_{Z,N}(\varrho) = \frac{(2\pi^2)^{5/3}}{10m_e\pi^2} \int d^3r\, \varrho^{5/3}(\boldsymbol{r}) - Ze^2 \int d^3r\, \frac{\varrho(\boldsymbol{r})}{r} + \frac{e^2}{2} \int d^3r\, d^3r' \frac{\varrho(\boldsymbol{r})\varrho(\boldsymbol{r}')}{|\boldsymbol{r}-\boldsymbol{r}'|} . \tag{14.36}$$

Making $E_{Z,N}(\cdot)$ stationary as a functional of ϱ, subject to the condition $\int \varrho(\boldsymbol{r})d^3r = N$, we straightforwardly obtain – after applying the Laplacian operator – (14.13) and (14.14). The constant C appears as the Lagrange multiplier. It has been shown that if $N \leq Z$, there exists a unique function $\varrho \geq 0$ in $L^1(\mathbb{R}^3) \cap L^{5/3}(\mathbb{R}^3)$, satisfying $\int \varrho\, d^3r = N$ and minimizing the functional $E_{Z,N}(\cdot)$. (This minimum does not exist if $N > Z$). Let $E^{\mathrm{TF}}_{Z,N}$ be this minimum, which is negative and equals the atom (or positive ion) energy in the Thomas-Fermi model and let $\varrho^{\mathrm{TF}}_{Z,N}$ be the unique associated function giving the electron density. The latter is a decreasing spherically symmetric function $\varrho^{\mathrm{TF}}_{Z,N}(r) \geq 0$, and for neutral atoms $\varrho^{\mathrm{TF}}_{Z,Z}$ is positive and C^∞ if $r > 0$, while for positive ions $\varrho^{\mathrm{TF}}_{Z,N}$ has compact support, and is C^∞ if $r > 0$ and $\varrho^{\mathrm{TF}}_{Z,N}(r) \neq 0$.

If $E^0_{Z,N}$ is the energy of the ground state for the Hamiltonian (14.14) (assuming the mass of the nucleus to be infinite) including the electron spins and incorporating the Fermi-Dirac statistics, and if $N^{-1}\varrho^0_{Z,N}(\boldsymbol{r})$ is the probability density for one electron in this state, it was possible to prove the following fundamental result

$$\begin{aligned} E^0_{Z,\lambda Z} &\underset{Z\to\infty}{\sim} E^{\mathrm{TF}}_{Z,\lambda Z} (= Z^{7/3} E^{\mathrm{TF}}_{1,\lambda}) , \\ \varrho^0_{Z,\lambda Z}(Z^{-1/3}\boldsymbol{r}) &\underset{Z\to\infty}{\sim} \varrho^{\mathrm{TF}}_{Z,\lambda Z}(Z^{-1/3}r) (= Z^2 \varrho^{\mathrm{TF}}_{1,\lambda}(r)) , \end{aligned} \tag{14.37}$$

maintaining $\lambda \in (0, 1]$ constant. The second relation holds in the sense of L^1_{loc}. The equalities in parentheses are a simple consequence of the transformation properties of the functional $E_{Z,N}(\varrho)$ under the scale change: $E_{\nu Z,\nu N}(\varrho') = \nu^{7/3}E_{Z,N}(\varrho)$ if $\varrho'(\boldsymbol{r}) = \nu^2\varrho(\nu^{1/3}\boldsymbol{r})$, $\nu > 0$. Compare this, for instance, with (14.29).

The Thomas-Fermi model, asymptotically correct according to (14.37), presents several drawbacks (which are not ameliorated by adding Dirac's term): It leads to a density which is infinite at the nucleus and does not decrease exponentially at large distances; it does not allow for stable negative ions, and it can also be shown that it does not lead to binding between two neutral atoms (Teller's theorem). These difficulties disappear if we include in (14.36) a term incorporating the correction to the kinetic energy due to the variation of ϱ with $\boldsymbol{r}$ (von Weizsäcker term). For a very detailed and rigorous study of these questions we refer the reader to *Lieb* [LI 81].

In particular, a picture for heavy atoms stems from this analysis. In general terms it consists of: (1) an internal part of radius $\propto Z^{-1/3}$, where $\varrho \propto Z^2$; (2) a shell surrounding the previous one with $\varrho(r) \propto r^{-6}$, independent of the value of Z, and very large in the scale of distances $Z^{-1/3}a_0$. These two parts contain practically 100% of the electrons for large Z; (3) an intermediate region whose description is very complex; (4) external shells whose thickness is not very sensitive to the value of Z and with $\propto Z^{2/3}$ electrons, as suggested by some estimates; and (5) the external region, where the density decreases exponentially.

The applications of the techniques based on density functionals [generalization of (14.36)] to problems in atomic, nuclear, molecular and solid state physics has become quite important following the pioneering works of [HK 64], [KS 65].

14.3 The Hartree-Fock Method

The Hartree-Fock method provides a way to determine the average potential acting on the electrons and the associated one-particle wave functions. This method was introduced by *Hartree* [HA 28] and improved by *Fock* [FO 30], who incorporated the fact that the wave function for a system of N electrons must be totally antisymmetric.

Following the arguments presented in Sect. 14.1, we take as the Hamiltonian for the N-electron system

$$H = \sum_{i=1}^{N} \left[-\frac{1}{2\mu} \Delta_i - \frac{Ze^2}{r_i} \right] + \sum_{1 \leq i < j}^{N} \frac{e^2}{r_{ij}} , \tag{14.38}$$

where we have neglected the Hughes-Eckart term and the spin-orbit coupling. Assuming each electron in the atom to be subject to an average potential due to the other electrons and the nucleus, we can describe the state of the electrons in terms of the single-particle wave functions of the bound states in the average potential considered. For these single-particle wave functions we use the notation $\phi_j(\xi)$, where ξ refers to the space and spin coordinates. These functions will be assumed to be orthonormalized:

$$\int d\xi \, \phi_j^*(\xi) \phi_i(\xi) = \delta_{ji} , \tag{14.39}$$

where the integration is understood as an integration over space coordinates and a sum over spin variables. With them we construct for the N-electron system the totally antisymmetric and normalized wave function (all indices $i_1, \dots, i_N$ must be different)

$$\Psi(\xi_1, \xi_2, \dots, \xi_N) = \frac{1}{\sqrt{N!}} \sum_{P \in S_N} (-1)^{\pi(P)} P[\phi_{i_1}(\xi_1) \phi_{i_2}(\xi_2) \dots \phi_{i_N}(\xi_N)] \, . \tag{14.40}$$

In the Hartee-Fock method we determine the single-particle wave functions using the wave function (14.40) as a trial function in a variational problem with the subsidiary condition (14.39). Therefore, we have to first compute the expectation value of H in the state described by the wave function (14.40). Taking into account the general methods explained in Chap. 13 we obtain

$$\langle\Psi|H|\Psi\rangle = \sum_{i=1}^{N}\int d\xi\, \phi_i^*(\xi)\left[-\frac{1}{2\mu}\Delta - \frac{Ze^2}{r}\right]\phi_i(\xi)$$
$$+\sum_{1\leq i<j}^{N}\int d\xi_1\, d\xi_2\, \phi_i^*(\xi_1)\phi_j^*(\xi_2)\frac{e^2}{r_{12}}\phi_i(\xi_1)\phi_j(\xi_2)$$
$$-\sum_{1\leq i<j}^{N}\int d\xi_1 d\xi_2\, \phi_i^*(\xi_1)\phi_j^*(\xi_2)\frac{e^2}{r_{12}}\phi_i(\xi_2)\phi_j(\xi_1)\,. \tag{14.41}$$

We must now stationarize $\langle\Psi|H|\Psi\rangle$, maintaining (14.39). Hence under infinitesimal variations of the single-particle wave functions ϕ_i, we should have

$$\delta\left[\langle\Psi|H|\Psi\rangle - \frac{1}{2}\sum_{i,j}\lambda_{ij}\{\langle\phi_i|\phi_j\rangle + \langle\phi_j|\phi_i\rangle\}\right] = 0\,, \tag{14.42}$$

where $\lambda_{ij} = \lambda_{ji} = \lambda_{ij}^*$ are Lagrange multipliers. We consider variations of the form $\phi_i(\xi) \to \phi_i(\xi)+\delta\phi_i(\xi)$ for each i, and hence $\phi_i^*(\xi) \to \phi_i^*(\xi)+\delta\phi_i^*(\xi)$, keeping the other wave functions fixed. In these variations, the real and imaginary parts of $\phi_i(\xi)$ can change arbitrarily. Thus we can take $\delta\phi_i(\xi)$ and $\delta\phi_i^*(\xi)$ as being independent. Varying only $\phi_i^*(\xi)$, we obtain

$$(H_\Psi\phi_i)(\xi_1) \equiv \left[-\frac{1}{2\mu}\Delta_1 - \frac{Ze^2}{r_1}\right]\phi_i(\xi_1) + \left[\sum_{j=1}^{N}\int d\xi_2\, |\phi_j(\xi_2)|^2\frac{e^2}{r_{12}}\right]\phi_i(\xi_1)$$
$$-\sum_{j=1}^{N}\left[\int d\xi_2\, \phi_j^*(\xi_2)\frac{e^2}{r_{12}}\phi_i(\xi_2)\right]\phi_j(\xi_1) = \sum_{j=1}^{N}\lambda_{ij}\phi_j(\xi_1)\,, \tag{14.43}$$

which are the so-called *Hartree-Fock equations*. It is however convenient to find an equivalent representation using the fact that the matrix with elements λ_{ij} is real and Hermitian and therefore diagonalizable by a unitary transformation (u_{ij}). If ε_i are the (obviously real) eigenvalues of the matrix (λ_{ij}), and defining

$$\phi_i(\xi) = \sum_{i=1}^{N} u_{ij}\phi'_j(\xi)\,, \tag{14.44}$$

it is easy to see that neither $\langle\Psi|H|\Psi\rangle$ nor H_Ψ change under $\phi_i \to \phi'_i$, and a simple computation brings (14.43) into the form

$$\boxed{\begin{aligned}&(H_\Psi\phi_i)(\xi_1)\\ &= \left[-\frac{1}{2\mu}\Delta_1 - \frac{Ze^2}{r_1}\right]\phi_i(\xi_1) + \left[\sum_{j=1}^{N}\int d\xi_2\, |\phi_j(\xi_2)|^2\frac{e^2}{r_{12}}\right]\phi_i(\xi_1)\\ &\quad -\sum_{j=1}^{N}\left[\int d\xi_2\, \phi_j^*(\xi_2)\,\frac{e^2}{r_{12}}\,\phi_i(\xi_2)\right]\phi_j(\xi_1) = \varepsilon_i\phi_i(\xi_1)\,,\end{aligned}} \tag{14.45}$$

where the primes in the wave functions have been omitted. Consequently, the average potential acting on the ith electron consists of two parts: 1) a *direct potential* defined as

$$\boxed{V_{i,\mathrm{d}}(\boldsymbol{r}_1) = -\frac{Ze^2}{r_1} + \sum_{i\neq j=1}^{N} \int d\xi_2\, |\phi_j(\xi_2)|^2 \frac{e^2}{r_{12}}} \tag{14.46}$$

and interpreted as the potential generated by the nucleus and the other electrons. 2) An *exchange potential* which is non-local, and defined by

$$\boxed{V_{i,\mathrm{ex}}(\xi_1, \xi_2) = -\sum_{i\neq j=1}^{N} \phi_j^*(\xi_2)\frac{e^2}{r_{12}}\phi_j(\xi_1)} \tag{14.47}$$

and whose origin is due to the antisymmetry of the original test function. If (as in the original method of Hartree) we had taken the trial wave functions simply as products of the single-particle wave functions, this term would not have appeared, with the other terms remaining unchanged. It is interesting to notice that even though the operators defined by (14.46) and (14.47) depend on the electron considered, their sum does not, as can be easily derived from (14.45).

If in the Hartree-Fock equation (14.45) we take the scalar product with ϕ_i, we obtain

$$\begin{aligned}\varepsilon_i = \int d\xi_1 \phi_i^*(\xi_1)\left[-\frac{1}{2\mu}\Delta_1 - \frac{Ze^2}{r_1}\right]\phi_i(\xi_1) + \sum_{j=1}^{N}\int d\xi_1 d\xi_2\, \phi_i^*(\xi_1)\phi_j^*(\xi_2)\frac{e^2}{r_{12}} \\ \times \phi_i(\xi_1)\phi_j(\xi_2) - \sum_{j=1}^{N}\int d\xi_1 d\xi_2\, \phi_i^*(\xi_1)\phi_j^*(\xi_2)\frac{e^2}{r_{12}}\phi_i(\xi_2)\phi_j(\xi_1)\,.\end{aligned} \tag{14.48}$$

To understand the physical meaning of ε_i it is necessary to recall that in the Hartree-Fock approximation the total energy E of the atom is given by (14.41). If after removing the ith electron from the atom the single-particle wave functions leading to a stationary $\langle H\rangle$ are still the same, then the total energy of the ion, E_{ion}, would be given by (14.41) after eliminating all terms containing $\phi_i(\xi)$:

$$E_{\mathrm{ion}} - E = -\varepsilon_i\,; \tag{14.49}$$

thus $(-\varepsilon_i)$ would in this approximation be the energy necessary to remove an electron in the state $\phi_i(\xi)$. This is the content of *Koopmans' theorem.* It is nonetheless worth while to point out that in general $E \neq \sum_{i=1}^{N} \varepsilon_i$, as follows easily from (14.41) and (14.48).

In principle the problem of finding the average potential and the associated single-particle wave functions is equivalent to solving the N integro-differential equations (14.45). In general this is not feasible, and a possible method is to proceed by means of a self-consistent approximation. We start with a set of approximate wave functions $\phi_i^{(1)}(\xi)$, obtained, for example, using the Thomas-Fermi-Dirac method. With them we compute the potentials appearing in (14.45), and then we solve these equations, and obtain a new set of single-particle wave functions $\phi_i^{(2)}(\xi)$. Following the same method and starting with $\phi_i^{(2)}(\xi)$ we compute $\phi_i^{(3)}(\xi)$ and so on, until $\phi_i^{(k+1)} \simeq \phi_i^{(k)}$ to the desired accuracy.

It is not easy to illustrate the Hartree-Fock method in realistic situations without turning to numerical calculations. In the following we present a simple model proposed by *Moshinsky* [MO 68], [CM 70]. The problem consists of two identical particles of spin $1/2$ in the presence of a common harmonic oscillator potential , and interacting via a harmonic potential which only depends on their relative distance. Using the notation of Chap. 4 and the units $\hbar = 2M = \omega/2 = 1$, the Hamiltonian is

$$\begin{aligned} H &= \sum_{i=1}^{2} H_0(i) + H_1 \,, \\ H_0(i) &= -\Delta_i + \boldsymbol{r}_i^2 \,, \quad H_1 = \lambda(\boldsymbol{r}_1 - \boldsymbol{r}_2)^2 \,, \quad \lambda \geq 0 \,. \end{aligned} \tag{14.50}$$

This problem is simple enough to admit an exact solution. Using as new coordinates

$$\boldsymbol{R} = \frac{1}{\sqrt{2}}(\boldsymbol{r}_1 + \boldsymbol{r}_2) \,, \quad \boldsymbol{r} = \frac{1}{\sqrt{2}}(\boldsymbol{r}_1 - \boldsymbol{r}_2) \,, \tag{14.51}$$

the Hamiltonian becomes

$$H = [-\Delta_R + \boldsymbol{R}^2] + [-\Delta_r + (1 + 2\lambda)\boldsymbol{r}^2] \,, \tag{14.52}$$

representing the energy of two uncoupled oscillators. Hence the ground state energy and normalized wave function are

$$\begin{aligned} E_0 &= 3(1 + \sqrt{1 + 2\lambda}) \,, \\ \Psi_0(\boldsymbol{r}_1, \boldsymbol{r}_2) &= \pi^{-3/2}(1 + 2\lambda)^{3/8} \exp[-R^2/2] \exp[-(1 + 2\lambda)^{1/2} r^2/2] \,. \end{aligned} \tag{14.53}$$

To obtain the total wave function we must simply multiply $\Psi_0(\boldsymbol{r}_1, \boldsymbol{r}_2)$ by the spin wave function. The latter must be one corresponding to a normalized singlet state, so that the total wave function is antisymmetric and of unit norm.

To solve this problem in the Hartree-Fock method, we start with a wave function which is a spin singlet, and whose spatial part is given by

$$\Psi(\boldsymbol{r}_1, \boldsymbol{r}_2) = \varphi(\boldsymbol{r}_1)\varphi(\boldsymbol{r}_2) \,. \tag{14.54}$$

In this case the exchange term vanishes, because the orbital single-particle wave functions are identical, and the interaction is spin independent. Then the two Hartree-Fock equations reduce to one:

$$[-\Delta + r^2]\varphi(\boldsymbol{r}) + \left\{ \int d^3r' \varphi^*(\boldsymbol{r}')\lambda(\boldsymbol{r} - \boldsymbol{r}')^2 \varphi(\boldsymbol{r}') \right\} \varphi(\boldsymbol{r}) = \varepsilon\varphi(\boldsymbol{r}) \,. \tag{14.55}$$

Taking the first approximation to the single-particle wave function to be the ground state for the non-interacting Hamiltonian

$$\varphi^{(1)}(\boldsymbol{r}) = \pi^{-3/4} \exp[-r^2/2] \,, \tag{14.56}$$

the effective potential in this approximation becomes

$$\int d^3r' \varphi^{(1)*}(\boldsymbol{r}') \lambda (\boldsymbol{r} - \boldsymbol{r}')^2 \varphi^{(1)}(\boldsymbol{r}') = \lambda r^2 + \frac{3\lambda}{2} \,. \tag{14.57}$$

Using (14.55, 57), the second approximation $\varphi^{(2)}(\boldsymbol{r})$ will be a solution of

$$[-\Delta + (1+\lambda) r^2] \varphi^{(2)}(\boldsymbol{r}) = \left[\varepsilon - \frac{3\lambda}{2}\right] \varphi^{(2)}(\boldsymbol{r}) \,, \tag{14.58}$$

and therefore the energy and single-particle wave function for the ground state are, to this order of iteration,

$$\begin{aligned} \varphi^{(2)}(\boldsymbol{r}) &= \pi^{-3/4}(1+\lambda)^{3/8} \exp[-(1+\lambda)^{1/2} r^2/2] \,, \\ \varepsilon^{(2)} &= 3\sqrt{1+\lambda} + \frac{3\lambda}{2} \,. \end{aligned} \tag{14.59}$$

It is easy to check that after repeating the procedure starting with $\varphi^{(2)}(\boldsymbol{r})$ we obtain $\varphi^{(3)}(\boldsymbol{r}) \equiv \varphi^{(2)}(\boldsymbol{r})$ and

$$\varepsilon^{(3)} = \frac{3}{2} \frac{2+3\lambda}{\sqrt{1+\lambda}} \tag{14.60}$$

and in the next iterations $\varepsilon^{(n)} = \varepsilon^{(3)}$ and $\varphi^{(n)}(\boldsymbol{r}) \equiv \varphi^{(2)}(\boldsymbol{r})$ for all $n \geq 3$. As a result the normalized orbital wave function and energy in the Hartree-Fock approximation are

$$\begin{aligned} \Psi_{\mathrm{HF}}(\boldsymbol{r}_1, \boldsymbol{r}_2) &= \pi^{-3/2}(1+\lambda)^{3/4} \exp[-(1+\lambda)^{1/2}(r_1^2 + r_2^2)/2] \,, \\ E_{\mathrm{HF}} &= 6\sqrt{1+\lambda} \,. \end{aligned} \tag{14.61}$$

To obtain E_{HF}, it is enough to take into account that $E_{\mathrm{HF}} = \langle \Psi_{\mathrm{HF}} | H | \Psi_{\mathrm{HF}} \rangle$, which is evaluated immediately after H is rewritten as

$$H = \sum_{i=1}^{2} [-\Delta_i + (1+\lambda) \boldsymbol{r}_i^2] - 2\lambda \boldsymbol{r}_1 \cdot \boldsymbol{r}_2 \,. \tag{14.62}$$

The comparison between the Hartree-Fock approximation and the exact result is given in Figs. 14.2 and 14.3, where we represent graphically as functions of the coupling parameter λ the quantities $(E_{\mathrm{HF}} - E_0)/E_0$ and $|\langle \Psi_{\mathrm{HF}} | \Psi_0 \rangle|^2$. The first is straightforwardly computed from (14.53) and (14.61):

$$\frac{E_{\mathrm{HF}} - E_0}{E_0} = -1 + \frac{2(1+\lambda)^{1/2}}{1 + (1+2\lambda)^{1/2}} \tag{14.63}$$

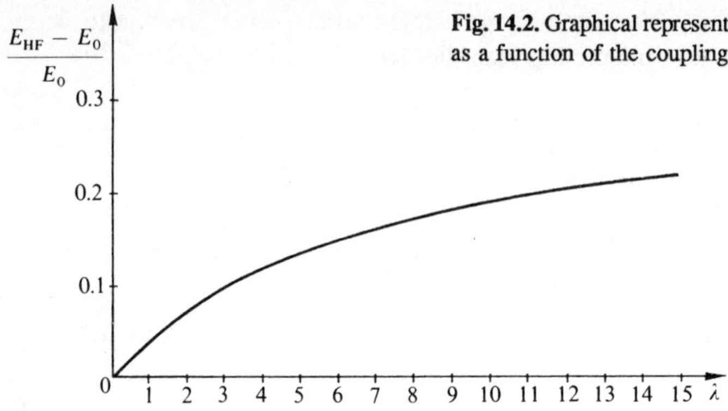

Fig. 14.2. Graphical representation of $(E_{HF} - E_0)/E_0$ as a function of the coupling parameter λ

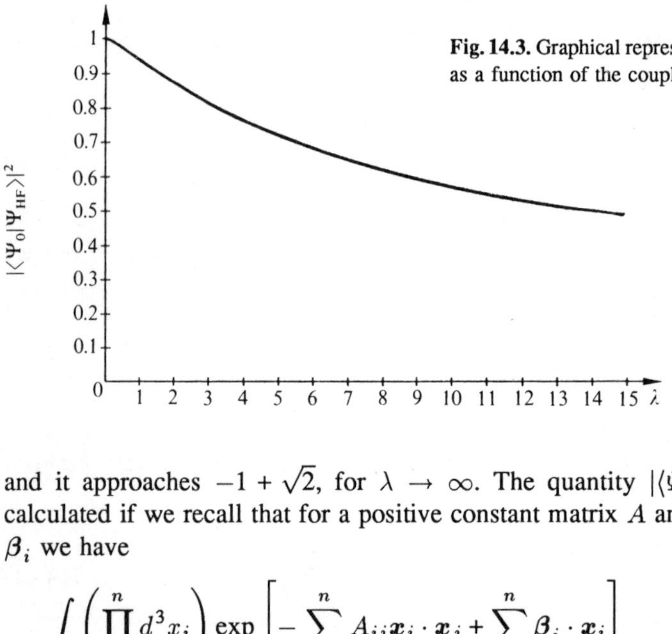

Fig. 14.3. Graphical representation of $|\langle\Psi_0|\Psi_{HF}\rangle|^2$ as a function of the coupling parameter λ

and it approaches $-1 + \sqrt{2}$, for $\lambda \to \infty$. The quantity $|\langle\Psi_{HF}|\Psi_0\rangle|^2$ is easily calculated if we recall that for a positive constant matrix A and constant vectors $\boldsymbol{\beta}_i$ we have

$$\int \left(\prod_{i=1}^{n} d^3x_i\right) \exp\left[-\sum_{i,j=1}^{n} A_{ij}\boldsymbol{x}_i \cdot \boldsymbol{x}_j + \sum_{i=1}^{n} \boldsymbol{\beta}_i \cdot \boldsymbol{x}_i\right]$$

$$= \frac{\pi^{3n/2}}{|\det A|^{3/2}} \exp\left[\frac{1}{4}\sum_{i,j=1}^{n}(A^{-1})_{ij}\boldsymbol{\beta}_i \cdot \boldsymbol{\beta}_j\right] . \tag{14.64}$$

Hence

$$|\langle\Psi_{HF}|\Psi_0\rangle|^2 = \frac{64(1+\lambda)^{3/2}(1+2\lambda)^{3/4}}{\{[1+(1+\lambda)^{1/2}][(1+\lambda)^{1/2}+(1+2\lambda)^{1/2}]\}^3} \tag{14.65}$$

and it approaches zero as $\lambda \to \infty$. Notice that for $\lambda = 1$, i.e., when the interaction potential has the same strength as the common potential acting on the particles, the exact value of the energy is 96.47% of its Hartree-Fock value and $|\langle\Psi_{HF}|\Psi_0\rangle|^2 = 0.9416$. Even though $(E_{HF} - E_0)/E_0 \lesssim 0.41$ always holds, Ψ_{HF} becomes orthogonal to the exact wave function in the limit $\lambda \to \infty$, and this supports the property of variational methods pointed out in Chap. 10, that they usually give a better approximation to the energy than to the wave function.

In the simple example we have presented, Hartree's iterative procedure converges rather quickly to the exact solution of the Hartree-Fock equations. This is not the case when the method is applied to more complex situations in atomic or nuclear physics. In these cases it is unknown in general whether the iterative method converges. [There are however classes of two-body potentials for which the Hartree-Fock equations admit a rapidly convergent iterative analysis based on the Newton-Kantorovich method [FM 73] and not on Hartree's original one].

In spite of the seemingly negative remarks of the previous paragraph, some general results have been obtained. It has been possible to show [LS 74], [LS 77] that whenever $N < Z+1$ the non-linear Hartree-Fock system (14.45) has some (unique?) solution $\phi_1, \dots, \phi_N$, consisting of orthonormal functions leading to a wave function Ψ, (14.40), which minimizes the expectation value of the Hamiltonian (14.38) for Slater determinant states constructed with single-particle wave functions whose gradients are also square integrable. Letting $E^{\mathrm{HF}}_{Z,N} \equiv \langle \Psi | H | \Psi \rangle$ be the Hartree-Fock approximation to the exact energy $E^0_{Z,N}$, it is possible to show that: (1) the ionization energy is positive ($E^{\mathrm{HF}}_{Z,N} < E^{\mathrm{HF}}_{Z,N-1}$); (2) as in the Thomas-Fermi model, $E^{\mathrm{HF}}_{Z,\lambda Z} \sim E^0_{Z,\lambda Z}$ if $Z \to \infty$, with λ fixed in $(0,1]$; (3) the functions ϕ_i decrease exponentially far away from the nucleus; and (4) the eigenvalues ε_i, $1 \le i \le N$, not all necessarily different, are negative and they give the lowest eigenvalues of H_Ψ. [Identical conclusions hold for Hartree's old model and also, except for (2), for the Eqs. (14.72) where the single-particle test functions have been conveniently restricted].

14.4 The Central Field Approximation

It is very difficult to solve the Hartree-Fock equations in atomic physics without resorting to new approximations. Among these, the *central field approximation* stands out. To exhibit its features, consider the trial functions $\phi(\xi) = (1/r)u_{nl}(r)Y_l^m(\hat{\boldsymbol{r}})\alpha$, with u_{nl} real and the spinor α normalized. We are going to show that for atoms with closed shells, i.e., atoms where the shell (nl) is either empty, or completely filled with $(4l+2)$ electrons, these wave functions give rise to a central potential. Hence the angular part of the problem can be separated, and the Hartree-Fock equations considerably simplified.

Let us see what each of the terms in the Hartree-Fock equations (14.45) become for atoms with closed shells and in the central approximation.

a) For the first term we immediately obtain

$$\left[-\frac{1}{2\mu}\Delta_1 - \frac{Ze^2}{r_1}\right]\frac{u_{nl}(r_1)}{r_1}Y_l^m(\hat{\boldsymbol{r}}_1)\alpha_\sigma = \frac{1}{r_1}\left[-\frac{1}{2\mu}\frac{d^2}{dr_1^2} + \frac{l(l+1)}{2\mu r_1^2} - \frac{Ze^2}{r_1}\right]u_{nl}(r_1)Y_l^m(\hat{\boldsymbol{r}}_1)\alpha_\sigma\,. \tag{14.66}$$

b) The second term is

$$A \equiv 2 \sum_{n'l'm'} \int d^3r_2 \frac{e^2}{r_1 r_2^2 r_{12}} \times u_{n'l'}^2(r_2) u_{nl}(r_1) Y_{l'}^{m'*}(\hat{\boldsymbol{r}}_2) Y_{l'}^{m'}(\hat{\boldsymbol{r}}_2) Y_l^m(\hat{\boldsymbol{r}}_1) \alpha_\sigma \,, \tag{14.67}$$

where the factor 2 comes from the presence of two electrons for each n', l', m', and these quantum numbers correspond to closed shells, with the usual ranges for one-electron atoms. Using (A.39) we can perform the sum over m', and afterwards, applying (A.135) together with (A.41), we can easily evaluate the integral over the $\boldsymbol{r}_2$ directions, yielding

$$A = \sum_{n'l'} 2(2l'+1) \int_0^\infty dr_2 \frac{e^2}{r_>} u_{n'l'}^2(r_2) \frac{u_{nl}(r_1)}{r_1} Y_l^m(\hat{\boldsymbol{r}}_1) \alpha_\sigma \,. \tag{14.68}$$

c) For the last term we have

$$B \equiv -\sum_{n'l'm'} \int d^3r_2 \frac{e^2}{r_1 r_2^2 r_{12}} \times u_{n'l'}(r_2) u_{nl}(r_2) u_{n'l'}(r_1) Y_{l'}^{m'*}(\hat{\boldsymbol{r}}_2) Y_l^m(\hat{\boldsymbol{r}}_2) Y_{l'}^{m'}(\hat{\boldsymbol{r}}_1) \alpha_\sigma \,. \tag{14.69}$$

Substituting for $1/r_{12}$ the same expansion as in the previous case and carrying out the angular integration with the help of (A.44) we obtain

$$B = -\sum_{n'l'm'} \sum_{L=0}^{\infty} \sqrt{\frac{4\pi(2l+1)}{(2l'+1)(2L+1)}} C(lLl'; m, m'-m) C(lLl'; 00) \times \int_0^\infty dr_2 \frac{e^2}{r_1} \frac{r_<^L}{r_>^{L+1}} u_{n'l'}(r_2) u_{nl}(r_2) u_{n'l'}(r_1) Y_{l'}^{m'}(\hat{\boldsymbol{r}}_1) Y_L^{m'-m*}(\hat{\boldsymbol{r}}_1) \alpha_\sigma . \tag{14.70}$$

Using (A.43) and the orthogonality and symmetry properties of the Clebsch-Gordan coefficients (14.70) reduces to

$$B = -\sum_{n'l'} \sum_{L=0}^{\infty} C^2(lLl'; 00) \times \int_0^\infty dr_2 \frac{e^2}{r_1} \frac{r_<^L}{r_>^{L+1}} u_{n'l'}(r_2) u_{nl}(r_2) u_{n'l'}(r_1) Y_l^m(\hat{\boldsymbol{r}}_1) \alpha_\sigma \,. \tag{14.71}$$

Summing up all the partial results, the Hartree-Fock equations for closed shells can be written as

$$\left[-\frac{1}{2\mu} \frac{d^2}{dr_1^2} + \frac{l(l+1)}{2\mu r_1^2} - \frac{Ze^2}{r_1} - \varepsilon_{nl} + \sum_{n'l'} 2(2l'+1) \int_0^\infty dr_2 \frac{e^2}{r_>} u_{n'l'}^2(r_2) \right] u_{nl}(r_1)$$
$$= \sum_{n'l'} \sum_{L=0}^{\infty} C^2(lLl'; 00) e^2 \left[\int_0^\infty dr_2 \frac{r_<^L}{r_>^{L+1}} u_{n'l'}(r_2) u_{nl}(r_2) \right] u_{n'l'}(r_1) \,, \tag{14.72}$$

where the sum over $n'l'$ ranges over all closed shells. This proves that under the conditions stated previously, the potential (direct + exchange) is spherically symmetric.

In the computations of atomic structures it is preferable to use central potentials averaging over the incomplete shells with certain weights. This is the central field approximation to the Hartree-Fock equations. In most cases this approximation is quite good, and this is because in general there is only one open shell. Also, in the ground state the electron spins tend to be oriented parallel to each other in a manner compatible with Pauli's exclusion principle: to the state with highest spin corresponds a symmetric spin wave function, so that the orbital wave function will be antisymmetric, reducing the electrostatic repulsion between the electrons (Hund's rule [HU 28]). Thus when there are $2l+1$ electrons in a shell, their spins will all be parallel, and repeating the previous computations we easily see that the average potential is still central; so the cases where the approximation may worsen are those with l electrons in an incomplete (nl) shell. With the exception of $l = 0$, which does not generate any problems, all orbitals of the known atoms correspond to $l = 1, 2, 3$, as mentioned in Sect. 14.2. For the case $l = 1$ with a single electron in the open shell, the average potential acting on it is still central, because in Eq. (14.45) the contribution due to this electron in the direct potential is cancelled by the contribution coming from the exchange term. For $l = 2$ and 2 electrons, only the interaction between them breaks the symmetry, and it is to be expected that the approximation still holds good. For $l = 3$ and 3 electrons in this shell, it still seems that the approximation is acceptable, for this case only appears for values of $Z \gtrsim 38$. All the above make us expect that the central approximation is reasonably good. Detailed computations using this method show that it is possible to reproduce the energies for extracting an electron from different atomic shells with an error of less than 10% in general. For details concerning the computations in the Hartree-Fock approximation, the reader is referred to [HA 57], [HA 63] and [BJ 68].

The presence in (14.72) of the exchange potential makes it more difficult to find a solution. For this reason it is often replaced by a local central potential obtained with the Thomas-Fermi method or with other approximations [BJ 68].

In the following, and accepting the central field approximation, we specify the distribution of atomic electrons with the standard notation $n_1 l_1^{k_1} n_2 l_2^{k_2} \ldots$, indicating the presence of k_i electrons in the shell $n_i l_i$. For example, the ground state of chlorine corresponds to $1s^2 2s^2 2p^6 3s^2 3p^5$. All shells are filled except for the last one, containing five electrons. When a state is described giving the number of electrons in each shell, we say that its *configuration* is given.

The filling order of the shells is determined by Pauli's exclusion principle and by energetic considerations. It can be shown that the filling order is given by the following rule: The shells are filled in increasing order of $n + l$, and for the same $n + l$, in increasing order of n. Therefore the normal filling order for atomic shells is

$$1s;\ 2s,\ 2p;\ 3s,\ 3p;\ 4s,\ 3d,\ 4p;\ 5s,\ 4d,\ 5p;\ 6s,\ 4f,\ 5d,\ 6p;\ 7s,\ 5f,\ 6d,\ 7p;\ \ldots, \tag{14.73}$$

where we have placed a semicolon to separate those shells that when filled provide particularly stable configurations. This occurs for $Z = 2, 10, 18, 36, 54, 86$, in other words, for all noble gases. In Table 14.2 we give the exceptions to the normal filling rule. One should keep in mind that the shells previous to those given are all filled. Notice that most of the atoms with an anomalous filling of their shells are either rare earths or transuranic elements.

Table 14.2. Atoms with an anomalous filling of their shells [CW 66]

	$4s$	$3d$
Cr	1	5
Cu	1	10

	$4f$	$5d$
La	0	1
Gd	7	1

	$6s$	$4f$	$5d$
Pt	1	14	9
Au	1	14	10

	$5s$	$4d$
Nb	1	4
Mo	1	5
Tc	1	6
Ru	1	7
Rh	1	8
Pd	0	10
Ag	1	10

	$5f$	$6d$
Ac	0	1
Th	0	2
Pa	2	1
U	3	1
Cm	7	1
Bk	8	1

14.5 Perturbative Calculations

Consider now some given configuration, and let nl^k be a term in it. The k electrons can be distributed among the $(4l + 2)$ possible states, so that this term will contribute a factor $\binom{4l+2}{k}$ to the degeneracy of the configuration. Obviously, a complete or closed shell does not introduce any degeneracy, and the total degeneracy of a configuration is the product of the degeneracies of each incomplete shell. For example, for carbon, whose ground state corresponds to $1s^2 2s^2 2p^2$, the degeneracy is 15, whereas the corresponding degeneracy for a configuration $1s^2 2s^2 2p^6 3s^2 3p^6 4s^2 3d^2 4p^1$ would be $45 \times 6 = 270$.

If the total Hamiltonian were given by (14.7) with an average central potential $V(r_i)$ common to all electrons and computed by the Hartree-Fock method in the central approximation after the exchange interaction had been replaced by a central effective potential, then all spectra would be relatively simple. However, this is not the case, because the interaction terms H_1, H_2, and H_3 given respectively in (14.8), (14.9) and (14.10) tend to break the configuration degeneracy, and the spectra of many-electron atoms will in general be very complex. Let us analyze how these corrections can be treated.

Denote by $\Phi_\alpha, \Phi_\beta, \ldots,$ an orthonormal basis of totally antisymmetric wave functions of the type (14.40) corresponding to a given configuration; their number equals the degeneracy g of the configuration. If we assume that $H_1+H_2+H_3$ can be treated as a small perturbation to H_0, the energy E_0 of the configuration will split due to the perturbation into a set of energy levels with energies $E_0 + E_i$, $1 \le i \le g$, where the shifts E_i are the roots of the equation

$$\det |\langle \Phi_\beta | H_1 + H_2 + H_3 | \Phi_\alpha \rangle - E\delta_{\beta\alpha}| = 0 \,. \tag{14.74}$$

This equation can be simplified if we take into account that H_1 is a single-particle operator given by a local central potential $W(r) = -Ze^2/r - V(r)$. This makes its general expression as a single-particle operator

$$H_1 = \sum_{nlm\sigma} \sum_{n'l'm'\sigma'} \langle n'l'm'\sigma' | W | nlm\sigma \rangle a^\dagger_{n'l'm'\sigma'} a_{nlm\sigma} \,, \tag{14.75}$$

to be reduced to

$$H_1 = \sum_{nn'} \sum_{lm\sigma} \langle n'lm\sigma | W | nlm\sigma \rangle a^\dagger_{n'lm\sigma} a_{nlm\sigma} \,. \tag{14.76}$$

From here it is clear that since all Φ_α belong to the same configuration, then

$$\begin{aligned} \langle \Phi_\beta | H_1 | \Phi_\alpha \rangle &= \delta_{\alpha\beta} \sum_{nl} I(nl) \,, \\ I(nl) &\equiv \int_0^\infty dr\, |u_{nl}(r)|^2 W(r) \,, \end{aligned} \tag{14.77}$$

where the sum runs over all the occupied electronic states in the configuration considered [if in the shell (nl) there are k electrons, the term $I(nl)$ appears k times in the sum]. Hence it is clear that H_1 alone is unable to break the degeneracy and that the original level will split into levels with energies of the form

$$E_i = E_0 + \sum_{nl} I(nl) + E_i' \,, \tag{14.78}$$

where the E_i' are the roots of the equation

$$\det |\langle \Phi_\beta | H_2 + H_3 | \Phi_\alpha \rangle - \delta_{\beta\alpha} E'| = 0 \,. \tag{14.79}$$

Equation (14.79) is still too complicated to be analyzed directly.

In the following we discuss those cases where for a given configuration $|\langle \Phi_\beta | H_2 | \Phi_\alpha \rangle| \ll |\langle \Phi_\beta | H_3 | \Phi_\alpha \rangle|$ is satisfied, as is the case for light atoms and the ground state and first excited states of some heavy atoms. We will also give some ideas on how to treat the case $|\langle \Phi_\beta | H_3 | \Phi_\alpha \rangle| \ll |\langle \Phi_\beta | H_2 | \Phi_\alpha \rangle|$, present mainly in heavy atoms. The case where both terms are of the same order of magnitude will not be studied. For details see [CS 57], [KU 62], among others.

14.6 Russell-Saunders or LS Coupling

Consider a configuration satisfying $|\langle\Phi_\beta|H_2|\Phi_\alpha\rangle| \ll |\langle\Phi_\beta|H_3|\Phi_\alpha\rangle|$. Assuming for the moment $H_2 = 0$, the total Hamiltonian in the notation of Sect. 14.1 is $H = H_0 + H_1 + H_3$. Let $\boldsymbol{L}$ be the total orbital angular momentum of the electrons; it is obvious that $[\boldsymbol{L}, H_0 + H_1] = 0$, and on the other hand, a direct computation shows that $[\boldsymbol{L}_i+\boldsymbol{L}_j, f(r_{ij})] = 0$, where $f(r_{ij})$ is an arbitrary function of $r_{ij} = |\boldsymbol{r}_i - \boldsymbol{r}_j|$. Thus

$$[\boldsymbol{L}, H] = [\boldsymbol{L}, H_3] = 0 . \tag{14.80}$$

Similarly the total spin angular momentum operator $\boldsymbol{S}$ also satisfies

$$[\boldsymbol{S}, H] = [\boldsymbol{S}, H_3] = 0 . \tag{14.81}$$

This result is a consequence of the fact that H does not contain any spin operators, but this argument, even though applicable to the individual spin operators, does not have new physical consequences due to the antisymmetry of the N-electron wave function.

Since H and H_3 commute with $\boldsymbol{L}^2, L_3, \boldsymbol{S}^2$ and S_3, it is convenient to choose the states corresponding to a given configuration as simultaneous eigenstates of these four operators. This coupling scheme is known as the *Russell-Saunders* or *LS coupling* [RS 25]. These states will be denoted by $|LMS\Lambda\alpha\rangle$, where Λ is the third component of the total spin, and α stands for all other quantum numbers necessary to completely specify the states. The advantage of this basis, is that (14.79) becomes (with $H_2 = 0$):

$$\begin{aligned} \det |\langle LMS\Lambda\beta|H_3|LMS\Lambda\alpha\rangle\delta_{LL'}\delta_{MM'}\delta_{SS'}\delta_{\Lambda\Lambda'} \\ - E'\delta_{LL'}\delta_{MM'}\delta_{SS'}\delta_{\Lambda\Lambda'}\delta_{\beta\alpha}| = 0 , \end{aligned} \tag{14.82}$$

and the problem is reduced to solving the equations

$$\det |\langle LMS\Lambda\beta|H_3|LMS\Lambda\alpha\rangle - E'\delta_{\beta\alpha}| = 0 \tag{14.83}$$

for all values of L, S in the configuration considered. The states with the same values of L and S but different values of M and Λ give rise to the same equation, because H_3 commutes with $\boldsymbol{L}$ and $\boldsymbol{S}$, and it cannot break the degeneracy with respect to the possible values of M and Λ.

To compute the matrix elements appearing in (14.83) it is convenient to express the state $|LMS\Lambda\alpha\rangle$ as a linear combination of Slater determinant states for which we use the notation $|(n_1l_1m_1\sigma_1,\ldots,n_Nl_Nm_N\sigma_N)\rangle$, with the occupied states explicitly indicated. The states are ordered so that the first $N-n$ correspond to the electrons in closed shells and the last n to those in open or incomplete shells. Keeping in mind that the electrons in closed shells couple in such a way that $L = S = 0$, the relation between the two bases is

$$|LMS\Lambda\alpha\rangle = \sum_{\{n_il_im_i\sigma_i\}} c_\alpha(\{n_il_im_i\sigma_i\})|(n_1l_1m_1\sigma_1,\ldots,n_Nl_Nm_N\sigma_N)\rangle , \tag{14.84}$$

where the sum ranges over the electrons in open shells; the coefficients $c_\alpha(\{n_i l_i m_i \sigma_i\})$ are more or less complicated, and can be obtained as combinations of Clebsch-Gordan coefficients, but their explicit form will not be needed here.

To find out the allowed values of L and S in a given configuration, let us consider, for instance, the fundamental configuration of carbon, $1s^2 2s^2 2p^2$, whose degeneracy is 15 due to the two electrons in the $2p$ shell. Taking into account that these two electrons cannot be in the same state, we can immediately construct Table 14.3, where all the possible different states of these electrons are given with the notation $|m_5^{\text{"sign"}\sigma_5}, m_6^{\text{"sign"}\sigma_6}\rangle$, and we also list the values of M and Λ. According to this table, and using the standard spectroscopic notation ^{2S+1}L, we obtain as possible states ^1D, ^3P, ^1S, with degeneracies 5, 9 and 1, respectively, whose sum gives the total degeneracy of the initial configuration. Notice that in this case the matrix determining their energies is already diagonal, and barring accidental degeneracies we can say that the original level has split due to the action of H_3 into three levels ^1D, ^3P and ^1S.

With similar arguments – or, more conveniently, using group theory techniques – one shows that for incomplete shells of the form np^k or nd^k the possible values of L and S are those appearing in Table 14.4. The necessity to include additional quantum numbers α is clear for nd^3, nd^4 and nd^5 in the table, and this in general means that solving (14.83) is not a trivial problem.

Table 14.3. Different possible states for two electrons in the configuration np^2

np^2		Λ		
		1	0	-1
	2		$\|1^+1^-\rangle$	
	1	$\|1^+0^+\rangle$	$\|1^+0^-\rangle, \|1^-0^+\rangle$	$\|1^-0^-\rangle$
M	0	$\|1^+-1^+\rangle$	$\|1^+-1^-\rangle, \|1^--1^+\rangle, \|0^+0^-\rangle$	$\|1^--1^-\rangle$
	-1	$\|0^+-1^+\rangle$	$\|0^+-1^-\rangle, \|0^--1^+\rangle$	$\|0^--1^-\rangle$
	-2		$\|-1^+-1^-\rangle$	

Table 14.4. Possible values of L and S for open shells np^k and nd^k. Those for np^{6-k} and nd^{10-k} are the same as those for np^k and nd^k

np	$k=1$	^2P
	$k=2$	$^1S, ^3P, ^1D$
	$k=3$	$^4S, ^2P, ^2D$
nd	$k=1$	^2D
	$k=2$	$^1S, ^3P, ^1D, ^3F, ^1G$
	$k=3$	$^2P, ^4P, ^2D, ^2D, ^2F, ^4F, ^2G, ^2H$
	$k=4$	$^1S, ^1S, ^3P, ^3P, ^1D, ^1D, ^3D, ^5D, ^1F, ^3F, ^3F, ^1G, ^1G, ^3G, ^3H, ^1J$
	$k=5$	$^2S, ^6S, ^2P, ^4P, ^2D, ^2D, ^2D, ^4D, ^2F, ^2F, ^4F, ^2G, ^2G, ^4G, ^2H, ^2J$

To calculate the spectrum of $H_0+H_1+H_3$ to first order in perturbation theory, we need to know the matrix elements appearing in (14.83). In other words, from (14.18), we need the quantities

$$A_k \equiv \langle (n_N l_N m'_N \sigma'_N, \ldots, n_1 l_1 m'_1 \sigma'_1)| \times \sum_{1\leq i<j}^{N} \frac{e^2}{r_{ij}} |(n_1 l_1 m_1 \sigma_1, \ldots, n_N l_N m_N \sigma_N)\rangle \ . \tag{14.85}$$

Here obviously $m_i = m'_i$ and $\sigma_i = \sigma'_i$ for $i \leq N-n$, the indices corresponding to electrons in closed shells. The subindex k indicates that there are k final single-particle states that are different from the initial ones. Since H_3 is an operator containing only two-body operators, we have $A_k = 0$ if $k \geq 3$ and it suffices to consider A_0, A_1, and A_2: We use the abbreviation $i \equiv \{n_i l_i m_i \sigma_i\}$, $i' \equiv \{n_i, l_i m'_i \sigma'_i\}$.

1) A_2. Suppose $i \neq i'$ and $j \neq j'$; in this case both i and j must correspond to open shells, and using the general techniques for the computation of matrix elements of the type (14.85), we obtain

$$A_2 = \langle (j'i')| \frac{e^2}{r_{12}} |(ij)\rangle = \langle i'j'| \frac{e^2}{r_{12}} |ij\rangle - \langle j'i'| \frac{e^2}{r_{12}} |ij\rangle \ . \tag{14.86}$$

Using (A.41) and (A.135) we have

$$\langle i'j'| \frac{1}{r_{12}} |ij\rangle = \delta_{\sigma_i\sigma'_i}\delta_{\sigma_j\sigma'_j} \sum_{L=0}^{\infty} \sum_{M=-L}^{+L} \frac{4\pi}{2L+1} F^L(n_i l_i; n_j l_j) \times \int d\Omega_1\, Y_{l_i}^{m'_i *}(\hat{\boldsymbol{r}}_1) Y_L^{M*}(\hat{\boldsymbol{r}}_1) Y_{l_i}^{m_i}(\hat{\boldsymbol{r}}_1) \times \int d\Omega_2\, Y_{l_j}^{m'_j *}(\hat{\boldsymbol{r}}_2) Y_L^{M}(\hat{\boldsymbol{r}}_2) Y_{l_j}^{m_j}(\hat{\boldsymbol{r}}_2) \ , \tag{14.87}$$

where

$$F^L(n_i l_i; n_j l_j) \equiv \int_0^\infty dr_1 \int_0^\infty dr_2\, u^2_{n_i l_i}(r_1) u^2_{n_j l_j}(r_2) \frac{r_<^L}{r_>^{L+1}} \ . \tag{14.88}$$

The integrals in (14.87) can be carried out with the help of (A.44) so that

$$\langle i'j'| \frac{1}{r_{12}} |ij\rangle = \delta_{\sigma_i\sigma'_i}\delta_{\sigma_j\sigma'_j}\delta_{m_i+m_j, m'_i+m'_j} \sum_{L=0}^{\infty} F^L(n_i l_i; n_j l_j) \times C(l_i L l_i; m'_i, m_i - m'_i) C(l_i L l_i; 00) \times C(l_j L l_j; m_j, m'_j - m_j) C(l_j L l_j; 00) \ . \tag{14.89}$$

With the notation

$$C^L(lm; l'm') \equiv (-1)^{m+m'} \sqrt{\frac{2l+1}{2l'+1}}\, C(lLl'; m, m'-m) C(lLl'; 00) \tag{14.90}$$

the previous result becomes

$$\langle i'j'|\frac{1}{r_{12}}|ij\rangle = \delta_{\sigma_i\sigma_i'}\delta_{\sigma_j\sigma_j'}\delta_{m_i+m_j,m_i'+m_j'} \times \sum_{L=0}^{\infty} C^L(l_i m_i'; l_i m_i) C^L(l_j m_j; l_j m_j') F^L(n_i l_i; n_j l_j)\,. \quad (14.91)$$

Similarly, we obtain

$$\langle j'i'|\frac{1}{r_{12}}|ij\rangle = \delta_{\sigma_i\sigma_j'}\delta_{\sigma_j\sigma_i'}\delta_{m_i+m_j,m_i'+m_j'} \times \sum_{L=0}^{\infty} C^L(l_i m_i'; l_j m_j) C^L(l_i m_i; l_j m_j') G^L(n_i l_i; n_j l_j)\,, \quad (14.92)$$

where

$$G^L(n_i l_i; n_j l_j) \equiv \int_0^\infty dr_1 \int_0^\infty dr_2\, u_{n_i l_i}(r_1) u_{n_j l_j}(r_1) u_{n_i l_i}(r_2) u_{n_j l_j}(r_2) \frac{r_<^L}{r_>^{L+1}}\,. \quad (14.93)$$

Combining (14.91) and (14.92), we conclude the evaluation of A_2.

2) A_1. In this case the only orbital different in the initial and final states must belong to an open shell. If $j \neq j'$ then

$$A_1 = \sum_i \left[\langle i'j|\frac{e^2}{r_{12}}|ij\rangle - \langle j'i|\frac{e^2}{r_{12}}|ij\rangle \right]\,. \quad (14.94)$$

From (14.91) and (14.92) we obtain $A_1 = 0$, after we take into account $i = i'$, $j \neq j'$.

3) A_0. In this case we can write:

$$A_0 = \frac{1}{2}\sum_i \sum_j \left[\langle ij|\frac{e^2}{r_{12}}|ij\rangle - \langle ji|\frac{e^2}{r_{12}}|ij\rangle \right]\,. \quad (14.95)$$

Denoting by C the orbitals belonging to complete shells and by NC those for incomplete shells:

$$A_0 = \frac{1}{2}\sum_i \left[\sum_{j\in C} + \sum_{j\in NC} \right] \left[\langle ij|\frac{e^2}{r_{12}}|ij\rangle - \langle ji|\frac{e^2}{r_{12}}|ij\rangle \right]\,. \quad (14.96)$$

In order to obtain a simple expression for the sum over complete shells, we use the relations

$$C^0(lm;l'm') = \delta_{ll'}\delta_{mm'} ,$$
$$\sum_{m=-l}^{+l} C^L(lm;lm) = \delta_{L0}(2l+1) , \tag{14.97}$$
$$\sum_{m'=-l'}^{+l'} C^L(lm;l'm')C^L(lm;l'm') = C^2(lLl';00) .$$

The second identity is a consequence of a representation of the Clebsch-Gordan coefficients as an integral over spherical harmonics using (A.44) and (A.39). The other two are trivial to prove. These relations together with (14.91) and (14.92) lead to

$$\sum_{\sigma_j}\sum_{m_j=-l_j}^{+l_j} \langle ij|\frac{1}{r_{12}}|ij\rangle = 2(2l_j+1)F^0(n_il_i;n_jl_j) ,$$
$$\sum_{\sigma_j}\sum_{m_j=-l_j}^{+l_j} \langle ji|\frac{1}{r_{12}}|ij\rangle = \sum_{L=0}^{\infty} C^2(l_iLl_j;00)G^L(n_il_i;n_jl_j) , \tag{14.98}$$

so that

$$A_0 = \frac{e^2}{2}\sum_i\left\{\sum_{nlC}\left[(4l+2)F^0(n_il_i;nl) - \sum_{L=0}^{\infty} C^2(l_iLl;00)G^Ln_il_i;nl)\right]\right.$$
$$\left. + \sum_{j\in NC}\left[\langle ij|\frac{1}{r_{12}}|ij\rangle - \langle ji|\frac{1}{r_{12}}|ij\rangle\right]\right\} , \tag{14.99}$$

where the sum over nlC is a sum over all values of n and l corresponding to closed shells. Separating in $\sum_i$ the terms corresponding to complete shells from the others, we obtain

$$A_0 = \frac{e^2}{2}\sum_{nlC}\sum_{n'l'C}(4l+2)(4l'+2)$$
$$\times\left[F^0(n'l';nl) - \sum_{L=0}^{\infty}\frac{C^2(l'lL;00)}{2(2L+1)}G^L(n'l';nl)\right]$$
$$+ e^2\sum_{nlC}\sum_{n'l'NC}(4l+2)k'\left[F^0(n'l';nl) - \sum_{L=0}^{\infty}\frac{C^2(l'lL;00)}{2(2L+1)}G^L(n'l';nl)\right]$$
$$+\frac{e^2}{2}\sum_{i\in NC}\sum_{j\in NC}\left[\langle ij|\frac{1}{r_{12}}|ij\rangle - \langle ji|\frac{1}{r_{12}}|ij\rangle\right] . \tag{14.100}$$

Taking now into account the relations (14.84) and the matrix elements we have just computed explicitly, one can analyze the equation (14.83) for the different configurations. When none of the states ^{2S+1}L appear repeated in the configuration given, the quantum numbers α are unnecessary, and the energy levels can be computed using a sum rule due to *Slater* [SL 29]. To establish it, we simply notice that the transformation (14.84) is unitary, and consequently it leaves invariant the trace of a matrix. Hence the trace of H_3 in the space of states $|LMS\Lambda\rangle$, for M and Λ fixed, is

$$\sum_{\{n_i l_i m_i \sigma_i\}} \langle (n_N l_N m_N \sigma_N, \dots, n_1 l_1 m_1 \sigma_1)|H_3|(n_1 l_1 m_1 \sigma_1, \dots, n_N l_N m_N \sigma_N)\rangle$$

$$= \sum_{L\geq|M|} \sum_{S\geq|\Lambda|} E(LMS\Lambda) , \qquad (14.101)$$

$$\sum_{i=N-n+1}^{N} m_i = M , \qquad \sum_{i=N-n+1}^{N} \sigma_i = \Lambda ,$$

where the E's are the energies of interest, and the variation in the quantum numbers m_i, σ_i is subject to the last two equations.

To understand how this sum rule is applied, consider again the configuration $1s^2 2s^2 2p^2$. According to (14.100) the contribution of each term in the right-hand side of (14.101) can be written as a constant $\mathcal{E}$ depending only on the configuration considered [first two terms in (14.100)] plus a sum over the occupied orbitals in the open shells. In this particular example, each term in the right-hand side of (14.101), which for abbreviation will be denoted by $\langle (m_6\sigma_6, m_5\sigma_5)|H_3|(m_5\sigma_5, m_6\sigma_6)\rangle$, can be written as

$$\langle (m_6\sigma_6, m_5\sigma_5)|H_3|(m_5\sigma_5, m_6\sigma_6)\rangle$$
$$= \mathcal{E} + e^2 \sum_{L=0}^{\infty} [C^L(1m_5; 1m_5) C^L(1m_6; 1m_6) - \delta_{\sigma_5\sigma_6}$$
$$\times C^L(1m_5; 1m_6) C^L(1m_5; 1m_6)] F^L(21; 21) , \qquad (14.102)$$

using (14.91), and (14.92) and the fact that $G^L(nl; nl) = F^L(nl; nl)$ – easily derived from their definitions. From (14.92) and the relation $C(111; 00) = 0$ it is obvious that L can take only the values $L = 0$ and $L = 2$, and the necessary values of C^L are listed in Table 14.5.

Table 14.5. Values of $C^L(1m_1; 1m_2)$ with $L = 0, 2$

$C^0(1m_1; 1m_2)$

m_2 \ m_1	1	0	-1
1	1	0	0
0	0	1	0
-1	1	0	1

$C^2(1m_1; 1m_2)$

m_2 \ m_1	1	0	-1
1	$-\frac{1}{5}$	$-\frac{\sqrt{3}}{5}$	$-\frac{\sqrt{6}}{5}$
0	$\frac{\sqrt{3}}{5}$	$\frac{2}{5}$	$\frac{\sqrt{3}}{5}$
-1	$-\frac{\sqrt{6}}{5}$	$-\frac{\sqrt{3}}{5}$	$-\frac{1}{5}$

Taking into account that in this example only ^1D, ^3P and ^1S states appear, it is easy to obtain from (14.101) by successively taking $(M = 2, \Lambda = 0)$, $(M = 1, \Lambda = 1)$ and $(M = 0, \Lambda = 0)$ the following results

$$\begin{aligned} E(^1D) &= \mathcal{E} + e^2F^0(21;21) + \frac{e^2}{25}F^2(21;21)\,, \\ E(^3P) &= \mathcal{E} + e^2F^0(21;21) - \frac{e^2}{5}F^2(21;21)\,, \\ E(^1S) + E(^3P) + E(^1D) &= 3\mathcal{E} + 3e^2F^0(21;21) + \frac{6e^2}{25}F^2(21;21)\,, \end{aligned} \tag{14.103}$$

which provide $E(^1S)$, $E(^3P)$ and $E(^1D)$ after carrying out the radial integrals. Notice that independently of the values of $\mathcal{E}$, F^0 and F^2, the following relation holds:

$$\frac{E(^1S) - E(^1D)}{E(^1D) - E(^3P)} = \frac{3}{2}\,. \tag{14.104}$$

From the way this result was obtained, it is obvious that it holds for all configurations with a single incomplete shell of the form np^2. For similar formulae in other cases of interest, see [CS 57], [KU 62] and [SL 60]. In Table 14.6 we compare the relation between (14.104) and some experimental data. The agreement is qualitatively good, but not very satifactory quantitatively, possibly because we have not taken into account the possibility of configuration mixing, i.e., the possibility that the ground state considered be a superposition of several configurations with appropriate weights.

Table 14.6. Comparison of (14.104) with experimental data

Atom	Configuration	$\frac{E(^1S) - E(^1D)}{E(^1D) - E(^3P)}$
C	$2p^2$	1.13
N^+	$2p^2$	1.14
O^{++}	$2p^2$	1.14
Si	$3p^2$	1.48
Ge	$4p^2$	1.50
Sn	$5p^2$	1.39

Before concluding, we would like to add, that as a consequence of numerous calculations similar to the one presented, and supported by substantial experimental evidence, we have Hund's rule [HU 25]: the term with lowest energy of a given configuration is the one with the largest value of S; and if there are several of these, the one among them with the largest value of L.

So far we have neglected the spin-orbit interaction H_2, with the assumption that its matrix elements are very small compared with those of H_3. We now discuss the effect of H_2. When this term is included $\boldsymbol{L}$ and $\boldsymbol{S}$ stop being conserved quantities independently, although $\boldsymbol{J}$ certainly remains a constant of the motion.

Fortunately, the assumption that the effect of H_2 is small implies that to first order its non-diagonal elements in the LS coupling scheme can be neglected, with the result that a specific level ${}^{2S+1}L$ (for a given α not written explicitly) will split under the influence of H_2 into a set of levels corresponding to the possible values of $J = |L - S|, \ldots, L + S$. Their energies, referred to the energy of the original term ${}^{2S+1}L$, are the roots of the equation

$$\det |\langle LM'S\Lambda'|H_2|LMS\Lambda\rangle - E\delta_{M'M}\delta_{\Lambda'\Lambda}| = 0 , \tag{14.105}$$

where the matrix elements are

$$\langle LM'S\Lambda'|H_2|LMS\Lambda\rangle = \langle LM'S\Lambda'|\sum_{i=1}^{N} \xi(r_i)\boldsymbol{L}_i \cdot \boldsymbol{S}_i|LMS\Lambda\rangle . \tag{14.106}$$

Using (B.74) we obtain

$$\begin{aligned}\langle LM'S\Lambda'|H_2|LMS\Lambda\rangle = &\left[\sum_{i-1}^{N} \xi_{n_i l_i} \frac{\langle L\|(\boldsymbol{L}\cdot\boldsymbol{L}_i)\|L\rangle\langle S\|(\boldsymbol{S}\cdot\boldsymbol{S}_i)\|S\rangle}{L(L+1)S(S+1)}\right]\\ &\times \langle LM'S\Lambda'|\boldsymbol{L}\cdot\boldsymbol{S}|LMS\Lambda\rangle ,\end{aligned} \tag{14.107}$$

where

$$\xi_{n_i l_i} \equiv \frac{1}{2m_e^2}\int_0^\infty dr\, u_{n_i l_i}^2(r)\frac{1}{r}\frac{dV(r)}{dr} . \tag{14.108}$$

Obviously, for $L = 0$ or $S = 0$, we have to take as zero the value of (14.107). If we choose as average potential $V(r) = -Ze^2/r$, then $\xi_{n_i l_i} > 0$, and this is also true for more realistic central potentials. Taking into account that $2\boldsymbol{L}\cdot\boldsymbol{S} = \boldsymbol{J}^2 - \boldsymbol{L}^2 - \boldsymbol{S}^2$, and introducing a representation where $\boldsymbol{J}^2$ and J_3 are diagonal, we obtain for the desired energies the expression

$$\begin{aligned}E(LSJ) = &\left[\sum_{i=1}^{N} \xi_{n_i l_i} \frac{\langle L\|(\boldsymbol{L}\cdot\boldsymbol{L}_i)\|L\rangle\langle S\|(\boldsymbol{S}\cdot\boldsymbol{S}_i)\|S\rangle}{L(L+1)S(S+1)}\right]\\ &\times \tfrac{1}{2}[J(J+1) - L(L+1) - S(S+1)] .\end{aligned} \tag{14.109}$$

If all the shells are closed except for one (nl) with k electrons, using that for closed shells the orbital and spin angular momenta vanish, and that the operator H_2 is a single-particle operator, it can be shown [BJ 68] that the electrons in closed shells give a vanishing contribution to $E(LSJ)$, and therefore

$$\begin{aligned}E(LSJ) &= \xi_{nl}\tfrac{1}{2}A(LS)[J(J+1) - L(L+1) - S(S+1)] ,\\ A(LS) &\equiv \sum_{i=1}^{k} \frac{\langle L\|(\boldsymbol{L}\cdot\boldsymbol{L}_i)\|L\rangle\langle S\|\boldsymbol{S}\cdot\boldsymbol{S}_i\|S\rangle}{L(L+1)S(S+1)} .\end{aligned} \tag{14.110}$$

The coefficient $A(LS)$ can be computed using (14.84), but in many cases it is calculated using sum rules [KU 62]. Ordinarily $A(LS) \geq 0$ for $k < 2l + 1$ and

$A(LS) \leq 0$ if $k > 2l+1$. When $k = 2l+1$ it can be proved that $A(LS) = 0$, and therefore no splitting appears to first order. From (14.110) we obtain

$$E(LSJ) - E(LSJ-1) = \xi_{nl} A(LS) J \,. \tag{14.111}$$

This is the Landé interval rule [LA 23] according to which the splitting between two consecutive values of J for a given LS multiplet is proportional to the largest one of the two. The multiplets with $A(LS) > 0$ are called regular, and for them the lowest level is $J = |L - S|$; the multiplets with $A(LS) < 0$ are called inverted, and the lowest one corresponds to $J = L + S$. These multiplets constitute the fine structure of the spectrum.

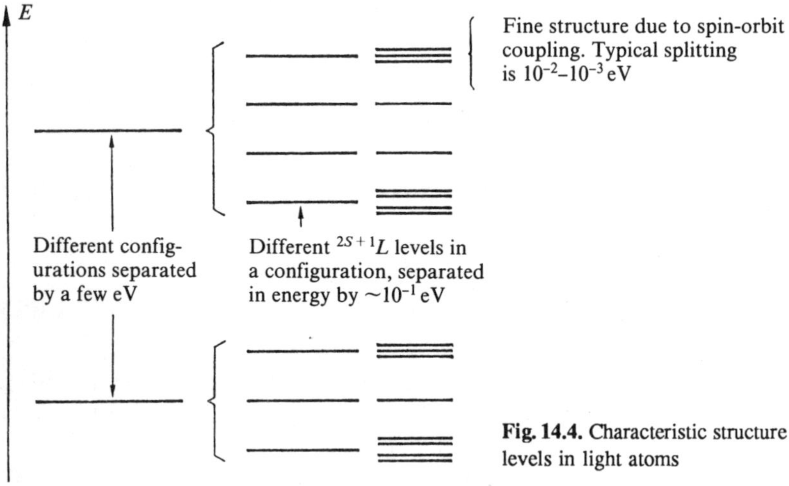

Fig. 14.4. Characteristic structure of energy levels in light atoms

Notice that all these results are justified only if the LS coupling scheme is a good approximation, and the only spin-orbit coupling has the form given in (14.9). In other words, each spin interacts with its own orbit. This is, however, just an approximation. It seems that the spin-spin interactions are also important in many light, and in several heavy atoms [KU 62].

The orders of magnitude for the different terms considered for light atoms are represented in Fig. 14.4.

14.7 *jj* Coupling

The previous section contains some ideas on how to perform spectrum calculations in the extreme case that the spin-orbit interaction is very small compared to the electrostatic repulsion between electrons; this is true in general for light elements. The other extreme case, which enters the study of some configurations in heavy atoms, appears when the spin-orbit coupling dominates the electrostatic interaction. In this situation we can use a different coupling scheme, known as *jj coupling*.

In this new scheme, to a first approximation the term H_3 is neglected in (14.79), and the basis we work with is the one obtained by coupling the orbital and spin angular momenta $\boldsymbol{L}_i$ and $\boldsymbol{S}_i$; the states obtained now are $|(n_1 l_1 j_1 m_1, \ldots, n_N l_N j_N m_N)\rangle$, where j_i is the total angular momentum of the ith electron, and m_i its third component. They diagonalize the spin-orbit interaction, and the energy levels obtained, corresponding to a given configuration, are

$$E(n_1 l_1 j_1 m_1, \ldots, n_N l_N j_N m_N) = \sum_{i=1}^{N} \xi_{n_i l_i} \frac{1}{2} \left[j_i(j_i+1) - l_i(l_i+1) - \frac{3}{4} \right] . \qquad (14.112)$$

Next we consider the effect of the term H_3, which is diagonalized in the basis given by the states $|(n_1 l_1 j_1 m_1, \ldots, n_N l_N j_N m_N)\rangle$ with fixed values of $j_1, \ldots, j_N$, so that a new splitting of the energy levels appears, according to the possible values of the total angular momentum $\boldsymbol{J}$.

It should be pointed out that even though this coupling scheme is relatively unimportant in atoms, it is the normal coupling scheme for nuclei, except for some of the lightest.

15. Quantum Theory of Radiation

15.1 Introduction

The understanding of radiative atomic and nuclear transitions makes it necessary to develop a quantum theory of radiation, and this in turn requires the concept of a quantized field. In a rigorous treatment of this problem, one starts with equations covariant under the Poincaré group, describing the interaction of matter and light, and subsequently one proceeds to quantize them in a covariant manner. A treatment along these lines is far beyond the scope of this book, and the interested reader is referred to the books by *Akhiezer* and *Berestestky* [AB 65], *Bogolyubov* and *Shirkov* [BS 59], *Heitler* [HE 54], *Jauch* and *Rohrlich* [JR 55], *Thirring* [TH 55] and *Itzykson* and *Zuber* [IZ 80] among others. In order to make the presentation as simple as possible, we follow a method similar to the ones used in the study of the quantum theory of many particle systems. The price to pay for this simplification, is the loss of apparent relativistic invariance of the theory.

It is known [JA 75] that in classical electrodynamics, and in the absence of sources, one can describe the E.M. field in the Coulomb or transverse gauge, where the scalar potential is identically zero, and the vector potential satisfies the equations

$$\boxed{\begin{array}{ll} \Box \boldsymbol{A}(\boldsymbol{r};t) \equiv \left[\dfrac{\partial^2}{\partial t^2} - \Delta\right] \boldsymbol{A}(\boldsymbol{r};t) = 0 & (\textit{d'Alembert equation}) \\ \nabla \cdot \boldsymbol{A}(\boldsymbol{r};t) = 0 & (\textit{transversality condition}) \end{array}} \tag{15.1}$$

The electric and magnetic fields are

$$\boldsymbol{E}(\boldsymbol{r};t) = -\frac{\partial \boldsymbol{A}(\boldsymbol{r};t)}{\partial t}\,, \qquad \boldsymbol{B}(\boldsymbol{r};t) = \nabla \times \boldsymbol{A}(\boldsymbol{r};t)\,. \tag{15.2}$$

As in Chap. 12, we adopt the unrationalized gaussian system of units, where $e^2 = \alpha \simeq (137)^{-1}$. From (15.1) and (15.2), it immediately follows that

$$\boxed{\Box \boldsymbol{E}(\boldsymbol{r};t) = 0\,, \quad \Box \boldsymbol{B}(\boldsymbol{r};t) = 0}\,. \tag{15.3}$$

It is also worthwhile to recall at this moment that the energy, momentum and total angular momentum of the electromagnetic field are respectively given by

$$\boxed{\begin{aligned} E &= \frac{1}{8\pi}\int d^3r\,[|\boldsymbol{E}(\boldsymbol{r};t)|^2 + |\boldsymbol{B}(\boldsymbol{r};t)|^2] \\ \boldsymbol{P} &= \frac{1}{4\pi}\mathrm{Re}\int d^3r\,[\boldsymbol{E}(\boldsymbol{r};t)\times\boldsymbol{B}^*(\boldsymbol{r};t)] \\ \boldsymbol{J} &= \frac{1}{4\pi}\mathrm{Re}\int d^3r\,\{\boldsymbol{r}\times[\boldsymbol{E}(\boldsymbol{r};t)\times\boldsymbol{B}^*(\boldsymbol{r};t)]\} \end{aligned}} \tag{15.4}$$

where complex notation is used. Although only the real parts of the fields are physically significant, the use of complex notation simplifies greatly the mathematical computations. In (15.4) the integrations range over all space, and due to the conservation laws, the result of the integration is independent of the instant of time t considered.

15.2 Plane Wave Expansions

To determine the most general solution to (15.1) we take into account that a complete set of solutions to the d'Alembert equation is given by the plane waves $\boldsymbol{\varepsilon}\exp(\mathrm{i}kx)$ and $\boldsymbol{\varepsilon}\exp(-\mathrm{i}kx)$, where $kx \equiv \omega t - \boldsymbol{k}\cdot\boldsymbol{r}$ and $\omega = |\boldsymbol{k}|$. Thus $\boldsymbol{A}(\boldsymbol{r};t)$ can be expanded in a (continuous) linear combination of them. For fixed $\boldsymbol{k}$ we define a basis for $\boldsymbol{\varepsilon}$ consisting of three real vectors $\boldsymbol{\varepsilon}(\boldsymbol{k},i)$, $i = 1,2,3$, such that

$$\boldsymbol{\varepsilon}(\boldsymbol{k},i)\cdot\boldsymbol{\varepsilon}(\boldsymbol{k},j) = \delta_{ij}\ , \quad \boldsymbol{\varepsilon}(\boldsymbol{k},i)\times\boldsymbol{\varepsilon}(\boldsymbol{k},j) = \varepsilon_{ijl}\boldsymbol{\varepsilon}(\boldsymbol{k},l)\ . \tag{15.5}$$

These vectors form a positively oriented orthonormal triad. A convenient choice for

$$\boldsymbol{k} \equiv (\omega\sin\theta\cos\phi, \omega\sin\theta\sin\phi, \omega\cos\theta) \tag{15.6}$$

is

$$\begin{aligned} \boldsymbol{\varepsilon}(\boldsymbol{k},1) &\equiv (\cos\theta\cos\phi, \cos\theta\sin\phi, -\sin\theta)\ , \\ \boldsymbol{\varepsilon}(\boldsymbol{k},2) &\equiv (-\sin\phi, \cos\phi, 0)\ , \\ \boldsymbol{\varepsilon}(\boldsymbol{k},3) &\equiv \hat{\boldsymbol{k}} = \boldsymbol{k}/\omega = (\sin\theta\cos\phi, \sin\theta\sin\phi, \cos\theta)\ . \end{aligned} \tag{15.7}$$

The vector $\boldsymbol{\varepsilon}$ in the plane waves $\boldsymbol{\varepsilon}\exp(\pm\mathrm{i}kx)$ describes the direction of $\boldsymbol{E}$, i.e., the wave *polarization.*

From all this one finds the most general real solution to the d'Alembert equation to be

$$\boldsymbol{A}(\boldsymbol{r};t) = \frac{1}{2\pi}\sum_{i=1}^{3}\int\frac{d^3k}{\sqrt{\omega}}\,[\boldsymbol{\varepsilon}(\boldsymbol{k},i)a(\boldsymbol{k},i)\mathrm{e}^{-\mathrm{i}kx} + \boldsymbol{\varepsilon}(\boldsymbol{k},i)a^*(\boldsymbol{k},i)\mathrm{e}^{\mathrm{i}kx}]\ , \tag{15.8}$$

where $a(\boldsymbol{k},i)$ $i = 1,2,3$, are three arbitrary functions. Imposing the transversality condition for all t gives $a(\boldsymbol{k},3) = 0$, and therefore the sum in (15.8) can be restricted to $i = 1,2$. The fact that for every value of $\boldsymbol{k}$ there are only two independent amplitudes is due to the well known fact that light has only two independent polarizations.

Instead of the two *linear* polarization vectors $\boldsymbol{\varepsilon}(\boldsymbol{k}, i)$, $i = 1, 2$, it is useful to introduce *right* and *left handed* polarizations

$$\boldsymbol{\varepsilon}(\boldsymbol{k}, \pm 1) = \mp \frac{1}{\sqrt{2}} [\boldsymbol{\varepsilon}(\boldsymbol{k}, 1) \pm \mathrm{i}\boldsymbol{\varepsilon}(\boldsymbol{k}, 2)] \tag{15.9}$$

satisfying ($\lambda = \pm 1$)

$$\begin{aligned} &\boldsymbol{\varepsilon}^*(\boldsymbol{k}, \lambda) \cdot \boldsymbol{\varepsilon}(\boldsymbol{k}, \lambda') = \delta_{\lambda\lambda'} , \quad \boldsymbol{\varepsilon}(\boldsymbol{k}, \lambda) \times \boldsymbol{\varepsilon}^*(\boldsymbol{k}, \lambda') = (2\mathrm{i}\omega)^{-1}(\lambda + \lambda')\boldsymbol{k} , \\ &\boldsymbol{k} \times \boldsymbol{\varepsilon}(\boldsymbol{k}, \lambda) = -\mathrm{i}\lambda\omega\boldsymbol{\varepsilon}(\boldsymbol{k}, \lambda) . \end{aligned} \tag{15.10}$$

Defining

$$a(\boldsymbol{k}, \pm 1) = \mp \frac{1}{\sqrt{2}} [a(\boldsymbol{k}, 1) \mp \mathrm{i} a(\boldsymbol{k}, 2)] \tag{15.11}$$

we obtain for the vector potential

$$\boldsymbol{A}(\boldsymbol{r}; t) = \frac{1}{2\pi} \sum_{\lambda=\pm 1} \int \frac{d^3 k}{\sqrt{\omega}} [\boldsymbol{\varepsilon}(\boldsymbol{k}, \lambda) a(\boldsymbol{k}, \lambda) \mathrm{e}^{-\mathrm{i}kx} + \boldsymbol{\varepsilon}^*(\boldsymbol{k}, \lambda) a^*(\boldsymbol{k}, \lambda) \mathrm{e}^{\mathrm{i}kx}] , \tag{15.12}$$

and using (15.2) and (15.10), the electric and magnetic fields are given by

$$\boxed{\begin{aligned} \boldsymbol{E}(\boldsymbol{r}; t) &= \frac{\mathrm{i}}{2\pi} \sum_{\lambda=\pm 1} \int d^3 k \sqrt{\omega} \\ &\quad \times [\boldsymbol{\varepsilon}(\boldsymbol{k}, \lambda) a(\boldsymbol{k}, \lambda) \mathrm{e}^{-\mathrm{i}kx} - \boldsymbol{\varepsilon}^*(\boldsymbol{k}, \lambda) a^*(\boldsymbol{k}, \lambda) \mathrm{e}^{\mathrm{i}kx}] , \\ \boldsymbol{B}(\boldsymbol{r}; t) &= \frac{1}{2\pi} \sum_{\lambda=\pm 1} \int d^3 k \sqrt{\omega} \lambda \\ &\quad \times [\boldsymbol{\varepsilon}(\boldsymbol{k}, \lambda) a(\boldsymbol{k}, \lambda) \mathrm{e}^{-\mathrm{i}kx} + \boldsymbol{\varepsilon}^*(\boldsymbol{k}, \lambda) a^*(\boldsymbol{k}, \lambda) \mathrm{e}^{\mathrm{i}kx}] . \end{aligned}} \tag{15.13}$$

From (15.4) for any time t, for instance $t = 0$, we can compute the energy and momentum by direct substitution of (15.13). The $\boldsymbol{r}$ integration is trivial, and with the resulting delta functions we can eliminate one of the momentum integrations. Using the defining relations (15.10) we finally obtain

$$E = \int d^3 k \, \omega \sum_{\lambda=\pm 1} |a(\boldsymbol{k}, \lambda)|^2 , \quad \boldsymbol{P} = \int d^3 k \, \boldsymbol{k} \sum_{\lambda=\pm 1} |a(\boldsymbol{k}, \lambda)|^2 . \tag{15.14}$$

Thus using the language of photons, we can say that $|a(\boldsymbol{k}, \lambda)|^2$ represents the average number of photons with energy ω, momentum $\boldsymbol{k}$, and right ($\lambda = +1$) or left ($\lambda = -1$) polarization present in the field. In the same way that for plane waves $\boldsymbol{\varepsilon} \exp(-\mathrm{i}kx)$ the energy and momentum densities are proportional to ω and $\boldsymbol{k}$, for waves of finite extension in directions transverse to $\boldsymbol{k}$ it can be shown

[JA 75] that λ measures the photon's helicity, i.e., the component of the total angular momentum of the wave along the direction of propagation. Summarizing, $a(\boldsymbol{k}, \lambda)$ can be interpreted as the probability amplitude for a photon of energy ω, momentum $\boldsymbol{k}$ and helicity λ to be present in the field.

Implicitly we have assumed that E.M. waves occupy the whole of space $\mathbb{R}^3$. We will continue to do so throughout the chapter. If this were not true, the boundary conditions would alter the possible solutions to the Maxwell equations by modifying the normal modes of the E.M. field. The influence on the interaction matter-radiation of the boundary conditions, for instance in the mean lifetime of an excited atomic state inside a metallic cavity, has been observed experimentally [GR 83].

15.3 Quantization of the E.M. Field

So far the results obtained are purely classical. To obtain the quantum version, we will replace the *amplitudes* $a(\boldsymbol{k}, \lambda)$ and $a^*(\boldsymbol{k}, \lambda)$ appearing in (15.12) and (15.13) by *operators* $a(\boldsymbol{k}, \lambda)$ and $a^\dagger(\boldsymbol{k}, \lambda)$ which respectively annihilate and create a photon of energy ω, momentum $\boldsymbol{k}$ and helicity λ with wave function $(2\pi)^{-3/2}\boldsymbol{\varepsilon}(\boldsymbol{k}, \lambda)\exp(-ikx)$.

Since the photons are bosons, these operators satisfy the commutation relations

$$\boxed{\begin{aligned} &[a(\boldsymbol{k}, \lambda), a(\boldsymbol{k}', \lambda')] = [a^\dagger(\boldsymbol{k}, \lambda), a^\dagger(\boldsymbol{k}', \lambda')] = 0 \\ &[a(\boldsymbol{k}, \lambda), a^\dagger(\boldsymbol{k}', \lambda')] = \delta_{\lambda\lambda'}\delta(\boldsymbol{k} - \boldsymbol{k}') \end{aligned}} \; . \tag{15.15}$$

The operators $\boldsymbol{A}, \boldsymbol{E}, \boldsymbol{B}$ so obtained are self-adjoint.

We would like to remark that the field $\boldsymbol{A}$ is represented as a "sum" over momenta and helicities of $\sqrt{4\pi/2\omega}\,[a_\alpha f_\alpha + a_\alpha^\dagger f_\alpha^*]$ where $\alpha \equiv (\boldsymbol{k}, \lambda)$ and f_α is the wave function of the photon that $a(\boldsymbol{k}, \lambda)$ annihilates. (The factor in front would have equalled unity had we used rationalized gaussian units and a Lorentz covariant normalization for the states). For later reference we write

$$\begin{aligned} \boldsymbol{A}(\boldsymbol{r}; t) = &\int d^3k \sqrt{\frac{4\pi}{2\omega}} \sum_{\lambda=\pm 1} \\ &\times \left[\frac{1}{(2\pi)^{3/2}}\,\boldsymbol{\varepsilon}(\boldsymbol{k}, \lambda)e^{-ikx}a(\boldsymbol{k}, \lambda) + \frac{1}{(2\pi)^{3/2}}\,\boldsymbol{\varepsilon}^*(\boldsymbol{k}, \lambda)e^{ikx}a^\dagger(\boldsymbol{k}, \lambda)\right] . \end{aligned} \tag{15.16}$$

If for the energy operator which from now on will be denoted by H_{rad} we take the first equation in (15.4), a computation analogous to the derivation of (15.14) yields

$$H_{\text{rad}} = \frac{1}{2}\int d^3k\,\omega \sum_{\lambda=\pm 1} [a^\dagger(\boldsymbol{k}, \lambda)a(\boldsymbol{k}, \lambda) + a(\boldsymbol{k}, \lambda)a^\dagger(\boldsymbol{k}, \lambda)] \, , \tag{15.17}$$

and thus its expectation value in the vacuum state is not zero as one would expect, but rather it is infinite. This difficulty is resolved if for H_{rad} we use instead

$$H_{\rm rad} = \frac{1}{8\pi} \int d^3r : [\boldsymbol{E}^2(\boldsymbol{r};t) + \boldsymbol{B}^2(\boldsymbol{r};t)] : \tag{15.18}$$

where : ... : stands for *Wick's normal ordering*. This is a prescription dictating that in any expression where both creation and annihilation operator appear, each monomial should be reordered as though they were commuting until the annihilation operators appear to the right of the creation operators. Consequently

$$\boxed{H_{\rm rad} = \int d^3k\, \omega \sum_{\lambda=\pm 1} a^\dagger(\boldsymbol{k},\lambda) a(\boldsymbol{k},\lambda)\,.}$$

Replacing (15.17) by (15.19) is physically equivalent to a renormalization of the origin of energies. We take the vacuum as the origin of the energy scale. The same prescription will be used when we associate operators to those classical quantities appearing in the description of the E.M. field, and in particular for those in (15.4).

Since the components of the E.M. field are now operators, we should study their commutation relations. As an illustration, let us consider in detail the commutator of two components of the electric field. From (15.13) and (15.15), we obtain

$$\begin{aligned}[E_j(\boldsymbol{r};t), E_l(\boldsymbol{r}';t')] &= \frac{1}{4\pi^2} \sum_{\lambda=\pm 1} \int d^3k\, \omega \\ &\times [\varepsilon_j(\boldsymbol{k},\lambda)\varepsilon_l^*(\boldsymbol{k},\lambda) e^{-ik(x-x')} - \varepsilon_j^*(\boldsymbol{k},\lambda)\varepsilon_l(\boldsymbol{k},\lambda) e^{ik(x-x')}]\,, \end{aligned} \tag{15.20}$$

and using

$$\sum_{\lambda=\pm 1} \varepsilon_j(\boldsymbol{k},\lambda)\varepsilon_l^*(\boldsymbol{k},\lambda) = \delta_{jl} - \frac{k_j k_l}{\omega^2} \tag{15.21}$$

we may write

$$[E_j(\boldsymbol{r};t), E_l(\boldsymbol{r};t)] = -\frac{i}{2\pi^2} \int \frac{d^3k}{\omega} [\omega^2 \delta_{jl} - k_j k_l] e^{i\boldsymbol{k}\cdot(\boldsymbol{r}-\boldsymbol{r}')} \sin\omega(t-t')\,. \tag{15.22}$$

Defining the function

$$D(\boldsymbol{r},t) \equiv -\frac{1}{(2\pi)^3} \int \frac{d^3k}{\omega} e^{i\boldsymbol{k}\cdot\boldsymbol{r}} \sin\omega t \tag{15.23}$$

one easily obtains

$$[E_j(\boldsymbol{r};t), E_l(\boldsymbol{r};t)] = -4\pi i \left[\delta_{jl} \frac{\partial^2}{\partial t^2} - \frac{\partial^2}{\partial x_j \partial x_l}\right] D(\boldsymbol{r}-\boldsymbol{r}', t-t')\,. \tag{15.24}$$

Similarly

$$[B_j(\boldsymbol{r};t), B_l(\boldsymbol{r};t)] = -4\pi\mathrm{i}\left[\delta_{jl}\frac{\partial^2}{\partial t^2} - \frac{\partial^2}{\partial x_j \partial x_l}\right] D(\boldsymbol{r}-\boldsymbol{r}', t-t') ,$$
$$[E_j(\boldsymbol{r};t), B_l(\boldsymbol{r};t)] = 4\pi\mathrm{i}\varepsilon_{jls}\frac{\partial^2}{\partial t\,\partial x_s} D(\boldsymbol{r}-\boldsymbol{r}', t-t') , \tag{15.25}$$

where in the derivation of the last equation we have used

$$\sum_{\lambda=\pm 1} \lambda\varepsilon_j(\boldsymbol{k},\lambda)\varepsilon_l^*(\boldsymbol{k},\lambda) = -\mathrm{i}\varepsilon_{jls}\frac{k_s}{\omega} . \tag{15.26}$$

These commutation relations were originally derived by *Jordan* and *Pauli* [JP 28].

Some important properties of the function $D(\boldsymbol{r},t)$, easy consequences of its definition (15.23), are

$$D(\boldsymbol{r},0) = 0 , \quad D(-\boldsymbol{r},-t) = -D(\boldsymbol{r},t) , \quad \Box D(\boldsymbol{r},t) = 0 ,$$
$$\left[\frac{\partial D(\boldsymbol{r},t)}{\partial t}\right]_{t=0} = -\delta(\boldsymbol{r}) , \quad \left[\frac{\partial D(\boldsymbol{r},t)}{\partial x_k}\right]_{t=0} = 0 . \tag{15.27}$$

Integrating in (15.23) over $\hat{\boldsymbol{k}}$ yields

$$D(\boldsymbol{r},t) = -\frac{1}{2\pi^2 r}\int_0^\infty d\omega\ \sin\omega r\ \sin\omega t . \tag{15.28}$$

Writing the sine function as a sum of exponentials, and taking into account the properties of the Dirac delta functions, we can write

$$D(\boldsymbol{r},t) = \frac{1}{4\pi r}[\delta(r+t) - \delta(r-t)] . \tag{15.29}$$

Hence $D(\boldsymbol{r},t)$ has its support on the light cone $|\boldsymbol{r}| = |t|$. If two space-time points cannot be connected by means of a light signal each component of the E.M. field at one of those points is an observable compatible with each component of the field at the other point. Actually the singular nature of the field commutator reveals that the notion of a field at a point is a pure mathematical idealization, and that only fields averaged with test functions of $\boldsymbol{r}$ and t are meaningful. For such averages, the commutation relations are finite quantities and they yield uncertainty relations between field components averaged over space-time regions. The physical explanation of these uncertainty relations appear in the classic work by *Bohr* and *Rosenfeld* [BR 33,50], who gave the following argument: the measurement of the average field in a space-time region I_1 can be performed by observing the action on a charged macroscopic body placed in that region. The position and momentum of this probe are subject to Heisenberg's uncertainty relations, and the uncertainty $\Delta\boldsymbol{x}$ originates an electric dipole moment and the corresponding field, whereas the uncertainty $\Delta\boldsymbol{p}$ generates a current which produces a magnetic field. This uncontrollable E.M. field produced by the probe affects the average field in some other region of space-time I_2 , as long as I_2 is within the causal region of influence of I_1. It is possible to show that the uncertainty produced by the existence of this perturbing field agrees with the indeterminacy predicted by the commutation relations (15.24) and (15.25).

The uncertainty relations for the average values of the fields are unimportant when the number of photons present in the state in a volume $\sim \lambda^3$ is large, where λ is the photon wavelength. In this case the quantum fluctuations are negligible and the electromagnetic field admits a classical description. Finally from (15.24, 25) and the properties of $D(\boldsymbol{r}, t)$ given in (15.27), one obtains the equal-time commutation relations

$$\begin{aligned} &[E_j(\boldsymbol{r};t), E_l(\boldsymbol{r}';t)] = [B_j(\boldsymbol{r};t), B_l(\boldsymbol{r}';t)] = 0 \, , \\ &[E_j(\boldsymbol{r};t), B_l(\boldsymbol{r}';t)] = -4\pi \mathrm{i} \varepsilon_{jls} \frac{\partial}{\partial x_s} \delta(\boldsymbol{r}-\boldsymbol{r}') \, . \end{aligned} \tag{15.30}$$

It is easy to show that the variable $A_j(\boldsymbol{r};t)$ and its conjugate momentum given in our units by $\pi_j(\boldsymbol{r};t) \equiv (4\pi)^{-1} \partial A_j(\boldsymbol{r};t)/\partial t$, satisfy the commutation relations

$$\begin{aligned} &[A_j(\boldsymbol{r};t), A_l(\boldsymbol{r}';t)] = [\pi_j(\boldsymbol{r};t), \pi_l(\boldsymbol{r}';t)] = 0 \, , \\ &[A_j(\boldsymbol{r};t), \pi_l(\boldsymbol{r}';t)] = \mathrm{i}\delta_{jl}\delta(\boldsymbol{r}-\boldsymbol{r}') \, , \end{aligned} \tag{15.31}$$

entirely analogous to the commutation relations among the classical canonical variables q_j and p_l.

It is often found in the literature that instead of presenting the quantization of the electromagnetic field along the lines followed here one starts by postulating the commutators (15.31), and using (15.16), one then derives the commutation relations (15.15).

15.4 Multipole Waves

In Sect. 15.2, we wrote the classical fields as superpositions of photon wave functions with energy ω, momentum $\boldsymbol{k}$, and helicity λ. Even though the plane waves are very useful in scattering problems, for the development of the theory of radiation it is more convenient to write the fields as superpositions of photon wave functions with well defined energy ω, total angular momentum J, third component M and parity Π, because the atomic and nuclear states emitting and absorbing photons have generally well defined values for these quantum numbers. We first address the problem of determining a new expansion basis for the fields consisting of photon wave functions with the previous quantum numbers specified.

Since $\boldsymbol{A}(\boldsymbol{r};t)$ is a polar vector field, under parity it transforms according to

$$\boldsymbol{A}(\boldsymbol{r};t) \xrightarrow{P} \boldsymbol{A}'(\boldsymbol{r};t) = -\boldsymbol{A}(-\boldsymbol{r};t) \, , \tag{15.32}$$

so that

$$\begin{aligned} &\boldsymbol{E}(\boldsymbol{r};t) \xrightarrow{P} \boldsymbol{E}'(\boldsymbol{r};t) = -\boldsymbol{E}(-\boldsymbol{r};t) \, , \\ &\boldsymbol{B}(\boldsymbol{r};t) \xrightarrow{P} \boldsymbol{B}'(\boldsymbol{r};t) = \boldsymbol{B}(-\boldsymbol{r};t) \, . \end{aligned} \tag{15.33}$$

If $\boldsymbol{A}$ has a well defined parity Π, in other words, if

$$\boldsymbol{A}'(\boldsymbol{r};t) = \Pi \boldsymbol{A}(\boldsymbol{r};t)\,, \tag{15.34}$$

the same occurs with $\boldsymbol{E}$ and $\boldsymbol{B}$:

$$\boldsymbol{E}'(\boldsymbol{r};t) = \Pi \boldsymbol{E}(\boldsymbol{r};t)\,, \quad \boldsymbol{B}'(\boldsymbol{r};t) = \Pi \boldsymbol{B}(\boldsymbol{r};t)\,. \tag{15.35}$$

These three fields have therefore the same parity, and this is the parity of the associated photon. Usually one defines the parity of the photon as that of its magnetic field; this is because a magnetic field is an axial vector, and thus, it transforms under P as a scalar function component by component.

According to their parity transformations, we introduce two types of electromagnetic radiations:

1) EJ radiation or 2^J-polar electric radiation: its photons have energy ω, angular momentum J and parity $(-1)^J$.
2) MJ radiation or 2^J-polar magnetic radiation: its photons have energy ω angular momentum J and parity $(-1)^{J+1}$.

The fact that the energy is ω means that the time dependence of the photon wave function is of the form $\exp(-\imath\omega t)$ and for the time being we will omit this factor in our discussion. For MJ radiation the vector potential (or more precisely the vector potential amplitude associated to the photon), $\boldsymbol{A}_{JM}(\boldsymbol{r};\omega;\mathrm{m})$, (where m $\equiv$ magnetic), has parity $(-1)^{J+1}$ and according to the results of Sect. B.6, its most general form is

$$\boldsymbol{A}_{JM}(\boldsymbol{r};\omega;\mathrm{m}) = C_J(\mathrm{m})\xi_J(\omega r)\boldsymbol{X}_{JM}(\hat{\boldsymbol{r}})\,, \tag{15.36}$$

where $C_J(\mathrm{m})$ is a normalization constant to be fixed later. Using the d'Alembert equation we have

$$(\Delta + \omega^2)\boldsymbol{A}_{JM}(\boldsymbol{r};\omega;\mathrm{m}) = 0\,, \tag{15.37}$$

so that $\xi_J(x)$ satisfies the equation

$$\frac{d^2\xi_J(x)}{dx^2} + \frac{2}{x}\frac{d\xi_J(x)}{dx} + \left[1 - \frac{J(J+1)}{x^2}\right]\xi_J(x) = 0\,, \tag{15.38}$$

typical of spherical Bessel functions. The regularity condition as $r \to 0$ requires $\xi_J(\omega r) \propto j_J(\omega r)$, hence

$$\boldsymbol{A}_{JM}(\boldsymbol{r};\omega;\mathrm{m}) = C_J(\mathrm{m})j_J(\omega r)\boldsymbol{X}_{JM}(\hat{\boldsymbol{r}})\,. \tag{15.39}$$

The transversality condition (15.1) is automatically satisfied as clearly shown by (B.108).

For the EJ radiation the vector potential $\boldsymbol{A}_{JM}(\boldsymbol{r};\omega;\mathrm{e})$ (e $\equiv$ electric) has parity $(-1)^J$; proceeding as before and using the results of Sect. B.6 we find that its most general form is

$$\begin{aligned}
&\boldsymbol{A}_{JM}(\boldsymbol{r};\omega;\mathrm{e}) \\
&\quad = C_{J+1}(\mathrm{e})j_{J+1}(\omega r)\mathcal{Y}_{JM}^{J+1}(\hat{\boldsymbol{r}}) + C_{J-1}(\mathrm{e})j_{J-1}(\omega r)\mathcal{Y}_{JM}^{J-1}(\hat{\boldsymbol{r}}) \,. \qquad (15.40)
\end{aligned}$$

To find the relation between $C_{J+1}(\mathrm{e})$ and $C_{J-1}(\mathrm{e})$ we must impose the transversality condition. Using (B.108) we obtain

$$\begin{aligned}
&\nabla \cdot \boldsymbol{A}_{JM}(\boldsymbol{r};\omega;\mathrm{e}) \\
&\quad = -C_{J+1}(\mathrm{e})\sqrt{\frac{J+1}{2J+1}}\left[\frac{dj_{J+1}(\omega r)}{dr} + \frac{J+2}{r}j_{J+1}(\omega r)\right] Y_J^M(\hat{\boldsymbol{r}}) \\
&\quad + C_{J-1}(\mathrm{e})\sqrt{\frac{J}{2J+1}}\left[\frac{dj_{J-1}(\omega r)}{dr} - \frac{J-1}{r}j_{J-1}(\omega r)\right] Y_J^M(\hat{\boldsymbol{r}}) \,, \qquad (15.41)
\end{aligned}$$

and taking into account (A.128) and (A.129):

$$\begin{aligned}
\nabla \cdot \boldsymbol{A}_{JM}(\boldsymbol{r};\omega;\mathrm{e}) = & -\omega j_J(\omega r) Y_J^M(\hat{\boldsymbol{r}}) \\
& \times \left[\sqrt{\frac{J+1}{2J+1}}C_{J+1}(\mathrm{e}) + \sqrt{\frac{J}{2J+1}}C_{J-1}(\mathrm{e})\right] \,. \qquad (15.42)
\end{aligned}$$

Hence, the most general vector field (15.40) satisfying (15.2) is

$$\begin{aligned}
&\boldsymbol{A}_{JM}(\boldsymbol{r};\omega;\mathrm{e}) = C_J(\mathrm{e}) \\
&\quad \times \left[-\sqrt{\frac{J}{2J+1}}j_{J+1}(\omega r)\mathcal{Y}_{JM}^{J+1} + \sqrt{\frac{J+1}{2J+1}}j_{J-1}(\omega r)\mathcal{Y}_{JM}^{J-1}(\hat{\boldsymbol{r}})\right] \,, \qquad (15.43)
\end{aligned}$$

where $C_j(\mathrm{e})$ is a normalization constant to be fixed later.

For photons with a well defined energy ω, the Maxwell equations lead to

$$\nabla \cdot \boldsymbol{B} = 0 \,, \quad \nabla \cdot \boldsymbol{E} = 0 \,, \quad \nabla \times \boldsymbol{B} = -\mathrm{i}\omega \boldsymbol{E} \,, \quad \nabla \times \boldsymbol{E} = \mathrm{i}\omega \boldsymbol{B} \,, \qquad (15.44)$$

where (the amplitudes of) the fields $\boldsymbol{E}, \boldsymbol{B}$ associated to the photon are given by

$$\boldsymbol{E} = \mathrm{i}\omega \boldsymbol{A} \,, \quad \boldsymbol{B} = \nabla \times \boldsymbol{A} \,. \qquad (15.45)$$

Starting now with (15.39, 43, 45), and (B.110), we obtain for the fields

$$\begin{aligned}
&\boldsymbol{E}_{JM}(\boldsymbol{r};\omega;\mathrm{m}) = \mathrm{i}\omega C_J(\mathrm{m}) j_J(\omega r)\boldsymbol{X}_{JM}(\hat{\boldsymbol{r}}) \,, \\
&\boldsymbol{B}_{JM}(\boldsymbol{r};\omega;\mathrm{m}) = \mathrm{i}\omega C_J(\mathrm{m}) \qquad (15.46) \\
&\quad \times \left[-\sqrt{\frac{J}{2J+1}}j_{J+1}(\omega r)\mathcal{Y}_{JM}^{J+1}(\hat{\boldsymbol{r}}) + \sqrt{\frac{J+1}{2J+1}}j_{J-1}(\omega r)\mathcal{Y}_{JM}^{J-1}(\hat{\boldsymbol{r}})\right] \,,
\end{aligned}$$

$$\begin{aligned}
&\boldsymbol{E}_{JM}(\boldsymbol{r};\omega;\mathrm{e}) = \mathrm{i}\omega C_J(\mathrm{e}) \\
&\quad \times \left[-\sqrt{\frac{J}{2J+1}}j_{J+1}(\omega r)\mathcal{Y}_{JM}^{J+1}(\hat{\boldsymbol{r}}) + \sqrt{\frac{J+1}{2J+1}}j_{J-1}(\omega r)\mathcal{Y}_{JM}^{J-1}(\hat{\boldsymbol{r}})\right] \,, \qquad (15.47) \\
&\boldsymbol{B}_{JM}(\boldsymbol{r};\omega;\mathrm{e}) = -\mathrm{i}\omega C_J(\mathrm{e}) j_J(\omega r)\boldsymbol{X}_{JM}(\hat{\boldsymbol{r}}) \,.
\end{aligned}$$

If $\boldsymbol{E}$ and $\boldsymbol{B}$ are solutions to (15.44) so are $\boldsymbol{B}_{\mathrm{d}} \equiv -\boldsymbol{E}$ and $\boldsymbol{E}_{\mathrm{d}} \equiv +\boldsymbol{B}$, the so-called dual fields. If we wish the MJ radiation field to be dual to the EJ radiation field, a possible choice of normalization constants is

$$C_J(\mathrm{e}) = \sqrt{\frac{2}{\pi}}\,, \quad C_J(\mathrm{m}) = -\sqrt{\frac{2}{\pi}}\,, \tag{15.48}$$

and recalling (B.96) and (A.134), we have

$$\int d^3r\, \boldsymbol{A}^*_{J'M'}(\boldsymbol{r};\omega';\lambda')\cdot \boldsymbol{A}_{JM}(\boldsymbol{r};\omega;\lambda) = \frac{1}{\omega^2}\delta(\omega-\omega')\delta_{\lambda\lambda'}\delta_{JJ'}\delta_{MM'}, \tag{15.49}$$

where $\lambda = \mathrm{e}, \mathrm{m}$. This finishes our construction of a complete orthogonal basis of multipolar states for the photon.

Before we continue, we would like to remark that (B.96) implies

$$\boldsymbol{r}\cdot \boldsymbol{X}_{JM}(\hat{\boldsymbol{r}}) = 0 \tag{15.50}$$

so that for 2^J-multipole electric (magnetic) radiation the magnetic (electric) field does not have a radial component.

We thus conclude that the second quantized form of the vector potential operator is

$$\begin{aligned} \boldsymbol{A}(\boldsymbol{r};t) = \int_0^\infty \omega^2 d\omega \sqrt{\frac{4\pi}{2\omega}} \sum_{JM}\sum_{\lambda} [&\boldsymbol{A}_{JM}(\boldsymbol{r};\omega;\lambda)\mathrm{e}^{-\mathrm{i}\omega t} a_{JM}(\omega;\lambda) \\ &+ \boldsymbol{A}^*_{JM}(\boldsymbol{r};\omega;\lambda)\mathrm{e}^{\mathrm{i}\omega t} a^\dagger_{JM}(\omega;\lambda)]\,, \end{aligned} \tag{15.51}$$

where $a_{JM}(\omega;\lambda)$ and $a^\dagger_{JM}(\omega;\lambda)$ are the multipole wave annihilation and creation operators respectively, and they satisfy the commutation relations

$$\begin{aligned} &[a_{JM}(\omega;\lambda), a_{J'M'}(\omega';\lambda')] = [a^\dagger_{JM}(\omega;\lambda), a^\dagger_{J'M'}(\omega';\lambda')] = 0\,, \\ &[a_{JM}(\omega;\lambda), a^\dagger_{J'M'}(\omega';\lambda')] = \frac{1}{\omega^2}\delta(\omega-\omega')\delta_{\lambda\lambda'}\delta_{JJ'}\delta_{MM'}\,. \end{aligned} \tag{15.52}$$

The validity of (15.51) and (15.52) is derived from (15.16), taking into account that the operator $a^\dagger_{JM}(\omega;\lambda)$ creates a photon with wave function $\boldsymbol{A}_{JM}(\boldsymbol{r};\omega;\lambda) \times \exp(-\mathrm{i}\omega t)$, and that $\sum[a_\alpha f_\alpha + a^\dagger_\alpha f^*_\alpha]$ is invariant under unitary changes of basis compatible with the quantum number ω.

A simple computation starting from (15.18) leads to

$$H_{\mathrm{rad}} = \int_0^\infty \omega^2 d\omega \left[\omega \sum_{JM}\sum_{\lambda} a^\dagger_{JM}(\omega;\lambda) a_{JM}(\omega;\lambda)\right] \tag{15.53}$$

as expected.

15.5 Interaction Between Matter and Radiation

So far we have only considered the electromagnetic field for free radiation, whose Hamiltonian in the formalism of quantum field theory is

$$H_{\rm rad} = \frac{1}{8\pi} \int d^3r : [\boldsymbol{E}^2(\boldsymbol{r};0) + \boldsymbol{B}^2(\boldsymbol{r};0)] : \; . \tag{15.54}$$

In order to study the interaction of this field with a system of N point particles we must take into account that the total Hamiltonian has the form

$$H = H_{\rm rad} + H_{\rm m} + H_1 \; , \tag{15.55}$$

where $H_{\rm m}$ is the matter Hamiltonian, i.e., the Hamiltonian of the N particles (for example, in the case of atoms, it will be the Hamiltonian for the Z electrons which interact among themselves and with the atomic nucleus, and for the case of nuclei it will describe the interactions among the A nucleons). The term H_1 describes the interaction between the N particles and the radiation field. In the non-relativistic approximation and hypothesizing the additivity of interactions of the constituent particles, (12.70) shows that H_1 can be written as

$$H_1 = \sum_{i=1}^{N} \left[-\frac{e_i}{m_i} \boldsymbol{A}(\boldsymbol{r}_i;0) \cdot \boldsymbol{p}_i - \tilde{\mu}_i g_i \boldsymbol{S}_i \cdot \boldsymbol{B}(\boldsymbol{r}_i;0) + \frac{e_i^2}{2m_i} : \boldsymbol{A}^2(\boldsymbol{r}_i;0) : \right] . \tag{15.56}$$

[Actually (12.70) refers to the case of a material system interacting with an external electromagnetic field. In writing (15.56), we have implicitly assumed that the form of the interaction is the same when we consider the electromagnetic field as quantized and since the matter-radiation system is assumed to be closed, the total energy H is constant in time. Thus in the Schrödinger picture, which is the one we will use to study the interaction, H takes the form (15.56) at $t = 0$. It is worth pointing out that the time dependence due to the presence of an external field is now automatically included in the wave function for the photon states].

In (15.56) e_i is the charge, m_i the mass, $\tilde{\mu}_i \equiv -\mu_{\rm B}$ if "i" is an electron, $\tilde{\mu}_i \equiv \mu_{\rm N}$ if "i" is a nucleon, g_i is the gyromagnetic factor and $\boldsymbol{S}_i$ is the spin of the ith particle. Equation (15.56) may be rewritten as

$$H_1 = \int d^3r \, \mathcal{H}_1(\boldsymbol{r};0) \; , \tag{15.57}$$

where the interaction energy density has the form

$$\mathcal{H}_1(\boldsymbol{r};0) = - : \boldsymbol{j}(\boldsymbol{r};0) \cdot \boldsymbol{A}(\boldsymbol{r};0) : \; , \tag{15.58}$$

$\boldsymbol{j}(\boldsymbol{r};0)$ being the current density operator

$$
\begin{aligned}
\boldsymbol{j}(\boldsymbol{r};0) &= \sum_{1}^{N} \frac{e_i}{2}[\boldsymbol{v}_i\delta(\boldsymbol{r}-\boldsymbol{r}_i)+\delta(\boldsymbol{r}-\boldsymbol{r}_i)\boldsymbol{v}_i] \\
&+ \nabla_r \left[\sum_{1}^{N} g_i\mu_i \boldsymbol{S}_i\delta(\boldsymbol{r}-\boldsymbol{r}_i)\right] .
\end{aligned} \tag{15.59}
$$

Notice that $\boldsymbol{j}(\boldsymbol{r};0) = \boldsymbol{j}_{\text{convection}}(\boldsymbol{r};0) + \boldsymbol{j}_{\text{spin}}(\boldsymbol{r};0)$, with

$$
\begin{aligned}
\boldsymbol{j}_{\text{convection}}(\boldsymbol{r};0) &= \sum_{1}^{N} \frac{e_i}{2m_i} \\
&\times [\boldsymbol{p}_i\delta(\boldsymbol{r}-\boldsymbol{r}_i)+\delta(\boldsymbol{r}-\boldsymbol{r}_i)\boldsymbol{p}_i - 2e_i\boldsymbol{A}(\boldsymbol{r},0)\delta(\boldsymbol{r}-\boldsymbol{r}_i)] , \\
\boldsymbol{j}_{\text{spin}}(\boldsymbol{r};0) &= \nabla_r \times \boldsymbol{M}_{\text{spin}}(\boldsymbol{r}) ,
\end{aligned} \tag{15.60}
$$

and $\boldsymbol{M}_{\text{spin}}(\boldsymbol{r};0)$ being the density of spin magnetic moment

$$
\boldsymbol{M}_{\text{spin}}(\boldsymbol{r};0) = \sum_{1}^{N} g_i\tilde{\mu}_i \boldsymbol{S}_i\delta(\boldsymbol{r}-\boldsymbol{r}_i) . \tag{15.61}
$$

Equation (15.57) has the advantage of being manifestly invariant under translations, and therefore those transitions generated by H_1 will not only conserve the energy but they will also preserve the momentum.

In order to study the radiative transitions we decompose the total Hamiltonian in two parts. The first one is the unperturbed Hamiltonian, H_0, which we take as $H_0 = H_{\text{rad}} + H_{\text{m}}$; the second piece, H_1, is the perturbation which we assume to be small so that its effects are amenable to perturbative methods. The eigenstates of H_0 are characterized by giving for the N-particle system its state vector $|\Psi\rangle$ with internal quantum numbers E, J, M and Π, and for the radiation we give the number of photons present in each of its multipolar states, these referred to a common origin that generally coincides with the center of mass of the matter system.

To first order of perturbation theory, the first two terms in H_1, which are of order e and linear in the creation and annihilation operators, relate states whose number of photons present differs by one unit and they correspond to emission and absorption processes of a single photon. The last term in H_1, of order e^2, relates states with either the same number of photons or such that the number of photons changes in two units. It contributes to processes where two photons are either absorbed or emitted simultaneously. This term may be of the same order of magnitude as those appearing in second order of perturbation theory for the first two terms in H_1. In the perturbative analysis, the expansion parameter for the transition probabilities is the fine structure constant $\alpha \simeq 1/137$. Since this is indeed a small number, we will restrict our study of radiative transitions to first order corrections and therefore we can safely omit the last term in H_1. As indicated in Sect. 12.5, this diamagnetic term cannot be omitted in processes where the transition amplitude is of second order in the coupling constant e. For instance in the computation of the scattering cross section of photons with free electrons (the Compton effect), in the computation of shifts of energy levels, in the decay of the $2s_{1/2}$ level of hydrogen, etc.

15.6 Transition Probabilities

We want to compute the transition probability from an initial state $|i\rangle$ to a final state $|f\rangle$. We limit our considerations to emission processes where the initial state contains no photons and it is characterized by a state vector $|\Psi_i\rangle$ of the matter system with well defined internal quantum numbers E_i, J_i, M_i, Π_i, whereas the final state contains a photon in a multipole state $|\omega JM\lambda\rangle$ of wave function $\boldsymbol{A}_{JM}(r;\omega;\lambda)$ and the matter system is described by $|\Psi_f\rangle$ with quantum numbers E_f, J_f, M_f, Π_f.

In the following we suppose that the initial matter system is at rest, and we will neglect its recoil. In other words, $|\Psi_i\rangle, |\Psi_f\rangle$ will be the normalized eigenstates of the matter system after one eliminates the motion of its center of mass. This is fully justified in atomic or molecular processes, where the recoil velocities are of the order $10^{-9}c$. In nuclear processes these are of order $10^{-3}c$, but their contributions are still negligible when compared to the errors generated by our lack of knowledge of the exact nuclear wave functions. If we insist in carrying out the computation using the current density $\boldsymbol{j}$ given in (15.59), when we take into account the global motion of the initial and final matter systems, we would obtain terms depending of their total momenta giving a vanishing contribution when the initial system is at rest. These terms, however, are important in the computation of Thomson scattering where one has to take into account intermediate matter states off the energy shell.

Since the energy must be conserved in this process (see Sect. 11.6), it is obvious that the transition will not take place unless

$$E_i = E_f + \omega \,. \tag{15.62}$$

Actually, the initial state whose decay we are studying, does not have a well defined energy according to the uncertainty principle. The very process considered in this section contributes to the width ΔE_i of the state. However, if all the decay channels of the initial system are due to electromagnetic or weak interactions, ΔE_i will be negligible with respect to experimental errors and we will assume that the states $|\Psi_i\rangle$ and $|\Psi_f\rangle$ are stationary with respect to H_{m}.

Using the results of Sect. 11.6, the emission probability $\Gamma_{i\to f}$ per unit time, of a photon of type $JM\lambda$ with any energy (necessarily we must have $\omega = E_i - E_f$) in the transition from $|\Psi_i\rangle$ to $|\Psi_f\rangle$, is given by

$$\boxed{\Gamma_{i\to f} = 2\pi|\langle f|H_1|i\rangle|^2\omega^2} \,, \tag{15.63}$$

where we have taken into account that the density of final states is ω^2, due to the normalization chosen for the multipole states.

Since H_1 is a scalar under rotations and parity invariant, it is obvious that in the transition $(J_i M_i \Pi_i) \to (J_j M_f \Pi_f)$ the only allowed multipole radiations $(JM\Pi)$ satisfy

$$\boxed{|J_i - J_f| \leq J \leq J_i + J_f\ , \quad M = M_i - M_f\ , \quad \Pi = \Pi_f \Pi_i}\ , \tag{15.64}$$

Form the general expression for $\boldsymbol{A}_{JM}(r;\omega;\lambda)$ given in (15.39) and (15.43) we conclude that $\boldsymbol{A}_{00}(r;\omega;\lambda) \equiv 0$, that is, there are no photons with $J = 0$. Taken this into account together with the constrains (15.64) we can easily determine what are the electric or magnetic multipole transitions of smallest J allowed for given values of $(J_i \Pi_i)$ and $(J_f \Pi_f)$. The results appear in Table 15.1.

Table 15.1. Lowest order multipoles allowed in the transitions $(J_i, \Pi_i) \to (J_f, \Pi_f)$

		Electric multipoles	Magnetic multipoles
$J_i \neq J_f$	$\Pi_i \Pi_f = (-1)^{J_i - J_f}$	$J = \lvert J_i - J_f \rvert$	$J = \lvert J_i - J_f \rvert + 1$ Does not exist if J_i or J_f vanish
	$\Pi_i \Pi_f = (-1)^{J_i - J_f + 1}$	$J = \lvert J_i - J_f \rvert + 1$ Does not exist if J_i or J_f vanish	$J = \lvert J_i - J_f \rvert$
$J_i = J_f \neq 0$	$\Pi_i \Pi_f = +1$	$J = 2$ Does not exist if $J_i = J_f = \frac{1}{2}$	$J = 1$
	$\Pi_i \Pi_f = -1$	$J = 1$	$J = 2$ Does not exist if $J_i = J_f = \frac{1}{2}$
$J_i = J_f = 0$	Any	Does not exist	Does not exist

We next find an explicit expression for the transition matrix element entering (15.63). Out of the two terms in the right hand side of (15.51), only the second one contributes to the emission process and the matrix element of the operator $\boldsymbol{A}(\boldsymbol{r};0)$ between the photon states considered is

$$\langle 1 \text{ photon } (\omega J M \lambda) | \boldsymbol{A}(\boldsymbol{r};0) | \text{ vacuum } \rangle = \sqrt{\frac{4\pi}{2\omega}} \boldsymbol{A}^*_{JM}(\boldsymbol{r};\omega;\lambda)\ . \tag{15.65}$$

We thus obtain

$$\langle f|H_1|i\rangle = \sqrt{\frac{4\pi}{2\omega}} \sum_{j=1}^{N} \int dV\, \Psi_f^\dagger$$
$$\times \left[-\frac{e_j}{m_j} \boldsymbol{A}^*_{JM}(\boldsymbol{r}_j;\omega;\lambda) \cdot \boldsymbol{p}_j - g_j \tilde{\mu}_j \boldsymbol{S}_j \cdot [\boldsymbol{\nabla}_j \times \boldsymbol{A}^*_{JM}(\boldsymbol{r}_j;\omega;\lambda)] \right] \Psi_i\ , \tag{15.66}$$

where dV represents the volume element of the $3N$ coordinates appearing as arguments of the initial and final wave functions Ψ_i, Ψ_f. [One should not forget that the previous expression was obtained assuming that the matter systems are "anchored" to the origin of coordinates, and hence there are no restrictions on the values of the $3N$ coordinates. Had we carried out the calculation without this hypothesis, the intrinsic wave functions would depend on $3(N-1)$ free coordinates; equations (15.63) and (15.66) would still be valid if the integration is limited to be over the $3(N-1)$ intrinsic coordinates (we assume that the transformation "$\boldsymbol{r}_1, \ldots, \boldsymbol{r}_N \to \boldsymbol{R} \equiv \sum m_j \boldsymbol{r}_j / \sum m_j, 3(N-1)$ intrinsic coordinates" has unit jacobian), the $3N$ coordinates $\boldsymbol{r}_1, \ldots, \boldsymbol{r}_N$ are subject to the constraints $\boldsymbol{R} = 0$ and Ψ_i, Ψ_f are now the intrinsic wave functions].

Even though the integration over each of the variables $\boldsymbol{r}_i$ extends to the whole space, the finite size of the matter system allows us for all practical purposes to limit the integration to a spherical region whose radius is of the order of the average radius R of the system considered. The nuclear radii are relatively well defined quantities [PR 62] and they follow the law $R \simeq 1.2 A^{1/3}$ fm, where A is the number of nucleons. In atoms, and within the framework of the Thomas-Fermi model, the radius R of the sphere containing all but one of the electrons is practically independent of Z, $R \simeq 2.5 \times 10^{-8}$ cm. Since in most nuclear radiative transitions $\omega \sim 1$ MeV while for atomic ones $\omega \sim 1$ eV (with the important exception of X-ray emission), it follows that for nuclei $\omega R \sim 10^{-2} A^{1/3}$ and for atoms $\omega R \sim 10^{-3}$. Within our approximations, these estimates permit the replacement of the spherical Bessel functions appearing in $\boldsymbol{A}_{JM}(\boldsymbol{r}; \omega; \lambda)$ by their asymptotic behavior when $\omega R \to 0$ given in (A.124).

The computation of the various terms appearing in (15.66) is lengthy and awkward, and the reader uninterested in the details can proceed directly to Eqs. (15.83) which summarize the results obtained. We begin by calculating the quantities

$$
\begin{aligned}
R(\lambda) &\equiv \sqrt{\frac{\pi}{2}} \sum_{j=1}^{N} \frac{e_j}{m_j} \int dV\, \Psi_f^\dagger \boldsymbol{A}^*_{JM}(\boldsymbol{r}_j; \omega; \lambda) \cdot \boldsymbol{p}_j \Psi_i \,, \\
R'(\lambda) &\equiv \sqrt{\frac{\pi}{2}} \sum_{j=1}^{N} g_j \tilde{\mu}_j \int dV\, \Psi_f^\dagger \boldsymbol{S}_j \cdot [\boldsymbol{\nabla}_j \times \boldsymbol{A}^*_{JM}(\boldsymbol{r}_j; \omega; \lambda)] \Psi_i \,,
\end{aligned}
\tag{15.67}
$$

recalling that the $3N$ coordinates are supposed to be free.

1) R(m). Using (15.36, 48) and (B.95)

$$
R(\mathrm{m}) = \frac{\mathrm{i}}{\sqrt{J(J+1)}} \sum_{j=1}^{N} \frac{e_j}{m_j} \int dV\, \Psi_f^\dagger j_J(\omega r_j) [\boldsymbol{L}_j Y_J^M(\hat{\boldsymbol{r}}_j)]^\dagger \cdot \nabla_j \Psi_i \,. \tag{15.68}
$$

Taking into account that $\boldsymbol{L}$ is a Hermitian operator commuting with $j_J(\omega r)$ we obtain

$$R(\mathrm{m}) = \frac{\mathrm{i}}{\sqrt{J(J+1)}} \sum_{j=1}^{N} \frac{e_j}{m_j} \int dV\, Y_J^{M*}(\hat{\boldsymbol{r}}_j) j_J(\omega r_j) \boldsymbol{L}_j \cdot (\Psi_f^\dagger \nabla_j \Psi_i) \,. \quad (15.69)$$

Since for any vector field $\boldsymbol{B}$ one has $\boldsymbol{L} \cdot \boldsymbol{B} = \mathrm{i}\nabla \cdot (\boldsymbol{r} \times \boldsymbol{B})$ and, as explained before, we can replace $j_J(\omega r)$ by its asymptotic expression (A.124):

$$R(\mathrm{m}) = -\mathrm{i}\frac{\omega^J}{\sqrt{J(J+1)}} \frac{1}{(2J+1)!!} \times \sum_{j=1}^{N} \frac{e_j}{m_j} \int dV\, r_j^J Y_J^{M*}(\hat{\boldsymbol{r}}_j) \nabla_j \cdot (\Psi_f^\dagger \boldsymbol{L}_j \Psi_i) \,. \quad (15.70)$$

2) $R'(\mathrm{m})$. From (15.39, 48) and (B.95) we have

$$R'(\mathrm{m}) = -\frac{1}{\sqrt{J(J+1)}} \sum_{j=1}^{N} g_j \tilde{\mu}_j \times \int dV\, \Psi_f^\dagger \boldsymbol{S}_j \cdot \{\nabla_j \times j_J(\omega r_j)[\boldsymbol{L}_j Y_J^M(\hat{\boldsymbol{r}}_j)]^\dagger\} \Psi_i \,, \quad (15.71)$$

and since $\boldsymbol{L}$ is Hermitian and ∇ is antihermitian

$$R'(\mathrm{m}) = -\frac{1}{\sqrt{J(J+1)}} \sum_{j=1}^{N} g_j \tilde{\mu}_j \times \int dV\, j_J(\omega r_j) Y_J^{M*}(\hat{\boldsymbol{r}}_j) \boldsymbol{L}_j \cdot [\nabla_j \times (\Psi_f^\dagger \boldsymbol{S}_j \Psi_i)] \,. \quad (15.72)$$

Any vector field $\boldsymbol{B}$ satisfies $\boldsymbol{L} \cdot (\nabla \times \boldsymbol{B}) = \mathrm{i}[\Delta(\boldsymbol{r} \cdot \boldsymbol{B}) - (\boldsymbol{r} \cdot \nabla + 2)(\nabla \cdot \boldsymbol{B})]$. This together with (A.124), yields

$$R'(\mathrm{m}) = \frac{\mathrm{i}\omega^J}{\sqrt{J(J+1)}(2J+1)!!} \sum_{j=1}^{N} g_j \tilde{\mu}_j \int dV\, r_j^J Y_J^{M*}(\hat{\boldsymbol{r}}_j) \times [(\boldsymbol{r}_j \cdot \nabla_j + 2)\nabla_j \cdot (\Psi_f^\dagger \boldsymbol{S}_j \Psi_i) - \Delta_j(\boldsymbol{r}_j \cdot \Psi_f^\dagger \boldsymbol{S}_j \Psi_i)] \,. \quad (15.73)$$

Since Δ is Hermitian and it gives 0 when acting on $r^J Y_J^{M*}(\hat{\boldsymbol{r}})$ we may omit the last term in (15.73). In the first term $(2 + \boldsymbol{r} \cdot \nabla)$ can be replaced by $(2 + r\partial/\partial r)$, and performing an integration by parts in polar coordinates we will obtain the operator $[2 - (\partial/\partial r)r]$ acting on $r^{J+2} Y_J^{M*}(\hat{\boldsymbol{r}})$, with the result $-(J+1)r^{J+2} Y_J^{M*}(\hat{\boldsymbol{r}})$. Hence finally

$$R'(\mathrm{m}) = -\frac{\mathrm{i}\omega^J}{(2J+1)!!} \sqrt{\frac{J+1}{J}} \sum_{j=1}^{N} g_j \tilde{\mu}_j \times \int dV\, r_j^J Y_J^{M*}(\hat{\boldsymbol{r}}_j) \nabla_j \cdot (\Psi_f^\dagger \boldsymbol{S}_j \Psi_i) \,. \quad (15.74)$$

3) $R(\mathrm{e})$. From (15.44) and (15.45) we obtain $\boldsymbol{A} = \omega^{-2}\nabla \times \boldsymbol{B}$, and taking into account (15.47, 48) and (B.95):

$$R(\mathrm{e}) = -\frac{1}{\omega\sqrt{J(J+1)}}\sum_{j=1}^{N}\frac{e_j}{2m_j} \times \int dV\,[\nabla_j \times j_J(\omega r_j)\boldsymbol{L}_j Y_J^M(\hat{\boldsymbol{r}}_j)]^\dagger \cdot \Psi_f^\dagger \overleftrightarrow{\nabla}_j \Psi_i\,, \tag{15.75}$$

where we have used the transversality $\boldsymbol{A}$ to replace ∇_j by $(1/2)\overleftrightarrow{\nabla}_j$, with $f\overleftrightarrow{\nabla}g \equiv f\nabla g - (\nabla f)g$.

Due to the hermiticity of $\boldsymbol{L}$ and the antihermiticity of ∇

$$R(\mathrm{e}) = -\frac{1}{\omega\sqrt{J(J+1)}}\sum_{j=1}^{N}\frac{e_j}{2m_j} \times \int dV\,Y_J^{M*}(\hat{\boldsymbol{r}}_j) j_J(\omega r_j)\boldsymbol{L}_j \cdot [\nabla_j \times (\Psi_f^\dagger \overleftrightarrow{\nabla}_j \Psi_i)]\,, \tag{15.76}$$

and following the arguments of the previous case we arrive at

$$R(\mathrm{e}) = -\frac{\mathrm{i}\omega^{J-1}}{(2J+1)!!}\sqrt{\frac{J+1}{J}}\sum_{j=1}^{N}\frac{e_j}{2m_j} \times \int dV\,r_j^J Y_J^{M*}(\hat{\boldsymbol{r}}_j)\nabla_j \cdot (\Psi_f^\dagger \overleftrightarrow{\nabla}_j \Psi_i)\,. \tag{15.77}$$

The continuity equation (3.61) admits the following generalization

$$-\omega\Psi_f^\dagger\Psi_i = \sum_j \frac{1}{2m_j}\nabla_j \cdot (\Psi_f^\dagger \overleftrightarrow{\nabla}_j \Psi_i)\,. \tag{15.78}$$

Indeed, it suffices to use the equality

$$\mathrm{e}^{-\mathrm{i}\omega t}\Psi_f^\dagger(\boldsymbol{x}_1,\ldots,\boldsymbol{x}_N)\Psi_i(\boldsymbol{x}_1,\ldots,\boldsymbol{x}_N) = \Psi_f^\dagger(\boldsymbol{x}_1,\ldots,\boldsymbol{x}_N;t)\Psi_i(\boldsymbol{x}_1,\ldots,\boldsymbol{x}_N;t)\,, \tag{15.79}$$

following from the stationary character of Ψ_i, Ψ_f with respect to H_m. Differentiating both sides with respect to time, using the Schrödinger equation, and assuming that the non-kinetic part of H_m is a local potential, we immediately obtain (15.78).

Using (15.78) and (15.77) we get

$$R(\mathrm{e}) = \frac{\mathrm{i}\omega^J}{(2J+1)!!}\sqrt{\frac{J+1}{J}}\sum_{j=1}^{N} e_j \int dV\,r_j^J Y_J^{M*}(\hat{\boldsymbol{r}}_j)\Psi_f^\dagger\Psi_i\,. \tag{15.80}$$

4) $R'(\mathrm{e})$. From (15.47) and (15.48) together with (B.95) we obtain

$$R'(\mathrm{e}) = -\frac{\mathrm{i}\omega}{\sqrt{J(J+1)}} \sum_{j=1}^{N} g_j \tilde{\mu}_j \int dV\, j_J(\omega r_j)[\boldsymbol{L}_j Y_J^M(\hat{\boldsymbol{r}}_j)]^\dagger \Psi_f^\dagger \boldsymbol{S}_j \Psi_i \,. \quad (15.81)$$

The hermiticity of $\boldsymbol{L}$ and the relation $\boldsymbol{L}\cdot\boldsymbol{B} = \mathrm{i}\nabla\cdot(\boldsymbol{r}\times\boldsymbol{B})$ imply

$$R'(\mathrm{e}) = \frac{\omega^{J+1}}{\sqrt{J(J+1)}(2J+1)!!} \sum_{j=1}^{N} g_j \tilde{\mu}_j \times \int dV\, r_j^J Y_J^{M*}(\hat{\boldsymbol{r}}_j) \nabla_j \cdot [\Psi_f^\dagger (\boldsymbol{r}_j \times \boldsymbol{S}_j) \Psi_i] \,. \quad (15.82)$$

Replacing finally the expressions we have found for $R(\lambda)$ and $R'(\lambda)$ in (15.66) and (15.63), the emission rates for a $JM\lambda$-photon, which we write explicitly as $\Gamma_{fi}(JM;\omega;\lambda)$, become

$$\begin{aligned} \Gamma_{fi}(JM;\omega;\mathrm{e}) &= \frac{8\pi(J+1)}{J[(2J+1)!!]^2}\omega^{2J+1} |Q_{fi}(JM;\omega) + Q'_{fi}(JM;\omega)|^2 \\ \Gamma_{fi}(JM;\omega;\mathrm{m}) &= \frac{8\pi(J+1)}{J[(2J+1)!!]^2}\omega^{2J+1} |M_{fi}(JM;\omega) + M'_{fi}(JM;\omega)|^2 \end{aligned}, \quad (15.83)$$

where

$$\begin{aligned} Q_{fi}(JM;\omega) &= \sum_{j=1}^{N} e_j \int dV\, r_j^J Y_J^{M*}(\hat{\boldsymbol{r}}_j) \Psi_f^\dagger \Psi_i \\ Q'_{fi}(JM;\omega) &= \frac{-\mathrm{i}\omega}{J+1} \sum_{j=1}^{N} g_j \tilde{\mu}_j \int dV\, r_j^J Y_J^{M*}(\hat{\boldsymbol{r}}_j) \nabla_j \cdot [\Psi_f^\dagger (\boldsymbol{r}_j \times \boldsymbol{S}_j) \Psi_i] \end{aligned} \quad (15.84)$$

and

$$\begin{aligned} M_{fi}(JM;\omega) &= \frac{-1}{J+1} \sum_{j=1}^{N} \frac{e_j}{m_j} \int dV\, r_j^J Y_J^{M*}(\hat{\boldsymbol{r}}_j) \nabla_j \cdot (\Psi_f^\dagger \boldsymbol{L}_j \Psi_i) \\ M'_{fi}(JM;\omega) &= -\sum_{j=1}^{N} g_j \tilde{\mu}_j \int dV\, r_j^J Y_J^{M*}(\hat{\boldsymbol{r}}_j) \nabla_j \cdot (\Psi_f^\dagger \boldsymbol{S}_j \Psi_i) \,. \end{aligned} \quad (15.85)$$

The computations we just completed for $R(\lambda)$ and $R'(\lambda)$ are still valid in the case when only $3(N-1)$ coordinates are free, as long as we implement the replacements $\boldsymbol{r}_i \to \boldsymbol{r}_i - \boldsymbol{R}$, $\boldsymbol{p}_i = -i\nabla_{\boldsymbol{r}_i} \to \boldsymbol{p}_i - (m_i/\sum m_j)\boldsymbol{P}$, as expected, because these are the relative positions and momenta of the particles in their C.M.

In order to understand the relative magnitudes of the different terms appearing in (15.83) as well as the size of the different multipole transitions it is convenient to make some order of magnitude estimates. In these estimates we assume that the effect of the operator $\boldsymbol{\nabla}$ is of order R^{-1}, $\boldsymbol{L}$ of order 1, $g\boldsymbol{S}$ of order 1 and $\boldsymbol{r}$ of order R, and furthermore, that $J \sim 1$. Taking all of this into account

$$\begin{aligned} \left|\frac{Q'_{fi}(JM;\omega)}{Q_{fi}(JM;\omega)}\right| &\sim \frac{\omega}{m} \sim \begin{cases} 10^{-6} & \text{in atoms} \\ 10^{-3} & \text{in nuclei} \end{cases}, \\ \left|\frac{M'_{fi}(JM;\omega)}{M_{fi}(JM;\omega)}\right| &\sim 1\,. \end{aligned} \tag{15.86}$$

Hence Q'_{fi} will generally be negligible with respect to Q_{fi}, while M_{fi} and M'_{fi} will be of the same order of magnitude.

If for the transition $(J_iM_i\Pi_i) \to (J_fM_f\Pi_f)$ the EJ radiation is allowed, then the next allowed electric radiation, if it exists, will be a 2^{J+2}-multipole radiation, and the same applies for magnetic transitions. If we take $Y_J^M(\hat{\boldsymbol{r}}) \sim 1$ to make an order of magnitude estimate, we obtain

$$\frac{\Gamma_{fi}(J+2,M;\omega;\lambda)}{\Gamma_{fi}(J,M;\omega;\lambda)} \sim (\omega R)^4 \sim \begin{cases} 10^{-12} & \text{in atoms} \\ 10^{-9}A^{4/3} & \text{in nuclei} \end{cases}, \tag{15.87}$$

so that for both electric and magnetic radiation it will suffice to take into account the lowest order transition allowed.

On the other hand if in a given radiative transition the lowest 2^J-multipole emission permitted is electric (magnetic), the next one allowed in increasing order of J would be a 2^{J+1} multipole magnetic (electric) transition, and one easily obtains that

$$\begin{aligned} \frac{\Gamma_{fi}(J+1,M;\omega;\mathrm{m})}{\Gamma_{fi}(J,M;\omega;\mathrm{e})} &\sim \frac{\omega^2}{m^2} \sim \begin{cases} 10^{-12} & \text{in atoms} \\ 10^{-6} & \text{in nuclei} \end{cases}, \\ \frac{\Gamma_{fi}(J+1,M;\omega;\mathrm{e})}{\Gamma_{fi}(J,M;\omega;\mathrm{m})} &\sim \omega^2 m^2 R^4 \sim \begin{cases} 1 & \text{in atoms} \\ 10^{-3}A^{4/3} & \text{in nuclei} \end{cases}. \end{aligned} \tag{15.88}$$

From all these order of magnitude estimates we obtain the following general rules:

1) Atomic Systems. If the lowest order allowed 2^J-multipole transition is electric, then this is the dominating one, and its transition probability can be computed by means of (15.83) neglecting the term Q'_{fi}. All others are negligible. If the first allowed transition is magnetic, then the 2^{J+1}-electric transition (if not vanishing) is of comparable order of magnitude and it should be taken into account.

2) Nuclear Systems. If the lowest order allowed 2^J-multipole transition is electric, then this is the dominating one, and its transition probability can be computed by means of (15.83) neglecting the term Q'_{fi}. All other transitions are negligible. If the first one permitted is magnetic, all other multipole transitions will be negligible (with the exception perhaps of the electric transition at the next order, which for quite a few nuclei with $A \gg 1$ could be comparable with the magnetic one).

In many cases the initial matter system is not polarized and the polarization of the final state is not measured. Under these circumstances, the quantity of interest is

$$\Gamma_{fi}(J;\omega;\lambda) \equiv \frac{1}{2J_i+1} \sum_{M_i} \sum_{M_f} \Gamma_{fi}(JM;\omega;\lambda) \,. \tag{15.89}$$

The sum can be carried out easily when we realize that the amplitudes defined in (15.84) and (15.85) are matrix elements of tensor operators of order J and use is made of the Wigner-Eckart theorem as well as of the orthogonality properties of the Clebsch-Gordan coefficients. We obtain

$$\Gamma_{fi}(J;\omega;e) = \frac{8\pi(J+1)}{J[(2J+1)!!]^2} \frac{2J_f+1}{2J_i+1} \omega^{2J+1} |Q_{fi}(J;\omega) + Q'_{fi}(J;\omega)|^2 \tag{15.90}$$

where $Q_{fi}(J;\omega)$ and $Q'_{fi}(J;\omega)$ are respectively the reduced matrix elements for $Q_{fi}(JM;\omega)$ and $Q'_{fi}(JM;\omega)$. A similar expression is obtained for magnetic transitions.

From an experimental point of view, one is often interested in the angular distributions of the emitted photons. In the next section we analyze these distributions when the photons have a definite type of polarization. In principle, all the multipolar amplitudes add up coherently to produce the experimental angular distribution. The interference terms disappear only after integrating over all possible directions of the emitted photon. The computation of the angular distribution is carried out by taking into account that the emission probability of a photon with momentum $\boldsymbol{k}$ and helicity μ per unit time and solid angle is:

$$\frac{d\Gamma_{i\to f}(\boldsymbol{k},\mu)}{d\Omega} = 2\pi |\langle \boldsymbol{k},\mu; \Psi_f | H_1 | \Psi_i \rangle|^2 \omega^2 \,. \tag{15.91}$$

Introducing in the previous matrix elements the complete basis $|\omega JM\lambda\rangle$, and using (B.113) to compute the amplitudes $\langle \boldsymbol{k},\mu|\omega' JM\lambda\rangle$, it is not difficult to obtain

$$\frac{d\Gamma_{i\to f}(\boldsymbol{k},\mu)}{d\Omega} = \omega \left| \sum_J (-i)^J \sqrt{\frac{(J+1)(2J+1)}{J}} \frac{\omega^J}{(2J+1)!!} \right.$$
$$\left. \times \left[(Q+Q')_{fi}(J\mu,\omega,\hat{\boldsymbol{k}}) - \mathrm{i}\mu(M+M')_{fi}(J\mu,\omega,\hat{\boldsymbol{k}}) \right] \right|^2 , \tag{15.92}$$

where by $Q_{fi}(J\mu,\omega,\hat{\boldsymbol{k}})$, etc. ..., we mean the expression

$$Q_{fi}(J\mu,\omega,\hat{\boldsymbol{k}}) \equiv \sum_M D^J(\mathcal{R}^{-1})_{\mu M} Q_{fi}(JM;\omega) \,, \quad \text{etc.} \ldots \,, \tag{15.93}$$

$\mathcal{R}$ being the rotation matrix transforming χ_μ, χ_0 (B.87) to ε_μ, $\hat{\boldsymbol{k}}$. With the choice (15.7) of polarization vectors, and using (15.9), we have $D^J_{MM'}(\mathcal{R}) = D^J_{MM'}(\phi, \theta, 0)$.

Integrating (15.92) over all directions $\hat{\boldsymbol{k}}$, and summing over the two polarizations $\mu = \pm 1$, yields

$$\Gamma_{i\to f} = \sum_{JM\lambda} \Gamma_{fi}(JM;\omega;\lambda) . \tag{15.94}$$

The interference between different multipole disappears due to the orthogonality of their wave functions. Equation (15.92) reveals that if only one of the multipole transitions is large, the angular distribution will be dominated by this multipole.

15.7 Emission and Absorption of Photons

We just studied the *spontaneous emission* process of a photon by the de-excitation of a matter system. Initially we had the photon vacuum and in the final state a single photon appears. We can similarly deal with the case when there is E.M. radiation in the initial state, but now, together with emission process one can also have absorption of part of the initial radiation.

If initially the matter system is described by a state $|\Psi_i\rangle$, and the radiation consists of N photons in a state $|\varphi\rangle$, then since to first order in e there is a term in H_1 which annihilates a photon, and another one creating it, we have to this order one of the following processes:

$$\begin{aligned} &\Psi_i, N \to \Psi_f, N-1 \quad \text{(absorption)} ,\\ &\Psi_i, N \to \Psi_f, N+1 \quad \text{(emission)} ; \end{aligned} \tag{15.95}$$

one or the other takes place depending on the energies E_i, E_f of the matter system.

If $E_i < E_f$ we will have absorption. The amplitude for this process is proportional to

$$\begin{aligned} \langle \Psi_f, \underbrace{\varphi\ldots\varphi}_{N-1} | H_1 | \Psi_i, \underbrace{\varphi\ldots\varphi}_{N} \rangle &= \sqrt{N} \langle \Psi_f | H_1 | \Psi_i, \varphi\rangle \\ &= \sqrt{N} \sum_\lambda \int d^3k \, \langle \Psi_f | H_1 | \Psi_i, \boldsymbol{k}\varepsilon_\lambda\rangle \langle \boldsymbol{k}\varepsilon_\lambda | \varphi\rangle , \end{aligned} \tag{15.96}$$

and the presence of N photons in the initial state is reflected in the fact that the corresponding absorption probability is N times the absorption probability of a single initial photon. This factor of N has its origin in the commutation relations, and it is therefore characteristic of the bosonic nature of photons. Next let us consider the particular scattering process Ψ_i+(1 photon) $\to \Psi_f$. If the levels E_i, E_f were pointlike, that is, infinitely narrow, the absorption probability would vanish since there are no perfectly monocromatic photons. In practice however it is the finite width ΔE_f which makes the transition possible. The measured absorption cross section is the average of the ideal theoretical result over the band ΔE_f. The final states are of the form $|\Psi_f(E)\rangle$, with $E \in \Delta E_f$. We assume the band to be narrow enough so that in our computations we can make the approximation $|\Psi_f(E)\rangle \simeq |\Psi_f(E')\rangle \simeq |\Psi_f\rangle$. Using the results in Sects. 8.3, 9 this cross section integrated over ΔE_f is given by

$$\int dE\, \sigma_{\text{abs}}(i \to f) = \int dE\, d^2\varrho\, W(\Psi_i, \varphi_{\varrho} \to \Psi_f(E)) \,, \tag{15.97}$$

where ϱ is a vector transverse to the central momentum of the photon state $|\varphi\rangle$, this one assumed to be very concentrated in a region of momentum space around $\boldsymbol{k}_i$. The transition probability $W(\Psi_i, \varphi_{\varrho} \to \Psi_f(E))$, in the Born approximation we are considering, is:

$$\begin{aligned} W(\Psi_i, \varphi_{\varrho} \to \Psi_f(E)) \simeq (2\pi)^2 \sum_{\lambda\lambda'} \int & d^3k\, d^3k' \langle \Psi_f(E)|H_1|\Psi_i, \boldsymbol{k}\boldsymbol{\varepsilon}_\lambda\rangle \\ & \times \delta(\omega + E_i - E)\langle \boldsymbol{k}\boldsymbol{\varepsilon}_\lambda|\varphi_{\varrho}\rangle\langle\varphi_{\varrho}|\boldsymbol{k}'\boldsymbol{\varepsilon}'_{\lambda'}\rangle \\ & \times \delta(\omega' + E_i - E)\langle\Psi_i, \boldsymbol{k}'\boldsymbol{\varepsilon}'_{\lambda'}|H_1|\Psi_f(E)\rangle \,. \end{aligned} \tag{15.98}$$

Substituting (15.98) in (15.97), and arguing as in Sect. 8.9 leads to

$$\int dE\, \sigma_{\text{abs}}(i \to f) = (2\pi)^4 |\langle\Psi_f|H_1|\Psi_i, \boldsymbol{k}_i\boldsymbol{\varepsilon}_{\lambda_i}\rangle|^2 \,, \tag{15.99}$$

as long as all of the energies in $|\varphi\rangle$ are capable of producing the transition to the band ΔE_f, and their polarizations are practically $\boldsymbol{\varepsilon}_{\lambda_i}$. The product of $\langle\sigma_{\text{abs}}\rangle_{\text{average}}$ times the incident E.M. energy flux provides the amount of energy absorbed by the system per unit time.

Finally, the study of the emission process (15.95) ($E_f < E_i$) is carried out along the same steps. The difference now is the appearance in the amplitude of the factor $\sqrt{(N+1)}$ which is non-vanishing even when $N = 0$ (*spontaneous* emission already treated in the previous section). Therefore, the transition probability in this case contains the factor $N + 1$, and it has two contributions: the spontaneous emission and one proportional to N, or *induced* emission. One should not forget however that in the induced emission or absorption we have the matrix element $\langle \boldsymbol{k}\boldsymbol{\varepsilon}_\lambda|\varphi\rangle$, which selects among all the photons contained in the initial E.M. radiation those capable of producing the transition, and only when the state of each initial photon falls in the correct energy band, will the probability be proportional to N. In other cases we will have to replace N by some N_{average} representing the photon fraction of interest.

15.8 Angular Distribution of Multipole Radiation

According to (15.92) and (15.93) the angular distribution of the power radiated in an electric or magnetic 2^J-polar transition, normalized to unity, is given by

$$W_{JM}(\hat{\boldsymbol{r}}) = \frac{2J+1}{8\pi} \sum_{\mu=\pm 1} |D^J_{M\mu}(\phi,\theta,0)|^2 \,. \tag{15.100}$$

From (B.31) one immediately obtains

$$\begin{aligned} &C(J1J;0\mu)D^J_{M\mu}(\phi,\theta,0) \\ &\quad = \sum_{M_1} C(J1J;M_1,M-M_1)D^J_{M_1 0}(\phi,\theta,0)D^1_{M-M_1,\mu}(\phi,\theta,0)\,, \end{aligned} \tag{15.101}$$

and since $C(J1J;01) = -C(J1J;0-1) = -1/\sqrt{2}$ and $C(J1J;00) = 0$, (15.100) can be rewritten as:

$$\begin{aligned} W_{JM}(\hat{\boldsymbol{r}}) = \frac{2J+1}{4\pi} &\sum_{M_1 M_1'} C(J1J;M_1,M-M_1) \\ &\times C(J1J;M_1',M-M_1')D^J_{M_1 0}(\phi,\theta,0)D^{J*}_{M_1' 0}(\phi,\theta,0) \\ &\times \sum_{\mu=-1}^{+1} D^1_{M-M_1,\mu}(\phi,\theta,0)D^{1*}_{M-M_1',\mu}(\phi,\theta,0)\,. \end{aligned} \tag{15.102}$$

Using (B.32) and the orthogonality relations (B.30) we obtain

$$W_{JM}(\hat{\boldsymbol{r}}) = \sum_{M_1} C^2(J1J;M_1,M-M_1)|Y_J^{M_1}(\phi,\theta)|^2\,, \tag{15.103}$$

namely:

$$\boxed{\begin{aligned} &W_{JM}(\hat{\boldsymbol{r}}) \\ &= \frac{1}{2J(J+1)}\{[J(J+1)-M(M+1)]|Y_J^{M+1}(\hat{\boldsymbol{r}})|^2 \\ &+ 2M^2|Y_J^M(\hat{\boldsymbol{r}})|^2 + [J(J+1)-M(M-1)]|Y_J^{M-1}(\hat{\boldsymbol{r}})|^2\}\,. \end{aligned}} \tag{15.104}$$

Notice that $W_{JM}(\hat{\boldsymbol{r}}) = W_{J,-M}(\hat{\boldsymbol{r}})$. In particular for dipole and quadrupole transitions one has explicitly

$$\begin{aligned} W_{10}(\hat{\boldsymbol{r}}) &= \frac{3}{8\pi}\sin^2\theta\,, & W_{1,\pm1}(\hat{\boldsymbol{r}}) &= \frac{3}{16\pi}(1+\cos^2\theta)\,, \\ W_{20}(\hat{\boldsymbol{r}}) &= \frac{15}{8\pi}\sin^2\theta\cos^2\theta\,, & W_{2,\pm1}(\hat{\boldsymbol{r}}) &= \frac{5}{16\pi}(1-3\cos^2\theta+4\cos^4\theta), \\ W_{2,\pm2}(\hat{\boldsymbol{r}}) &= \frac{5}{16\pi}(1-\cos^4\theta)\,, \end{aligned} \tag{15.105}$$

represented graphically in Fig. 15.1.

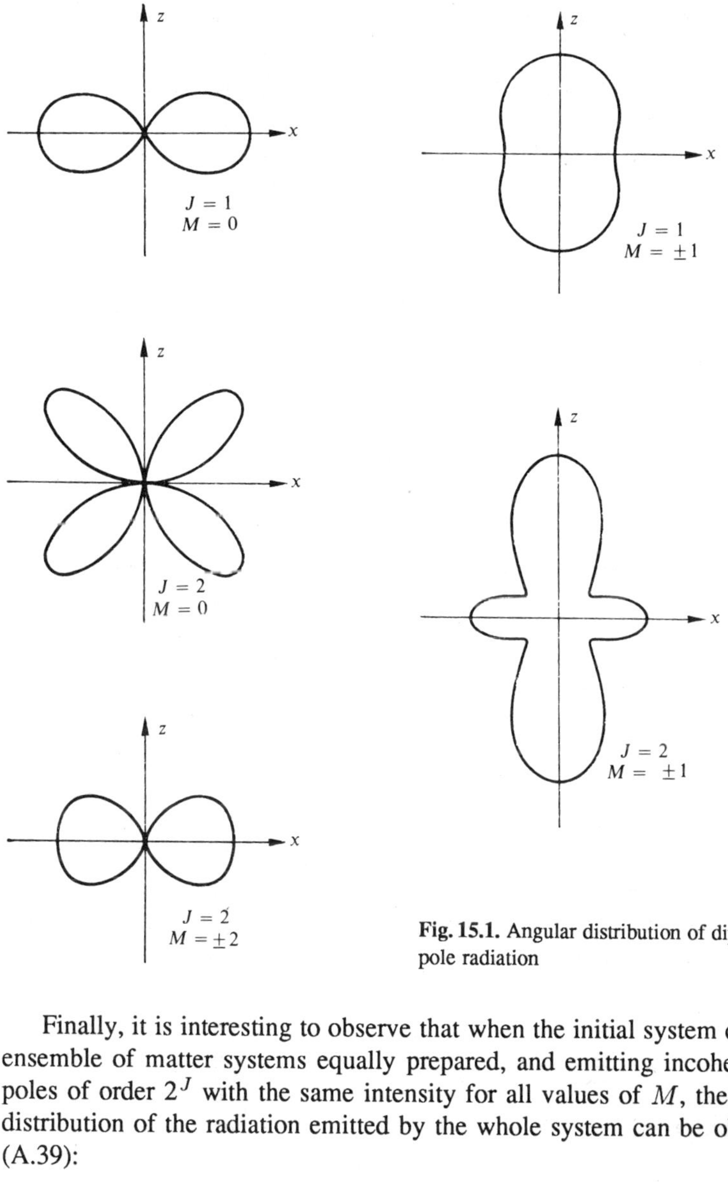

Fig. 15.1. Angular distribution of dipole and quadrupole radiation

Finally, it is interesting to observe that when the initial system consists of an ensemble of matter systems equally prepared, and emitting incoherently multipoles of order 2^J with the same intensity for all values of M, then the angular distribution of the radiation emitted by the whole system can be obtained using (A.39):

$$W_J(\hat{\boldsymbol{r}}) = \frac{1}{2J+1}\sum_M W_{JM}(\hat{\boldsymbol{r}}) = \frac{1}{4\pi}\ , \tag{15.106}$$

and it is therefore isotropic.

15.9 Electric Dipole Transitions in Atoms

Since this is a particularly interesting case, and as an application of the formulae derived previously, we now consider the spontaneous electric dipole transitions in atoms. According to Eq. (15.83) and taking into account that generally $|Q'_{fi}(JM;\omega)| \ll |Q_{fi}(JM;\omega)|$, the transition probability is

$$\Gamma_{fi}(1M;\omega;\mathrm{e}) = \frac{16\pi}{9}\omega^3 |\langle\Psi_f|\sum_{j=1}^{Z} er_j\, Y_1^{M*}(\hat{\boldsymbol{r}}_j)|\Psi_i\rangle|^2 . \tag{15.107}$$

Table 15.1 indicates that this transition is allowed only if $|J_i - J_f| = 0, 1, \Pi_i\Pi_f = -1$, and J_i and J_f do not vanish simultaneously. Equation (15.107) can be rewritten as

$$\Gamma_{fi}(1M;\omega;\mathrm{e}) = \frac{16\pi}{9}\omega^3 |\langle\Psi_i|\sum_{j=1}^{Z} er_j\, Y_1^{M}(\hat{\boldsymbol{r}}_j)|\Psi_j\rangle|^2 . \tag{15.108}$$

Using (A.37) and the definition of electric dipole moment:

$$\boldsymbol{D} = e\sum_{j=1}^{Z} \boldsymbol{r}_j , \tag{15.109}$$

it is obvious that (15.108) becomes

$$\boxed{\Gamma_{fi}(1M;\omega;\mathrm{e}) = \tfrac{4}{3}\omega^3 |\langle\Psi_i|D_M|\Psi_j\rangle|^2} . \tag{15.110}$$

Assuming the initial atomic system not to be polarized and that the final polarization is not measured, we have to average (15.108) over the initial states and sum over the polarization of the final states. These sums are easy to carry out using the Wigner-Eckart theorem and the orthogonality properties of Clebsch-Gordan coefficients. We obtain

$$\Gamma_{fi}^{\mathrm{E1}} = \frac{16\pi}{9}\omega^3 |\langle\Psi_i\|\sum_{j=1}^{Z} er_j\, Y_1(\hat{\boldsymbol{r}}_j)\|\Psi_j\rangle|^2 , \tag{15.111}$$

or equivalently

$$\boxed{\Gamma_{fi}^{\mathrm{E1}} = \tfrac{4}{3}\omega^3 |\langle\Psi_i\|D\|\Psi_j\rangle|^2} . \tag{15.112}$$

To proceed any further, it is necessary to specify the atomic wave functions. For simplicity, we perform the computation for a $nl_j \to n'l'_{j'}$ transition in hydrogen. Using (B.75), (15.111) becomes

$$\begin{aligned}\Gamma^{\mathrm{E1}}&(nlj \to n'l'j') \\ &= \frac{16\pi}{9}\omega^3 W^2(l'j'lj;1/2\,1)(2l+1)(2j'+1)|\langle nl\|er\,Y_1\|n'l'\rangle|^2 .\end{aligned} \tag{15.113}$$

The reduced matrix element appearing above can easily be evaluated from the corresponding matrix element with all the third components of angular momentum equal to 0 and using (A.44):

$$|\langle nl\|er\,Y_1\|n'l'\rangle|^2 = \alpha\frac{3(2l'+1)}{4\pi(2l+1)}C^2(l'1l;00)I^2\,, \tag{15.114}$$

where α is the fine structure constant and I is given by

$$I = \int_0^\infty dr\, r^3 R_{nl}(r)R_{n'l'}(r)\,. \tag{15.115}$$

Hence

$$\begin{aligned}&\Gamma^{\mathrm{E1}}(nlj \to n'l'j')\\ &\quad = \tfrac{4}{3}\alpha\omega^3(2l'+1)(2j'+1)W^2(l'j'lj;1/2\,1)C^2(l'1l;00)I^2\,,\end{aligned} \tag{15.116}$$

which can be simplified with (B.60):

$$\boxed{\Gamma^{\mathrm{E1}}(nlj \to n'l'j') = \tfrac{4}{3}\alpha\omega^3 C^2(j1j';1/2\,0)I^2}\,. \tag{15.117}$$

The step from (15.116) to (15.117) is justified only if $l+l'$ = odd and $l = j\pm 1/2$, $l' = j'\pm 1/2$. But this is precisely the case of interest here, for otherwise $\Gamma^{\mathrm{E1}}(nlj \to n'l'j') = 0$, as follows directly from (15.116).

If as it is often the case, one is only interested in

$$\Gamma^{\mathrm{E1}}(nlj \to n'l' \text{ all } j') = \sum_{j'}\Gamma^{\mathrm{E1}}(nlj \to n'l'j')\,, \tag{15.118}$$

the sum can be obtained from (15.116) after using (B.61). Neglecting the fine structure, the result, which does not depend on j and will be denoted by $\Gamma^{\mathrm{E1}}(nl \to n'l')$, is given by

$$\boxed{\Gamma^{\mathrm{E1}}(nl \to n'l') = \tfrac{4}{3}\alpha\omega^3 I^2 C^2(l1l';00)}\,. \tag{15.119}$$

Had we neglected the spin variables, we could have obtained this result directly from (15.108).

The agreement between the theory and the available experimental data [RS 85] on lifetimes $\tau_{nl} \equiv 1/\Gamma_{nl}$, with $\Gamma_{nl} \equiv \sum_{n'l'}\Gamma(nl \to n'l')$, and absorption probabilities is very good within the experimental errors (not smaller than 1%). Table 15.2 contains theoretical and experimental information on lifetimes.

Table 15.2. Electric dipole transitions in hydrogen, with $a_0 = (\mu\alpha)^{-1}$, and μ the reduced mass of the electron. The experimental values have typical errors of 1% [RS 85].

Transition	$\alpha^{-4}a_0\Gamma^{E1}(nl \to n'l')$	$\Gamma^{E1}_{\text{theory}}$ (in s^{-1})	$\tau^{E1}_{np\,\text{theory}}$ (in s)	$\tau_{np\,\exp}$ (in s)
$2p \to 1s$	$2^8 \cdot 3^{-8}$	6.265×10^8	1.596×10^{-9}	1.60×10^{-9}
$3p \to 2s$	$2^{13} \cdot 3^{-1} \cdot 5^{-9}$	2.245×10^7	5.271×10^{-9}	5.30×10^{-9}
$3p \to 1s$	$2^{-5} \cdot 3^{-1}$	1.673×10^8		
$4p \to 3s$	$2^9 \cdot 5^3 \cdot 7^{-13} \cdot 17^2$	3.065×10^6	12.305×10^{-9}	12.7×10^{-9}
$4p \to 2s$	$2^6 \cdot 3^{-12} \cdot 5$	9.668×10^6		
$4p \to 1s$	$2^9 \cdot 3^4 \cdot 5^{-10}$	6.819×10^7		
$4p \to 3d$	$2^{21} \cdot 7^{-13}$	3.475×10^5		
$5p \to 4s$	$2^{16} \cdot 3^{-37} \cdot 5 \cdot 13^4 \cdot 47^2$	7.372×10^5	23.789×10^{-9}	23.8×10^{-9}
$5p \to 3s$	$2^{-26} \cdot 5 \cdot 37^2$	1.638×10^6		
$5p \to 2s$	$2^{13} \cdot 3^6 \cdot 5 \cdot 7^{-13}$	4.948×10^6		
$5p \to 1s$	$2^{11} \cdot 3^{-14} \cdot 5$	3.438×10^7		
$5p \to 4d$	$2^{19} \cdot 3^{-31} \cdot 5^4 \cdot 127^2$	1.885×10^5		
$5p \to 3d$	$2^{-26} \cdot 5^4$	1.495×10^5		

The *oscillator strenghts* are some dimensionless quantities directly related with the E1 widths of the atomic levels. They played a very important role in the development of quantum mechanics, and they are still in use today. They are defined by

$$f_{a\to b} = 2\mu\omega_{ba}|\langle b|\sum_1^Z \boldsymbol{r}_i|a\rangle|^2 , \tag{15.120}$$

where $|a\rangle$, $|b\rangle$ are normalized atomic eigenstates of H_{m} with energies E_a, E_b, and where $\omega_{ba} \equiv (E_b - E_a)$ and μ is the reduced mass of the electron in the atom considered. As a function of these quantities, the E1 decay width is

$$\langle \Gamma^{\mathrm{E1}}_{a\to b}\rangle = -\frac{2}{3}\frac{\alpha}{\mu}\omega^2_{ba}\langle f_{a\to b}\rangle . \tag{15.121}$$

The symbol $\langle\ldots\rangle$ represents averaging over the initial polarizations and summing over the final ones. (Notice that $f_{a\to b}$ is positive for absorption and negative for emission processes.)

The oscillator strengths satisfy an important *sum rule*, nowadays known as the Thomas-Reiche-Kuhn sum rule, obtained by *Thomas* [TH 25], *Reiche-Thomas* [RT 25] and *Kuhn* [KU 25] using pre-quantum concepts and later on by *Heisenberg* [HE 25] from his commutation relations. This rule implies that

$$\boxed{\sum_b f_{a\to b} = 3Z} \tag{15.122}$$

independently of the initial atomic state.

The proof is rather simple: If $\boldsymbol{r} \equiv \Sigma \boldsymbol{r}_i$, $\boldsymbol{p} \equiv \Sigma \boldsymbol{p}_i$, we have

$$[r_i, p_j] = \mathrm{i}Z\delta_{ij} , \quad \boldsymbol{p} = \mathrm{i}\mu[H_\mathrm{m}, \boldsymbol{r}] , \tag{15.123}$$

where the last equation is obtained assuming the interactions between matter particles to be given by a local potential and neglecting the Hughes-Eckart terms associated to the recoil of the nucleus.

Hence

$$\mathrm{i}Z = \langle a|[r_j, p_j]|a\rangle = \sum_b [\langle a|r_j|b\rangle\langle b|p_j|a\rangle - \langle a|p_j|b\rangle\langle b|r_j|a\rangle] . \tag{15.124}$$

Since

$$\langle b|p_j|a\rangle = \mathrm{i}\mu\langle b|[H_\mathrm{m}, r_j]|a\rangle = \mathrm{i}\mu\omega_{ba}\langle b|r_j|a\rangle , \tag{15.125}$$

it follows that

$$2\mu\sum_b \omega_{ba}|\langle b|r_j|a\rangle|^2 = Z \tag{15.126}$$

and we obtain (15.122).

As an important application of the sum rule (15.122) the following relation is obtained easily using (15.99) and (15.120):

$$\int d\omega\, \sigma_\mathrm{abs}\,(\text{ground state + unpolarized photon of energy } \omega$$
$$\to \text{any atomic state}) = 6\pi^2 Z\alpha/\mu \tag{15.127}$$

valid for any atom in the E1 approximation.

15.10 Radiative Transitions in Nuclei

Since the strong interactions are invariant under the group $\mathrm{SU}_T(2)$ of isospin rotations and since these interactions dominate the electromagnetic ones, it is possible to use in the specification of nuclear states (Sect. 7.15) the quantum numbers T and T_3 which correspond, respectively, to the total and third component of isospin. The spontaneous emission probability of 2^J-polar radiation for a transition $jmTT_3 \to j'm'T'T_3$ is given by (15.83), and taking into account that $|Q'_{fi}| \ll |Q_{fi}|$, Eqs. (15.84) and (15.85) can be written in the isospin formalism as

$$Q = \langle j'm'T'T_3|\sum_{i=1}^{A}\frac{|e|}{2}(1-\tau_{3i})r_i^J Y_J^{m-m'*}(\hat{\boldsymbol{r}}_i)|jmTT_3\rangle , \tag{15.128}$$

$$M = \langle j'm'T'T_3| \sum_{i=1}^{A} \frac{|e|}{2m_N} \nabla_i [r_i^J Y_J^{m-m'*}(\hat{r}_i)] \cdot \Big\{ (1+\tau_{3i}) \frac{L_i}{J+1}$$

$$+ \tfrac{1}{2}[(\mu_p + \mu_n) + \tau_{3i}(\mu_p - \mu_n)]\sigma_i \Big\} |jmTT_3\rangle \,, \qquad (15.129)$$

where $m_N(\simeq m_p \simeq m_n)$ is the nucleon mass, $|e|$ is the proton charge and τ are the isospin matrices for which we use the same representation as for the Pauli matrices σ. The magnetic moments μ_p, μ_n (in units of the nuclear magneton $|e|/2m_p$) are related to the gyromagnetic factors by $\mu_p = g_p/2$, $\mu_n = g_n/2$.

The operators appearing in (15.128) and (15.129) can be decomposed in a isoscalar part and the third component of an isovector. Hence T_3' must equal T_3 for the transition to take place. This result is still valid when the contribution of Q'_{fi} is included, and it is always true for any E.M. transition between hadrons (strongly interacting particles). Moreover we obtain the following isospin selection rules:

a) The isoscalar interaction only allows $T = T'$ transitions.
b) The isovector part allows transitions $|T - T'| = 1$, and those with $T = T'$ if $T_3 \neq 0$. (Recall that $C(T1T; T_3 0 T_3)$ is proportional to T_3.)

If as it is often the case both interactions give contributions of similar orders of magnitude, the selection rule valid for all cases is $|T - T'| = 0, 1$. For dipole transitions it is useful to spell out the selection rules in more detail.

For E1 transitions, the isoscalar part Q^s of Q may be written as

$$Q^s = \sqrt{\frac{3}{4\pi}} \frac{|e|}{2} A \langle j'm'T'T_3 | R^*_{m-m'} | jmTT_3\rangle \,, \qquad (15.130)$$

where R is the nuclear center of mass position.

Equation (15.130) has been obtained using previous formulae where the $3A$ nuclear coordinates were assumed to be free. Actually, as explained in Sect. 15.6, these expressions are still valid using internal wave functions as long as we replace the r_i by coordinates relative to the C.M. so that for nuclei the operator R appearing in (15.130) should be replaced by 0. Hence $Q^s = 0$ and we can formulate the approximate rule of *Trainor* [TR 52] and *Gamba* et al. [GM 52]: the only E1 transitions allowed satisfy $|T - T'| = 1$ and $T = T'$ if $T_3 \neq 0$. As an application of this selection rule we consider the E1 transitions in $^{14}N(T_3 = 0)$ shown in Fig. 15.2. Since the E1 transition probabilities depend on the energy as ω^3, assuming that both transitions have nuclear matrix elements of the same order of magnitude we expect that $\Gamma_2(E1)/\Gamma_1(E1) \approx (5.685/3.373)^3 \approx 4.8$. According to the previous selection rule $\Gamma_2(E1) = 0$, and experimentally it is found [AJ 70] that $\Gamma_2(E1)/\Gamma_1(E1) = 0.62$.

For magnetic dipole transitions the isoscalar part M^s of M is

$$M^s = \frac{|e|}{4m_N} \sqrt{\frac{3}{4\pi}} \chi^*_{m-m'} \cdot \langle j'm'T'T_3 \rangle | \sum_{i=1}^{A}$$

$$\times \left[L_i + (\mu_p + \mu_n)\sigma_i \right] |jmTT_3\rangle \,, \qquad (15.131)$$

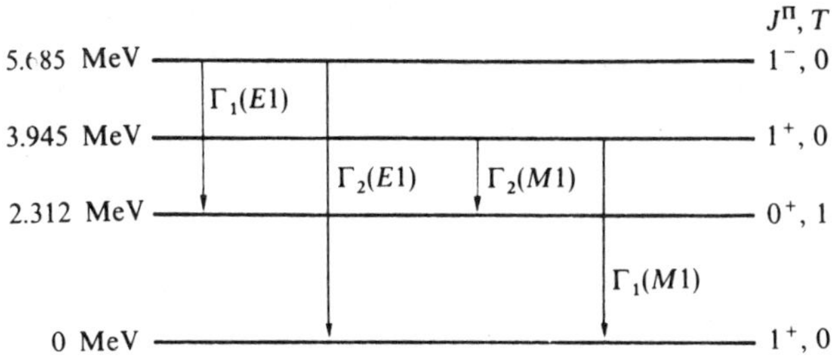

Fig. 15.2. Dipole transitions in $^{14}N(T_3 = 0)$. For each level of interest we give the binding energy, J^{Π} and T

as a consequence of (15.129) using $\nabla[rY^{m-m'}(\hat{\boldsymbol{r}})] = \sqrt{3/4\pi}\chi_{m-m'}$ where χ are the vectors introduced in (B.87). If $\boldsymbol{J}$ is the total angular momentum operator:

$$M^{\mathrm{s}} = \frac{|e|}{4m_{\mathrm{N}}}\sqrt{\frac{3}{4\pi}}\chi^*_{m-m'} \cdot \langle j'm'T'T_3|\boldsymbol{J} + \left(\mu_{\mathrm{p}} + \mu_{\mathrm{n}} - \tfrac{1}{2}\right)\sum_{i=1}^{A}\boldsymbol{\sigma}_i|jmTT_3\rangle\,, \tag{15.132}$$

and since the matrix element of $\boldsymbol{J}$ vanishes ($\boldsymbol{J}$ transforms the initial state into another state of the same energy, orthogonal to the final state):

$$M^{\mathrm{s}} = \frac{|e|}{4m_{\mathrm{N}}}\sqrt{\frac{3}{4\pi}}\left(\mu_{\mathrm{p}} + \mu_{\mathrm{n}} - \tfrac{1}{2}\right) \times \chi^*_{m-m'} \cdot \langle j'm'T'T_3|\sum_{i=1}^{A}\boldsymbol{\sigma}_i|jmTT_3\rangle\,. \tag{15.133}$$

A similar argument gives the isovector part M^{v} of M as

$$M^{\mathrm{v}} = \frac{|e|}{4m_{\mathrm{N}}}\sqrt{\frac{3}{4\pi}}\chi^*_{m-m'} \cdot \langle j'm'T'T_3|\sum_{i=1}^{A} \times [\boldsymbol{L}_i + (\mu_{\mathrm{p}} - \mu_{\mathrm{n}})\boldsymbol{\sigma}_i]\tau_{3i}|jmTT_3\rangle\,. \tag{15.134}$$

As $\mu_{\mathrm{p}} - \mu_{\mathrm{n}} \simeq 4.7$ and $\mu_{\mathrm{p}} + \mu_{\mathrm{n}} - 1/2 \simeq 0.38$, one should expect the isovector part to dominate in general the isoscalar part in M1 transitions. Since in nuclei with $T_3 = 0$ the transitions $T' = T$ can only proceed through the isoscalar part, it is likely that in such nuclei the transitions $T' = T$ are suppressed with respect to those with $T' \neq T$ by a factor of the order of $(0.38/4.7)^2 \simeq 7 \times 10^{-3}$ (*Morpurgo*'s rule [MO 58]).

As an application of this rule we study the M1 transition shown in Fig. 15.2. Since the M1 transition probabilities depend on the energy as ω^3, if the nuclear matrix elements of both transitions had the same order of magnitude, we would expect $\Gamma_1(\mathrm{M1})/\Gamma_2(\mathrm{M1}) \simeq (3.945/1.633)^3 \simeq 14$. According to Morpurgo's

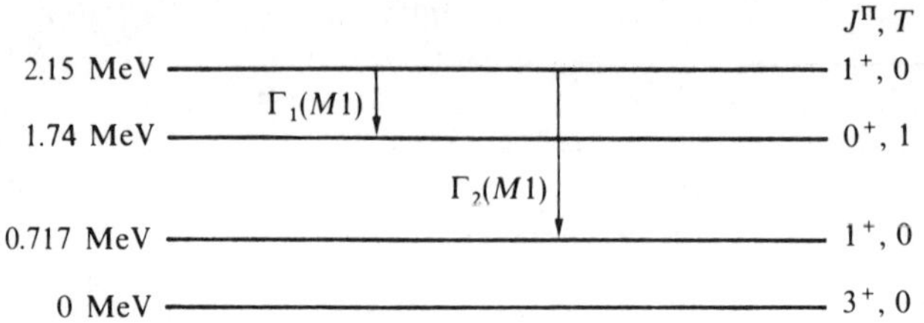

Fig. 15.3. M1 transitions in ^{10}B

rule the transition Γ_1(M1) should be strongly suppressed, and experimentally it is found that $\Gamma_1(\mathrm{M1})/\Gamma_2(\mathrm{M1}) = 0.042$ [GF 66].

Another example of the validity of Morpurgo's rule is the following: In Fig. 15.3 some of the energy levels of the nucleus $^{10}\mathrm{B}(T_3 = 0)$ are represented. If the nuclear matrix elements are of the same order of magnitude $\Gamma_2(\mathrm{M1})/\Gamma_1(\mathrm{M1}) \simeq (1.43/0.41)^3 \simeq 42$. According to the rule mentioned Γ_2(M1) should be strongly suppressed. Experimentally one finds $\Gamma_2(\mathrm{M1})/\Gamma_1(\mathrm{M1}) \simeq 0.77$.

An important drawback often encountered when performing computations with nuclear wave functions is that their intrinsic parts are almost always unknown. Adopting a nuclear model, like for instance the shell model where one assumes the nucleons to move under the action of an effective potential anchored at the origin of coordinates, leads to wave functions depending on the $3A$ coordinates, in such a way that their C.M. is not free. The corrections necessary in order to resolve this issue pose very difficult and sometimes not well defined problems in practical cases [LI 58], [FR 71]. For multipole electric transitions a simple solution to this problem can be outlined: The quantities to be computed Q_{fi} (15.84), with the replacement $\boldsymbol{r}_i \to \boldsymbol{r}_i - \boldsymbol{R}$, are expressed as a sum of terms of the form

$$\int dV_{\mathrm{int}}\, f(\boldsymbol{r}_j - \boldsymbol{R})\Psi_f^\dagger \Psi_i \ , \tag{15.135}$$

where dV_{int} denotes the volume element of the $3(A-1)$ intrinsic coordinates on which Ψ_i, Ψ_f depend.

Let us define the functions

$$\begin{aligned} \phi_i(\boldsymbol{r}_i, \dots, \boldsymbol{r}_A) &= g(\boldsymbol{R})\Psi_i \ , \\ \phi_f(\boldsymbol{r}_1, \dots, \boldsymbol{r}_A) &= g(\boldsymbol{R})\Psi_f \ , \end{aligned} \tag{15.136}$$

where $g(R)$ is normalized to unity. Then (15.135) is equivalent to

$$\int dV\, f(\boldsymbol{r}_j - \boldsymbol{R})\phi_f^\dagger(\boldsymbol{r}_1, \dots, \boldsymbol{r}_A)\phi_i(\boldsymbol{r}_1, \dots, \boldsymbol{r}_A) \ , \tag{15.137}$$

with the integration extended over the $3A$ coordinates. However, we have the exceptional circumstance that when the nuclear potential is a harmonic oscillator and the shells are occupied in increasing order of energy [ES 55], the intrinsic wave functions of the nuclei can always be written in the form (15.136), choosing $\phi_i(\boldsymbol{r}_1, \ldots, \boldsymbol{r}_A)$ and $\phi_f(\boldsymbol{r}_1, \ldots \boldsymbol{r}_A)$ as Slater determinants of one-particle wave functions in that potential. Even though this should be the case for the ground state of the nucleus in the shell model, it is not necessarily the case for excited states. This is why the computation of (15.135) by means of (15.137) with ϕ_f, ϕ_i given by Slater determinants requires corrections even after admitting the shell model with a harmonic nuclear potential as a good approximation. Such corrections (which in some cases can be of the order of 10%) affect the EJ transitions to the ground level only if J is odd [GU 77]. Neglecting these corrections we take (15.137) as the approximate starting point to begin our computations. The problem is still difficult, due to the dependence of the integrand on $\boldsymbol{R}$. Sometimes one simply neglects this dependence and replaces $\boldsymbol{r}_j - \boldsymbol{R}$ by $\boldsymbol{r}_j$. We now analyze how to correct the error in this substitution for E1 transitions. The amplitude to be calculated is

$$\sum_j e_j \int dV\,(\boldsymbol{r}_j - \boldsymbol{R})\phi_f^\dagger(\boldsymbol{r}_1, \ldots, \boldsymbol{r}_A)\phi_i(\boldsymbol{r}_1, \ldots, \boldsymbol{r}_A) \tag{15.138}$$

If we consider the nucleus as formed by an inert core of $(A-1)$ nucleons with only one nucleon taking part in the transition (the independent particle model), say nucleon number 1, the other nucleons $2, \ldots, A$, would be mere spectators in the process and only those terms in (15.138) depending on the coordinate $\boldsymbol{r}_1$ of the active particle would be relevant. Equation (15.138) coincides in this case with

$$\left[e_1\left(1-\frac{1}{A}\right)-\frac{1}{A}\sum_{j\neq 1} e_j\right]\int d^3r_1\,\boldsymbol{r}_1\xi_f^\dagger(\boldsymbol{r}_1)\xi_i(\boldsymbol{r}_1)\,, \tag{15.139}$$

where ξ_i, ξ_f are the initial and final single-particle states of nucleon number 1. This result is identical to what one would obtain if in (15.138) we replace $\boldsymbol{r}_j - \boldsymbol{R} \to \boldsymbol{r}_j$ and at the same time the charge e_1 of the particle undergoing the transition is replaced by an effective charge

$$e_1 \to e_1\left(1-\frac{1}{A}\right)-\frac{1}{A}\sum_{j\neq 1} e_j \tag{15.140}$$

which is $|e|(1-Z/A)$ if such particle is a proton and $-(Z/A)|e|$ if it is a neutron. For electric quadrupole transitions, it is possible to show that no correction for the charges is required [BM 69]. For magnetic and electric transitions with $J > 2$ the corrections are very sensitive to the model used and they can be estimated to be small [GW 55], [EG 70]. Here we choose to ignore them. Consequently, for the EJ transitions, in the independent particle model, the effects of nuclear recoil (the dependence of (15.138) on $\boldsymbol{R}$) are well simulated by assigning to the active nucleon an effective charge $|e|\varepsilon_j$, with

$$\begin{aligned} \varepsilon_J &= 1 - \frac{Z}{A}\delta_{J1} \quad \text{for protons}\,, \\ \varepsilon_J &= -\frac{Z}{A}\delta_{J1} \quad \text{for neutrons}\,. \end{aligned} \tag{15.141}$$

In this model therefore, the unpolarized EJ and MJ transition rates can be written, using (15.83, 128) and (15.129), as

$$\begin{aligned} \Gamma^{\mathrm{E}J}(nlj \to n'l'j') &= \frac{8\pi(J+1)}{J[(2J+1)!!]^2}\,\frac{\alpha\varepsilon_j^2\omega^{2J+1}}{2j+1} \\ &\quad \times \sum_{mm'} |\langle nl1/2jm|r^J Y_J^{m-m'}(\hat{\boldsymbol{r}})|n'l'1/2j'm'\rangle|^2\,, \\ \Gamma^{\mathrm{M}J}(nlj \to n'l'j') &= \frac{8\pi(J+1)}{J[(2J+1)!!]^2}\,\frac{\alpha\omega^{2J+1}}{m_N^2(2j+1)} \\ &\quad \times \sum_{mm'} |\langle nl1/2jm|O_{\mathrm{M}}|n'l'1/2j'm\rangle|^2\,, \\ O_{\mathrm{M}} &\equiv \nabla[r^J Y_J^{m-m'}(\hat{\boldsymbol{r}})]\cdot\left[a\frac{\boldsymbol{J}}{J+1} + \frac{b}{2}\boldsymbol{\sigma}\right]\,, \end{aligned} \tag{15.142}$$

where if the particle participating in the magnetic transition is a proton $a = 1$ and $b = \mu_{\mathrm{p}} - (J+1)^{-1}$, whereas $a = 0$ and $b = \mu_n$ if it is a neutron; and $nljm$ are the quantum numbers of the active particle in the shell model.

From now on we only consider those dominating transitions where $J = j' - j > 0$ $[J = j - j' > 0]$, and since $l = j \pm 1/2$, $l' = j' \pm 1/2$, then $l' - 1 = J - 1, J, J+1$ $[l - l' = J - 1, J, J+1]$. With respect to the electric transitions it is obvious from (15.142) that the only contributions in this case will satisfy $l' - l = J$ $[l - l' = J]$. On the other hand, using (B.105), we have

$$\nabla[r^J Y_J^{m-m'}(\hat{\boldsymbol{r}})] = \sqrt{J(2J+1)}\,r^{J-1}\boldsymbol{\mathcal{Y}}_{Jm-m'}^{J-1}(\hat{\boldsymbol{r}})\,, \tag{15.143}$$

and there will be a magnetic contribution only if $l' - l = J - 1$ $[l - l' = J - 1]$.

The calculation of $\Gamma^{\mathrm{E}J}$ via (15.142) can be easily carried out following the same method used in Sect. 15.9 to derive (15.116), and we obtain

$$\begin{aligned} &\Gamma^{\mathrm{E}J}(nlj \to n'l'j') \\ &\quad = \frac{2(J+1)}{J[(2J+1)!!]^2}\alpha\varepsilon_J^2\omega(\omega R)^{2J} R(nl;n'l';J) S_{\mathrm{e}}(jJj';ll')\,, \\ &R(nl;n'l';J) \equiv \left|\int_0^\infty dr\,r^2\left(\frac{r}{R}\right)^J R_{nl}(r)R_{n'l'}(r)\right|^2\,, \\ &S_{\mathrm{e}}(jJj';ll') \equiv (2J+1)(2j'+1)(2l'+1)C^2(l'Jl;00)W^2(l'j'lj;1/2J)\,. \end{aligned} \tag{15.144}$$

To estimate the order of magnitude of these quantities, one usually takes $R_{nl}(r) =$ const. for $r \le R$ and $R_{nl}(r) = 0$ for $r > R$. Normalizing

$$R_{nl}(r) = \begin{cases} \left(\frac{3}{R^3}\right)^{1/2}, & r \le R\,, \\ 0\,, & r > R\,, \end{cases} \tag{15.145}$$

one gets

$$R(nl;n'l';J) \equiv \left(\frac{3}{J+3}\right)^2 . \tag{15.146}$$

On the other hand, (B.60) implies that

$$S_e(jJj';ll') = (2J+1)C^2(jJj';1/2\,0\,1/2) , \tag{15.147}$$

and this quantity will be denoted from now on as $S_e(jJj')$. Since $j+J+j'$ is odd, use of (B.45) yields

$$S_e(jJj') = \frac{(2j_>+1)}{(2j+1)}\,\frac{(j_>-1/2)!}{(j_<-1/2)!}\,\frac{(2j_<)!!}{(2j_>)!!}\,\frac{(2J+1)!!}{J!} , \tag{15.148}$$

where $j_>$ and $j_<$ are respectively the larger and the lesser of the two quantities j and j'. Hence the EJ transition rate with $J=|j'-j|$ can be estimated using

$$\Gamma_e(J;j\to j') = \frac{2(J+1)}{J[(2J+1)!!]^2}\left(\frac{3}{J+3}\right)^2 \alpha\varepsilon_J^2(\omega R)^{2J}\omega S_e(jJj') . \tag{15.149}$$

For magnetic transitions, using (15.143) and (B.92), the matrix element becomes

$$\begin{aligned}\langle nl1/2jm|O_M|n'l'1/2j'm'\rangle &= \sqrt{J(2J+1)}\langle nl1/2jm|r^{J-1}\\ &\times \sum_\mu C(J-1\,1\,J;m-m'-\mu,\mu)Y_{J-1}^{m-m'-\mu}(\hat{\boldsymbol{r}})\\ &\times \left[\frac{a}{J+1}J_\mu + \frac{b}{2}\sigma_\mu\right]|n'l'1/2j'm'\rangle .\end{aligned} \tag{15.150}$$

Since the action of J_μ on $|n'l'1/2j'm'\rangle$ gives a linear combination of states all with the quantum number j', and since the tensor operator acting next is of order $(J-1)$, it cannot connect the resulting states on the right with $\langle nl1/2jm|$ due to the assumption $J=|j-j'|$. Therefore we can omit the J_μ contribution and write

$$\begin{aligned}&\Gamma^{MJ}(nlj\to n'l'j')\\ &\quad= \frac{2\pi(J+1)}{J[(2J+1)!!]^2}\,\frac{\alpha b^2\omega^{2J+1}}{m_N^2}|\langle nl1/2j\|A^J\|n'l'1/2j'\rangle|^2 ,\\ &A_M^J \equiv \sqrt{J(2J+1)}\sum_\mu C(J-1\,1\,J;M-\mu,\mu)r^{J-1}Y_{J-1}^{M-\mu}(\hat{\boldsymbol{r}})\sigma_\mu .\end{aligned} \tag{15.151}$$

Using (B.77), (B.72) and (B.73), we obtain

$$\begin{aligned}&|\langle nl1/2j\|A^J\|n'l'1/2j'\rangle|^2\\ &\quad= \frac{3}{2\pi}R^{2J-2}R(nl;n'l';J-1)(2J+1)^2J(2J-1)\\ &\quad\times(2l'+1)(2j'+1)C^2(l'J-1l;00)\\ &\quad\times X^2(l'1/2j';l1/2j;J-1\,1\,J) ,\end{aligned} \tag{15.152}$$

and therefore

$$
\begin{aligned}
&\Gamma^{\mathrm{M}J}(nlj \to n'l'j') \\
&\quad = \frac{2(J+1)}{J[(2J+1)!!]^2}\alpha\frac{b^2J^2}{m_{\mathrm{N}}^2R^2}(\omega R)^2\omega R(nl;n'l';J-1)S_{\mathrm{m}}(jJj';ll') \\
&S_{\mathrm{m}}(jJj';ll') \equiv \frac{3}{2J}(2J-1)(2J+1)^2(2l'+1)(2j'+1) \\
&\qquad\qquad \times C^2(l'J-1l;00)X^2(l'1/2j';l1/2j;J-1\,1J)\,.
\end{aligned}
\tag{15.153}
$$

(The presence of $J-1$ in $R(nl;n'l';J-1)$ in (15.153) makes the M1 transitions strongly suppressed in those nuclei where the independent particle model is a good approximation, due to the orthogonality of the radial wave functions.)

As explained previously, the magnetic transitions will take place when $\pm j' = J \pm j$ as long as $l = j \pm 1/2$ and $l' = j' \mp 1/2$. Taking this into account, and using (B.81) we conclude that $S_{\mathrm{m}} = S_{\mathrm{e}} \equiv S$. Thus, we can estimate the transition rates considered when the active particle is a proton using

$$
\begin{aligned}
\Gamma_{\mathrm{e}}(J;j \to j') &= \frac{2(J+1)}{J[(2J+1)!!]^2} \\
&\quad \times \left(\frac{3}{J+3}\right)^2 \alpha\varepsilon_J^2(\omega R)^{2J}\omega S(jJj')\,, \\
\Gamma_{\mathrm{m}}(J;j \to j') &= \frac{2(J+1)}{J[(2J+1)!!]^2} \\
&\quad \times \left(\frac{3}{J+3}\right)^2 \frac{\alpha}{(m_{\mathrm{N}}R)^2}[\mu_{\mathrm{p}} - (J+1)^{-1}]^2J^2(\omega R)^{2J}\omega S(jJj')\,,
\end{aligned}
\tag{15.154}
$$

where $S(jJj')$ is defined in (15.148) and it will be generically of order 1. If the nucleon participating in the transition is a neutron it suffices to replace $[\mu_{\mathrm{p}} - (J+1)^{-1}]^2J^2$ by $(\mu_{\mathrm{n}}J)^2$ in the second equation.

The measurement of widths of radiative transitions are often given in Weisskopf units, defined by

Table 15.3. Weisskopf units in eV, for $r_0 = 1.2\,\mathrm{fm}$, $\omega = 1\,\mathrm{MeV}$ and $A = 1$

J	$\dfrac{\Gamma_{\mathrm{W}}(\mathrm{E}J)}{A^{2J/3}(\omega/\mathrm{MeV})^{2J+1}}$ (in eV)	$\dfrac{\Gamma_{\mathrm{W}}(\mathrm{M}J)}{A^{2(J-1)/3}(\omega/\mathrm{MeV})^{2J+1}}$ (in eV)
1	6.75×10^{-2}	2.07×10^{-2}
2	4.79×10^{-8}	1.47×10^{-8}
3	2.23×10^{-14}	6.86×10^{-15}
4	7.02×10^{-21}	2.16×10^{-21}
5	1.58×10^{-27}	4.84×10^{-28}

$$\Gamma_W(EJ) = \frac{2(J+1)}{J[(2J+1)!!]^2}\left(\frac{3}{J+3}\right)^2 \alpha r_0^{2J} A^{2J/3}\omega^{2J+1} ,$$
$$\Gamma_W(MJ) = \frac{20(J+1)}{J[(2J+1)!!]^2}\left(\frac{3}{J+3}\right)^2 \frac{\alpha}{m_N^2} r_0^{2J-2} A^{2(J-1)/3}\omega^{2J+1} , \tag{15.155}$$

where $R = r_0 A^{1/3}$ has been assumed. In Table 15.3 we give the values for these units in eV, for $r_0 = 1.2\,\mathrm{fm}$, $\omega = 1\,\mathrm{MeV}$, and $A = 1$.

15.11 Low Energy Compton Scattering

As a final interesting application of the theory of matter-light interaction we will derive the cross section for the elastic scattering between soft photons and free electrons nearly at rest, the so-called low energy Compton scattering:

$$\gamma(\boldsymbol{k},\lambda) + e^-(\boldsymbol{p},\mu) \to e^-(\boldsymbol{p}',\mu') + \gamma(\boldsymbol{k}',\lambda') \tag{15.156}$$

where we specify the momentum and polarization of each of the particles involved. We only consider energies of the incident photon much smaller than the electron rest mass $\omega \ll m_e$. Taking into account the conservation laws of energy and momentum and neglecting terms of the order of ω/m_e it follows immediately that $\omega' = \omega$, i.e., there is no transfer of energy from the photon to the electron.

The interaction Hamiltonian H_1 for the process considered is given by (15.56) with $N = 1$. In first order of perturbation theory only the last term in H_1 contributes to the annihilation of the incident photon and the creation of the outgoing one, with a transition amplitude of the order of e^2. For this reason it is also necessary to consider second order terms in H_1 in the perturbation expansion which also contribute to this process to order e^2. Using the explicit form of H_1 and the properties of such expansion, it is possible to show that for frequencies ω much larger than the kinetic energies of the electrons, the latter contribution is negligible with respect to the diamagnetic term [SA 67].

To obtain under these assumptions the differential cross section in the Born approximation in a plain way, we should notice that when $\omega \ll m_e$ we can assume the electron to be located at the origin of coordinates with a wave function given simply by its polarization spinor. Using (8.64) corresponding to simple scattering the transition amplitude will be

$$\begin{aligned}\langle f|H_1|i\rangle &= \frac{\alpha}{2m_e}(\alpha^\dagger_{\mu'}\alpha_\mu)\langle \boldsymbol{k}'\varepsilon'_{\lambda'}| : \boldsymbol{A}^2(0) : |\boldsymbol{k}\varepsilon_\lambda\rangle \\ &= \frac{\alpha}{4\pi^2 m_e\omega}(\alpha^\dagger_{\mu'}\alpha_\mu)\boldsymbol{\varepsilon}(\boldsymbol{k}',\lambda')^*\cdot\boldsymbol{\varepsilon}(\boldsymbol{k},\lambda) ,\end{aligned} \tag{15.157}$$

and hence

$$\frac{d\sigma}{d\Omega} = \frac{\alpha^2}{m_e^2}|\boldsymbol{\varepsilon}(\boldsymbol{k}',\lambda')^*\cdot\boldsymbol{\varepsilon}(\boldsymbol{k},\lambda)|^2|\alpha^\dagger_{\mu'}\alpha_\mu|^2 . \tag{15.158}$$

Averaging over the polarizations of the initial electron and summing over the polarizations of the final electron we obtain

$$\frac{d\sigma}{d\Omega} = \frac{\alpha^2}{m_e^2}|\boldsymbol{\varepsilon}^*(\boldsymbol{k}',\lambda')\cdot\boldsymbol{\varepsilon}(\boldsymbol{k},\lambda)|^2 . \tag{15.159}$$

It is useful to notice that this formula multiplied by Z^4 and changing $m_e \to m$, describes the scattering at low energies of light on any target (free and at rest) of charge $\pm Ze$ and mass m. The structure of the target does not show up when the wave length of the incident photon is large compared to the size of the target.

If the polarization of the outgoing photon is not measured, the corresponding differential cross section is obtained by summing over the final polarizations in (15.159) and using (15.21):

$$\frac{d\sigma}{d\Omega} = \frac{\alpha^2}{m_e^2}\left[1 - \frac{1}{\omega^2}|\boldsymbol{k}'\cdot\boldsymbol{\varepsilon}(\boldsymbol{k},\lambda)|^2\right] . \tag{15.160}$$

Assuming $\boldsymbol{k} = (0,0,\omega)$, $\boldsymbol{k}' = (\omega\sin\theta\cos\phi, \omega\sin\theta\sin\phi, \omega\cos\theta)$ and that the initial polarization vector, orthogonal to $\boldsymbol{k}$, is $\boldsymbol{\varepsilon} = (\cos\psi, \sin\psi, 0)$, then

$$\frac{d\sigma}{d\Omega} = \frac{\alpha^2}{m_e^2}[1 - \sin^2\theta\cos^2(\theta - \psi)] , \tag{15.161}$$

in agreement with the equivalent formula derived in classical electrodynamics [JA 75]. If the initial radiation is not polarized, the cross section is obtained after averaging (15.161) over the angle ψ, or equivalently, averaging (15.160) over the initial polarizations:

$$\frac{d\sigma}{d\Omega} = \frac{\alpha^2}{m_e^2}\frac{1}{2}(1 + \cos^2\theta) . \tag{15.162}$$

Consequently the total cross section is

$$\boxed{\sigma = \frac{8\pi}{3}\left(\frac{\alpha}{m_e}\right)^2} , \tag{15.163}$$

known as the *Thomson* cross section. It is equal to $\sigma = 0.6652448(33)$ barn. The length $r_0 \equiv \alpha/m_e = 2.8179380(70)$ is called the *classical electron radius* because a classical charge distribution whose total charge equals e must have this radius if one wants its electrostatic energy to be of the order of the electron rest mass [JA 75].

In the case of electrons and to order α^2, it is possible to compute the cross section for any value of ω [AB 65]. The expression obtained is known as the *Klein-Nishina* formula. In particular one obtains

$$\sigma = \frac{8\pi}{3}\left(\frac{\alpha}{m_e}\right)^2\left[1 - \frac{2\omega}{m_e} + \mathrm{O}\left(\frac{\omega^2}{m_e^2}\right)\right] , \quad \omega \ll m_e . \tag{15.164}$$

One might expect that for a particle of mass m the deviations from Thomson's formula should only be important when $\omega \gtrsim m$ as is the case for electrons. This is not so. For hadrons, the strongly interacting particles such as the proton and the neutron, the deviations from Thomson's formula appear earlier than expected. For protons for example, when $\omega \simeq 140\,\mathrm{MeV}$ (approximately the pion mass), the total cross section grows quite rapidly with energy showing resonant structures at the same energies as those in πN scattering.

Bibliography

AA 87 Aharonov, A.; Anandan, J.: Phys. Rev. Lett. **58**, 1593 (1987)

AB 59 Aharonov, Y.; Bohm, D.: Phys. Rev. **115**, 485 (1959)

AB 65 Akhiezer, A.I.; Berestetskii, V.B.: *Quantum Electrodynamics* (translation of the Russian edition). Wiley, New York (1965)

AB 65a Arens, R.; Babbitt, D.: J. Math. Phys. **6**, 1071 (1965)

AC 72 Arecchi, F.; Courtens, E.; Gilmore, R.; Thomas, A.: Phys. Rev. A **6**, 2211 (1972)

AC 72a Amaldi, E.; Cabibbo, N.: "On the Dirac Magnetic Poles", in *Aspects of Quantum Theory*, Cambridge Univ. Press (1972)

AC 76 Aragone, C.; Chalbaud, E.; Salamó, S.: J. Math. Phys. **17**, 1963 (1976)

AC 84 Aguilar–Benítez, M.; Cahn, R.N.; Crawford, R.L.; Frosch, R.; Gopal, G.P.; Hendrick, R.E.; Hernández, J.J.; Höhler, G.; Losty, M.J.; Montanet, L.; Porter, F.C.; Rittenberg, A.; Roos, M.: Roper, L.D.; Shimada, T.; Shrock, R.E.; Törnqvist, N.A.; Trippe, T.G.; Trower, W.P; Walck, C.; Wohl, C.G.; Yost, G.P.; Armstrong, B.: Rev. Mod. Phys. **56**, no. 2, part II (1984)

AD 82 Aspect, A.; Dalibard, J.; Roger, G.: Phys. Rev. Lett. **49**, 1804 (1982)

AD 85 Adler, S.L.: Phys. Rev. Lett. **55**, 783 (1985)

AG 61 Akhiezer, N.A.; Glazman, I.M.: *Theory of Linear Operators in Hilbert Space* (two volumes), Ungar, New York (1961)

AG 74 Aragone, C.; Guerri, G.; Salomó, S.; Tani, J.L.: J. Phys. A **7**, L149 (1974)

AG 81 Aspect, A.; Grangier, P.; Roger, G.: Phys. Rev. Lett. **47**, 460 (1981)

AG 82 Aspect, A.; Grangier, P.; Roger, G.: Phys. Rev. Lett. **49**, 91 (1982)

AG 84 Aspect, A.; Grangier, P.: In Proceedings of the International Symposium *Foundations of Quantum Mechanics in the Light of New Technology*. Eds. S. Kamefuchi et al., Phys. Soc. of Japan (1984)

AG 84a Albeverio, S.; Gesztesy, F.; Høegh–Krohn, R.; Holden, H.: *Some exactly solvable models in Quantum Mechanics and the low energy expansions*, Preprint. University of Bielefeld (1984)

AG 88 Abellanas, L.; Galindo, A.: *Espacios de Hilbert (Geometría, Operadores, Espectros)*, Eudema, Madrid (1988)

AH 78 Avron, J.; Herbst, I.W.; Simon, B.: Duke Math. J. **45**, 847 (1978)

AJ 70 Ajzenberg–Selove, F.: Nucl. Phys. A **152**, 1 (1970)

AJ 77 Amrein, W.O.; Jauch, J.M.; Sinha, K.B.: *Scattering Theory in Quantum Mechanics*. W.A. Benjamin, Reading, Mass. (1977)

AL 59 Ajzenberg–Selove, F.; Lauritsen, T.: Nucl. Phys. **11**, 1 (1959)

AL 69 Allcock, G.R.: Ann. Phys. (N.Y.) **53**, 253, 286, 311 (1969)

AL 70 Ans, J.D'.; Lax, E.: *Taschenbuch für Chemiker und Physiker*. Springer, Berlin, Heidelberg (1970)

AL 72 Albeverio, S.: Ann. Phys. (N.Y.) **71**, 167 (1972)

AL 88 Alvarez, G.: Phys. Rev. A **37**, 4079 (1988)

AM 74 Amrein, W.O.: "Some Questions in Non-Relativistic Quantum Scattering Theory", in *Scattering Theory in Mathematical Physics*. J.A. La Vita and J.P. Marchand. Reidel, Dordrecht, Holland (1974)

AN 69 Anderson, R.F.V.: J. Functional Anal. **4**, 240 (1969)

AN 81 Andrews, M.: J. Phys. A **14**, 1123 (1981)

AR 65 Alfaro, V. de; Regge, T.: *Potential Scattering*. North-Holland, Amsterdam (1965)

AS 64 Abramowitz, M.; Stegun, I.A. (Eds.): *Handbook of Mathematical Functions with Formulas, Graphs and Mathematical Tables*. Dover, New York (1964)

AS 83 Austen, G.J.M.; Swart, J.J. de: Phys. Rev. Lett. **50**, 2039 (1983)

AS 87 Avron, J.E.; Seiler, R.; Yaffe, L.G.: Commun. Math. Phys. **110**, 33 (1987)

BA 49 Bargmann, V.: Rev. Mod. Phys. **21**, 488 (1949)

BA 49a Bargmann, V.: Phys. Rev. **75**, 301 (1949)

BA 52 Bargmann, V.: Proc. Nat. Acad. Sci. U.S.A. **38**, 961 (1952)

BA 52a Bargmann, V.: Ann. Math. **59**, 1 (1952)

BA 57 Bohm, D.; Aharonov, Y.: Phys. Rev. **108**, 1070 (1957)

BA 60 Bazley, N.W.: Phys. Rev. **120**, 144 (1960)

BA 64 Bargmann, V.: J. Math. Phys. **5**, 862 (1964)

BA 67 Bacry, H.: *Leçons sur la Théorie des Groups et les Symétries des Particules Elémentaires*. Gordon and Breach, Paris (1967)

BA 75 Baker, G.A.: *Essentials of Padé Approximants*. Academic Press, New York (1975)

BA 76 Banerjee, K.: Lett. Math. Phys. **1**, 323 (1976)

BB 66 Bohm, D.; Bub, J.: Rev. Mod. Phys. **38**, 453 (1966)

BB 78 Banerjee, K.; Bhatnagar, S.P.; Choudhry, V.; Kanwali, S.S.: Proc. Roy. Soc. London A **360**, 575 (1978)

BC 57 Bardeen, J.; Cooper, L.N.; Schrieffer, J.R.: Phys. Rev. **108**, 1175 (1957)

BD 71 Biswas, S.N.; Datta, K.; Saxena, R.P.; Srivastaka, P.K.; Varma, V.S.: Phys. Rev. D **4**, 3617 (1971)

BD 87 Bitter, T.; Dubbers, D.: Phys. Rev. Lett. **59**, 251 (1987)

BE 42 Bergmann, P.G.: *Introduction to the Theory of Relativity*. Prentice-Hall, Englewood Cliffs, N.J. (1942)

BE 64 Bell, J.S.: Physics **1**, 195 (1964)

BE 66 Bell, J.S.: Rev. Mod. Phys. **38**, 447 (1966)

BE 73 Bertrand, J.: C. R. Acad. Sci. **77**, 849 (1873)

BE 73a Belinfante, F.J.: *A Survey of Hidden-Variables Theories*. Pergamon, Oxford (1973)

BE 75 Bell, J.S.: Helv. Phys. Acta **48**, 93 (1975)

BE 84 Berry, M.V.: Proc. Roy. Soc. London A **329**, 45 (1984)

BG 84 Baumgartner, B.; Grosse, H.; Martin, A.: Phys. Lett. **146**B, 363 (1984)

BG 85 Baumgartner, B.; Grosse, H.; Martin, A.: *"Inequalities for scattering phase shifts"*. U.W. Th. Ph.-1985-6 (1985)

BH 85 Bhattacharya, S.K.: Phys. Rev. A **31**, 1991 (1985)

BI 61 Birman, M.S.: Mat. Sb. **55**, 125 (1961)

BJ 68 Bethe, H.A.; Jackiw, R.: *Intermediate Quantum Mechanics*, 2nd ed., Benjamin, New York (1968)

BK 61 Bromley, D.A.; Kuehner, J.A.; Almqvist, E.: Phys. Rev. **123**, 878 (1961)

BL 46 Bloch, F.: Phys. Rev. **70**, 460 (1946)

BL 73 Blin–Stoyle, R.J.: *Fundamental Interactions and the Nucleus*. North-Holland, Amsterdam (1973)

BL 77 Brezin, E.; Le-Gillou, J.; Zinn-Justin, J.: Phys. Rev. D **15**, 1544 and 1558 (1977)

BL 81 Biedenharn, L.C.; Louck, J.D.: *Angular Momentum in Quantum Physics*, Addison-Wesley, Reading, Mass. (1981)

BM 69 Bohr, A.; Mottelson, B.R.: *Nuclear Structure* (two volumes). Benjamin, New York (1969)

BM 72 Berry, M.V.; Mount, K.E.: Rep. Prog. Phys. **35**, 315 (1972)

BO 13 Bohr, N.: Phil. Mag. **26**, 1, 476 and 875 (1913)

BO 18 Bohr, N.: Kgl. Danske Vid. Selsk. Skr., nat.-math. Afd., 8 Raekke IV, 1 (1918)

BO 22 Bohr, N.: Z. Phys. **13**, 117 (1922)

BO 26 Born, M.: Z. Phys. **37**, 863 (1926) and **38**, 803 (1926)

BO 28 Bohr, N.: "The Quantum Postulate and the Recent Development of Atomic Theory", in *Atti del Congresso Internazionale dei Fisici*. N. Zanichelli, Bologna (1928)

BO 52 Bohm, D.: Phys. Rev. **85**, 166, 180 (1952)

BO 57 Bogolyubov, N.N.: J. Phys. USSR **9**, 23 (1957)
BO 58 Bogolyubov, N.N.: J.E.P.T. **7**, 41 (1958)
BO 67 Bogolyubov, N.N.: *Lectures on Quantum Statistics* (two volumes). Gordon and Breach, New York (1967)
BO 68 Bogolyubov, N.N. (Ed.): *The Theory of Superconductivity*. Gordon and Breach, New York (1968)
BO 77 Bender, C.M.; Olaussen, K.; Wang, P.S.: Phys. Rev. D **16**, 1740 (1977)
BP 79 Balian, R.; Parisi, G.; Voros, A.: "Quartic Oscillator" in *Feynman Path Integrals*. Lecture Notes in Physics, Vol. 106, Springer, Berlin, Heidelberg (1979)
BP 83 Balsa, R.; Plo, M.; Esteve, J.G.; Pacheco, A.F.: Phys. Rev. D **28**, 1945 (1983)
BR 23 Broglie, L. de: C. R. Acad. Sci. **177**, 507, 548 and 630 (1923)
BR 26 Brillouin, L.: C. R. Acad. Sci. **183**, 24 (1926)
BR 33 Bohr, N.; Rosenfeld, L.: Klg. Danske Vid. Sels. Math.-Fys. Medd. **12**, no. 8 (1933)
BR 50 Bohr, N.; Rosenfeld, L.: Phys. Rev. **78**, 794 (1950)
BR 59 Brueckner, K.A.: "Theory of Nucleon Structure", in *The Many Body Problem*. Lectures given at l'École d'Été de Physique Théorique. Les Houches. Session 1958. Wiley, New York (1959)
BR 65 Bremermann, H.: *Distributions, Complex Variables, and Fourier Transforms*. Addison-Wesley, Reading, Mass. (1965)
BS 59 Bogolyubov, N.N.; Shirkov, D.V.: *Introduction to the Theory of Quantized Fields*. Wiley, New York (1959)
BW 65 Born, M.; Wolf, E.: *Principles of Optics*, 3rd ed., Pergamon, Oxford (1965)
BW 69 Bender, C.M.; Wu, T.T.: Phys. Rev. **184**, 1231 (1969)

CA 67 Calogero, F.: *Variable Phase Approach to Potential Scattering*. Academic Press, New York (1967)
CA 85 Caves, C.M.: Phys. Rev. Lett. **54**, 2465 (1985)
CB 85 Comtet, A.; Bandrauk, A.D.; Campbell, K.: Phys. Lett. **150B**, 159 (1985)
CD 73 Cohen–Tannoudji, C.; Diu, B.; Laloë, F.: *Mecanique Quantique*, Hermann, Paris (1973)
CE 68 Cerny, J.: Ann. Rev. Nuc. Sci. **18**, 27 (1968)
CF 70 Capasso, V.; Fortunato, D.; Selleri, F.: Riv. Nuovo Cimento **11**, 149 (1970)
CF 81 Creutz, M.; Freedman, B.: Ann. Phys. (N.Y.) **132**, 427 (1981)
CG 80 Caliceti, E.; Graffi, R.; Maioli, M.: Commun. Math. Phys. **75**, 51 (1980)
CH 39 Chandrasekhar, S.: *An Introduction to the Study of Stellar Structure*. Dover, New York (1939)
CH 50 Cherry, T.M.: Trans. Am. Math. Soc. **68**, 224 (1950)
CH 60 Chambers, R.G.: Phys. Rev. Lett. **5**, 3 (1960)
CH 68 Chadan, K.: Nuovo Cimento **58**A, 191 (1968)
CM 70 Calles, A.; Moshinsky, M.: Am. J. Phys. **38**, 456 (1970)
CM 82 Chiu, C.B.; Misra, B.; Sudarshan, E.C.G.: Phys. Lett. **117**B, 34 (1982)
CN 68 Carruthers, P.; Nieto, M.M.: Rev. Mod. Phys. **40**, 411 (1968)
CO 21 Compton, A.H.: Phil. Mag. **41**, 749 (1921); Phys. Rev. **18**, 96 (1921) and **19**, 267 (1922)
CO 23 Compton, A.H.: Phys. Rev. **21**, 483, 715 (1923); ibid. **22**, 409 (1923)
CO 65 Cohn, J.H.E.: J. London Math. Soc. **40**, 523 (1965)
CO 75 Colella, R.; Overhauser, A.W.; Werner, S.A.: Phys. Rev. Lett. **34**, 1472 (1975)
CO 77 Coleman, S.: *The Uses of Instantons*. Proc. Int. School of Physics. Erice (1977)
CO 83 Coleman, S.: In *The Unity of the Fundamental Interactions*. Ed. A. Zichichi. Plenum, London (1983)
CO 85 Common, A.K.: J. Phys., A.: Math. Gen. **18**, 2219 (1985)
CP 48 Casimir, H.G.R.; Polder, D.: Phys. Rev. **73**, 360 (1948)
CP 63 Cerny, J.; Pehl, R.H.; Rivet, E.; Harvey, B.G.: Phys. Lett. **7**, 67 (1963)
CR 66 Conley, C.C.; Rejto, P.A.: "Spectral Concentration II-General Theory", in *Perturbation Theory and Its Applications in Quantum Mechanics*. Ed. C.H. Wilcox. Wiley, New York (1966)

CS 57 Condon, E.U.; Shortley, G.H.: *The Theory of Atomic Spectra*, 4th ed., Cambridge University Press, London (1957)

CS 77 Chadan, K.; Sabatier, P.C.: *Inverse Problems in Quantum Scattering Theory*. Springer, New York (1977)

CS 78 Clauser, J.F.; Shimony, A.: Rep. Progr. Phys. **41**, 1881 (1978)

CW 66 Cotton, F.A.; Wilkinson, G.: *Advanced Inorganic Chemistry*, Wiley, New York (1966)

CW 77 Cwikel, M.: Ann. Math. **106**, 93 (1977)

DA 61 Dalgarno, A.: "Stationary Perturbation Theory", in *Quantum Theory* (three volumes). Ed. D.R. Bates. Academic Press, New York (1961)

DG 27 Davisson, C.; Germer, L.H.: Phys. Rev. **30**, 705 (1927)

DI 28 Dirac, P.A.M.: Proc. Roy. Soc. London A **117**, 610 and A **118**, 351 (1928)

DI 30 Dirac, P.A.M.: Proc. Cambridge Phil. Soc. **26**, 376 (1930)

DI 31 Dirac, P.A.M.: Proc. Roy. Soc. London A **133**, 60 (1931)

DI 58 Dirac, P.A.M.: *The Principles of Quantum Mechanics*, 4th ed., Clarendon Press, London (1958)

DI 61 Dicke, R.H.: Phys. Rev. Lett. **7**, 359 (1961)

DK 75 Detwiler, L.C.; Klauder, J.R.: Phys. Rev. D **11**, 1436 (1975)

DK 79 Duru, I.H.; Kleinert, H.: Phys. Lett. **84**B, 185 (1979)

DL 62 Danieri, A.; Loinger, A.; Prosperi, G.M.: Nucl. Phys. **33**, 297 (1962)

DO 64 Dollard, J.D.: J. Math. Phys. **5**, 729 (1964)

DO 69 Dollard, J.D.: Commun. Math. Phys. **12**, 193 (1969)

DO 76 Dowker, J.S.: J. Math. Phys. **17**, 1873 (1976)

DP 84 Damburg, R.; Propin, R.; Martyschenko, V.: J. Phys. A **17**, 3493 (1984)

DR 81 Drummond, J.E.: J. Phys. A **14**, 1651 (1981); A **15**, 2321 (1982)

DS 64 Dunford, N.; Schwartz, J.T.: *Linear Operators* (two volumes). Wiley, New York (1964)

DS 78 Davies, E.B.; Simon, B.: Commun. Math. Phys. **63**, 277 (1978)

DT 79 Deift, P.; Trubowitz, E.: Comm. Pure Appl. Math. **32**, 121 (1979)

DU 32 Dunham, J.L.: Phys. Rev. **41**, 713 (1932)

DY 66 Dyson, F.J.: *Symmetry Groups in Nuclear and Particle Physics*. Benjamin, New York (1966)

DY 72 Dyson, F.J.: "Fundamental constants and their time variation", in *Aspects of Quantum Theory*. Eds. A. Salam and E.P. Wigner. Cambridge Univ. Press (1972)

EC 30 Eckart, C.: Rev. Mod. Phys. **2**, 205 (1930)

EC 30a Eckart, C.: Phys. Rev. **35**, 1303 (1930)

ED 57 Edmonds, A.R.: *Angular Momentum in Quantum Mechanics*. Princeton Univ. Press, Princeton, N.J. (1957)

EG 70 Eisenberg, J.M.; Greiner, W.: *Nuclear Theory* (three volumes). North-Holland, Amsterdam (1970)

EH 13 Ehrenfest, P.: Versl. Gewone Vergard. Wis. Natuurk. Afd. Kon. Akad. Wetensch. Amsterdam **22**, 586 (1913) and *Collected Scientific Papers*. Ed. M.J. Klein. North-Holland, Amsterdam (1959)

EI 05 Einstein, A.: Ann. Phys. **17**, 132 (1905)

EI 71 Einsenbud, L.: *The Conceptual Foundations of Quantum Mechanics*. Van Nostrand Reinhold, New York (1971)

EL 67 Endt, P.M.; Van der Leuen, C.: Nucl. Phys. A **105**, 1 (1967)

EM 63 Emch, G.: Helv. Phys. Acta **36**, 739 and 770 (1963)

EM 72 Emch, G.G.: Helv. Phys. Acta **45**, 1049 (1972)

EN 78 Enss, V.: Commun. Math. Phys. **61**, 285 (1978)

EN 79 Enss, V.: Ann. Phys. (N.Y.) **119**, 117 (1979)

EN 84 Enss, V.: Physica **124**A, 269 (1984)

EN 85 Enss, V.: "Quantum Scattering Theory for Two- and Three-Body Systems with Potentials of Short and Long Range", in *Schrödinger Operators*. Lecture Notes in Mathematics, Vol. 1159. Ed. S. Graffi. Springer, Berlin, Heidelberg (1985)

EP 16 Epstein, P.S.: Ann. Phys. (Leipzig) **50**, 489 (1916)

ER 35 Einstein, A.; Rosen, N.; Podolsky, B.: Phys. Rev. **47**, 777 (1935)

ER 53 Erdély, A. (Ed.): *Higher Transcendental Functions* (three volumes) and *Tables of Integral Transforms* (two volumes). McGraw-Hill, New York (1953)

ES 55 Elliot, J.P.; Skyrme, T.H.R.: Proc. Roy. Soc. London A **232**, 561 (1955)

ES 71 Espagnat, B.d' (Ed.): *Foundations of Quantum Mechanics*. Proceedings of the International School of Physics "Enrico Fermi". Curso IL. Academic Press, New York (1971)

ES 71a Espagnat, B.d': *Conceptual Foundations of Quantum Mechanics*. Benjamin, Menlo Park, Cal. (1971)

ES 80 Enss, V.; Simon, B.: Commun. Math. Phys. **76**, 177 (1980)

ET 74 Eberhard, P.H.; Tripp, R.D.; Déclais, Y.; Séguinot, J.; Baillon, P.; Bricman, C.; Ferro-Luzzi, M.; Perreau, J.M.; Ypsilantis, T.: Phys. Lett. **53B**, 121 (1974)

EV 57 Everett, H.: Rev. Mod. Phys. **29**, 454 (1957)

FA 57 Fano, U.: Rev. Mod. Phys. **29**, 74 (1957)

FA 59 Faddeev, L.D.: Usp. Mat. Nauk **14**, 57 (1959).[English translation J. Math. Phys. **4**, 72 (1963)]

FA 65 Faddeev, L.D.: *Mathematical Aspects of the Three Body Problem in the Quantum Theory of Scattering*. Israel Program of Scientific Translations, Jerusalem (1965)

FA 67 Faddeev, L.D.: Am. Math. Soc. Trans. **65**, 139 (1967)

FA 67a Faris, W.G.: Pac. J. Math. **22**, 47 (1967)

FA 75 Faris, W.G.: *Self-Adjoint Operators*. Springer, Berlin, Heidelberg (1975)

FA 78 Faris, W.G.: J. Math. Phys. **19**, 461 (1978)

FC 56 Feynman, R.P.; Cohen, M.: Phys. Rev. **102**, 1189 (1956)

FC 72 Freedman, S.J.; Clauser, J.F.: Phys. Rev. Lett. **28**, 938 (1972)

FD 70 Fröman, N.; Dammert, Ö.: Nucl. Phys. A **147**, 627 (1970)

FE 27 Fermi, E.: Rend. Accad. Naz. Lincei. **6**, 602 (1927); Z. Phys. **48**, 73 (1928)

FE 30 Fermi, E.: Z. Phys. **60**, 320 (1930)

FE 32 Feenberg, E.: Phys. Rev. **40**, 40 (1932)

FE 39 Feynman, R.P.: Phys. Rev. **56**, 340 (1939)

FE 48 Feynman, R.P.: Rev. Mod. Phys. **20**, 367 (1948)

FE 58 Feshbach, H.: Ann. Rev. Nucl. Sci. **8**, 49 (1958)

FE 69 Feenberg, E.: *Theory of Quantum Fluids*. Academic Press, New York (1969)

FF 65 Fröman, H.; Fröman, P.O.: *JWKB Approximation. Contribution to the Theory*. North-Holland, Amsterdam (1965)

FG 59 Feinberg, G.; Goldhaber, M.: Proc. Nat. Acad. Sci. U.S.A. **45**, 1301 (1959)

FG 70 Fonda, L.; Ghirardi, G.C.: *Symmetry Principles in Quantum Physics*. Marcel Dekker, New York (1970)

FH 14 Franck, J.; Hertz, G.: Verh. Deut. Phys. Ges. **16**, 457, 512 (1914); Phys. Z. **17**, 609 (1916); ibid. **20**, 132 (1919)

FH 65 Feynman, R.P.; Hibbs, A.R.: *Quantum Mechanics and Path Integrals*. McGraw-Hill, New York (1965)

FL 71a Frank, W.M.; Land, D.J.; Spector, R.M.: Rev. Mod. Phys. **43**, 36 (1971)

FL 74 Faris, W.G.; Lavine, R.B.: Commun. Math. Phys. **35**, 39 (1974)

FL 77 Feynman, R.P.; Leighton, R.B.; Sands, M.: *The Feynman Lectures on Physics*, Vol. 3. Addison-Wesley, Reading, Mass. (1977)

FM 73 Fonte, G.; Mignani, R.; Schiffrer, G.: Commun. Math. Phys. **33**, 293 (1973)

FM 76 Flato, M.; Maric, Z.; Milojevic, A.; Sternheimer, D.; Vigier, J.P.: *Quantum Mechanics, Determinism, Causality and Particles*. Reidel, Dordrecht, Holland (1976)

FO 30 Fock, V.: Z. Phys. **61**, 126 (1930); ibid. **62**, 795 (1930)

FO 76 Fonda, L.: *Preprint Trieste IC/76/20*. Lectures presented at the XIII Winter School of Theoretical Physics. Karpacz, Poland (Feb. 1976)

FP 66 Frankowski, K.; Pekeris, C.L.; Phys. Rev. **146**, 46 (1966)

FR 26 Frenkel, J.: Z. Phys. **37**, 243 (1926)

FR 50 Fröhlich, H.: Phys. Rev. **79**, 845 (1950); Proc. Roy. Soc. London A **215**, 291 (1952)

FR 55 Friedrichs, K.O.: *On the Adiabatic Theorem in Quantum Theory*. Report IMM NYU-218, New York (1955)

FR 59 Fano, U.; Racah, G.: *Irreducible Tensorial Sets*. Academic Press, New York (1959)

FR 71 Friar, J.L.: Nucl. Phys. A **173**, 257 (1971)

FR 72 Friedman, C.N.: J. Funct. Anal. **10**, 346 (1972)

FS 70 Feinberg, G.; Sucher, J.: Phys. Rev. A **2**, 2395 (1970)

GA 68 Galindo, A.: Anales de Física **64**, 141 (1968)

GA 70 Gari, M.: Phys. Lett. **31**B, 627 (1970)

GA 76 Galindo, A.: "Another Proof of the Kochen-Specker Paradox", in *Algunas cuestiones de Física Teórica*. Ed. GIFT, Zaragoza (1976)

GA 85 Grangier, P.; Aspect, A.; Vigué, J.: Phys. Rev. Lett. **54**, 418 (1985)

GC 34 Gorter, C.J.; Casimir, H.: Phys. Z. **35**, 963 (1974); Z. Tech. Phys. **15**, 539 (1934)

GC 62 Guelfand, J.M.; Chilov, G.E.: *Les Distributions* (vol. III). Dunod, Paris (1962)

GF 66 Gorodetzky, S.; Freeman, R.M.; Gallmann, A.; Haas, F.: Phys. Rev. **149**, 801 (1966)

GG 66 Gilbey, D.M.; Goodman, F.O.: Am. J. Phys. **34**, 143 (1966)

GG 67 Gardner, C.S.; Greene, J.M.; Kruskal, M.D.; Miura, R.M.: Phys. Rev. Lett. **19**, 1095 (1967)

GG 70 Graffi, S.; Grecchi, V.; Simon, B.: Phys. Lett. **32**B, 631 (1970)

GG 78 Graffi, S.; Grecchi, V.: J. Math. Phys. **19**, 1002 (1978)

GG 78a Graffi, S.; Grecchi, V.: Commun. Math. Phys. **62**, 834 (1978)

GH 78 Grosse, H.; Herjel, P.; Thirring, W.: Acta Phys. Austr. **49**, 89 (1978)

GL 51 Gelfand, I.M.; Levitan, B.M.: Izv. Akad. Nauk SSSR. Ser Mat. **15**, 309 (1951)

GL 57 Gleason, A.M.: J. Math. Mech. **6**, 885 (1957)

GM 52 Gamba, A.; Malvano, R.; Radicati, L.A.: Phys. Rev. **87**, 440 (1952)

GM 62 Galindo, A.; Morales, A.; Núñez-Lagos, R.: J. Math. Phys. **3**, 324 (1962)

GM 65 Greenberg, O.W.; Messiah, A.M.L.: Phys. Rev. **138**B, 1115 (1965)

GM 76 Glaser, V.; Martin, A.; Grosse, H.; Thirring, N.: "A family of optimal conditions for the absence of bound states in a potential", in *Studies in Mathematical Physics*. Eds. E.H. Lieb et al. Princeton Univ. Press, Princeton (1976)

GM 86 García-Alcaine, G.; Molera, J.M.: "Efectos de tipo Aharonov-Bohm", Tesina UCM

GN 64 Gell-Mann, M.; Ne'eman, Y.: *The Eightfold Way*. Benjamin, New York (1964)

GN 80 Gutschick, V.P.; Nieto, M.M.: Phys. Rev. D **22**, 403 (1980)

GO 56 Gombás, P.: *Handbuch der Physik*, Vol. 36 (1956)

GO 70 Goldstein, H.: *Classical Mechanics*. Addison-Wesley, Reading, Mass. (1970)

GP 67 Guardiola, R.; Pascual, P.: Anales de Física **63**, 279 (1967)

GP 71 Gudder, S.; Piron, C.: J. Math. Phys. **12**, 1583 (1971)

GP 75 Galindo, A.; Pascual, P.: Nuovo Cimento **30**A, 111 (1975)

GP 76 Galindo, A.; Pascual, P.: Nuovo Cimento **34**B, 155 (1976)

GP 77 Galindo, A.; Pascual, P.: Anales de Física **73**, 217 (1977)

GR 64 Greenberg, O.W.: Phys. Rev. **135**B, 1447 (1964)

GR 83 Greenberg, D.M.: Rev. Mod. Phys. **55**, 875 (1983)

GR 83a Goy, P.; Raimond, J.M:; Gross, M.; Haroche, S.: Phys. Rev. Lett. **50**, 1903 (1983)

GS 22 Gerlach, W.; Stern, O.: Z. Phys. **9**, 349 (1922)

GS 61 Galindo, A.; Sánchez del Río: Am. J. Phys. **29**, 582 (1961)

GS 85 Gislason, E.A.; Sabelli, N.H.; Wood, J.W.: Phys. Rev. A **31**, 2078 (1985)

GT 30 Gebauer, R.; Traubenberg, H.R.V.: Z. Phys. **62**, 289 (1930)

GU 66 Guardiola, R.: *Estudio Analítico de la Matriz S*. University of Valencia (1966)

GU 77 Guardiola, R.: *Recoil Corrections in Electric Gamma Transitions*. University of Granada (1977)

GW 55 Goldhaber, M.; Weneser, J.: Ann. Rev. Nucl. Sci. **5**, 1 (1955)

GW 64 Goldberger, M.L.; Watson, K.W.: *Collision Theory*. Wiley, New York (1964)

GY 69 Grotch, H.; Yennie, D.R.: Rev. Mod. Phys. **41**, 350 (1969)

HA 28 Hartree, D.R.: Proc. Cambridge Phil. Soc. **24**, 89, 111 (1928)

HA 57 Hartree, D.R.: *The Calculation of Atomic Structure*. Wiley, New York (1957)

HA 62 Hamermesh, M.: *Group Theory and Its Applications to Physical Problems*. Addison-Wesley, Reading, Mass. (1962)

HA 80 Harrell, E.M.: Commun. Math. Phys. **75**, 239 (1980)

HA 88 Hallwachs, W.: Ann. Phys. (Wiedemann) **33**, 301 (1888); ibid. **34**, 731 (1888)

HC 84 Hallin, A.L.; Calaprice, F.P.; MacArthur, D.W.; Piilonen, L.E.; Schneider, M.B.; Schreiber, D.F.: Phys. Rev. Lett. **52**, 337 (1984)

HE 25 Heisenberg, W.: Z. Phys. **33**, 879 (1925)

HE 27 Heisenberg, W.: Z. Phys. **43**, 172 (1927)

HE 30 Heisenberg, W.: *The Physical Principles of Quantum Theory*. University of Chicago Press, Chicago, Ill. (1930)

HE 32 Heisenberg, W.: Z. Phys. **77**, 1 (1932)

HE 35 Hellmann: Acta Physicochemica URSS **16**, 913 (1935); **IV2**, 225 (1936)

HE 54 Heitler, W.: *The Quantum Theory of Radiation*, 3rd ed. Oxford Univ. Press, Oxford (1954)

HE 62 Heisenberg, W.: *Physics and Philosophy*. New York (1962)

HE 69 Hepp, K.: Helv. Phys. Acta **42**, 425 (1969)

HE 69a Henley, E.M.: Ann. Rev. Nucl. Sci. **19**, 367 (1969)

HE 69b Henley, E.M.: "Charge Independence and Charge Symmetry of Nuclear Forces", in *Isospin in Nuclear Physics*. Ed. D.H. Wilkison. North-Holland, Amsterdam (1969)

HE 72 Hepp, K.: Helv. Phys. Acta **45**, 237 (1972)

HE 74 Herbst, I.W.: Commun. Math. Phys. **35**, 181 (1974)

HE 74a Hepp, K.: Commun. Math. Phys. **35**, 265 (1974)

HE 81 Herbst, I.W.: In *Rigorous Atomic and Molecular Physics*. Eds. G. Velo and A.S. Wightman. Plenum, New York (1981)

HE 85 Hegerfeldt, G.C.: Phys. Rev. Lett. **54**, 2395 (1985)

HE 87 Hertz, H.: Ann. Phys. (Wiedemann) **31**, 982 (1887)

HH 70 Hättig, H.; Hünchen, K.; Wäffler, H.: Phys. Rev. Lett. **25**, 941 (1970)

HI 77 Hill, R.N.: Phys. Rev. Lett. **38**, 643 (1977)

HI 77a Hill, R.N.: J. Math. Phys. **18**, 2316 (1977)

HI 82 Hirsbrunner, B.: Helv. Phys. Acta **55**, 295 (1982)

HK 60 Harting, D.; Kluyver, J.C.; Kusumegi, A.; Rigopoulos, R.; Sachs, A.M.; Tibell, G.; Vanderhaeghe, G.; Weber, G.: Phys. Rev. **119**, 1716 (1960)

HK 64 Hohemberg, P.; Kohn, W.: Phys. Rev. **136B**, 864 (1964)

HL 51 Hulthén, L; Laurikainen, K.V.: Rev. Mod. Phys. **23**, 1 (1951)

HO 71 Hochstadt, H.: *The Functions of Mathematical Physics*. Wiley, New York (1971)

HO 85 Hansen, T.O.; Østgaard, E.: Can. J. Phys. **63**, 1022 (1985)

HR 60 Hughes, V.W.; Robinson, H.G.; Beltram-López, V.: Phys. Rev. Lett. **4**, 342 (1960)

HR 80 Hegerfeldt, G.C.; Ruijsenaars, N.M.: Phys. Rev. D **22**, 377 (1980)

HS 63 Herman, F.; Skillman, S.S.: *Atomic Structure Calculations*. Prentice Hall, Englewood Cliffs, N.J. (1963)

HS 78 Herbst, I.W.; Simon, B.: Phys. Lett. **78B**, 304 (1978)

HT 56 Hanbury-Brown, R.; Twiss, R.Q.: Nature **157**, 27 (1956); Proc. Roy. Soc. London A **242**, 300 (1957); A **243**, 291 (1957)

HU 25 Hund, F.: Z. Phys. **33**, 345 (1925)

HU 42 Hulthén, L.: Arkiv. Mat. Astron. Fys. **28A**, no. 5 (1942); ibid. **29B**, no. 1 (1942)

HU 66 Hunziker, W.: Helv. Phys. Acta **39**, 451 (1966)

HU 68 Hunziker, W.: "Mathematical Theory of Multi-particle Quantum Systems", in *Lectures in Theoretical Physics*. Vol. X-A. Eds. A.O. Barut and W.E. Brittin. Gordon and Breach, New York (1968)

HU 79 Hunziker, W.: *I.A.M.P. Proc. Lausanne*. Springer, Berlin, Heidelberg (1979)

HW 61 Henshaw, D.G.; Woods, A.D.B.: Phys. Rev. **121**, 1266 (1961)

HY 30 Hylleraas, E.: Z. Phys. **65**, 209 (1930)

HY 37 Hylleraas, E.A.: Z. Phys. **107**, 258 (1937)

IH 28 Ishida, Y.; Himaya, S.: *Inst. Phys. Chem. Res. Tokyo. Sci. Paper,9*. Vol. 152 (1928)
IH 51 Infeld, L.; Hull, T.E.: Rev. Mod. Phys. **23**, 21 (1951)
IK 60 Ikebe, T.: Arch. Rat. Mech. Anal. **5**, 1 (1960)
IK 65 Ikebe, T.: Pacific J. Math. **15**, 511 (1965)
IS 15 Ishiwara, J.: Tokyo Sugaku Buturigakkawi Kizi **8**, 106 (1915)
IZ 80 Itzykson, C; Zuber, J.B.: *Quantum Field Theory*. McGraw-Hill, New York (1980)

JA 62 Jacobson, N.: *Lie Algebras*. Wiley, New York (1962)
JA 66 Jammer, M.: *The Conceptual Development of Quantum Mechanics*. McGraw-Hill, New York (1966)
JA 68 Jauch, J.M.: *Foundations of Quantum Mechanics*. Addison-Wesley, Reading, Mass. (1968)
JA 73 Jauch, J.M.: In *The Physicist's Conception of Nature*. Ed. J. Mehra. Reidel, Dordrecht, Holland (1973)
JA 74 Jammer, M.: *The Philosophy of Quantum Mechanics*. Wiley, New York (1974)
JA 75 Jackson, J.D.: *Classical Electrodynamics*. 2nd ed., Wiley, New York (1975)
JE 05 Jeans, J.H.: Phil. Mag. **10**, 91 (1905)
JE 24 Jeffreys, H.: Proc. London Math. Soc. **23**, 428 (1924)
JK 52 Jost, R.; Kohn, W.: Phys. Rev. **87**, 979 (1952); ibid. **88**, 382 (1952)
JL 72 Jauch, J.M.; Lavine, R.; Newton, R.G.: Helv. Phys. Acta **45**, 325 (1972)
JO 75 Joachain, C.J.: *Quantum Collision Theory*. North-Holland, Amsterdam (1975)
JP 28 Jordan, P.; Pauli, W.: Z. Phys. **47**, 151 (1928)
JP 63 Jauch, J.M.; Piron, C.: Helv. Phys. Acta **36**, 827 (1963)
JR 55 Jauch, J.M.; Rohrlich, F.: *The Theory of Photons and Electrons*. Addison-Wesley, Reading, Mass. (1955)
JS 72 Jauch, J.M.; Sinha, K.: Helv. Phys. Acta **45**, 580 (1972)
JS 72a Jauch, J.M.; Sinha, K.B.; Misra, B.N.: Helv. Phys. Acta **45**, 398 (1972)
JW 59 Jacob, M.; Wick, G.C.: Ann. Phys. (N.Y.) **7**, 404 (1959)

KA 11 Kamerlingh-Onnes, H.: *Commun. Phys. Lab. Univ. Leiden*, 119b, 120b, 133b (1911)
KA 38 Kapitza, P.L.: Nature **141**, 74 (1938)
KA 49 Kato, T.: Progr. Theor. Phys. **4**, 514 (1949)
KA 50 Kato, T.: J. Phys. Soc. Jap. **5**, 435 (1950)
KA 50a Kac, M.: In *Proc. 2nd Berk. Symp. Math. Statist. Probability*, p. 189 (1950)
KA 51 Kato, T.: J. Fac. Sci. Univ. Tokio, Sect. I, **6**, 145 (1951)
KA 59 Kac, M.: *Probability and Related Topics in the Physical Sciences*. Wiley, New York (1959)
KA 80 Kato, T.: *Perturbation Theory for Linear Operators*. Springer, Berlin, Heidelberg (1980)
KA 84 Kamefuchi, S. et al. (Eds.): Proceedings of the International Symposium *Foundations of Quantum Mechanics in the Light of New Technology*, Phys. Soc. of Japan (1984)
KE 35 Kemble, E.C. Phys. Rev. **48**, 549 (1935)
KE 37 Kemble, F.: *The Fundamental Principles of Quantum Mechanics with Elementary Applications*. Dover, New York (1937)
KG 84 Kaiser, H.; George, E.A.; Werner, S.A.: Phys. Rev. A **29**, 2276 (1984)
KH 84 Kolbas, R.M.; Holonyak, N.: Am. J. Phys. **52**, 431 (1984)
KI 57 Kinoshita, T.: Phys. Rev. **105**, 1490 (1957)
KI 60 Kirchhoff, G.R.: Ann. Phys. (Poggendorf) **109**, 275 (1860)
KI 66 Kittel, C.: *Introduction to Solid State Physics*. Wiley, New York (1966)
KM 55 Kobayashi, S.; Matsukuma, T.; Nagai, S.; Umeda, K.: J. Phys. Soc. Japan **10**, 759 (1955)
KO 60 Köhler, H.S.: Phys. Rev. **118**, 1345 (1960)
KO 67 Kolos, W.: Int. J. Quantum Chem. **1**, 169 (1967)
KR 26 Kramers, H.A.: Z. Phys. **39**, 828 (1926)
KR 30 Kramers, H.A.: Proc. Acad. (Amsterdam) **33**, 959 (1930)
KS 65 Kohn, W.; Sham, J.L.: Phys. Rev. **140**A, 1133 (1965)
KS 67 Kochen, S.; Specker, E.P.: J. Math. Mech. **17**, 59 (1967)

KS 68 Klauder, J.R.; Sudarshan, E.C.G.: *Fundamentals of Quantum Optics*. Benjamin, New York (1968)

KU 25 Kuhn, W.: Z. Phys. **33**, 408 (1925)

KU 62 Kuhn, H.G.: *Atomic Spectra*. Longmans, Green and Co., London (1962)

KU 71 Kujawski, E.: Am. J. Phys. **39**, 1248 (1971)

LA 23 Landé, A.: Z. Phys. **15**, 189 (1923); ibid. **19**, 112 (1923)

LA 37 Langer, R.E.: Phys. Rev. **51**, 669 (1937)

LA 41 Landau, L.D.: J. Phys. USSR **5**, 71 (1941); ibid. **8**, 1 (1944)

LA 55 Latter, R.: Phys. Rev. **99**, 510 (1955)

LA 74 Labonté, G.: Commun. Math. Phys. **36**, 59 (1974)

LB 39 London, F.W.; Bauer, E.: *La théorie de l'observation en mécanique quantique*. Hermann, Paris (1939)

LE 26 Lewis, G.N.: Nature **118**, 874 (1926)

LE 49 Levinson, H.: Kgl. Danske. Videnskab. Selskab., Mat.-Fys. Medd. **25** (9) (1949)

LE 74 Levy-Leblond, J.M.: Riv. Nuovo Cimento **4**, 99 (1974)

LE 83 Leinfelder, H.: J. Operator Theory **9**, 163 (1983)

LE 84 Leggett, A.J.: Contemp. Phys. **25**, 6 (1984)

LE 99 Lenard, P.: Wien Ber. **108**, 1649 (1899); Ann. Phys. (Leipzig) **2**, 359 (1900); **8**, 149 (1902)

LI 58 Lipkin, H.J.: Phys. Rev. **110**, 1395 (1958)

LI 73 Lipkin, H.J.: *Quantum Mechanics. New Approaches to Selected Topics*. North-Holland, Amsterdam (1973)

LI 76 Lieb, E.: Bull. Amer. Math. Soc. **82**, 751 (1976)

LI 77 Lipatov, L.: Sov. Phys. JEPT **72**, 411 (1977)

LI 81 Lieb, E.H.: Rev. Mod. Phys. **53**, 603 (1981)

LI 84 Lieb, E.H.: Phys. Rev. A **29**, 3018 (1984)

LL 35 London, F.; London, H.: Proc. Roy. Soc. London A **149**, 71 (1935); Physica (Utrecht) **2**, 341 (1935)

LL 65 Landau, L.D.; Lifshitz, E.M.: *Quantum Mechanics: Non-Relativistic Theory*. Pergamon Press, Oxford (1965)

LL 80 Landau, L.D.; Lifshitz, E.M.: *Statistical Physics*. 3rd ed., Pergamon Press, London (1980)

LM 69 Loeffel, J.J.; Martin, A.; Simon, B.; Wightman, A.S.: Phys. Lett. **30B**, 656 (1969)

LM 70 Loeffel, J.J.; Martin, A.: Proc. R.C.P. 25 May (1970)

LO 30 London, F.: Z. Phys. **63**, 245 (1930)

LO 63 Lousiell, W.: Phys. Lett. **7**, 60 (1963)

LO 71 Loebel, E.M. (Ed.): *Group Theory and Its Applications*. Academic Press, New York (1968)

LO 97 Lorentz, H.A.: Ann. Phys. (Wiedemann) **63**, 278 (1897)

LP 72 Lautrup, B.E.; Peterman, A.; Rafael, E. de: Phys. Rep. **3C**, 193 (1972)

LP 97 Lummer, O.; Pringsheim, E.: Ann. Phys. (Wiedemann) **63**, 395 (1897); Verh. Deut. Phys. Ges. **2**, 163 (1900)

LS 73 Lieb, E.H.; Simon, B.: Phys. Rev. Lett. **31**, 681 (1973)

LS 74 Lieb, E.H.; Simon, B.: J. Chem. Phys. **61**, 735 (1974)

LS 77 Lieb, E.H.; Simon, B.: Commun. Math. Phys. **53**, 185 (1977)

LS 77a Lieb, E.H.; Simon, B.: Adv. Math. **23**, 22 (1977)

LS 81 Leinfelder, H.; Simader, C.G.: Math. Z. **176**, 1 (1981)

LT 85 Lieb, E.H.; Thirring, W.E.: *The Universal Nature of van der Waals Forces for Coulomb Systems*, preprint. Vienna (1985)

LU 61 Ludwig, G.: In *Werner Heisenberg und die Physik unserer Zeit*, Vieweg, Braunschweig (1961)

LY 56 Lee, T.D.; Yang, C.N.: Phys. Rev. **104**, 254 (1956)

LY 62 Lynton, E.A.: *Superconductivity*. Wiley, New York (1962)

MA 50 Marchenko, V.A.: Dokl. Akad. Nauk SSSR **72**, 457 (1950)

MA 55 Marchenko, V.A.: Dokl. Akad. Nauk SSSR **104**, 695 (1955)

MA 63 Mackey, G.W.: *The Mathematical Foundations of Quantum Mechanics*. Benjamin, New York (1963)

MA 68 Marchand, J.P.: "Rigorous Results in Scattering Theory", in *Quantum Theory and Statistical Physics*. Eds. A.O. Barut and W.E. Brittin. Gordon and Breach, New York (1968)

MA 72 Maslov, V.P.: *Théorie des Perturbations et Méthodes Asymptotiques*. Dunod, Paris (1972)

MA 75 Martin, A.: Ref. *TH 2085-CERN* (1975)

MA 79 Martin, A.: Commun. Math. Phys. **69**, 89 (1979)

MA 80 Martin, A.: Commun. Math. Phys. **73**, 79 (1980)

ME 59 Messiah, A.: *Mécanique Quantique* (two volumes). Dunod, Paris (1959)

ME 65 Meyer-Abich, K.M.: *Korrespondenz, Individualität und Komplementarität*. Vol. 5 of the series "Boethius-Texte und Abhandlungen zur Geschichte der exakten Wissenschaften". Eds. J.E. Hofmann, F. Klemm and B. Sticker. Franz Steiner, Wiesbaden (1965)

MG 53 Miller, S.C.; Good, R.H.: Phys. Rev. **91**, 174 (1953)

MI 16 Millikan, R.A.: Phys. Rev. **7**, 356 (1916)

MI 54 Miller, S.C.: Phys. Rev. **94**, 1345 (1954)

MI 64 Mikhlin, S.G.: *Variational Methods in Mathematical Physics*. Pergamon, Oxford (1964)

MI 65 Midtdal, J.: Phys. Rev. **138A**, 1010 (1965)

MI 87 Michelson, W.: J. Phys. **6**, 467 (1887)

MI 91 Michelson, A.A.: Phil. Mag. **31**, 338 (1891); ibid. **34**, 280 (1892)

MK 84 Meitzler, C.R.; Khalil, A.E.; Robbins, A.B.; Temmer, G.M.: Phys. Rev. C **30**, 1105 (1984)

ML 67 Mendt, P.; Van der Leon, C.: Nucl. Phys. A **105**, 1 (1967)

MM 65 Mott, N.F.; Massey, H.S.N.: *The Theory of Atomic Collisions*. 3rd ed., Clarendon Press, Oxford (1965)

MO 29 Morse, P.M.: Phys. Rev. **34**, 57 (1929)

MO 58 Morpurgo, G.: Phys. Rev. **110**, 721 (1958)

MO 68 Moshinsky, M.: Am. J. Phys. **36**, 52 and 763 (1968)

MP 79 Marshall, J.T.; Pell, J.L.: J. Math. Phys. **20**, 1297 (1979)

MR 69 Marshak, R.E.; Riazuddin; Ryan, C.P.: *Theory of Weak Interactions in Particle Physics*. Wiley, New York (1969)

MR 85 Manson, J.R.; Ritchie, R.H.: Phys. Rev. Lett. **54**, 785 (1985)

MT 45 Mandelstam, L.; Tamm, I.G.: J. Phys. USSR **9**, 249 (1945)

MZ 70 Mulherin, D.; Zinnes, I.I.: J. Math Phys. **11**, 1402 (1970)

NA 04 Nagaoka, H.: Bull. Math. Phys. Soc. Tokyo **2**, 140 (1904); Nature **69**, 392 (1904); Phil. Mag. **7**, 445 (1904)

NA 68 Naimark, M.A.: *Linear Differential Operators*. Ungar, New York (1968)

NE 32 Von Neumann, J.: *Mathematische Grundlagen der Quantenmechanik*. Springer, Berlin (1932)

NE 59 Nelson, E.: Ann. Math. **70**, 572 (1959)

NE 60 Newton, R.G.: J. Math. Phys. **1**, 319 (1960)

NE 64 Nelson, E.: J. Math. Phys. **5**, 332 (1964)

NE 76 Newton, R.G.: Am. J. Phys. **44**, 639 (1976)

NE 82 Newton, R.G.: *Scattering Theory of Waves and Particles*. 2nd ed., Springer, New York (1982)

NI 80 Nieto, M.M.: Phys. Rev. D **22**, 391 (1980)

NS 79 Nieto, M.M.; Simmons, L.M.: Phys. Rev. D **20**, 1321, 1332, 1342 (1979)

NW 29 Von Neumann, J.; Wigner, E.P.: Phys. Z. **30**, 465 (1929)

OP 28 Oppenheimer, J.R.: Phys. Rev. **31**, 66 (1928)

OP 85 Olariu, S.; Popescu, I.I.: Rev. Mod. Phys. **57**, 339 (1985)

PA 16 Paschen, F.: Ann. Phys. (Leipzig) **50**, 901 (1916)

PA 25 Pauli, W.: Z. Phys. **31**, 765 (1925)

PA 67 Papaliolios, C.: Phys. Rev. Lett. **18**, 622 (1967)

PA 79 Pascual, P.: Anales de Física **75**, 77 (1979)

PA 97 Paschen, F.: Ann. Phys. (Wiedemann) **60**, 662 (1897)
PD 88 Particle Data Group: Phys. Lett. **204**B, 1 (1988)
PE 01 Perrin, J.: Rev. Sci. **15**, 449 (1901)
PE 62 Pekeris, C.L.: Phys. Rev. **126**, 1470 (1962)
PE 72 Peterman, A.: Phys. Lett. **38**B, 330 (1972)
PE 75 Pearson, D.B.: Commun. Math. Phys. **40**, 125 (1975)
PI 53 Pippard, A.B.: Proc. Roy. Soc. London, A**216**, 547 (1953)
PI 76 Piron, C.: *Foundations of Quantum Physics*. Benjamin, Reading, Mass (1976)
PI 78 Pipkin, F.M.: *Advances in Atomic and Molecular Physics*. Eds. P.R. Bates and B. Bederson. Academic Press, New York (1978)
PL 00 Planck, M.: Verh. Deut. Phys. Ges. **2**, 202, 237 (1900)
PL 01 Planck, M.: Sitzungsber. K. Preuss. Akad. Wiss. Berlin **25**, 544 (1901)
PL 12 Planck, M.: "La Théorie du Rayonnement et les Quanta", in *Rapports et Discussion de la Réunion Tenue à Bruxelles 1911*. Eds. P. Langevin and M. de Broglie. Gauthier-Villars, Paris (1912)
PR 62 Preston, M.A.: *Physics of the Nucleus*. Addison-Wesley, Reading, Mass. (1962)
PR 67 Prugovečki, E.: Can. J. Phys. **45**, 2173 (1967)
PR 71 Prugovečki, E.: *Quantum Mechanics in Hilbert Space*. Academic Press, New York (1971)
PR 80 Privman, V.: Phys. Rev. A **22**, 1833 (1980)
PR 84 Preskill, J.: Ann. Rev. Nuc. Part. Sci. **34**, 461 (1984)
PR 97 Preston, T.: Sci. Trans. Roy. Dublin Soc. **6**, 385 (1897)
PS 68 Prepost, R.; Simons, R.M.; Wiik, B.H.: Phys. Rev. Lett. **21**, 1271 (1968)
PS 84 Pak, N.K.; Sökmen, I.: Phys. Lett. **100**A, 327 (1984)
PT 84 Pascual, P.; Tarrach, R.: *Q.C.D.: Renormalization for the Practitioner*. Springer, Berlin, Heidelberg (1984)
PU 56 Purcell, E.M.: Nature **178**, 1449 (1956)
PU 67 Putnam, C.R.: *Commutation Properties of Hilbert Space Operators and Related Topics*. Springer, Berlin, Heidelberg (1967)
PW 34 Paley, R.; Wiener, N.: *Fourier Transforms in the Complex Domain*. American Math. Soc. Providence, RI (1934)
PW 99 Paschen, F.; Wanner, H.: Sitzungsber. K. Preuss. Akad. Wiss. Berlin **2**, 5 (1899)

RA 00 Lord Rayleigh: Phil. Mag. **49**, 539 (1900)
RA 37 Rabi, I.I.: Phys. Rev. **51**, 652 (1937)
RA 42 Racah, G.: Phys. Rev. **62**, 438 (1942)
RA 43 Racah, G.: Phys. Rev. **63**, 367 (1943)
RA 75 Radin, C.: J. Math. Phys. **16**, 544 (1975)
RB 59 Rotenberg, M.; Bivins, R.; Metropolis, N.; Wooten, J.K.: *The 3-j and 6-j Symbols*. Crosby Lockwood, London (1959)
RG 70 Rogers, F.J.; Graboske, H.C.; Harwood, D.J.: Phys. Rev. A **1**, 1577 (1970)
RH 63 Robinson, P.D.; Hirschfelder, J.O.: J. Math. Phys. **4**, 338 (1963)
RI 08 Ritz, W.: Phys. Z. **9**, 521 (1908); Astrophys. J. **28**, 237 (1908)
RI 65 Rickayzen, G.: *Theory of Superconductivity*. Wiley, New York (1965)
RK 01 Rubens, H.; Kurlbaum, F.: Ann. Phys. (Leipzig) **4**, 649 (1901)
RK 68 Rosenzweig, C.; Krieger, J.B.: J. Math. Phys. **9**, 849 (1968)
RM 72 Rau, A.R.P.; Mueller, R.O.; Spruch, L.: Comm. Atom. Mol. Phys. **3**, 87 (1972)
RO 57 Rose, M.E.: *Elementary Theory of Angular Momentum*. Wiley, New York (1957)
RO 71 Robinson, D.W.: *The Thermodynamic Pressure in Quantum Statistical Mechanics*. Springer, Berlin, Heidelberg (1971)
RO 72 Rosenbljum, G.V.: Dokl. Akad. Nauk SSSR **202**, 1012 (1972)
RR 73 Rauch, J.; Reed, M.: Commun. Math. Phys. **29**, 105 (1973)
RS 25 Russell, H.N.; Saunders, F.A.: Astrophys. J. **61**, 38 (1925)
RS 55 Riesz, F.; Sz-Nagy, B.: *Functional Analysis*. Ungar, New York (1955)
RS 70 Robiscoe, R.T.; Shyn, T.W.: Phys. Rev. Lett. **24**, 559 (1970)

RS 72 Reed, M.; Simon, B.: *Functional Analysis*. Academic Press, New York (1972)
RS 75 Reed, M.; Simon, B.: *Fourier Analysis, Self-Adjointness*. Academic Press, New York (1975)
RS 78 Reed, M.; Simon, B.: *Analysis of Operators*. Academic Press, New York (1978)
RS 79 Reed, M.; Simon, B.: *Scattering Theory*. Academic Press, New York (1979)
RS 85 Radzig, A.A.; Smirnov, B.M.: *Reference Data on Atoms, Molecules and Ions*, Springer Series in Chemical Physics, Vol. 31. Springer, Berlin, Heidelberg (1985)
RT 25 Reiche, F.; Thomas, W.: Z. Phys. **34**, 510 (1925)
RU 00 Rubens, H.: Verh. Deut. Phys. Ges. **3**, 263 (1900)
RU 11 Rutherford, E.: Phil. Mag. **21**, 669 (1911)
RU 71 Ruelle, D.: "Equilibrium Statistical Mechanics of Infinite Systems", in *Statistical Mechanics of Quantum Field Theory*. Eds. C. de Witt and R. Stora. Gordon and Breach, New York (1971)
RZ 75 Rauch, H.; Zeilinger, A.; Badurek, G.; Wilfing, A.; Bauspiess, W.; Bonse, U.: Phys. Lett. **54**A, 425 (1975)

SA 64 Sakurai, J.J.: *Invariance Principles and Elementary Particles*. Princeton Univ. Press, Princeton, N.J. (1964)
SA 67 Sakurai, J.J.: *Advanced Quantum Mechanics*. Addison-Wesley, Reading, Mass. (1967)
SA 84 Sakaki, H.: Proceedings of the International Symposium *Foundations of Quantum Mechanics in the Light of New Technology*. Eds. S. Kamefuchi et al., Phys. Soc. of Japan (1984)
SB 82 Summhammer, J.; Badurek, G.; Rauch, H.; Kischko, U.: Phys. Lett. **90**A, 110 (1982)
SB 88 Samuel, J.; Bhandari, R.: Phys. Rev. Lett. **60**, 2339 (1988)
SC 14 Schwarzschild, K.: Verh. Deut. Phys. Ges. **16**, 20 (1914)
SC 26 Schrödinger, E.: Ann. Phys. (Leipzig) **79**, 361, 489 (1926); **80**, 437 (1926); **81**, 109 (1926)
SC 28 Schlapp, R.: Proc. Roy. Soc. London A **119**, 313 (1928)
SC 57 Schwartz, L.: *Théorie des Distributions* (two volumes). Hermann, Paris (1957)
SC 61 Schwinger, J.: Proc. Nat. Acad. Sci. **47**, 122 (1961)
SC 64 Schrieffer, J.R.: *Theory of Superconductivity*. Benjamin, New York (1964)
SC 65 Schwartz, L.: *Méthodes Mathématiques pour les Sciences Physiques*. 2nd ed., Hermann, Paris (1965)
SC 68 Schwartz, J.: *W^*-Algebras*. Nelson, London (1968)
SC 68a Schiff, L.I.: *Quantum Mechanics*, 3rd ed., McGraw-Hill, New York (1968)
SC 81 Schechter, M.: *Operator Methods in Quantum Mechanics*. North-Holland, New York (1981)
SC 81a Schulman, L.S.: *Techniques and Applications of Path Integration*. Wiley, New York (1981)
SG 61 Stueckelberg, E.C.G.; Guenin, M.: Helv. Phys. Acta **34**, 621 (1961)
SG 61a Stueckelberg, E.C.G.; Guenin, M.; Piron, C.; Ruegg, H.: Helv. Phys. Acta **34**, 675 (1961)
SG 62 Stueckelberg, E.C.G.; Guenin, M.: Helv. Phys. Acta **35**, 673 (1962)
SH 59 Shewell, J.R.: Am. J. Phys. **27**, 16 (1959)
SI 68 Siegbahn, K. (Ed.): *Alpha-, Beta- and Gamma-Ray Spectroscopy*. North-Holland, Amsterdam (1968)
SI 70 Simon, B.: Helv. Phys. Acta **43**, 607 (1970)
SI 70a Simon, B. (with an appendix by A. Dicke): Ann. Phys. (N.Y.) **58**, 76 (1970)
SI 71 Simon, B.: *Quantum Mechanics for Hamiltonians Defined as Quadratic Forms*. Princeton Univ. Press, Princeton N.J. (1971)
SI 71a Simon, B.: Commun. Math. Phys. **23**, 37 (1971)
SI 72 Simon, B.: In *Mathematics of Contemporary Physics*. Academic Press, New York (1972)
SI 72a Sinha, K.: Helv. Phys. Acta **45**, 619 (1972)
SI 73 Simon, B.: Ann. Math. **97**, 247 (1973)
SI 74 Simon, B.: Proc. Amer. Math. Soc. **42**, 395 (1974)
SI 76 Simon, B.: "On the Number of Bound States of Two Body Schrödinger Operators: A Review", in *Studies in Mathematical Physics*. Eds. E.H. Lieb, B. Simon and A.S. Wightman, Princeton Univ. Press, Princeton, N.J. (1976)
SI 76a Simon, B.: Ann. Phys. **97**, 279 (1976)
SI 78 Sienert, C.E.: J. Math. Phys. **19**, 434 (1978)

SI 79 Simon, B.: *Functional Integration and Quantum Physics*. Academic Press, New York (1979)

SI 81 Silverman, J.N.: Phys. Rev. A **23**, 441 (1981)

SI 82 Simon, B.: Bull. Am. Math. Soc. **7**, 447 (1982)

SI 82a Simon, B.: Int. J. Quant. Chem. **21**, 3 (1982)

SI 83 Simon, B.: Phys. Rev. Lett. **51**, 2167 (1983)

SI 84 Simon, B.: "Some Aspects of the Theory of Schrödinger Operators", in *C.I.M.E. Summer School on Schrödinger Operators*. Como, Italy (Sept. 1984)

SI 84a Simon, B.: Ann. Math. **120**, 89 (1984)

SI 85 Silverstone, H.J.: Phys. Rev. Lett. **55**, 2523 (1985)

SL 29 Slater, J.C.: Phys. Rev. **34**, 1293 (1929)

SL 39 Slater, J.C.: *Introduction to Chemical Physics*. McGraw-Hill, New York (1939)

SL 60 Slater, J.C.: *Quantum Theory of Atomic Structure* (two volumes). McGraw-Hill, New York (1960)

SN 85 Silverstone, H.J.; Nakai, S.; Harris, J.G.: Phys. Rev. A **32**, 1341 (1985)

SO 16 Sommerfeld, A.: Ann. Phys. (Leipzig) **51**, 1, 125 (1916)

SO 19 Sommerfeld, A.: *Atombau und Spetrallinien*, 7th ed., Vieweg, Braunschweig (1951)

SP 67 Spector, R.M.: J. Math. Phys. **8**, 2357 (1967)

SS 87 Sigal, I.M.; Soffer, A.: Ann. Math. **126**, 35 (1987)

ST 13 Stark, J.: Sitzungsber. K. Preuss. Akad. Wiss. Berlin **47**, 932 (1913)

ST 60 Stueckelberg, E.C.G.: Helv. Phys. Acta **33**, 727 (1960)

ST 66 Stillinger, F.H.: J. Chem. Phys. **45**, 3623 (1966)

ST 68 Stakgold, I.: *Boundary Value Problems of Mathematical Physics*. Macmillan, New York (1968)

ST 81 Selleri, F.; Tarozzi, G.: Riv. Nuovo Cimento **4**, 1 (1981)

ST 88 Stoletow, A.: C. R. Acad. Sci. **106**, 1149, 1593 (1888); **107**, 81 (1888); **108**, 1241 (1889); J. Phys. **9**, 468 (1889)

SV 84 Seetharaman, M.; Vasan, S.S.: J. Phys. A **17**, 2485 (1984)

SW 64 Scadron, M.; Weinberg, S.: Phys. Rev. **133**B, 1589 (1964)

SW 64a Scadron, M.; Weinberg, S.; Wright, J.: Phys. Rev. **135**B, 202 (1964)

SW 64b Streater, R.F.; Wightman, A.S.: *PCT, Spin, Statistics, and All That*. Benjamin, New York (1964)

SW 68 Swanson, C.A.: *Comparison and Oscillation Theory of Linear Differential Equations*. Academic Press, New York (1968)

SZ 79 Seznec, R.; Zinn-Justin, J.: J. Math. Phys. **20**, 1398 (1979)

TA 72 Taylor, J.R.: *Scattering Theory*. Wiley, New York (1972)

TB 84 Trelle, R.P.; Birkhäuser, J.; Hinterberger, F.; Kuhn, S.; Prashun, D.; von Rossen, P.: Phys. Lett. **134**B, 34 (1984)

TC 86 Tomita, A.; Chiao, R.Y.: Phys. Rev. Lett. **57**, 937 (1986)

TE 28 Temple, G.: Proc. Roy. Soc. London A **119**, 276 (1928)

TG 29 Traubenberg, H.R.V.; Gebauer, R.: Z. Phys. **54**, 307 (1929); ibid. **56**, 254 (1929)

TG 30 Traubenberg, H.R.V.; Gebauer, R.; Lewin, G.: Naturwissenschaften **18**, 418 (1930)

TH 03 Thomson, J.J. Phil. Mag. **6**, 673 (1903), and **7**, 237 (1904)

TH 25 Thomas, W.: Naturwissenschaften **13**, 627 (1925)

TH 26 Thomas, L.H.: Nature **117**, 514 (1926)

TH 26a Thomas, L.H.: Proc. Cambridge Phil. Soc. **23**, 542 (1926)

TH 55 Thirring, W.: *Einführung in die Quantenelektrodynamik*. Deuticke, Vienna (1955)

TH 81 Thirring, W.: *Quantum Mechanics of Atoms and Molecules*. Springer, New York (1981)

TH 83 Thirring, W.: *Quantum Mechanics of Large Systems*. Springer, New York (1983)

TI 63 Tisza, L.: Rev. Mod. Phys. **35**, 151 (1963)

TR 27 Thomson, G.P.; Reid, A.: Nature **119**, 890 (1927)

TR 52 Trainor, L.E.H.: Phys. Rev. **85**, 962 (1952)

UG 25 Uhlenbeck, G.E.; Goudsmit, S.: Naturwissenschaften **13**, 953 (1925)

VI 72 Vinh Mau, N.: *Non Conservation de la Parité dans les Noyaux*. Report GIFT 20/72 (1972)
VL 26 Van Vleck, J.H.: Proc. Nat. Acad. Sci. U.S.A. **12**, 662 (1926)
VP 66 Vessot, R.; Peters, H.; Vanier, J.; Bechler, R.; Halford, D.; Harrach, R.; Allan, D.; Glaze, D.; Snider, C.; Barnes, J.; Cutler, L.; Bodily, L.: IEEE Trans. Instr. Meas., IM-**15**, 165 (1966)
VW 81 Velo, G.; Wightman, A.S. (Eds.): *Rigorous Atomic and Molecular Problems*. Plenum, New York (1981)

WA 26 Waller, I.: Z. Phys. **38**, 635 (1926)
WA 57 Wu, C.S.; Ambler, E.; Hayward, R.W.; Hoppes, D.D.; Hudson, R.P.: Phys. Rev. **105**, 1413 (1957)
WA 72 Wagner, D.: *Introduction to the Theory of Magnetism*. Pergamon, Oxford (1972)
WC 75 Werner, S.A.; Colella, R.; Overhauser, A.W.; Eagen, C.F.: Phys. Rev. Lett. **35**, 1053 (1975)
WE 26 Wentzel, G.: Z. Phys. **38**, 518 (1926)
WE 63 Wei, J.: J. Math. Phys. **4**, 1337 (1963)
WE 63a Weinberg, S.: Phys. Rev. **130**, 776 (1963)
WE 63b Weinberg, S.: Phys. Rev. **131**, 440 (1963)
WE 64 Weinberg, S.: Phys. Rev. **133B**, 232 (1964)
WE 67 Weidmann, J.: Bull. Am. Math. Soc. **73**, 452 (1967)
WE 67a Weidmann, J.: Math. Zeit. **98**, 268 (1967)
WE 70 Weisskopf, V.F.: CERN 70-8 (1970)
WG 73 Witt, B. de; Graham, N.: *The Many-Worlds Interpretation of Quantum Mechanics*. Princeton University Press, Princeton (1973)
WI 15 Wilson, W.: Phil. Mag. **29**, 795 (1915)
WI 23 Wiener, N.: J. Mathematical and Physical Sci. **2**, 132 (1923)
WI 31 Wigner, E.P.: *Gruppentheorie und ihre Anwendung auf die Quantenmechanik der Atomspektren*. Vieweg, Braunschweig (1931). English translation: *Group Theory and Its Application to the Quantum Mechanics of Atomic Spectra*. Academic Press, New York (1959)
WI 41 Wintner, A.: *The Analytical Foundations of Celestial Mechanics*. Princeton University Press, Princeton, N.J. (1941)
WI 55 Wigner, E.P.: Phys. Rev. **98**, 145 (1955)
WI 60 Wigner, E.P.: J. Math. Phys. **1**, 409 and 414 (1960)
WI 62 Wightman, A.S.: Rev. Mod. Phys. **34**, 845 (1962)
WI 63 Wigner, E.P.: Am. J. Phys. **31**, 6 (1963)
WI 69 Wilkinson, D.H. (Ed.): *Isospin in Nuclear Physics*. North-Holland, Amsterdam (1969)
WI 72 Wigner, E.P.: In *Aspects of Quantum Theory*. Eds. A. Salam and E.P. Wigner. Cambridge Univ. Press, London (1982)
WI 81 Witten, E.: Nucl. Phys. B **188**, 513 (1981)
WI 82 Wilczek, F.: Phys. Rev. Lett. **48**, 1144 (1982)
WI 84 Will, C.M.: Phys. Rep. **113C**, 345 (1984)
WI 94 Wien, W.: Ann. Phys. (Wiedemann) **52**, 132 (1894)
WI 96 Wien, W.: Ann. Phys. (Wiedemann) **58**, 662 (1896)
WR 68 Witsch, W. von; Richter, A.; Bretano, P. von: Phys. Rev. **169**, 923 (1968)
WS 55 Wightman, A.S.; Schweber, S.: Phys. Rev. **98**, 812 (1955)
WS 72 Weinstein, A.; Stenger, W.: *Intermediate Problems for Eigenvalues*. Academic Press, New York (1972)
WU 84 Wu, Y.S.: Phys. Rev. Lett. **53**, 111 (1984)
WW 30 Weisskopf, V.; Wigner, E.: Z. Phys. **63**, 54 (1930)
WW 52 Wick, G.C.; Wightman, A.S.; Wigner, E.P.: Phys. Rev. **88**, 101 (1952)
WW 69 Whittaker, E.T.; Watson, G.N.: *A Course of Modern Analysis*, 4th ed., Cambridge Univ. Press, Cambridge, Mass. (1969)
WZ 83 Wheeler, J.A.; Zurek, W.H.: *Quantum Theory and Measurement*. Princeton Univ. Press, Princeton (1983)

YB 78 Yaris, R.; Bendler, J.; Lovett, A.; Bender, C.M.; Fedders, P.A.: Phys. Rev. A **18**, 1816 (1978)

YU 84 Yuen, H.P.: Phys. Rev. Lett. **52**, 1730 (1984)

ZE 65 Zemanian, A.H.: *Distribution Theory and Transform Analysis*. McGraw-Hill, New York (1965)

ZE 97 Zeeman, P.: Phil. Mag. **43**, 226 (1897); **44**, 55 and 255 (1897); Versl. Gewone Vergad. Wis. Natuurk. Afd. Kom. Akad. Wetensch. Amsterdam **6**, 13, 99 and 260 (1897)

ZO 80 Zorbas, J.: J. Math. Phys. **21**, 840 (1980)

Subject Index

The numbers I or II preceding the page numbers refer to the respective volume of **Quantum Mechanics**

Absorption cross section II 84
– photon II 319
Accidental degeneracy I 15, 240, 247, 250
Action I 123
Action variables I 13
Active interpretation I 270
Addition of angular momenta I 209, 326
Adiabatic approximation II 186, 190
Adiabatic invariance I 13
Adiabatic perturbation II 169, 180, 186
Adiabatic switch-on of perturbations II 180
Adiabatic theorem II 186
Agmon-Kato-Kuroda theorem I 222
Aharonov-Bohm effect II 209
Amplitude
– probability I 9, 27, 89
– reflection I 154
– transmission I 154
Analyticity properties of scattering amplitudes I 158, 160; II 56, 67
Angle variables I 13
Angular distribution of radiation II 317, 320
Angular momentum of the electromagnetic field II 298
Angular momentum(a) I 189, 322
– addition of I 209, 326
– conservation of I 279
– eigenvalues of I 196
– operator of I 189, 195, 322
– orbital I 200
– spin I 195
– standard basis for the irreducible representations of I 199
– total I 195
Annihilation operator(s) I 146; II 246
– for bosons II 246
– for fermions II 250
Anomalous Zeeman effect II 225
Anticommutation relations *or* rules
– for fermions II 250
– Fock representation of II 250
– irreducible representations of II 250
Antiunitary operator I 263, 274, 284, 386
Approximants, Padé II 129
Approximation
– adiabatic II 186,190
– Born *see* Born approximation
– central field II 283
– sudden II 169, 185
Associated Legendre function I 303
Asymptotes *in*, *out* II 10
Asymptotic completeness II 15, 78
Asymptotic series
Atom(s) II 268
– Bohr I 10, 15
– electric dipole transitions in II 322
– hydrogen I 10, 14, 242
– hydrogenic *or* hydrogen-like *or* one-electron I 10, 14, 242
– hydrogen-like, in parabolic coordinates I 252
– many-electron II 268, 271, 277, 286
– muonic I 249
– Nagaoka I 10
– one-electron *see* Hydrogen-like atoms
– Perrin I 10
– perturbative calculations II 287
  jj coupling II 296
  Russell-Saunders *or* *LS* coupling II 288
– Thomson I 10
– two-electron *see* Two-electron atoms
Atomic radius II 274, 312
Atomic units I 245
Auger effect II 270
Azimuthal quantum number I 14

Background phase shift II 65
Balance, detailed II 85
Balmer series I 246
Band structure I 178
Bargmann potentials II 63
Bargmann strip II 57
Bargmann's bound I 232
Bargmann's superselection rule I 292
Bargmann's theorem I 190
Bargmann-Segal space I 151

Barn II 4
Barrier
– centrifugal I 225
– potential I 152
– square I 173, 175
Baryonic charge *or* baryonic number I 87
BCS theory II 263
Bell's inequality I 382
Berry phase II 192
Bertrand's theorem I 252
Bessel functions I 312
Binding energy, deuteron II 64
Birkhoff's oscillation theorem I 177
Birman-Schwinger bound I 223
Black hole II 256
Blackbody I 1
Blackbody radiation II 243
Bloch states I 205
Bloch's theorem I 177
Bogolyubov canonical transformation II 259, 264
Bohr atom I 10, 15
Bohr complementarity I 7
Bohr magneton II 214
Bohr radius I 11, 243
Bohr-Peierls-Placzek relation II 27
Bohr-Rosenfeld argument II 303
Boltzmann constant I 2
Borel summability II 128, 134, 146
Born approximation II 33, 81
– in Coulomb scattering II 71
– for partial amplitudes II 52
– for scattering amplitudes II 33, 81
  in a spherical square well II 36
  in a Yukawa potential II 38
– for transition probabilities II 175
– unitarity violation in II 35
Born series II 31, 33, 51
– for partial amplitudes II 51
Bose gas II 243
Bose-Einstein statistics II 231, 238, 246, 256
Boson annihilation operators II 246
Boson commutation relations II 247
Boson creation operators II 246
Bosons II 231, 238, 246, 256
Bound states II 58
Bound(s) on number of bound states I 134, 223, 232
– Bargmann I 232
– Birman-Schwinger I 223
– Calogero I 233
– Cwikel-Lieb-Rosenbljum I 224
– Martin I 224
Bra I 39
Brackett series I 246
Breaking channel II 77
Breit-Wigner formula I 164; II 66, 164, 170
Brillouin zone I 179

Calogero's bound I 233
Campbell-Hausdorff formula I 82
Canonical quantization I 74, 77
Canonical transformation, Bogolyubov II 259, 264
Canonically cut plane II 32
Center of mass I 220
– correction due to recoil of I 245; II 328
Center-of-mass frame II 6
Central field approximation II 283
Central potential(s) I 220; II 44, 118
– Bargmann II 63
– Hylleraas II 119
– short range II 47
– square II 36, 83
– Yukawa II 38, 47, 57, 157
– Yukawian II 67
Centrifugal barrier I 225
Change LAB-C.M. II 6
Channel II 2
– breaking II 77
– (de)-excitation II 77
– elastic II 2, 77
– pick up II 77
Charge conservation II 200
Charge(s)
– baryonic I 87
– leptonic I 87
Chemical potential II 237
Chromodynamics I 281
Clapeyron's equation II 241
Classical electron radius II 334
Classical limit of the Schrödinger equation I 113
Clebsch-Gordan coefficients I 212, 326, 339
Clebsch-Gordan series I 211
Closure relation I 49
Coefficients X I 333
Coefficient(s)
– Clebsch-Gordan I 212, 326, 339
– diffusion I 157
– Racah I 329, 346
– reflection I 153
– reflection and transmission II 104
– transmission I 153
Coherences I 61, 64
Coherent spaces I 88, 266

Coherent states I 148
Collapse of the state I 65, 370
Collapse, gravitational II 255
Collision *see* Scattering
Combination principle I 10
Commutation relations *or* rules
– for angle-angular momentum I 201
– for angular momenta I 190, 195, 322
– for bosons II 250
– for electromagnetic field II 301, 302
– Fock representation of II 250
– irreducible representations of II 250
– for position-momentum I 78
– for velocity II 207
– in Weyl's form I 78, 205
Commutator I 53
Compatible observables I 55
– complete set of I 55
Complementarity principle I 7
Completeness, asymptotic II 15
Compton effect I 5
Compton scattering I 5; II 333
Compton wavelength I 6
Condition
– Lorentz gauge II 200
– transversality II 298
Condon-Shortley convention I 198, 212, 323
Configuration II 285
Confluent hypergeometric function I 317
Connection formulae II 90
Conservation
– angular momentum I 279
– charge II 200
– energy II 18, 80, 176
– flux I 153
– isospin I 295
– momentum I 278
– parity I 280
Conservation laws I 277
– and invariances I 277
Constant
– Boltzmann I 2
– fine structure I 11
– Planck I 3
– reduced Planck I 3
– renormalization II 125
– Stefan-Boltzmann II 245
Constant perturbations II 176
Constants of the motion I 73, 277
Continuity equation I 99, 100; II 200
Convection current II 309
Convention of Condon-Shortley I 198, 212, 323
Convolution product I 368
Cooper pair II 266
Coordinates
– principal I 103
– relative I 220
Copenhagen interpretation I 32, 34, 376
Correlation function(s) II 252
– two-boson II 256
– two-fermion II 254
– two-particle II 253
Correspondence principle I 15
Coulomb potential I 242
Coulomb scattering II 69
– amplitude of II 71
– Born approximation for II 71
– differential cross section for II 71
– Möller operators for II 69
– modified free evolution operator for II 69
– phase shifts for II 71
– Rutherford formula for II 71
– S matrix for II 69
Coulomb wave functions I 320
Coupling
– jj II 296
– LS *or* Russell-Saunders II 288
– minimal (electromagnetic) II 212
– spin-orbit II 218, 270
Creation operator(s) I 146; II 246
– for bosons II 246
– for fermions II 250
Cross section II 3
– absortion *see* Absortion cross section
– differential *see* Differential cross section
– elastic *see* Elastic cross section
– inelastic II 41
– invariance of II 6
– Thomson II 334
Current
– convection II 309
– probability II 203
– spin II 309
Cwikel-Lieb-Rosenbljum bound I 224
Cyclotron frequency II 216

D'Alembert equation II 298
Darwin term II 217
Davisson-Germer experiment I 23
De Broglie wavelength I 23
(De)-excitation channel II 77
Debye law II 243
Decay law I 173; II 179, 193
Decay rate II 179

Decomposition
– partial wave I 225
– spectral I 45, 350
Decomposition, electromagnetic field
– in multipole waves II 304
– in plane waves II 299
Deficiency indices I 43, 225, 348; II 131
Degeneracy
– accidental I 15, 240, 247, 250
– exchange II 228
– and invariance I 277
– Kramers I 286
– total II 241
Degenerate Fermi gas II 241
Delay time, Wigner formula for II 66
Delta function I 365
Delta-function potential I 130, 159, 236
Density
– of final states II 179
– probability I 90, 97; II 203
– probability current I 99; II 203
Density matrix *or* density operator I 58; II 74
Detailed balance II 85
Detailed balance theorem I 287
Determinant, Slater II 236
Determinism I 31
Deuteron binding energy II 64
Diamagnetic enhancement II 206
Diamagnetic term I 19; II 213, 220, 225, 333
Differential cross section II 9, 20, 27
– change LAB-C.M. II 6
– Coulomb II 71
– forward II 22
Diffusion coefficient I 157
Diffusion equation I 109
Dipole transition, electric II 322
Dirac delta function I 365
Dirac monopole II 208
Dirac picture II 172
Dirac quantization condition II 208
Dirac string II 208
Direct integral II 152
Direct potential II 279
Dispersion relations II 68
Dispersion-free states I 54
Displacement law of Wien I 1; II 244
Distribution(s) *or* generalized function(s) I 364
– differentiation of I 365
– support of I 366
– temperate I 364
Dunham formula II 97, 119
Dynamical momentum II 201
Dyson time ordering operator I 72

Eckart potential I 254, 257
Effect
– Aharonov-Bohm II 209
– Auger II 270
– Compton I 5
– Hall, quantum II 216
– Handbury-Brown and Twiss II 257
– Lamb I 248
– Paschen-Back II 224
– photoelectric I 4
– Stark I 21; *see also* Stark effect
– tunnel I 174
– Zeeman I 18; *see also* Zeeman effect
Effective range I 298; II 55, 62
Effective range expansion II 62
Ehrenfest's theorem I 101
Eigenfunction I 47
Eigenvalue I 44, 348
Eigenvalues for large coupling constants II 157
– for Yukawa potential II 159
Eigenvector I 44, 47, 348
Einstein-Rosen-Podolsky argument I 377
Elastic channel II 2, 77
Elastic cross section II 22
– in multichannel scattering II 84
Elastic scattering II 2
– two-body II 73
of identical particles II 75
Electric and magnetic multipoles II 305
Electric dipole transitions II 322
– in atoms II 322
– in hydrogen II 323
Electromagnetic coupling, minimal II 212
Electromagnetic field
– amplitudes of II 301
– angular momentum of II 209, 298
– classical Lagrangian of a particle in II 201
– commutation relations for II 301, 302
– conjugate momentum of II 304
– energy of II 298
– force on a particle in II 201
– Hamiltonian for a particle in II 201, 213
– momentum of II 298
– multipole decomposition of II 304
– particle in II 200
– plane wave decomposition of II 299
– quantization of II 301

– Schrödinger equation for a particle in II 202, 211
– uncertainty relations for II 303
– uncertainty relations for a particle in II 206
Electromagnetic interaction I 280
Electromagnetic potentials II 200
Electromagnetic transitions, selection rules for II 311
Electron configuration II 285
Electron radius, classical II 334
Electroweak model I 280
Elementary magnetic flux II 210
Elementary particles II 232
Emission
– induced II 319
– of radiation I 245; II 245
– spontaneous II 318
Energy
– conservation of II 18, 80, 176
– Coulomb, of a nucleus I 297
– deuteron binding II 64
– electromagnetic field II 298
free II 244
– radial kinetic I 225
– resonant I 164
– rotational kinetic I 225
– threshold II 2
Energy functional II 276
Energy shell I 188; II 19
Entropy I 371
Equation(s)
– d'Alembert II 298
– Clapeyron II 241
– continuity I 99, 100; II 200
– diffusion I 109
– Hamilton-Jacobi I 24, 113
– Lippmann-Schwinger *see* Lippmann-Schwinger equations
– Lorentz II 201
– Marchenko I 183
– Maxwell II 200
– Pauli II 213
– phase II 50
– radial I 225
– reduced radial I 226
– Schrödinger I 25, 26, 68
classical limit of I 113
for a particle in an electromagnetic field II 202, 211
stationary states of I 25
time-dependent I 26
Euclidean group I 190, 193
Euler angles I 206
Euler constant I 311
Euler gamma function I 311
Evolution
– stellar II 254
– time, of scattering wave packets II 41
Evolution operator *or* unitary propagator I 69, 70, 71
Evolution, types of I 370
Exchange degeneracy II 228
Exchange integral II 152
Exchange potential II 279
Exclusion principle II 233
– formal extension to nucleons II 234
Expansion
– effective range II 62
– multipole wave II 304
– partial wave II 44
– plane wave II 299
Expansion in eigenvectors I 48, 185
Expectation values, time evolution law for I 73
Expectation *or* mean value I 52, 59
Exponential decay law II 180
Exponential potential I 259

Factor(s)
– gyromagnetic II 214, 308
– inelasticity II 83
Family
– holomorphic self-adjoint
of type (A) I 356
of type (B) I 360
– spectral I 45, 351
Fast decrease functions I 363
Fermat's principle I 22
Fermi gas II 239
– degenerate II 241
Fermi momentum II 241
Fermi sea II 241
Fermi's golden rule II 179
Fermi-Dirac statistics II 231, 237, 250, 253
Fermion annihilation operators II 250
Fermion anticommutation relation II 250
Fermion creation operators II 250
Fermions II 231, 237, 250, 253
Feynman-Kac formula I 128
Field operators II 249
Final states, density of II 179
Fine structure II 217, 270, 297
– of hydrogen-like atoms II 217
Fine structure constant I 11
Floquet's theorem I 177

Flux conservation I 153
Fock representation of the (anti)commutation relations II 250
Fock space II 246
Force on a particle in an electromagnetic field II 201
Forced harmonic oscillator II 164
– adiabatic turning on II 169
– sudden turning on II 169
Form, quadratic I 140, 353
Formula(e)
– Breit-Wigner I 164; II 66, 164, 170
– Campbell-Hausdorff I 82
– connection II 90
– Dunham II 97, 119
– Feynman-Kac I 128
– Hellmann-Feynman II 127
– Klein-Nishina II 334
– mass I 299
– Mehler I 125
– Mott II 76
– Planck II 244
– Rodrigues I 302, 307, 308, 309
– Rutherford II 71
– Schläfli I 302
– Tiktopoulos I 359, 360
– Wigner II 66
Forward differential cross section II 22
Fourier transform I 367
Fourier transformation I 367
Fractional statistics II 231
Frames LAB and C.M. II 6
Franck-Hertz experiment I 12
Free energy II 244
Frequency
– cyclotron II 216
– Larmor I 20
– threshold I 4
Friedrichs extension I 140, 226, 348, 354
Frobenius theorem I 38
Function(s)
– associated Legendre I 303
– Bessel I 312
– confluent hypergeometric I 317
– correlation *see* Correlation function(s)
– Coulomb wave I 320
– Dirac delta I 365
– Euler gamma I 311
– fast decrease I 363
– Hamilton's characteristic I 24, 114
– Hamilton's principal I 24, 113
– Hankel I 313
– Hermitian analytic II 31
– Jost I 160; *see also* Jost functions
– Neumann I 315
– Racah triangle I 330
– radial wave I 225
– slow growth I 364
– spherical Bessel I 314
– spherical Hankel I 315
– step I 71
– test I 363
– triangle I 327
– wave I 24, 26, 88
– Weber I 312
– work I 5
Functional calculus with self-adjoint operators I 46, 350

Galilean invariance I 288, 294
Galilean transformations I 288
Galilei group I 288
Gamma function I 311
Gas
– Bose II 243
– Fermi *see* Fermi gas
Gauge condition, Lorentz II 200
Gauge independent observables II 202
Gauge invariance II 202, 204
Gauge transformation II 200
– of first kind II 205
– of second kind II 205
– of wave functions II 204
Generator(s) I 71, 269
– rotation I 272
– translation I 270
Gleason's theorem I 61
Golden rule, Fermi II 179
Grand unification models I 281
Gravitational collapse II 255
Gravitational interaction I 280
Group
– Euclidean I 190, 193
– Galilei I 288
– isospin I 296
– one-parameter unitary I 70, 78, 353
– rotation *or* SO(3) I 190, 271
– of space translations I 91, 269
– SU(2) I 193, 296
– of symmetry transformations I 268
– time evolution I 70, 71
– unitary abelian I 78
– velocity of I 25, 28, 180
Gyromagnetic factor I 249
Gyromagnetic factors *or* ratios II 214, 308

H_0-bounded I 356
H_0-infinitesimal I 354
h_0-infinitesimal I 358
H_0-small I 354
h_0-small I 358
Hadrons I 281
Hall effect, quantum II 216
Hamilton's characteristic function I 24, 114
Hamilton's principal function I 24, 113
Hamilton-Jacobi equation I 24, 113
Hamiltonian I 68
– in arbitrary coordinates I 98
– in Cartesian coordinates I 98
– for particle in electromagnetic field II 201, 213
– in spherical coordinates I 225
– in terms of annihilation and creation operators II 248, 301, 307
Handbury-Brown and Twiss effect II 257
Hankel functions I 313
Harmonic oscillator
– analytical treatment of I 142, 239
– classical equations for I 106, 143; II 167
– forced *see* Forced harmonic oscillator
– Hamiltonian of, in terms of annihilation and creation operators I 145
– in the Heisenberg picture I 147
– one-dimensional I 142
– operator treatment of I 145
– three-dimensional I 239
– W.B.K. levels for
  isotropic II 119
  one-dimensional II 98
– with x^3 perturbation II 131
  complex scaling techniques II 133
  lifetimes II 133
– with x^4 perturbation II 127
  Borel summability II 128
  Padé approximants II 129
Hartree-Fock equations II 278
– Moshinsky example II 280
Hartree-Fock method II 277
– central field approximation II 283
– direct potential II 279
– exchange potential II 279
– rigorous results II 283
Heat, specific II 242
Heisenberg picture I 84
Helicity I 217
Helicity, photon II 301
^{4}He, superfluidity of II 262
Hellmann-Feyman formula II 127
Hermite polynomials I 306
Hermitian analytic function II 31
Hidden variables I 31, 376
– local I 381
High-T_c superconductivity II 267
Hilbert space I 37, 39, 347
– rigged I 48
Hilbert-Schmidt norm I 354; II 22
Hölder continuity I 223; II 15, 194
Holomorphic self-adjoint family
– of type (A) I 356
– of type (B) I 360
Hughes-Eckart term II 269
Hulthén potential I 255
Hund's rule II 153, 285, 294
Hunziker-Van Winter-Zhislin theorem II 76, 269
Hydrogen atom I 10, 14, 242
Hydrogen, electric dipole transitions in II 323
Hydrogenic atom *or* hydrogen like atom *or* one-electron atom I 10, 14, 242
Hydrogen-like atom(s) II 119, 155
– fine structure of II 217
– in a magnetic field II 220
  ionization energy II 226
  Rayleigh-Schrödinger series II 226
– in parabolic coordinates I 252
– variational method II 155
– W.B.K. method II 119
Hylleraas potential I 256; II 119

Ideal measurement I 64
Identical particles II 227
– non-interacting II 235
  bosons II 238
  fermions II 237
Identity
– Jacobi I 74
– spectral resolution of I 45, 351
Ikebe class of potentials I 223; II 15
Ikebe conditions II 15
Impact parameter II 3
In and out states II 10, 13
Incoherent target II 5
Incomplete problems II 132
Indistinguishability II 6, 74, 227
Induced emission II 319
Inelastic cross section II 41
Inelastic scattering, phase shifts in II 83
Inelasticity factors II 83
Inequality
– Kato II 206
– Lieb II 269
– Temple II 149

Instanton II 117
Integral
– direct, in Hilbert spaces I 48, 361
– path I 123
Integral, direct and exchange II 152
Intelligent states I 206
Interaction between matter and radiation II 243, 308
Interaction picture *or* Dirac picture I 86; II 172
Interactions
– electromagnetic I 280
– gravitational I 280
– strong I 280
– weak I 280
Internal symmetries I 295
Interpretation(s)
– active and passive, of a symmetry transformation I 270
– Copenhagen I 32, 34, 376
Intertwining property II 18, 79
Interval rule of Landé
Intrinsic parity I 273
Invariance(s)
– adiabatic I 13
– and conservation laws I 277
– of cross section II 6
– Galilean I 288, 294
– gauge II 202, 204
– isospin I 295
– parity I 280
– rotation I 279
– time-reversal I 284; II 43, 84
– translation I 278
Inverse spectral problem I 180
Ion H^- II 136
Ion H^{--} II 269
Irreducibility of a set of operators I 268
Irreducible representation I 193, 277
Irreducible set of observables I 268
Irreducible tensor operator I 215
Irreducible vector tensors *or* vector spherical harmonics I 335
Isomultiplet *or* isospin multiplet I 296
Isospin I 295; II 325
– conservation of I 295
– group of I I 296
– invariance under I 295
– space of I 296
Isotopic shift II 269

jj coupling II 296
Jacobi identity I 74
Jost function(s) I 160; II 56
– zeroes and bound states II 58

Kato class of potentials I 221, 356
Kato's inequality II 206
Kato's perturbation theory II 138
Kato-Birman theorem I 355
Kato-Lax-Milgram-Lions-Nelson theorem I 358
Kato-Rellich criterion for the stability of self-adjointness I 354
Ket *or* state vector I 37, 39
Kirchhoff's law II 245
Klein-Nishina formula II 334
Koopmans' theorem II 279
Kramers' degeneracy I 286
Kramers' relation I 245, 310
Kramers' theorem I 286, 388
Kronig-Penney potential I 178

Laguerre generalized polynomials I 309
Laguerre polynomials I 308
Lamb effect I 248
Lamb shift I 248
Landé interval rule II 296
Landau levels II 215
Larmor frequency I 20
Law
– Debye II 243
– decay I 173; II 179, 193
– displacement I 1; II 244
– Kirchhoff II 245
– Planck I 2
– Rayleigh-Jeans I 2; II 244
– Wien I 2
Least action principle I 22
Left handed polarization vector II 300
Legendre polynomials I 301
Lehmann's ellipse II 47
Length, scattering II 53, 62
Leptonic charges *or* leptonic numbers I 87
Leptons I 280
Levinson's theorem I 166; II 60
Lieb's inequality II 269
Lifetime II 133, 180, 196, 323
Light, duality I 7
Limit, classical, of the Schrödinger equation I 113
Linear polarization vector II 300
Linear superposition principle I 86
Lippmann-Schwinger equations II 28, 31, 80
– in operator form II 31
Local hidden variables I 381

Localizability I 78, 81
Lorentz equation II 201
Lorentz gauge condition II 200
Lyman series I 246
LS coupling II 288

Magnetic field
– hydrogen atom in I 220
– one-electron atoms in *see* Hydrogen-like atoms
– particle in a constant uniform I 215
Magnetic flux, elementary II 210
Magnetic moment I 19, 249, 295; II 214
Magnetic monopoles *see* Monopoles
Magnetic quantum number I 14, 201
Magneton
– Bohr I 214
– nuclear I 214
Mandelstam representation II 68
Mandelstam variables II 6
Many-electron atoms II 268, 271, 277, 286
Marchenko integral equation I 183
Martin's bound I 224
Mass
– effective I 180
– formula of I 299
– reduced I 10, 220
– superselection rule associated with I 292
Matrix(ces)
– density I 58; II 74
time evolution of I 69
– Pauli I 67, 200, 323
– rotation I 190
– *S* I 156, 188; *see also* *S* matrix
– scattering I 156; *see also* *S* matrix
– *T* II 19, 30, 81; *see also* *T* matrix
Matter waves I 22
Maupertuis principle I 22
Maxwell equations II 200
Measure
– spectral I 351
– Wiener I 128
Measurement(s)
– filter I 64
– of the first kind I 35, 63
– ideal I 64
– problem of I 370
– results of I 50, 61
– of the second kind I 35, 63
Mechanical momentum II 201
Mehler formula I 125
Method
– Hartree-Fock *see* Hartree-Fock method
– Miller-Good 110
– quasi-particle II 38, 260, 264
– Rayleigh-Ritz II 149
– Thomas-Fermi *see* Thomas-Fermi method
– Thomas-Fermi-Dirac II 275
– variational *see* Variational method
– W.B.K. *see* W.B.K. method
Micro-states I 379
Microreversibility principle I 285; II 44, 175
Miller-Good method II 110
Min-max principle I 135, 223, 231, 360; II 148
Minimal (electromagnetic) coupling II 212
Mixed symmetries II 229
Möller operators I 188, 355; II 10, 14, 78
– in Coulomb scattering II 69
Moment, magnetic I 19, 249, 295; II 214
Momentum(a)
– conservation of I 278
– dynamical II 201
– of electromagnetic field II 298
– Fermi II 241
– mechanical II 201
– operator for I 41, 78, 90, 92
– relative I 220
– representation associated with I 90
Momentum transfer II 35
Monopoles II 208, 209
– Dirac II 208
– 't Hooft-Polyakov II 209
Morpurgo's rule II 327
Morse potential I 259
Mott's formula II 76
Multi-periodic systems I 13, 16
Multichannel scattering II 2, 76
– absorption cross section for II 84
– asymptotic completeness of II 78
– detailed balance in II 85
– elastic cross section for II 84
– inelasticity factors in II 83
– Möller operators for II 78
– rotational invariance of II 83
– *S* matrix for II 79
– symmetries of II 83
– time reversal invariance of II 84
– total cross section for II 84
Multiplicity I 44
Multiplier I 364
Multipole radiation, angular distribution of II 320
Multipole wave decomposition of an electromagnetic field II 304
Multipole waves II 304

Multipoles, electric and magnetic II 305
Muonic atom I 249

Nagaoka atom I 10
Natural units II 268
Natural width I 249
Nelson phenomenon I 229
Neumann functions I 315
Neutron interferometry I 23
Neutron stars II 256
Norm
– Hilbert-Schmidt I 354; II 22
– Rollnik II 32
– trace I 355
– trace class II 22
Normal modes, number of I 2
Normal ordering II 302
Normal Zeeman effect II 225
Nuclear magneton II 214
Nuclear radius I 297
Nuclear spin resonance II 162
Nucleon I 295
Number operator II 246
Number(s)
– azimuthal quantum I 14
– baryonic I 87
– of bound states I 134, 223, 232
– leptonic I 87
– magnetic quantum I 14, 201
– operator for I 146
– principal quantum I 11, 14, 243
– radial quantum I 14, 242

Observable(s)
– with classical analog I 79
– compatible I 55
  complete set of I 55
– irreducible set of I 268
– and operators I 40, 77
– superselection I 86
One-dimensional problem(s) I 130
– asymptotic behaviors in I 131
– energy spectrum in I 131
– mathematical conditions for I 183
– number of bound states in I 134
One-electron atoms *see* hydrogen-like atoms
Operator(s)
– adjoint I 348
– angular momentum I 189, 195, 322
– annihilation I 146; *see also* Annihilation operators
– antiunitary I 263, 274, 284, 386
– closable I 347
– closed I 347
– compact I 44, 355
– corresponding to classical observables I 78
– creation I 146; *see also* Creation operators
– decomposable I 362
– density I 58
– diagonal I 362
– evolution I 69, 70, 71
– field II 249
– Hamiltonian I 68, 97, 225
– Hermitian I 348
– irreducible tensor I 215
– kinetic energy I 140
– Möller I 188, 355; *see also* Möller operators
– momentum I 41, 78, 90, 92
– number I 146; II 246
– one-parameter unitary group of I 70, 71, 78, 353
– parity I 273
– position I 41, 78, 83, 91
– quasiparticle II 261
– radial kinetic energy I 225
– radial momentum I 225
– resolvent I 347
– rotation I 190, 271
– rotational kinetic energy I 225
– Runge-Lenz I 250
– scalar I 216
– scattering I 156, 188; *see also* Scattering operator
– self-adjoint I 40, 348
  essentially I 131, 221, 348
  functional calculus with I 46, 350
  perturbation of I 354
  semi-bounded *or* bounded from below I 348
  spectral theory of I 350
  spectrum of I 352
– spectral projection I 351
– symmetric I 41, 53, 348
– time ordering I 72
– time reversal I 274
– trace class I 355
– transition II 19, 30, 81
– translation I 91, 269
– unitary I 70, 71, 263, 351
– univalence I 209
– vector I 216
Optical potential II 85
Optical theorem I 157; II 26, 40, 47, 82

Orbitals II 235
Oscillator strengths II 324

Packets
- wave I 27
  minimal I 95, 105
Padé approximants II 129
Pair, Cooper II 266
Pair, Schrödinger I 80
Parabolic coordinates, hydrogen-like atoms in I 252
Paramagnetic term II 213
Parameters, normal I 269
Parastatistics II 230
Parity I 272
- conservation of I 280
- intrinsic I 273
  invariance of scattering operator under II 43
- invariance under I 280
- operator for I 273
Partial wave decomposition I 225
- of plane waves I 219, 302
Partial wave resonances II 65
Partial waves II 45
- expansion of scattering amplitudes in II 44
Partial (wave) scattering amplitude(s) II 46
- analyticity properties of II 56
- Born approximation to II 52
- Born series for II 51
- low energy behavior of II 62
Particle(s)
- under the action of a constant force I 105
- in a constant uniform magnetic field II 215
- in an electromagnetic field II 200, 211
  classical Lagrangian for II 201
  force on II 201
  Hamiltonian of II 201, 213
  Schrödinger equation for II 202
  uncertainty relations for II 206
- elementary I 139
- identical *see* Identical particles
Paschen series I 246
Paschen-Back effect II 224
Passive interpretation I 270
Path
- integral I 123
- integration I 120
Pauli equation II 213
Pauli matrices I 66, 200, 323
Periodic perturbations II 182
Periodic potentials I 175
Permutation group II 227
Perrin atom I 10
Perturbation
- of self-adjoint forms I 358
- of self-adjoint operators I 354
Perturbation(s)
- adiabatic II 169, 180, 186
- of close levels II 126
- constant II 176
- and energy conservation II 176
- of harmonic oscillator see Harmonic oscillator
- periodic II 182
- regular *see* Regular perturbations
- stationary *see* Time-independent perturbations
- sudden II 169, 184
- time-dependent *see* Time-dependent perturbations
- time-independent *see* Time-independent perturbations
Perturbative calculation of atoms II 287
- jj coupling II 296
- Russell-Saunders *or* LS coupling II 288
Perturbative calculation of atoms II 287
Peter-Weyl theorem I 196
Pfund series I 246
Phase, Berry II 192
Phase equation II 50
Phase shift(s) II 44, 46
- background II 65
- behavior near a resonance of II 65
- for central square well II 53
- computation of II 48
- Coulomb II 71
- for delta function potentials I 160
- in inelastic scattering II 83
- and Levinson theorem I 166; II 60
- low energy behavior of I 298; II 62
- for one-dimensional potentials I 157
- ordering of II 72
- properties of II 48
- resonance behavior of I 165
- resonant II 65
- W.B.K. approximation to II 120
Phase space I 4
Phase velocity I 25, 115
Phonons II 263
Photoelectric effect I 4
Photon I 5
Photon absorption cross section II 319
Photon emission *or* absorption II 318
Photon helicity II 301

Pick up channel II 77
Picture(s)
– Dirac *or* interaction I 85; II 172
– Heisenberg I 84
– interaction *or* Dirac I 85; II 172
– Schrödinger I 84
– time evolution I 84
Planck constant I 3
Planck constant, reduced I 3
Planck formula II 244
Planck's law I 2
Plane wave I 39, 51, 100
Plane wave expansion II 299
Platinum rule II 182
Poisson bracket I 74
Polarization I 66
Polarization (vector) II 299
– linear II 300
– right and left II 300
Pole structure of the S matrix II 66
Pole structure of the transmission amplitude I 167
Polynomials
– Hermite I 306
– Laguerre I 308
– Laguerre generalized I 309
– Legendre I 301
Populations I 61
Position(s)
– operator for I 41, 78, 83, 91
– representation associated with I 90
Potential(s)
– $a|x|$ II 98
– barrier II 104, 113
– central I 220; *see also* Central potentials
– chemical II 237
– Coulomb I 242
– delta function I 130, 159, 236
– direct II 279
– Eckart I 254, 257
– electromagnetic II 200
– exactly solvable, in s waves I 254
– exchange II 279
– exponential I 259
– harmonic oscillator I 142, 239
– Hulthén I 255
– Hylleraas I 256
– Ikebe class of I 223; II 15
– Kato class of I 221, 356
– Kronig-Penney I 178
– Morse I 259
– optical II 85
– Periodic I 175
– r^{-2} I 229
– r^{-4} I 254
– Rollnik class of I 222, 359
– soliton II 109
– square well I 136, 234
– standard II 15
– thermodynamic II 237
– Von Neumann-Wigner I 233
– $V_0 \cosh^{-2} ax$ II 108
– $|x|^\alpha$ II 101
– $-1/r$ at the origin II 157
Preparation I 35, 63
– maximal I 36
Principal quantum number I 11, 14, 243
Principle
– combination I 10
– complementarity I 7
– correspondence I 15
– exclusion *see* Exclusion principle
– Fermat I 22
– least action I 22
– linear superposition I 86
– Maupertuis I 22
– microreversibility I 285; II 44, 175
– min-max I 135, 223, 231, 360; II 148
– symmetrization II 231
– uncertainty I 29, 53
Probability amplitude I 9, 27, 89
Probability current density I 99; II 203
Probability density I 90, 97; II 203
Probability, relative I 51
Probability, transition II 174, 310
Problem(s)
– inverse spectral I 180
– one-dimensional I 130
  asymptotic behaviors in I 131
  energy spectrum in I 131
  mathematical conditions I 183
  number of bound states in I 134
Product of convolution I 368
Projectiles II 1
Projection operator, spectral I 351
Propagation of wave packets I 103, 106
Propagator
– advanced I 112
– free retarded I 109
– retarded I 107
– unitary evolution I 69, 70, 71, 121
Property(ies)
– analyticity, of scattering amplitudes II 56, 67
– intertwining II 18, 79
– of phase shifts II 48

Proposition *or* yes-no experiment I 63
Propositional calculus I 63

Quadratic form(s) I 140, 353
- closable I 353
- closed I 353
- positive I 353
- self-adjoint I 353
  semi-bounded *or* bounded from below I 353
- symmetric I 353
Quantization
- canonical I 74, 77
- of the electromagnetic field II 301
- rules of I 77
- second II 249
- Sommerfeld-Wilson-Ishiwara rules of I 4, 13
- space I 15, 21
Quantum chromodynamics I 281
Quantum theory of radiation II 298
- transition probabilities in II 310
Quarks I 280, 300
Quasiparticle method II 38, 260, 264
Quasiparticle operators II 261
Quasiparticle states II 260
Quaternions I 38

Racah
- coefficients of I 329, 346
- triangle function of I 330
Radial equation I 225
Radial momentum operator I 225
Radial quantum number I 14, 242
Radial wave function I 225
Radiation I 10; II 243, 298
- angular distribution of II 317, 320
- emission of II 245
- quantum theory of II 298
  transition probabilities in II 310
- 2^J-polar II 305
Radiation emission I 245
Radiative transitions in nuclei II 325
- nucleon effective charge for II 329
Radius
- atomic II 274, 312
- Bohr I 11, 243
- classical electron II 334
- nuclear I 297
Range, effective I 298; II 55, 62
Ratios, gyromagnetic II 214, 308
Rayleigh-Jeans law I 2; II 244
Rayleigh-Ritz method II 149
Rayleigh-Schrödinger series II 123
Reciprocity relation I 153
Reciprocity theorem I 287; II 44
Reduced mass I 10, 220
Reduced matrix element I 216
Reduced radial equation I 226
Reduction of the wave packet I 65, 370
Reflection amplitude I 154
Reflection coefficient I 153; II 104
Regular perturbations II 143
- convergence radius II 143
Regularity condition I 227, 229
Relation(s) *or* rule(s)
- anticommutation *see* Anticommutation relations
- Bohr-Peierls-Placzek II 27
- closure I 49
- commutation, *see also* Commutation relations
  for angle-angular momentum I 201
  for angular momenta I 190, 195, 322
  for position-momentum I 78
  in Weyl's form I 78, 205
- dispersion II 68
- Kramers I 245, 310
- reciprocity I 153
- uncertainty I 29, 53
Relative probability I 51
Renormalization constant II 125
Representation
- associated with a complete set of compatible observables I 88
- irreducible I 193, 277
- Mandelstam II 68
- matrix
  of J I 199
  of the rotation operators I 206, 324
- momentum I 90
- position I 90
- relation between position and momentum I 93
- spectral I 350
  of the identity I 45, 351
Resolvent operator I 347
Resolvent set I 46, 347
Resonance width I 164
Resonance(s) I 164; II 63, 64, 164
- nuclear spin II 162
- zero energy II 59, 63
Resonant phase shift II 65
Resonant state II 65
Results of measurements I 50, 61
Riesz-Fréchet theorem I 39

Rigged Hilbert space I 48
Right handed polarization vector II 300
Rodrigues' formula I 302, 307, 308, 309
Rollnik norm II 32
Rotation *or* SO(3) group I 190, 271
Rotation matrix I 190
Rotation operator I 190, 271
Rotational invariance I 279
– in multichannel scattering II 83
– in simple scattering II 44
Rotational kinetic energy I 225
Rotations I 271
– double-valued representations of I 208
Rule(s)
– anticommutation *see* Anticommutation rules
– commutation *see* Commutation rules
– Fermi's golden II 179
– Hund's II 153, 285, 294
– Landé interval II 296
– Morpurgo II 327
– platinum II 182
– quantization I 77
of Sommerfeld-Wilson-Ishiwara I 4, 13
– selection I 87; *see also* Selection rules
– Slater sum II 293
– superselection I 39, 85, 209, 292
– Thomas-Reiche-Kuhn sum II 324
– Trainor-Gamba II 326
Runge-Lenz operator I 250
Russell-Saunders coupling II 288
Rutherford formula II 71
Rydberg I 246

S matrix I 156, 188; II 16, 79
– for Coulomb scattering II 69
– pole structure of II 66
– poles of II 66
– symmetries of II 43, 83
– unitarity of II 16
Scalar operator I 216
Scattering
– by central potentials II 44
– by central square wells II 53
– Compton I 5; II 333
– into cones II 25, 69
– Coulomb *see* Coulomb scattering
– elastic *see* Elastic scattering
– exo-energetic II 2
– general description of II 1
– of identical particles II 6, 75
– multichannel *see* Multichannel scattering
– single-channel *or* simple *see* single-channel scattering
– two-body elastic *see* Elastic scattering
Scattering amplitude II 27
– analyticity properties of I 158, 160; II 56, 67
– Born approximation *see* Born approximation
– computation II 28
– Coulomb II 71
– low energy behavior II 62
– multichannel
– partial wave expansion of II 44
– symmetries of
– two-body II 80
Born approximation II 81
Scattering length, nucleon-nucleon I 298
Scattering lengths II 53, 62
Scattering matrix I 156; *see also* *S* matrix
Scattering operator I 156, 188; *see also* *S* matrix
Scattering wave packet, time evolution of II 41
Schläfli's formula I 302
Schrödinger equation I 25, 26, 68
– classical limit of I 113
– in an electromagnetic field II 202, 211
– one-dimensional I 130
– reduced radial I 226
– for the relative motion I 221
– for stationary states I 25
– time-dependent I 26
– for two-particle systems I 220
Schrödinger pair I 80
Schrödinger picture I 84
Sea, Fermi II 241
Second quantization II 249
Selection rule(s) I 87
– for electromagnetic transitions II 311
– Trainor-Gamba II 326
Self-adjoint operator I 40, 348
Self-ionization II 194, 270
Series
– asymptotic II 128
– Balmer I 246
– Born II 31, 33, 51
– Brackett I 246
– Clebsch-Gordan I 211
– Lyman I 246
– Paschen I 246
– perturbative II 123, 140, 173, 174
– Pfund I 246
– Rayleigh-Schrödinger II 123
– spectral I 246
Shift, isotopic II 269
Short range potentials II 47
Simple scattering *see* Single-channel scattering

Single-channel scattering II 2
– classical case II 9
– quantum case II 13
– space-time description of II 39
Slater determinant II 236
Slater sum rule II 293
Slow growth function I 364
Soliton I 182
Soliton potential II 109
Sommerfeld-Wilson-Ishiwara quantization rules I 4, 13
Space(s)
– Bargmann-Segal I 151
– coherent I 86, 266
– Fock II 246, 251
– Hilbert I 37, 39, 347
– isospin I 296
– phase I 4
– Stummel I 119
Space quantization I 15, 21
Space translation I 91, 269, 278
Specific heat II 242
Spectral concentration in Stark effect II 146
Spectral decomposition I 45, 350
Spectral family I 45, 351
Spectral measure I 351
Spectral projection operator I 351
Spectral representation I 350
Spectral series I 246
Spectral theory of self-adjoint operators I 350
Spectrum I 43, 347
– absolutely continuous I 184, 352
– of the angular momentum operator I 196
– continuous I 43, 348
– discrete I 352
– essential I 184, 222, 352, 355
– of H I 131, 222
– of the momentum operator I 83
– point I 43, 348
– of the position operator I 83
– pure point I 352
– residual I 348
– of a self-adjoint operator I 352
– singular continuous I 352
Spherical Bessel functions I 314
Spherical Hankel functions I 315
Spherical harmonics I 201, 304
Spherical square well, Born approximation for II 36
Spin I 21, 95, 192
Spin 1/2 particles in an electromagnetic field II 211
Spin current II 309
Spin magnetic moment density II 309
Spin-orbit coupling II 218, 270
Spin-statistics theorem II 231
Spinor I 95; II 162
Spontaneous emission II 318
Spreading of wave packets I 105, 111
Square barrier I 173, 175
Square well potentials I 136, 234
Standard potentials II 15
Stark effect I 21; II 144
– for hydrogen-like atoms II 144
– Oppenheimer formula for II 146
– perturbative series in II 146
– quadratic II 145
– spectral concentration in II 146
Stars
– neutron II 256
– white dwarf II 242, 255
State(s) I 34
– Bloch I 205
– bound I 73; II 58
– coherent I 148
– collapse of I 65, 370
– dispersion-free I 54
– excited I 136
– ground I 129, 136
– *in* and *out* I 156; II 10, 13
– intelligent I 206
– mixed I 34, 36, 57
evolution equation for I 69
– pure I 34, 36, 37
evolution equation for I 68
– quasiparticle II 260
– resonant I 164, 171; II 65
– scattering I 73
– stationary I 10, 72
– vacuum II 246
– virtual I 162; II 64
– width of II 134, 146, 179
State vector *or* ket I 37, 39
Stationary perturbations *see* Time-independent perturbations
Stationary states, Schrödinger equation for I 25
Statistics
– Bose-Einstein II 231, 238, 246, 256
– Fermi-Dirac II 231, 238, 250, 253
– fractional II 231
– parastatistics II 230
Stefan-Boltzmann constant II 245
Stellar evolution II 254
Step function I 71
Stern-Gerlach experiment I 20, 66, 196

Stone's theorem I 71, 353
String, Dirac II 208
Strong interaction I 280
Structure
– fine I 16, 247; *see also* Fine structure
  constant of I 11
– hyperfine I 249
Stummel space I 119
Sturm's theorem I 135, 186, 230
Sudden approximation II 169, 185
Sudden perturbations II 184
Sum rule(s)
– Slater II 293
– Thomas-Reiche-Kuhn II 324
Summability, Borel II 128, 134, 146
Superconductivity II 263
– critical temperature for II 263
– energy gap in II 266
– high-T_c II 267
– superconducting solution in II 266
Superfluidity II 258
– of ^{4}He II 262
Superposition, principle of linear I 86
Superselection observables I 86
Superselection rule(s) I 39, 85, 209, 292
– Bargmann I 292
– mass I 292
– univalence I 209
Superstrings I 281
Supersymmetry I 281; II 98, 143
Support of a distribution I 366
Symmetries of the scattering operator
  II 43, 83
Symmetrization of wave functions II 227
Symmetrization principle II 231
Symmetry transformations I 262

't Hooft-Polyakov monopole II 209
T matrix II 19, 30, 81
Targets II 1, 5
– incoherent II 5
Teller's theorem II 276
Temperate distribution I 364
Temple's inequality II 149
Tensor
– irreducible I 215, 331
– irreducible vector I 334
Term(s)
– Darwin II 217
– diamagnetic II 213, 220, 225, 333
– Hughes-Eckart II 269
– paramagnetic II 213
– Von Weizsäcker II 276
Test function I 363
Theorem
– adiabatic II 186
– Agmon-Kato-Kuroda I 222
– Bargmann I 190
– Bertrand I 252
– Birkhoff I 177
– Bloch I 177
– detailed balance I 287; II 85
– Ehrenfest I 101
– Floquet I 177
– Frobenius I 38
– Gleason I 61
– Hunziker-Van Winter-Zhislin II 76, 269
– Kato-Birman I 355
– Kato-Lax-Milgram-Lions-Nelson I 358
– Koopmans II 279
– Kramers I 286, 388
– Levinson I 166; II 60
– optical I 157; II 26, 40, 47, 82
– Peter-Weyl I 196
– reciprocity I 287; II 44
– Riesz-Fréchet I 39
– spin-statistics II 231
– Stone I 71, 353
– Sturm I 135, 186, 230
– Teller II 276
– virial I 118
– Von Neumann
  on complete sets of compatible
  observables I 57
  on hidden variables I 378
– Wigner I 190, 263
– Wigner-Eckart I 215, 332
Thermodynamic potential II 237
Thomas-Fermi method II 271
– energy functional in II 276
Thomas-Fermi-Dirac model II 275
Thomas-Reiche-Kuhn sum rule II 324
Thomson atom I 10
Thomson cross section II 334
Threshold energy II 2
Threshold frequency I 4
Tiktopolous formula I 359, 360
Time delay I 154
Time evolution group I 71
Time evolution law for expectation values
  I 73
Time evolution of scattering wave packets
  II 41
Time evolution pictures I 84
Time evolution, deterministic *or* causal I 370
Time ordering operator I 72

Time reversal I 274
– invariance under I 284
– operator for I 274
Time reversal invariance II 43, 84
Time-dependent perturbations II 161
– adiabatic turning-on of II 180
– forced harmonic oscillator and II 164
– interaction *or* Dirac picture and II 172
– nuclear spin resonance II 162
– transition probabilities in II 174
Born approximation to II 175
Time-dependent Schrödinger equation I 26
Time-independent perturbations II 122
– ion H^- and II 136
– Kato's theory of II 138
convergence radius II 143
degenerate level II 141
non-degenerate level II 141
regularity conditions II 143
– non-degenerate case II 122
Rayleigh-Schrödinger series II 123
– Stark effect II 144
– two-electron atoms II 134
– Van der Waals forces II 136
dipole-dipole approximation II 137
Total degeneracy II 241
Trace class I 355
Trace class norm II 22
Trace class operator II 19
Trace norm I 355
Trainor-Gamba selection rule II 326
Trajectories and the uncertainty principle I 31
Transfer, momentum II 35
Transformation
– Bogolyubov canonical II 259, 264
– Fourier I 367
– Galilean I 288
– gauge II 200
of wave functions II 204
– symmetry I 262
Transformation of operators I 267
Transition operator II 19, 30, 81
Transition probability II 174, 310
– for constant perturbations II 176
– per unit time II 24
Transition rate(s)
– radiative II 310
– 2^J-polar
in atomic systems II 316
in nuclear systems II 316
Transition, electric dipole II 322
Translation operator I 91, 269
Translation, space I 91, 269, 278
Translational invariance I 278
Transmission amplitude I 154
Transmission coefficient I 153; II 104
Transversality condition II 298
Trial functions *or* states II 147
Triangle function I 327
Tunnel effect I 174
Turning points II 89
Two-body elastic scattering II 73
– differential cross section II 74
– indistinguishability II 75
– separation of C.M. motion II 73
– spin dependent transition amplitude II 73
– unpolarized beams II 74
Two-body scattering
– amplitude for II 80
– Born approximation II 81
– differential cross section II 80
Lippmann-Schwinger equations II 80
Two-boson correlation function II 256
Two electron atoms II 134, 150, 152
– Hund's rule in II 153
Two-fermion correlation function II 254
Two-particle correlation function II 253
2^J-polar radiation II 305
2^J-polar transition rates *see* Transition rates

Ultraviolet catastrophe I 2
Uncertainty principle I 29, 53
Uncertainty relation(s) I 29, 53
– for angle-angular momentum I 201
– for energy-time I 29, 75
– for position-momentum I 29, 95
– in presence of electromagnetic field II 206
Uncertainty *or* standard deviation I 52
Unitarity
– limit of II 47
– of the S matrix II 16
– violation of II 35, 105
Unitarity bound I 157
Unitary propagator *or* evolution operator I 69, 70, 71
Units
– atomic I 245
– natural II 268
– Weisskopf II 332
Univalence superselection rule I 209
Unstable quantum systems, decay law for II 193

Vacuum state II 246
Van der Waals forces II 136, 154

Variational method II 122, 146
– helium atoms II 152
– hydrogen-like atoms II 155
– lower bounds II 149
– one-electron atoms II 155
– Temple's inequality II 149
– trial functions *or* states II 147
– two-electron atoms II 150
– Van der Waals forces II 154
Vector operator I 216
Vector spherical harmonics *or* irreducible vector tensors I 335
Velocity
– group I 25, 28, 180
– phase I 25, 115
Velocity commutation rules II 207
Virial theorem I 118
Virtual states II 64
Von Neumann's theorem on complete sets of compatible observables I 57
Von Neumann's theorem on hidden variables I 378
Von Neumann-Wigner potential I 233
Von Weizsäcker term II 276

W^*-algebra I 362
– abelian I 362
– commutant of I 362
W.B.K. approximation II 90
W.B.K. method II 88
– basic solutions II 89
– bound state energies II 95, 115, 118
  Dunham formula II 97
  for several turning points II 115
– central potentials II 118
  Coulomb II 119
  Hylleraas potential II 119
  isotropic harmonic oscillator II 119
  Langer modification II 118
  phase shifts II 120
– connection formulae II 90
– one-dimensional problems II 88
  harmonic oscillator II 98
  potential barrier II 104, 113
  transmission and reflection coefficients II 105
  transmission and reflection coefficients by double potential barriers II 113
– turning points II 89
Wave function I 24, 26, 88
Wave function symmetrization II 227
Wave packet reduction I 65, 370
Wave packet spreading I 105, 111
Wave packets I 27
Wave(s)
– expansion in plane II 299
– matter I 22
– multipole II 304
– partial II 45
– plane I 39, 51, 100
Wavelength
– Compton I 6
– De Broglie I 23
Weak interaction I 280
Weber function I 312
Weisskopf units II 332
Well(s)
– one-dimensional square I 136
– three-dimensional square I 234
Weyl's alternative I 184
Weyl's form of the commutation relations I 78, 205
White dwarfs II 242, 255
Wick's normal ordering II 302
Width(s)
– radiative transitions II 310, 315
– of a resonance I 164
– of state II 134, 146, 180
Wien's displacement law I 2; II 244
Wiener measure I 128
Wigner formula II 66
Wigner's friend I 373
Wigner's theorem I 190, 263
Wigner-Eckart theorem I 215, 332
Work function I 5

Yes-no experiment *or* proposition I 63
Young double-slit experiment I 7, 9
Yukawa cut II 57, 64
Yukawa potential II 38, 47, 57, 157
– Born approximation to scattering by II 38
– energy levels in II 159
Yukawian potentials II 67

Zeeman effect I 18; II 224
– anomalous II 225
– normal II 225
Zero energy resonance II 59, 63
Zone, Brillouin I 179

Contents of **Quantum Mechanics I**

1. The Physical Basis of Quantum Mechanics ... 1
1.1 Introduction ... 1
1.2 The Blackbody ... 1
1.3 The Photoelectric Effect ... 4
1.4 The Compton Effect ... 5
1.5 Light: Particle or Wave? ... 7
1.6 Atomic Structure ... 10
1.7 The Sommerfeld-Wilson-Ishiwara (SWI) Quantization Rules ... 13
1.8 Fine Structure ... 16
1.9 The Zeeman Effect ... 18
1.10 Successes and Failures of the Old Quantum Theory ... 21
1.11 Matter Waves ... 22
1.12 Wave Packets ... 27
1.13 Uncertainty Relations ... 29

2. The Postulates of Quantum Mechanics ... 33
2.1 Introduction ... 33
2.2 Pure States ... 34
2.3 Observables ... 40
2.4 Results of Measurements ... 50
2.5 Uncertainty Relations ... 53
2.6 Complete Sets of Compatible Observables ... 55
2.7 Density Matrix ... 57
2.8 Preparations and Measurements ... 63
2.9 Schrödinger Equation ... 68
2.10 Stationary States and Constants of the Motion ... 72
2.11 The Time-Energy Uncertainty Relation ... 75
2.12 Quantization Rules ... 77
2.13 The Spectra of the Operators X and P ... 83
2.14 Time Evolution Pictures ... 84
2.15 Superselection Rules ... 85

3. The Wave Function ... 88
3.1 Introduction ... 88
3.2 Wave Functions ... 88
3.3 Position and Momentum Representations ... 90

3.4 Position-Momentum Uncertainty Relations ... 95
3.5 Probability Density and Probability Current Density ... 97
3.6 Ehrenfest's Theorem ... 101
3.7 Propagation of Wave Packets (I) ... 103
3.8 Wave Packet Propagation (II) ... 106
3.9 The Classical Limit of the Schrödinger Equation ... 113
3.10 The Virial Theorem ... 118
3.11 Path Integration ... 120

4. One-Dimensional Problems ... 130
4.1 Introduction ... 130
4.2 The Spectrum of H ... 131
4.3 Square Wells ... 136
4.4 The Harmonic Oscillator ... 142
4.5 Transmission and Reflection Coefficients ... 152
4.6 Delta Function Potentials ... 159
4.7 Square Potentials ... 168
4.8 Periodic Potentials ... 175
4.9 Inverse Spectral Problem ... 180
4.10 Mathematical Conditions ... 183

5. Angular Momentum ... 189
5.1 Introduction ... 189
5.2 The Definition of Angular Momentum ... 189
5.3 Eigenvalues of Angular Momentum Operators ... 196
5.4 Orbital Angular Momentum ... 200
5.5 Angular Momentum Uncertainty Relations ... 201
5.6 Matrix Representations of the Rotation Operators ... 206
5.7 Addition of Angular Momenta ... 209
5.8 Clebsch-Gordan Coefficients ... 212
5.9 Irreducible Tensors Under Rotations ... 215
5.10 Helicity ... 217

6. Two-Particle Systems: Central Potentials ... 220
6.1 Introduction ... 220
6.2 The Radial Equation ... 225
6.3 Square Wells ... 234
6.4 The Three-Dimensional Harmonic Oscillator ... 239
6.5 The Hydrogen Atom ... 242
6.6 The Hydrogen Atom: Corrections ... 247
6.7 Accidental Degeneracy ... 250
6.8 The Hydrogen Atom: Parabolic Coordinates ... 252
6.9 Exactly Solvable Potentials for s-Waves ... 254

7. Symmetry Transformations 262
7.1 Introduction 262
7.2 Symmetry Transformations: Wigner's Theorem 262
7.3 Transformation Properties of Operators 267
7.4 Symmetry Groups 268
7.5 Space Translations 269
7.6 Rotations 271
7.7 Parity 272
7.8 Time Reversal 274
7.9 Invariances and Conservation Laws 277
7.10 Invariance Under Translations 278
7.11 Invariance Under Rotations 279
7.12 Invariance Under Parity 280
7.13 Invariance Under Time Reversal 284
7.14 Galilean Transformations 288
7.15 Isospin 295

Appendix A: Special Functions 301
A.1 Legendre Polynomials 301
A.2 Associated Legendre Functions 303
A.3 Spherical Harmonics 304
A.4 Hermite Polynomials 306
A.5 Laguerre Polynomials 308
A.6 Generalized Laguerre Polynomials 309
A.7 The Euler Gamma Function 311
A.8 Bessel Functions 312
A.9 Spherical Bessel Functions 314
A.10 Confluent Hypergeometric Functions 317
A.11 Coulomb Wave Functions 320

Appendix B: Angular Momentum 322
B.1 Angular Momentum 322
B.2 Matrix Representation of the Rotation Operators 324
B.3 Clebsch-Gordan Coefficients 326
B.4 Racah Coefficients 329
B.5 Irreducible Tensors 331
B.6 Irreducible Vector Tensors 334
B.7 Tables of Clebsch-Gordan and Racah Coefficients 338

Appendix C: Summary of Operator Theory 347
C.1 Notation and Basic Definitions 347
C.2 Symmetric, Self-Adjoint, and Essentially Self-Adjoint Operators 348
C.3 Spectral Theory of Self-Adjoint Operators 350
C.4 The Spectrum of a Self-Adjoint Operator 352

C.5 One-Parameter Unitary Groups 353
C.6 Quadratic Forms 353
C.7 Perturbation of Self-Adjoint Operators 354
C.8 Perturbation of Semi-Bounded Self-Adjoint Forms 358
C.9 Min-Max Principle 360
C.10 Direct Integrals in Hilbert Spaces 361

Appendix D: Elements of the Theory of Distributions 363
D.1 Spaces of Test Functions 363
D.2 Concept of a Distribution or Generalized Function 364
D.3 Operations with Distributions 364
D.4 Examples of Distributions 365
D.5 Fourier Transformation 367

Appendix E: On the Measurement Problem Quantum Mechanics 370
E.1 Types of Evolution 370
E.2 Sketch of a Measurement Process 371
E.3 Solutions to the Dilemma 374

Appendix F: Models for Hidden Variables. (A Summary 376
F.1 Motivation 376
F.2 Impossibility Theorems 379
F.3 Hidden Variables of the First Kind and of the Second Kind (or Local Hidden Variables) 381
F.4 Conclusions 384

Appendix G: Properties of Certain Antiunitary Operators 386
G.1 Definitions and Basic Properties 386
G.2 Canonical Form of Antiunitary V with $(V^2)_{\text{p.p.}} = V^2$ 387

Bibliography 389

Subject Index 405

Contents of **Quantum Mechanics II** 415